T0180193

Advanced Textbooks in Control and Signal Processing

Series editors

More information about this series at http://www.springer.com/series/4045

László Keviczky · Ruth Bars
Jenő Hetthéssy · Csilla Bányász

Control Engineering

László Keviczky
Institute for Computer Science
and Control
Hungarian Academy of Sciences
Budapest, Hungary

Jenő Hetthéssy
Department of Automation
and Applied Informatics
Budapest University of Technology
and Economics
Budapest, Hungary

Ruth Bars
Department of Automation
and Applied Informatics
Budapest University of Technology
and Economics
Budapest, Hungary

Csilla Bányász
Institute for Computer Science
and Control
Hungarian Academy of Sciences
Budapest, Hungary

ISSN 1439-2232 ISSN 2510-3814 (electronic)
Advanced Textbooks in Control and Signal Processing
ISBN 978-981-13-4114-4 ISBN 978-981-10-8297-9 (eBook)
https://doi.org/10.1007/978-981-10-8297-9

Printed on acid-free paper

This Springer imprint is published by the registered company Springer Nature Singapore Pte Ltd.
The registered company address is: 152 Beach Road, #21-01/04 Gateway East, Singapore 189721, Singapore

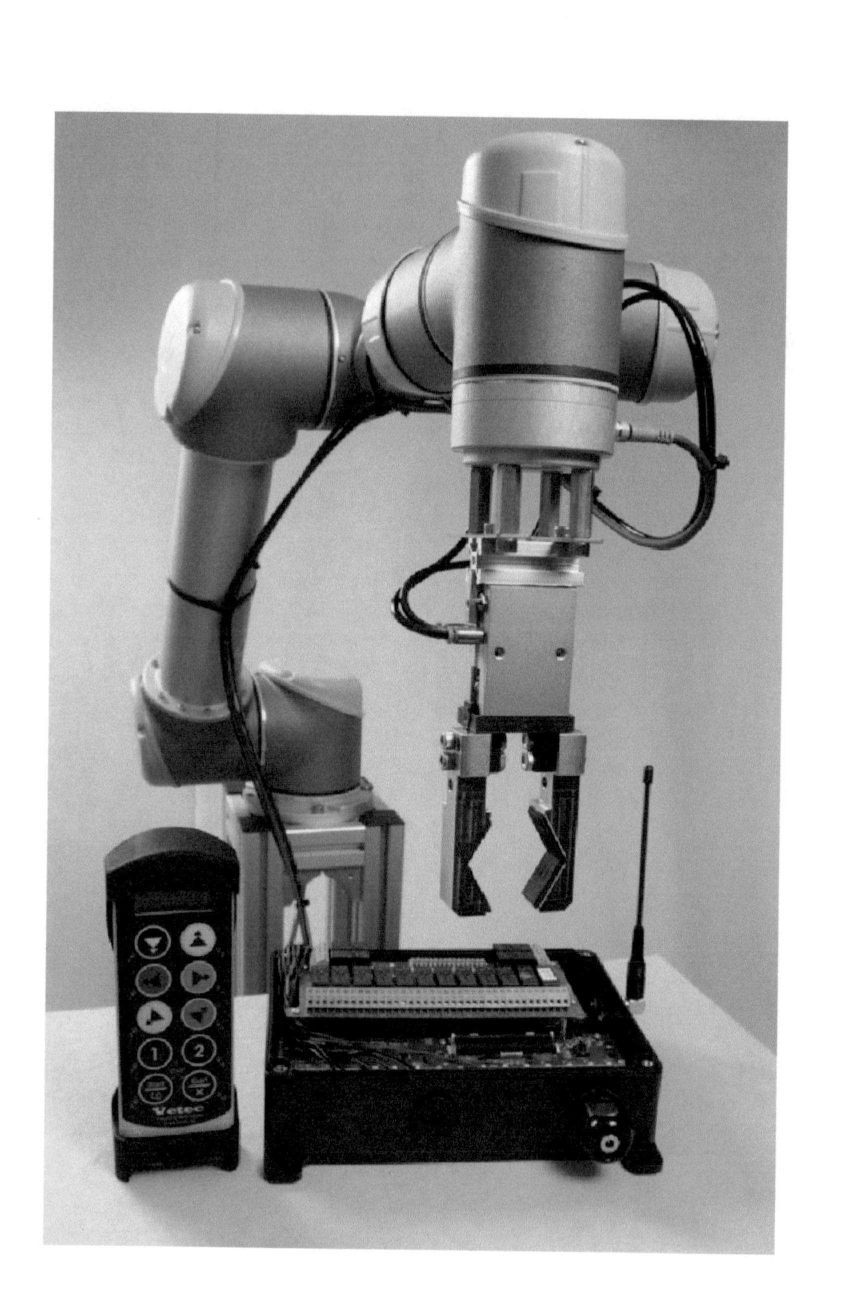
Vetec

Frigyes Csáki
(1921–1977)

This textbook is devoted to the memory of Frigyes Csáki, who was the first professor of control in Hungary

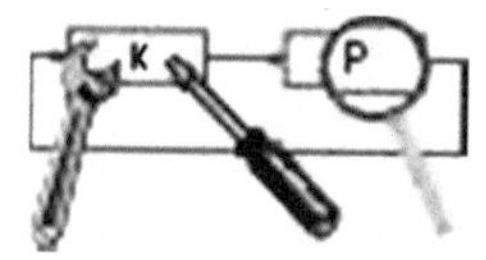

Foreword

The *Advanced Textbooks in Control and Signal Processing* series is designed as a vehicle for the systematic textbook presentation of both fundamental and innovative topics in the control and signal processing disciplines. It is hoped that prospective authors will welcome the opportunity to publish a more rounded and structured presentation of some of the newer emerging control and signal processing technologies in this textbook series. However, it is useful to note that there will always be a place in the series for contemporary presentations of foundational material in these important engineering areas.

It is currently quite a challenge to compose and write a new introductory textbook for control courses. One issue is that the electrical engineering discipline has grown and evolved immeasurably over the years. It now encompasses the fields of power systems technology, telecommunications, signal processing, electronics, optoelectronic and control systems engineering all served with a smattering of computer science. The undergraduates and postgraduates are faced with the unenviable task of selecting which subjects to study from this smorgasbord of topics.

Many academic institutions have introduced a modular semester structure to their engineering courses. This has the advantage of allowing undergraduates and postgraduates to study a set of basic modules from each of the disciplines before specializing through a selection of advanced subject modules. This means the student obtains a good foundational grounding in the electrical engineering discipline. Such an approach requires an introductory control course textbook of sufficient depth to be useful but not so advanced as to leave students bewildered given that the subject of control has a substantial mathematical content.

Other institutions have managed to retain an Automatic Control Department or Group where the main course is a first degree in control engineering per se. Such departments are also likely to offer master and Ph.D. postgraduate qualifications in the control discipline too. In these departments, the requirements of control systems theory for mathematics can be met by specific control mathematics course modules. An introductory control engineering textbook in this context can have considerably more analytical depth too.

There is one more consideration to add into this discussion of introductory control systems engineering course textbooks. The spectrum of control involves systems theory, systems modeling, control theory, control design techniques, system identification methods, system simulation and validation, controller implementation techniques, control hardware, sensors, actuators, and system instrumentation. Quite how much of each area to include in an introductory control course is something usually decided by the course lecturer, the institutional resources available, the academic level of the course, and the time available for the student to study control. But these issues will also have a considerable influence on the type, level, and structure of any introductory course textbook that is proposed.

László Keviczky, Ruth Bars, Jenö Hetthéssy, Csilla Bányász form a team of control academics who have worked in various Hungarian higher educational institutions, primarily the Department of Automation and Applied Informatics at the Budapest University of Technology and Economics, Hungary, and latterly with the Computer and Automation Research Institute of the Hungarian Academy of Science. Their introductory control course textbook presented here has evolved and been refined through many years of teaching practice. The textbook focuses on the control and systems theory, control design techniques, system simulation and validation part of the control curriculum and is supported by a substantial volume of MATLAB® exercises (ISBN 978-981-10-8320-4).

The textbook can be used by undergraduates in a first control systems course. The technical content is self-contained and provides all the signals and systems material that would be needed for a first control course. This is an obvious advantage for the student reader and also the lecturer as it avoids the need for a supplementary mathematical textbook or course. The use of the Youla parameterization approach is a distinctive feature of the text, and this approach will also be of interest to graduate students. The Youla parameterization approach has the advantage of unifying a number of control design methods.

Many popular undergraduate texts give cursory space to the PID controller yet it is a controller that is widely used in industry. In this control textbook, there is a good chapter on PID control and this will chime well with the more industrially orientated undergraduate and academic lecturer. Also valuable is the material presented in Chapter 13 on the tuning of discrete PID controllers. To close the textbook, the authors present an outlook chapter, Chapter 16, that directs the reader toward more advanced topics.

Industrial Control Centre
Glasgow, Scotland, UK
January 2017

M. J. Grimble
M. A. Johnson

Preface

"Navigare necesse est", i.e., the ship must be navigated, said the Romans in Antiquity. "Controlare necesse est", i.e. systems must be controlled, we have been saying since the technological revolution of the nineteenth century. Really, in our everyday life, or in our environment, one can hardly find equipment that does not contain at least one or more control tasks solved by automation instead of by us, or, more importantly, for our comfort.

In an iron, a temperature control system is operated by a relay, in a gas-heating system the temperature is also controlled, and in more sophisticated systems the temperature of the environment is also taken into consideration. In our homes, modern audio-visual systems contain dozens of control tasks, e.g., the regulation of the speed of the tape recorders, the start and stop operation of the equipment; similar operation modes of the CD and DVD systems; the temperature control of the processor in our PC, the positioning of the hard disks' heads, etc. In cars, the quantity of petrol used and the harmonized operation of the brakes are all controlled by automatic controllers. An aircraft could not fly without controllers, since its operation is a typical example of an unstable system. The number of control tasks in modern aircraft is more than one hundred. The universe could not have been investigated by humankind without the automatic control and guidance systems used at launching rockets, satellites, and ballistic missiles. In the recent Mars explorers, sophisticated high-level, so-called intelligent components, have been employed.

In complex, industrial processes the number of tasks to be solved is over a thousand or ten thousand. The quantity and quality of the products, as well as the safety of the environment, could not be guaranteed without these automatically operated systems. Launching products in the market requires the accurate control of a number of variables.

In almost all assembly factories—from simple production beltways to robots—automatic control is applied.

With the development of medical biology, it was discovered that in any organ, and so in human beings, dozens of basic control processes are at work (i.e., the control of the blood pressure, the body temperature, the level of the blood-sugar content, the level of hormones) and the present techniques are approaching the level when some of these tasks can be taken over in case of illnesses or some problems.

Several basic processes of economics (e.g., supply and demand, storage–inventory, macro- and micro-balance) afford possibilities for automatic control.

The everyday person hardly meets directly with the concept of automatic control, even though they operate several pieces of equipment by pushing buttons, switches, or using instrument panels. That is why control is often considered to be a hidden technology. This phenomenon used to be the reason for the ignorant opinion that there is no need for studying the theory of control and regulation, since it comes embedded in the equipment. But do not forget that such equipment has to be designed and produced, and brought to the market. Only those countries can be considered "developed" ones, that are in the front ranks in the development of these kinds of instruments and processes.

In the modern technologies of the twenty-first century, the basic processing, evaluating and decision-making tasks are executed by computers. The observation of the signals and characteristics of real-time processes, the transfer of executive commands, are made by digital communication. The above three areas (***Control–Computation–Communication*** = C^3) are often considered to be in close synergy.

The goal of this book is to summarize the knowledge required in the introductory courses of university education in these subjects. Each chapter, of course, can have different priorities, but they try to provide useful, basic knowledge in order to continue studies of the higher levels of control theory.

This textbook deals with single variable (single input, single output), linear, constant parameter systems, so, with the simplest systems. Multivariable, nonlinear, varying parameters, stochastic systems are not considered. (Similarly, the theory of the modern adaptive, optimal, and robust controllers is not discussed.) It has to be admitted that the real world is more complex, i.e., multivariable, nonlinear; thus, the material of this textbook is only the first step in studying the control methods of real systems. It also has to be mentioned though that several practical tasks can be solved with quite good results by applying these simplified approaches.

In this book, relatively great attention is devoted to the subject of "Signals and Systems" essential in the basic courses of control theory. In the Appendices, important mathematical fundamentals are summarized. The reason for this is to provide a comprehensive source for students and readers, not requiring additional textbooks to understand this textbook. If anyone's knowledge of certain fields is doubtful, it can be refreshed in the corresponding chapters.

There are many formulas in this textbook. This subject area, this field requires them, which sometimes is threatening to students. The complexity of the necessary computations, however, never exceeds the complexity of engineering computations, but where it cannot be performed by hand, the necessary computational resources and softwares are referred to. It has to be noted that this level is a basic requirement for the engineers employed by companies working for international markets. It has to be added, however, that the theoretical knowledge can really become useful only with many years of practical experience.

Nothing is more practical than a good theory!

The authors believe that this textbook provides a suitable basis for the basic level (B.Sc.) education of those faculties, where control theory is to be taught, and where the goal is to prepare a master's level (M.Sc.) education.

This textbook has been written by a working group of the Department of Automation and Applied Informatics, Budapest University of Technology and Economics. The group is headed by László Keviczky. This material is based on, long experience and textbooks used by the department, but, of course, it is not comparable with those in goals and coverage. The following members of the group played primary roles in writing the different chapters:

Chapter 1. Ruth Bars
Chapter 2. Ruth Bars
Chapter 3. László Keviczky
Chapter 4. Ruth Bars
Chapter 5. Ruth Bars
Chapter 6. László Keviczky and Ruth Bars
Chapter 7. László Keviczky
Chapter 8. László Keviczky and Ruth Bars
Chapter 9. László Keviczky
Chapter 10. László Keviczky
Chapter 11. Jenő Hetthéssy
Chapter 12. László Keviczky and Csilla Bányász
Chapter 13. László Keviczky and Jenő Hetthéssy
Chapter 14. László Keviczky
Chapter 15. László Keviczky
Chapter 16. László Keviczky and Csilla Bányász
Appendix. László Keviczky, Ruth Bars, Jenő Hetthéssy and Csilla Bányász

In the typographical preparation of this textbook, Csilla Bányász had the determining role. The figures were prepared partly with the help of the Ph.D. students Ágnes Bogárdi-Mészöly, Zoltán Dávid, and Gábor Somogyi.

An essential part of this textbook is the practical laboratory material published in a separate volume (MATLAB® Exercises), as well as several examples, helping the students in a good preparation for exams.

Budapest, Hungary

László Keviczky
Ruth Bars
Jenő Hetthéssy
Csilla Bányász

Contents

Notations

H	Transfer functions of continuous-time systems
G	Transfer functions of discrete-time systems
C	Controller transfer function
P	Process transfer function
G (or P_{d})	Discrete-time process pulse transfer function
S	Sensitivity function
T	Complementary sensitivity function
L	Transfer function of an open control loop
K	Gain of a control loop
k	Transfer coefficient of a control loop
Q	YOULA parameter
(t)	Continuous time
$[k]$	Discrete time
$\mathcal{L}\{\ldots\}$	LAPLACE transform
$\mathcal{F}\{\ldots\}$	FOURIER transform
$\mathcal{Z}\{\ldots\}$	z-transform
s	Complex variable ($\mathcal{L}$ transformation)
z	Complex variable ($\mathcal{Z}$ transformation)
r (or y_{r})	Reference signal
y	Controlled variable
e	Error signal
u	Actuating signal (or output of the regulator)
y_{ni}	Input noise
y_{n} (or y_{no})	Output noise
y_{z}	Measurement noise
$\boldsymbol{a}, \boldsymbol{b}, \boldsymbol{c}, \ldots$	Vector
$\boldsymbol{a}^{\mathrm{T}}, \boldsymbol{b}^{\mathrm{T}}, \boldsymbol{c}^{\mathrm{T}}, \ldots$	Row vector
$\boldsymbol{A}, \boldsymbol{B}, \boldsymbol{C}, \ldots$	Matrix
$\boldsymbol{A}^{\mathrm{T}}$	Transpose of a matrix
$\mathbf{adj}(\boldsymbol{A})$	Adjunct of a matrix

$\det(\boldsymbol{A})$ (or $\lvert\boldsymbol{A}\rvert$)	Determinant of a matrix
$\boldsymbol{x}$	State variable
$\boldsymbol{A}, \boldsymbol{b}, \boldsymbol{c}, d$	Parameters of the state equation (continuous)
$\boldsymbol{F}, \boldsymbol{g}, \boldsymbol{h}, d$ (or $\boldsymbol{F}, \boldsymbol{g}, \boldsymbol{c}, d$)	Parameters of the state equation (discrete)
$\mathbf{diag}[a_{11}, a_{22}, \ldots, a_{nn}]$	Diagonal matrix
$\boldsymbol{I} = \mathbf{diag}[1, 1, \ldots, 1]$	Unit matrix
T_{s}	Sampling time
T_{d}	Dead time (continuous)
d	Time delay (discrete)
T_{h}	Additional time delay
$v(t)$	Step response function
$w(t)$	Weighting function
ω	Frequency
ω_{c}	Crossover (cut-off) frequency
$F(j\omega)$	Frequency spectrum of a continuous signal
$F^*(j\omega)$	Frequency spectrum of a sampled signal series
$G(j\omega)$ (or $P_{\mathrm{d}}(j\omega)$)	Frequency spectrum of a discrete-time model
$\mathcal{A}, \mathcal{B}, \mathcal{C}, \mathcal{D}, \mathcal{G}, \mathcal{F}, \mathcal{R}, \mathcal{X}, \mathcal{Y}, \mathcal{V}$	Polynomials
$\deg\{\mathcal{A}\}$	Order of a polynomial
$\mathcal{A}(s) = 0$	Characteristic equation
$\mathbb{U}$	Limit of the control output
$\mathbf{grad}\,[f(\boldsymbol{x})]$	Gradient vector
$\forall\omega$	For all ω
$\angle$ (or $\mathrm{arc}(\ldots)$)	Angle of a complex number or functions
$e^{(\ldots)}$ (or $\exp(\ldots)$)	Exponential function
$\ln(\ldots)$	Natural logarithm
$\lg(\ldots)$	Base 10 logarithm
$E\{\ldots\}$	Expected value
$\mathrm{plim}\{\ldots\}$	Probability limit value
e^{A}	Matrix exponential
$\ln(\boldsymbol{A})$	Matrix logarithm
CT	Continuous time
DT	Discrete time
SRE	Step response equivalent
PFE	Partial fractional expansion
■	End of example

Chapter 1
Introduction

Control means a specific action to reach the desired behavior of a system. In the control of industrial processes generally technological processes, are considered, but control is highly required to keep any physical, chemical, biological, communication, economic, or social process functioning in a desired manner.

Control methods should be used whenever some quantity must be kept at a desired value. For example, control is used to maintain the temperature of our flat at a comfortable specific value both in winter and summer. Controlling an aircraft, the pilot (or the robot pilot) has to execute extremely diverse control tasks to keep the speed, the direction, and the altitude of the aircraft at desired values. Control systems are all around us, in the household (e.g., setting the program of a washing machine, ironing by on-off temperature control, air conditioning, etc.), in transportation, space research, communication, industrial manufacturing, economics, medicine, etc. A lot of control systems do operate in living organisms as well.

Control systems are everywhere in our surroundings. A control system is realized e.g., when taking a shower, where the temperature of the shower is to be kept at a comfortable value (Fig. 1.1). If the temperature sensed by our body differs from its desired value, we intervene by opening the cold tap or the warm tap. After being mixed, the water goes through the shower pipe. The effect of the change takes place after a delay. The effect of the delay has to be considered when deciding on a possible newer execution. The control process taking place is symbolized by the block-diagram shown in Fig. 1.2.

Figure 1.3 shows schematically a control system for room temperature control.

Figure 1.4 illustrates some processes which require control to ensure appropriate performance. The speed or angular position of the motor, as well as the level of the tank, is to be kept at a constant value. The temperature of the liquid flowing through the heat exchanger has to be maintained. In the chemical reactor, the quality and quantity of the materials being created during the chemical reaction have to be maintained. In the distillation column the individual components of the crude oil are

L. Keviczky et al., *Control Engineering*, Advanced Textbooks in Control and Signal Processing, https://doi.org/10.1007/978-981-10-8297-9_1

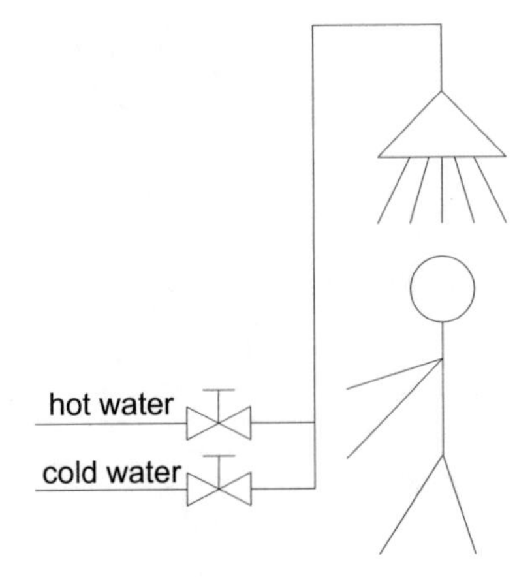

Fig. 1.1 Shower-bath as a control task

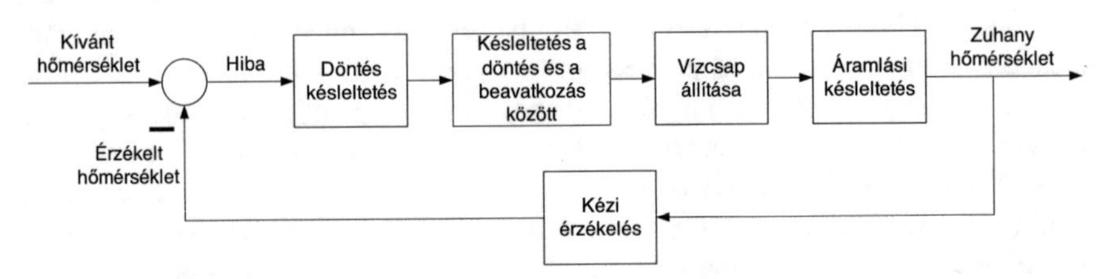

Fig. 1.2 Control block-scheme of the shower-bath

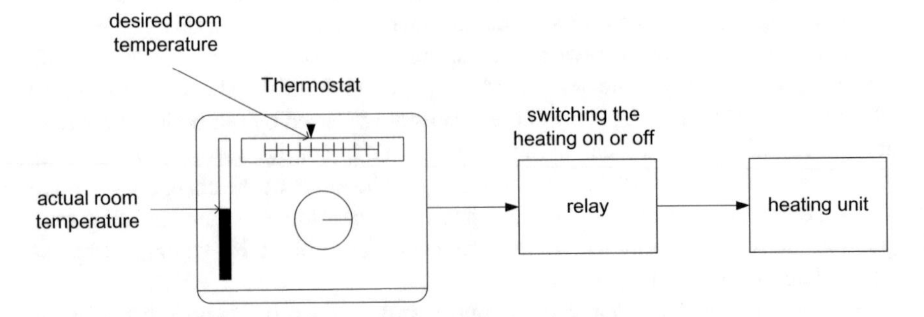

Fig. 1.3 Room temperature control

to be separated. For this purpose, the temperatures of the plates in the column have to be appropriately controlled relative to each other. Furthermore, in everyday practice in the household and in a variety of production processes different control tasks have to be solved.

In what follows, the control processes of technological systems will be discussed. The control of industrial processes plays a significant role in ensuring better product quality, minimizing energy consumption, increasing safety and decreasing environmental pollution.

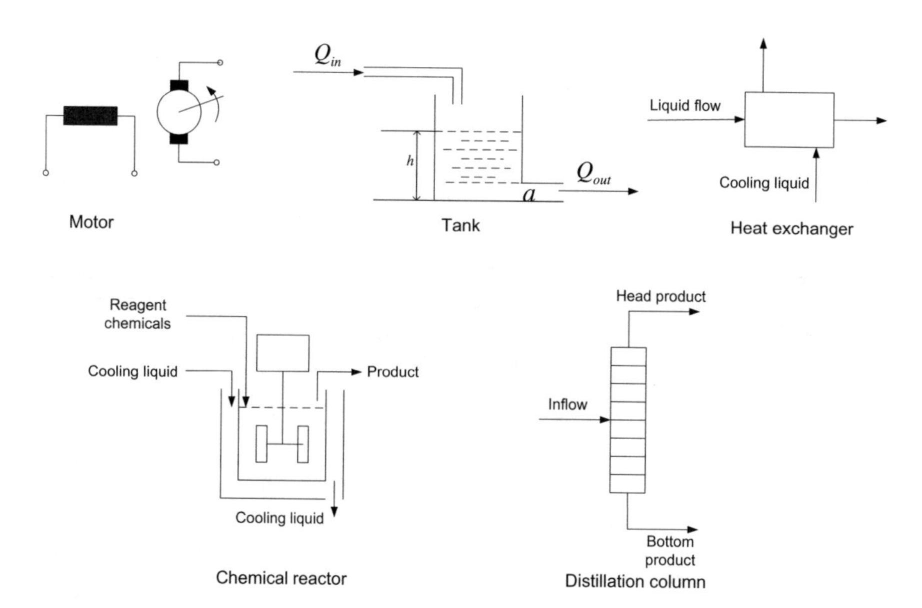

Fig. 1.4 Some typical control tasks

In the manufacturing production processes of material goods, mass and energy conversion takes place. Appropriate control is to be applied to ensure the suitable starting, maintenance and stopping of these processes. For example, in a thermal power station the chemical energy of the coal is converted to heat energy by burning. The heat is then used to produce steam. The steam drives the turbine, creating mechanical rotation energy. The turbine rotates the rotor of a synchronous generator in the magnetic field of the stator. This creates electric energy. All these processes must be operated in a prescribed way. The processes have to be started, and their performance has to be ensured according to the given technological prescriptions. For example, in electrical energy production, it has to be ensured that the voltage and frequency be kept at prescribed constant values within a given accuracy in spite of load changes during the day. Stopping the processes has to be executed safely.

To maintain the processes in a desired manner means keeping different physical quantities at constant values or altering them according to given laws. Such physical quantities could be, for instance, the temperature or pressure of a medium, the composition of a material, the speed of a machine, the angular position of an axe, the level in a tank, etc.

A process is a system which is connected to its environment in many ways. For example, a thermal power station converts the chemical energy of the fuel to electrical energy. The system consists of several pieces of interconnected equipment (furnace, turbine, synchronous generator, auxiliary equipment). The system converts the input quantity (fuel) to the output quantity (electrical energy), while it has multi-faceted relations with its environment (it produces waste material, transfers

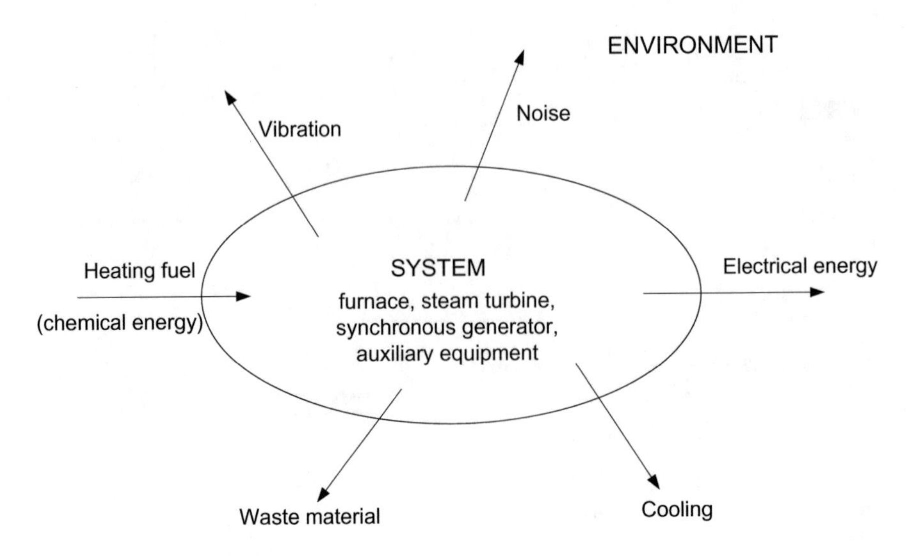

Fig. 1.5 The system and its environment

heat into the environment, produces mechanical vibration and noise, etc.). Figure 1.5 illustrates the relation of the system and its environment. If the operation of the turbine is investigated, then the relation of the system and its environment is considered in a different way (Fig. 1.6). In this case the system is the turbine, which converts the thermal energy of the steam into electrical energy.

The quantities going from the environment into the system are the inputs, while the quantities going from the system into the environment are the outputs. With control—by appropriately manipulating the input quantities—the output quantities are to be maintained according to the given requirements.

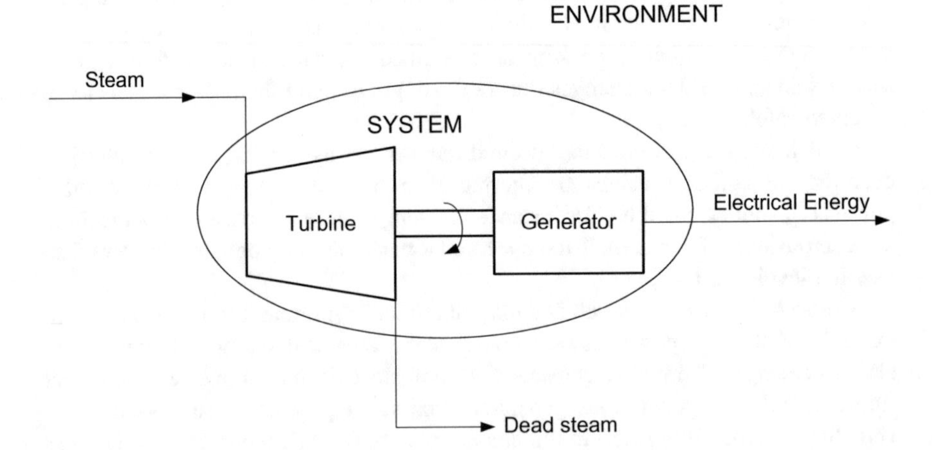

Fig. 1.6 The system and its environment (as a detailed part of the system in Fig. 1.5)

1.1 Basic Concepts

Control means the specific actions to influence a process in order to start it, to appropriately maintain it, and to stop it.

Control is based on information obtained from the process and its environment through measurements. Measuring instruments are needed to measure the different physical quantities involved in the control. Based on the knowledge of the control's aim and on the information obtained from the process and its environment, a decision is made about the appropriate manipulation of the process input. It is characteristic for control that high energy processes are influenced by low energy causes.

The methodology of control is that specifically designed external equipment is connected to the process and then, based on data obtained by measurements or in other ways, it directly modifies the input variables and in that way influences indirectly the output variables. The control system is the joint system made up of the interconnected plant to be controlled and the control equipment.

Control can be performed manually or automatically. In manual control the operator makes a decision and manipulates the input quantity of the process based on the observed output quantity. In automatic control automatic devices execute the functions of decision making and executing the manipulation. Taking a shower is a case of manual control (Fig. 1.1), Fig. 1.7 also illustrates manual control. The operator observes the level of the liquid in the tank and sets the required level by the valve position of the tap influencing the amount of the outlet liquid. Figure 1.8 shows an automatic level control in a tank. The level of the liquid is sensed by a floating sensor. If the level differs from its required value, the valve influencing the input flow will be opened more or less.

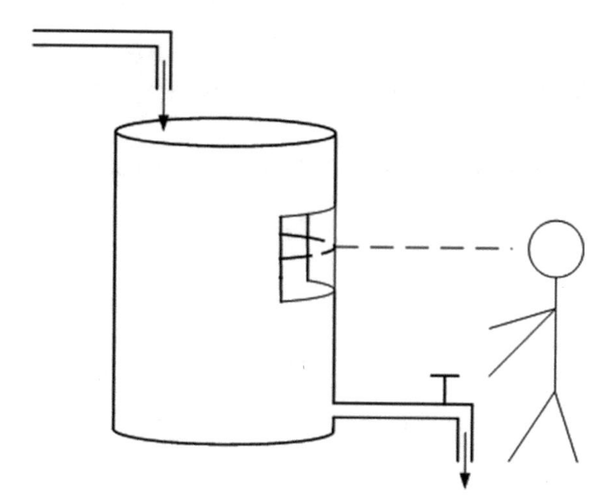

Fig. 1.7 Level control by hand

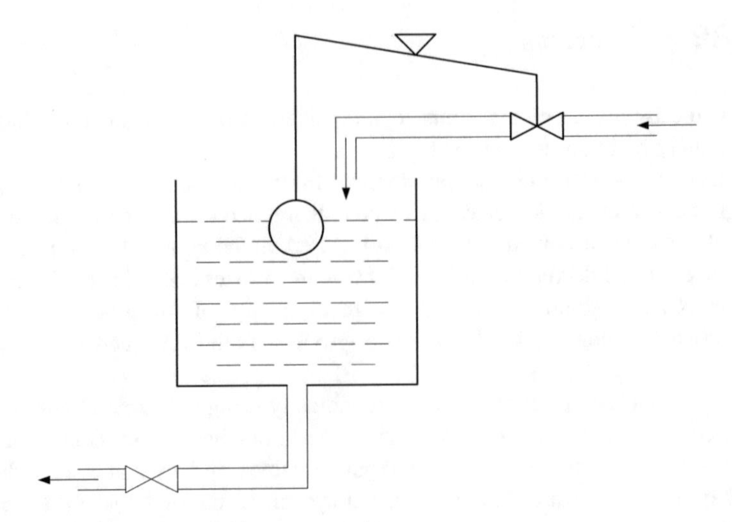

Fig. 1.8 Automatic level control in a water tank

Control engineering deals with the properties and behavior of control systems, with the methods for their analysis and design, and with the question of their realization.

1.1.1 The Basic Elements of a Control Process

A control process consists of the following operations (Fig. 1.9):

Sensing: gaining information about the process to be controlled and its environment
Decision making: processing the information and, based on the aim of the control taking decisions about the necessary manipulations
Disposition: giving a command for manipulation
Signal processing: determine the characteristics of intervention, acting
Intervention, Acting: the modification of the process input according to the disposition.

The individual operations are executed by the appropriate functional units.

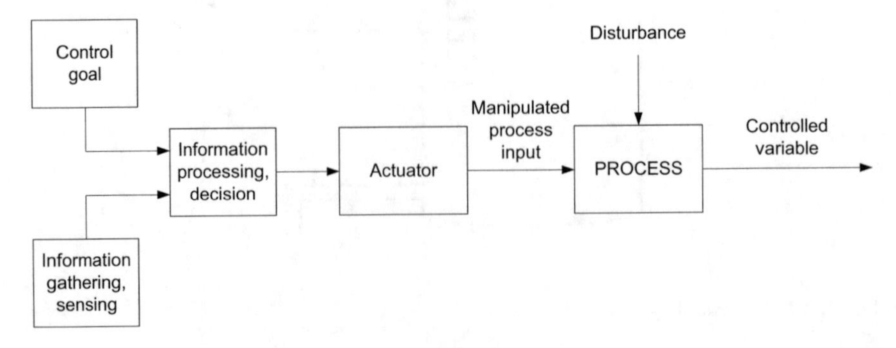

Fig. 1.9 Functional diagram of a control system

1.1.2 Signals and Their Classification

To control a process it is required to measure its changes. Changes of the process occur as consequences of external and internal effects. The features of the process which manifest its motion, and also the external and internal effects, are represented by signals. The signal is a physical quantity, or a change in a physical quantity, which carries information. The signal is capable of acquiring, transferring, as well as storing information. Signals can be observed by measurement equipment. Signals have a physical form (e.g., current, voltage, temperature, etc.)—this is the carrier of the signal. Signals also have informational content—which shows the effect represented by the signal (e.g., change of the current versus time).

Signals can be classified in different ways.

According to its temporal evolution

a signal is *continuous* if it is continuously maintained without interruption over a given range of time,

a signal is *discrete-time or sampled* if it provides information only at determined points in time in a given duration of time.

According to its set of value

a signal is *contiguous* if its set of value is contiguous,

a signal is *fractional* if its set of value is non contiguous and can take only definite values.

According to the form of representation of the information

a signal is *analog* if the value of the signal carrier directly represents the infor mation involved,

a signal is *digital* if the information is represented by digits which are the coded digital values of the signal carrier.

According to the definiteness of the signal value

a signal is *deterministic* if its value can definitely be given by a function of time,

a signal is *stochastic* if its evolution is probabilistic, which can be described using statistical methods.

The characteristic signals of a process are its inputs, outputs, and internal signals. Those input signals which are supposed to be used as inputs modifying the output of the process are called manipulated variables or control variables. The other input variables are disturbances.

1.1.3 Representation of System Engineering Relationships

The various parts of a control system are in interaction with each other. The relations of the individual parts can be represented by different diagrams. As was mentioned earlier, a piece of equipment which performs some control task is called functional unit (e.g., sensor, actuator, etc.). The symbols for the functional units also appear in the diagrams characterizing the connections of the elements of the control system.

A *structural diagram* gives an overview of the pieces of equipment forming the system and shows their connections. First of all it highlights those parts of the system which are substantial from the control viewpoint. Generally, a structural diagram uses the standard notation of the specific field under consideration.

Considering the performance of a control system, what is of primary interest is not the operation of the individual functional units, but rather the spreading effect of the information induced by their operation. An *operational block diagram* shows the connection and interaction of the individual control units disregarding their physical characteristics. In a block diagram the units are represented by rectangles. A line supplied with an arrow directed to a rectangle symbolizes the input signal, while a directed line going out of a rectangle represents the output signal. The direction of the arrow is also the direction of the flow of information. In the rectangles the functions of the structural units are indicated (e.g., sensor, actuator element, controller, etc.).

When realizing a control system, the requirements for the process and the aim of the control have to be formulated first. Then, to solve the control problem, the individual structural control units are chosen. These units are connected to the process and to each other according to the control structure. It has to be analyzed whether the control system meets the quality specifications. To do this it is required to examine the signal transfer properties of the individual elements and also the signal transfer in the interconnected system. In a *block diagram* the individual elements of the operational diagram are described by their signal transfer properties, i.e., by the mathematical formula giving the relationships between the outputs and inputs. These relationships can be mathematical equations, tables, characteristics, operation commands, etc. The signal transfer properties of the individual elements can be given by a mathematical description of the physical operation of the element, where the values of the parameters involved in the equations are also given. To indicate some frequently used operations, accepted symbols are written inside the rectangles (e.g., the symbol of integration). The symbols of summation and subtraction are shown in Fig. 1.10. A *chain of effect* is a set of connected elements along a given direction.

A block diagram can be considered as the mathematical model of the control system. In this model, mainly the signal transfer properties of the system are kept in view, other properties are ignored.

The static and dynamic behavior of the control system can be investigated based on the block diagram. The block diagram also provides the basis for the design of the control system.

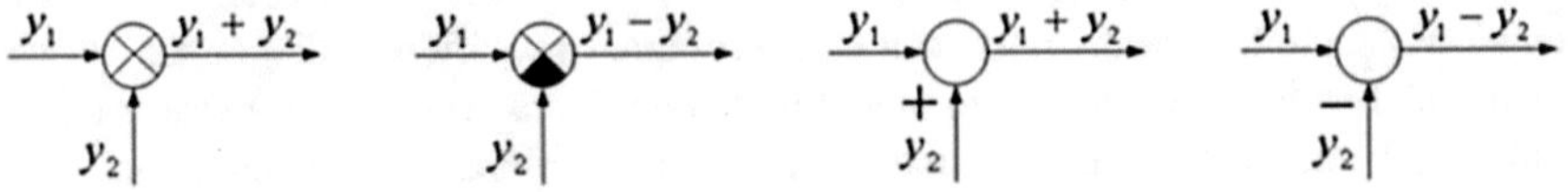

y_1 $y_1 + y_2$ y_2 y_1 $y_1 - y_2$ y_2 y_1 $y_1 + y_2$ $+$ y_2 y_1 $y_1 - y_2$ $-$ y_2

Fig. 1.10 Symbols of summation and subtraction

Of course, when the control system is actually implemented, in addition to its signal transfer properties, other aspects should also be taken into account (e.g., energy constraints, standardized solutions, etc.).

1.1.4 Open- and Closed-loop Control, Disturbance Elimination

If the information is not gained directly from the measurement of the controlled signal, an open-loop control is realized. If the information is derived by directly measuring the controlled signal, a closed-loop control or feedback control is obtained. Figure 1.11 gives the operational block diagram of a closed-loop control system.

An example of an open-loop control system is the control of a washing machine according to a time schedule of executing consecutive operations (rinsing, washing, spin drying). The output signal (the cleanness of the cloths) is not measured. An open-loop control is realized also if the heating of a room is set depending on the external temperature.

In the case of a closed-loop (feedback) control the controlled signal itself is measured. The control error, i.e., the deviation between the actual and the desired value of the controlled signal, influences the input of the process. The functional units are the sensor (measuring equipment), the unit providing the reference signal, the subtraction unit, the amplifier and signal forming unit, and the executing and actuator unit. The characteristic signals of the processes are measured by sensors. The measuring instruments provide signals which are proportional to the different physical quantities measured. The requirements set for the sensors are the following:

- reliable operation in the range of the measurements
- linearity in the range of the measurements
- accuracy

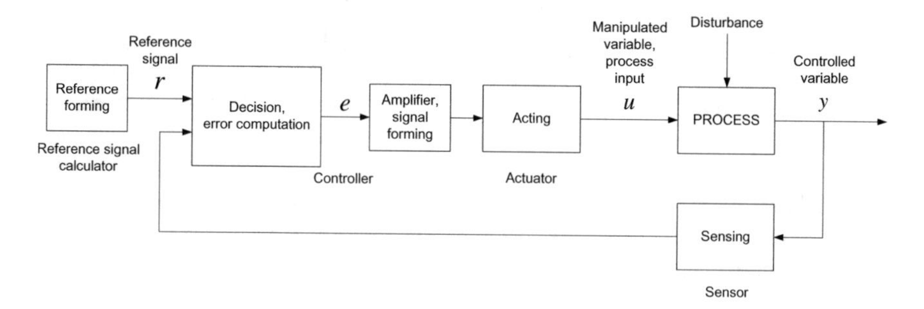

Fig. 1.11 Operational block diagram of the closed-loop control system

– small dead-time compared to the time constants of the process
– low measurement noise.

A sensor measures the physical quantity which is to be controlled and transforms it to another physical quantity which is proportional to the actual value of the controlled signal, and can be compared to the reference signal provided by the reference unit. The error signal operates the controller. The output signal of the controller is amplified, formed and operates the acting element (actuator) which provides the input signal (manipulated variable) for the process. The error signal gives the deviation of the actual output signal from its desired value. If it is different from zero, the system input is to be modified to eliminate the error.

The different functional units are selected according to practical considerations. The control system is built from the individual control elements (sensors which measure the given physical variables in the required range, controllers, actuators, miscellaneous elements) available on the market.

The basis of a closed-loop control system is *negative feedback*. The command for modifying the input of the process is performed based on comparing the reference signal and the actual value of the output signal to be controlled. (There are different schemes for realizing control systems, but all of them are based on negative feedback.)

Because of the dynamics of the plant and the individual elements of the control system, signals need time to go through the control loop. A well designed controller takes the dynamics of the closed-loop system into consideration and ensures the fulfillment of the quality specifications imposed on the control system.

Comparison of open-loop and closed-loop control

If the relationship between the control signal (manipulated variable) and the controlled signal (process variable) is known and reliable information is available on all the elements and all the disturbances in the control circuit, then open-loop control can ensure good control performance. But if our knowledge about the plant and about the disturbances is inaccurate, then the performance of the open-loop control will not be satisfactory. Open-loop control provides a cheap control solution, as it does not apply expensive sensors to measure the controlled quantity, but instead it uses apriori information or information gained about external physical quantities for decision making. In open-loop control there are no stability problems.

Closed-loop control is more expensive than open-loop control. The controlled variable is measured by sensor equipment, and manipulation of the input signal of the plant is executed based on the deviation between the reference signal and the measured output signal. Closed-loop control is able to track the reference signal and to reject the effect of the disturbances. As the actual value of the controlled signal is influenced by the disturbances, closed-loop control rejects the effect of the disturbances which are not known in advance, and also compensates the effect of the parameter uncertainties of the process model. If any kind of effect has caused the difference between the output signal and its required value, the closed-loop control is activated to eliminate the deviation. But because of the negative feedback stability problems may occur, oscillations may appear in the system. The stability of the control system can be ensured by the appropriate design of the controller.

If the disturbance is measurable, then closed-loop control is often supplemented by *feedforward* using the measured value of the disturbance. A block diagram of the feedforward principle is shown in Fig. 1.12. A signal depending on the measured disturbance variable is fed forward to some appropriate summation point of the control loop. This means an open-loop path which relieves the closed-loop control in disturbance rejection. This forward path tries to compensate the effect of the disturbance. This manipulation works in open-loop, the disturbance variable influences the controlled variable, but the manipulation does not affect the disturbance variable.

A classical example of feedforward compensation is the compound excitation of a direct current (DC) generator (Fig. 1.13). The armature voltage is the controlled variable, the excitation is the control (manipulated) variable. The load current (disturbance variable) decreases the armature voltage of the generator. With compound excitation, part of the excitation is created by the load current itself, thus the disturbance variable directly produces the effect of eliminating itself. In this way the armature voltage of the generator is greatly stabilized. For more accurate voltage control, an additional closed-loop configuration can be applied.

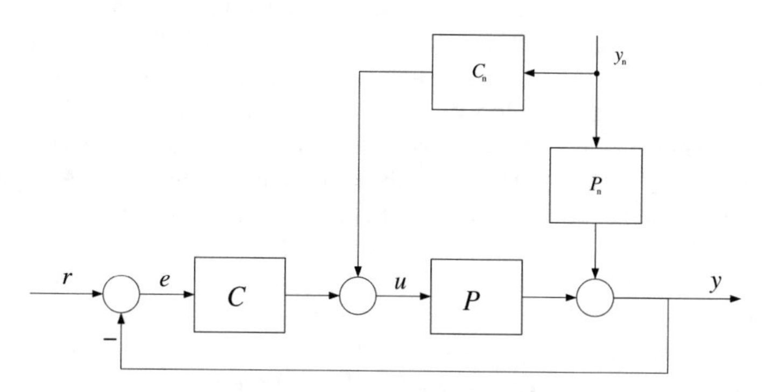

Fig. 1.12 Feedforward control (disturbance compensation)

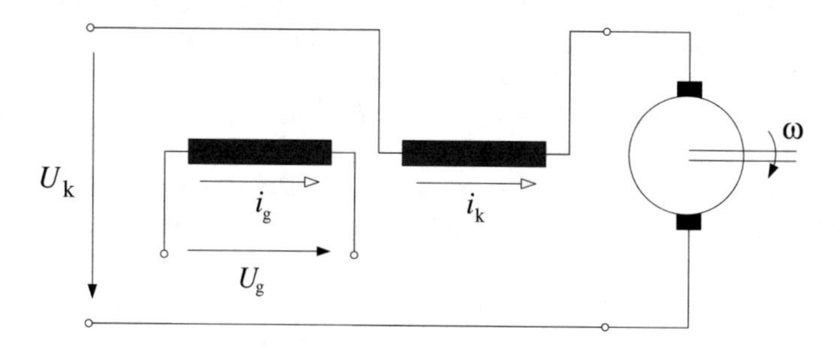

Fig. 1.13 DC generator with compound excitation

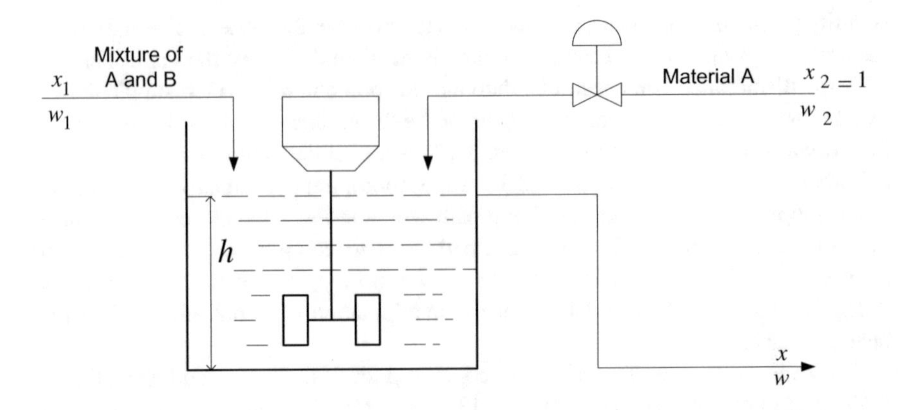

Fig. 1.14 Stirring tank

Let us consider the stirring tank in Fig. 1.14, where w_1 is the inflow quantity of the mixture of materials A and B flowing into the tank. In the mixture the partial rate of material A is x_1. w_2 is the inflow quantity of the pure material A, $x_2 = 1$. w denotes the amount of the outflow material of partial rate x. It is supposed that w_1 is constant, x_2 is constant, and the mixing process in the tank works ideally. The control aim is to keep the composition x of the outflow material (the controlled variable) at a prescribed value in spite of the variations in x_1 (disturbance variable). Manipulations can be executed by modifying the inflow quantity w_2 (control or manipulated variable) by setting the position of the valve. The control is realized by a closed-loop control, if x is measured and w_2 is set depending on this measurement (Fig. 1.15). An open-loop control is built if the composition x_1 of the inflow mixture material is measured, and the inflow amount w_2 is modified accordingly (Fig. 1.16). Figure 1.17 shows a feedforward solution, where both the composition x of the outflow material and the composition x_1 of the inflow mixture are measured, and the inflow quantity w_2 is set according to both measured values (In the figures, the standard symbols for the sensors, controllers, and valves are employed, see Appendix A.3).

The next example shows the speed control of a motor with open-loop and closed-loop control. In a CD player the disc has to be rotated at steady speed. A DC motor can be used as actuator. The angular velocity is proportional to the terminal voltage of the motor. Figure 1.18 shows the solution of the task in open-loop control. The terminal voltage of the motor is provided by a direct current power supply through an amplifier. The velocity is proportional to the terminal voltage. Figure 1.19 schematically presents the solution using closed-loop control. Figure 1.19a gives the structural diagram, while 1.19b shows the operational diagram. The speed of the motor is measured with a tachometer generator, whose output voltage is proportional to the velocity. The measured voltage is compared to the reference signal voltage set by the power supply, which is proportional to the prescribed value of the speed. The error signal operates the actuator DC motor.

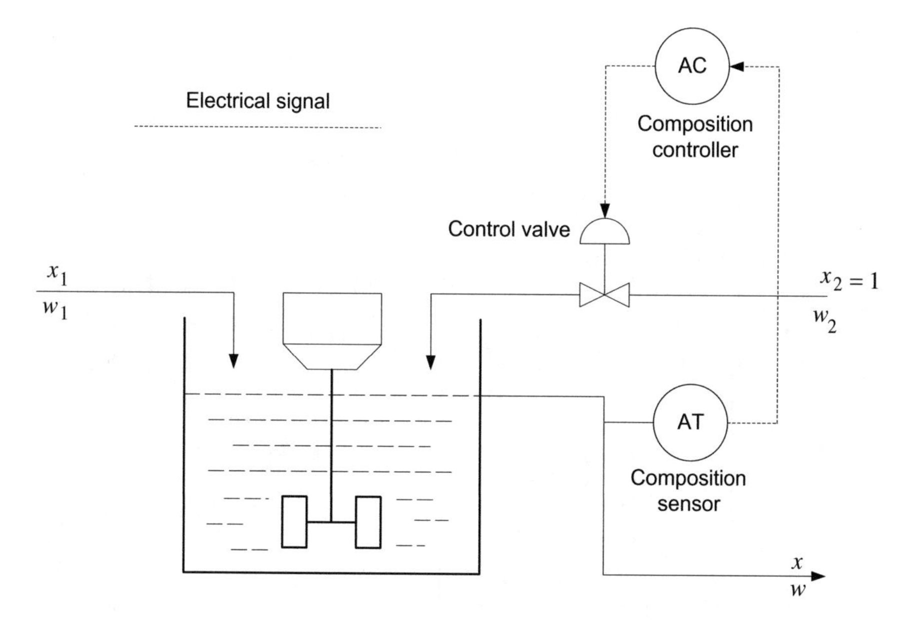

Fig. 1.15 Closed-loop composition control of the liquid in a tank

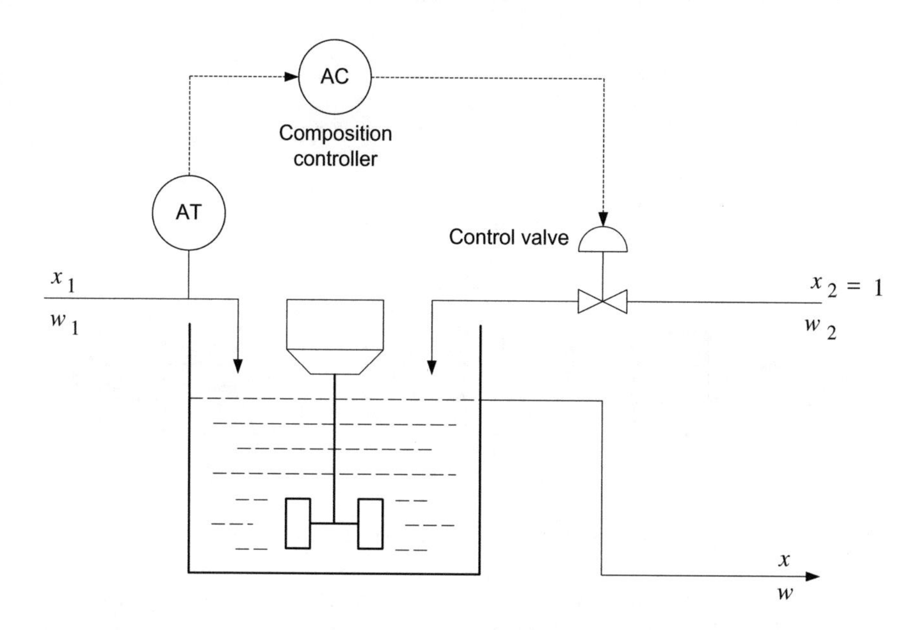

Fig. 1.16 Open-loop composition control of a liquid in a tank

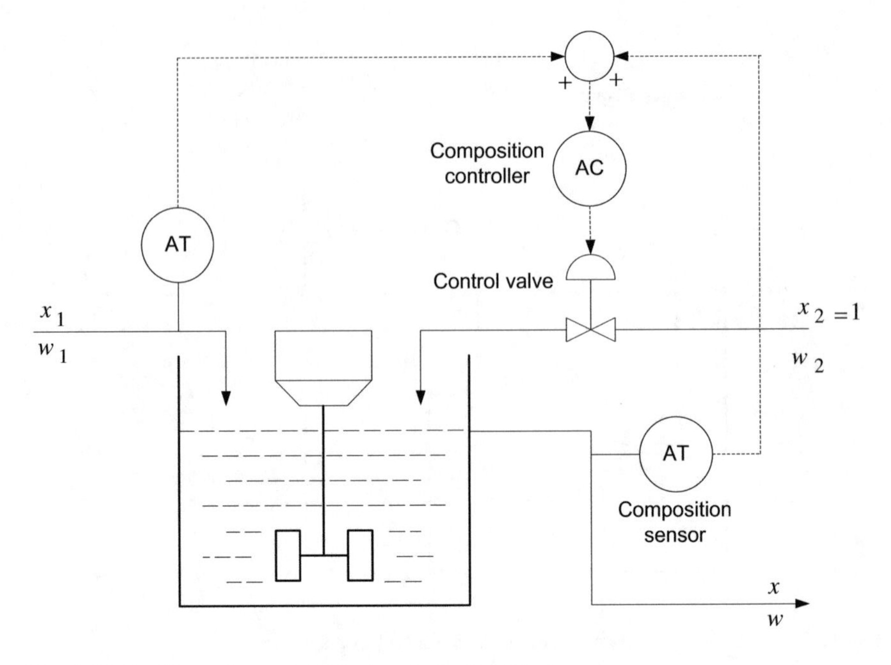

Fig. 1.17 Feedforward composition control of a liquid in a tank

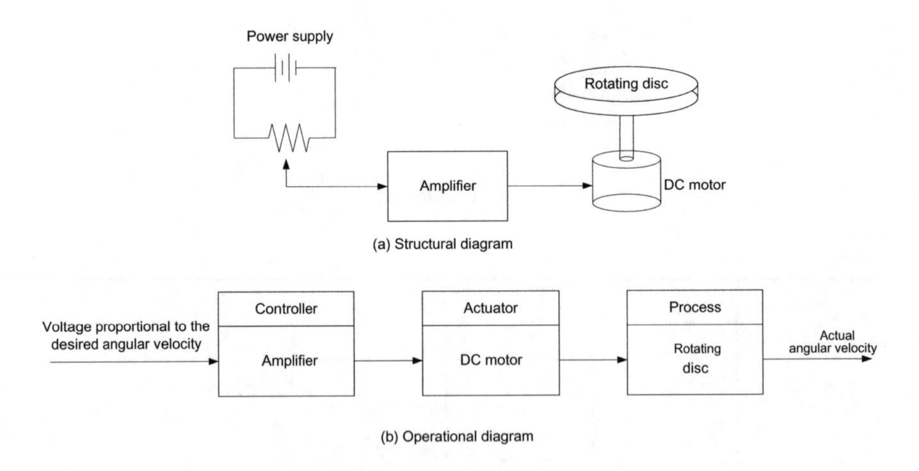

Fig. 1.18 Open-loop angular velocity control of a CD player

With closed-loop control more accurate and more reliable operation can be reached. Closed-loop control ensures not only reference signal tracking, but eliminates speed changes resulting from possible changes in the load, as well.

In practice, besides closed-loop control, open-loop control systems are also given an important role. When starting and stopping a complex system, a series of complex open-loop control operations has to be executed. Generally, intelligent

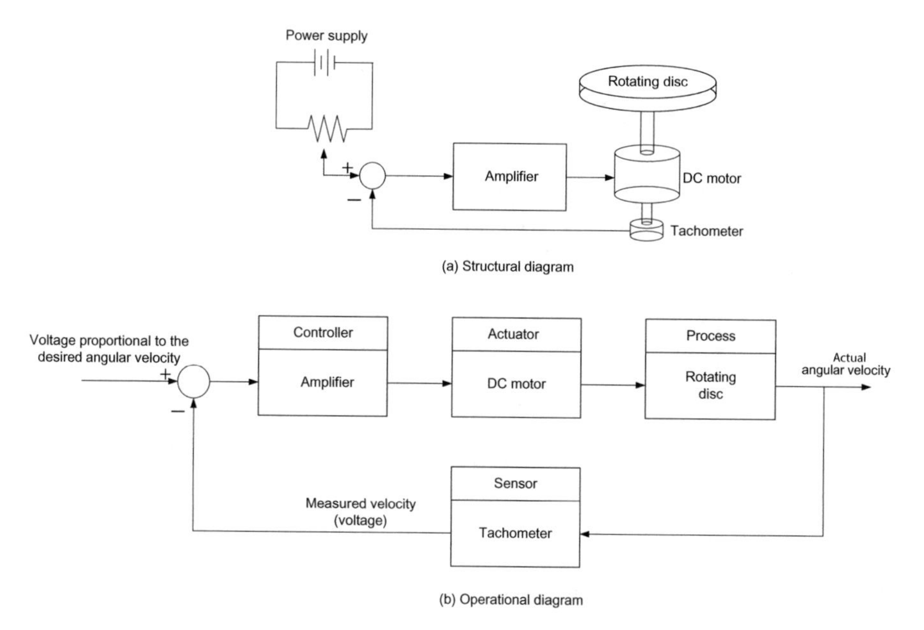

Fig. 1.19 Closed-loop angular velocity control of a CD player

Programmable Logic Controller (PLC) equipment is used to realize the open-loop control. To keep various physical quantities at their required constant values closed-loop control systems are applied.

1.1.5 General Specifications for Closed-Loop Control Systems

The main goal of a closed-loop control system is to track the reference signal and to reject the effect of the disturbances. Regarding the quality of the performance of the control system static and dynamic requirements are prescribed.

First of all a closed-loop control has to be stable, i.e., oscillations of steady or increasing amplitude in the loop variables are not allowed. After the change of the input signals a new balance state has to be reached. The problem of instability comes from the negative feedback realizing the closed-loop control. As after the appearance of the control error the manipulation of the process input can be executed only in a delayed fashion, it may occur that undesired transients do appear in the system (e.g., in Fig. 1.1 when taking a shower the water can be too hot or too cold, the desired temperature is not settled.) Stable behavior can be ensured by appropriate controller design. (The stability of a control system will be discussed in detail in Chap. 5).

Static specifications give the allowed maximum value of the steady error of the reference signal tracking, and the allowed remaining steady deviation in the output

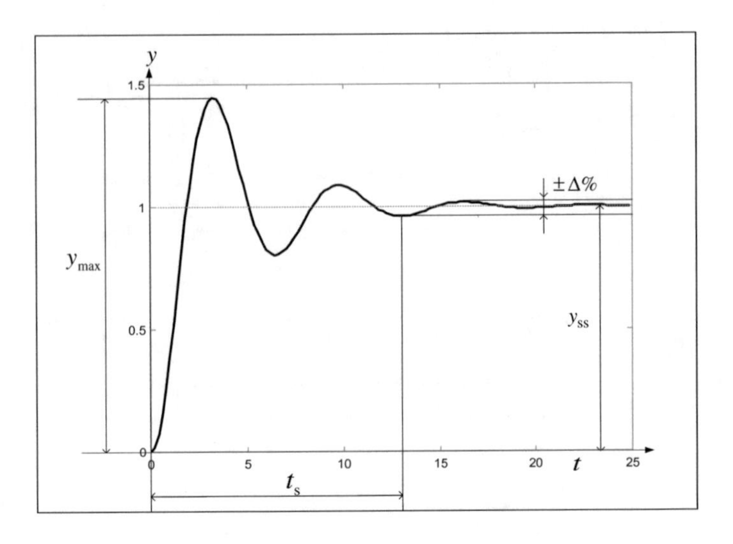

Fig. 1.20 Dynamic quality specifications

signal occurring as the effect of the disturbances, after deceasing of the transients, in steady state. It depends on the technology and on the process to be controlled whether deviations can be allowed at all, and if so, what their maximum possible value can be.

Dynamic specifications give prescriptions for the course of the transients. Let us consider the step response of the closed-loop control system (Fig. 1.20) with the indicated maximum value y_{max} and steady-state value $y_{ss} = y_{steady-state}$. The overshoot σ in percentages is expressed by

$$\sigma = \frac{y_{max} - y_{ss}}{y_{ss}} \cdot 100\%$$

There are processes where aperiodic performance is required (e.g., machine tools, landing of an airplane, etc.), while in other processes often an overshoot of 5–10% is tolerable.

The settling time t_s specifies the time it takes for the step response of the closed-loop control system to settle down within an accuracy of $\pm\Delta\%$ (generally $\pm(1–2)\%$) of its steady state value. Usually the number of allowed oscillations within the settling time is also prescribed.

The control signal in the control system is the output signal of the actuator. The control signal (or manipulated variable) can only take a restricted value corresponding to its physical realization (e.g., a valve setting the inflow liquid quantity in a tank can provide a maximum amount of liquid passing through in its totally open state, and is not able to provide more, in spite of possibly receiving such a command.). If a higher value were to be forced, the actuator would be saturated at only releasing its maximum possible amount, thus temporarily “opening” the control loop. The phenomenon of possible saturation of the manipulated variable (control signal) should be considered already in the control design phase, and it has to be ensured that the manipulated variable be within its specified range, or if nevertheless it exceeds it, effects substantially distorting the normal operation of the control loop should be avoided.

A control system is designed for the process to be controlled ensuring the quality specifications. The model of the process describing its signal transfer properties is obtained by mathematical description reflecting its physical operation. The values of the parameters in the equations are determined generally by measurements. Thus in their values uncertainties may occur. The closed-loop control has to operate appropriately (in a robust way) even if the actual parameters of the process and the parameters considered in its model do differ to some extent.

The requirements set for the closed-loop control system have to be realistic. For example, extremely fast settling can not be required from a slow heating process, as this would result in extremely high control signals. Instead, it is necessary to relax the strictness of the prescriptions in order to get a realizable solution.

Chapter 4 deals in more detail with the quality specifications set for a closed-loop control system.

1.1.6 Simple Control Examples

Next, some examples of closed-loop control will be presented.

Temperature control

Figure 1.21 shows a schematic structural diagram of a device producing warm water with a prescribed temperature. The water is circulating in tubes located in the stokehold of a furnace. The coal used for firing is delivered from the coal container to the heating equipment by a conveyor driven by an electrical motor. The velocity of the conveyor and thus the amount of the transported coal is controlled by the speed of the motor. On the basis of the difference between the prescribed temperature of the warm water and its measured actual value the controller sets the terminal voltage determining the speed of the electrical motor through a preamplifier and a power amplifier. Figure 1.22 shows a block diagram of the temperature control.

Speed control

Figure 1.23 shows the structural diagram of the speed control of a direct current (DC) motor with constant external excitation. The speed of the motor can be changed by the terminal voltage (manipulated variable). The machine driven by the motor produces a changing load for the motor (disturbance), and produces variation in the speed. The terminal voltage of the motor can be changed by an electronic unit

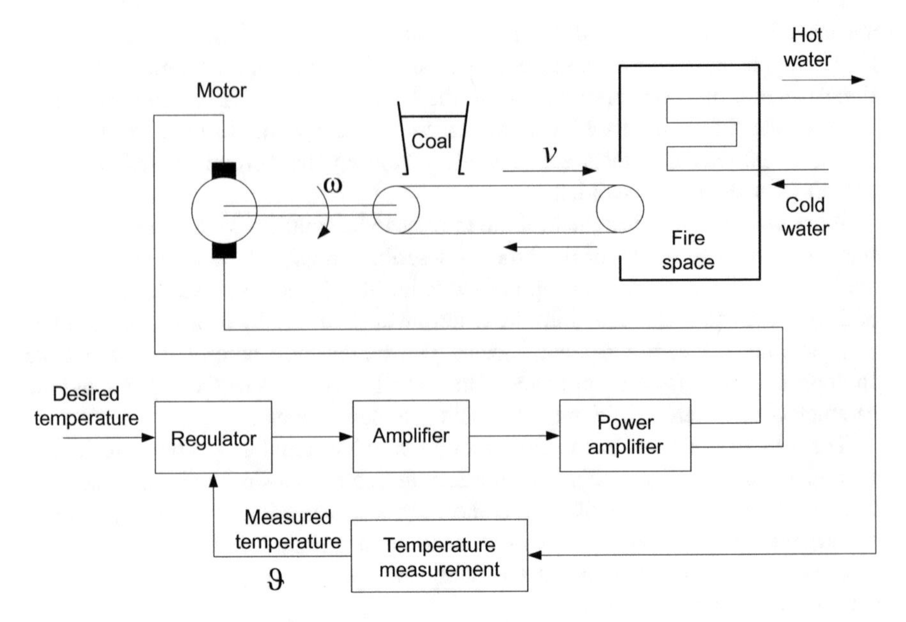

Fig. 1.21 Schematic structural diagram of temperature control

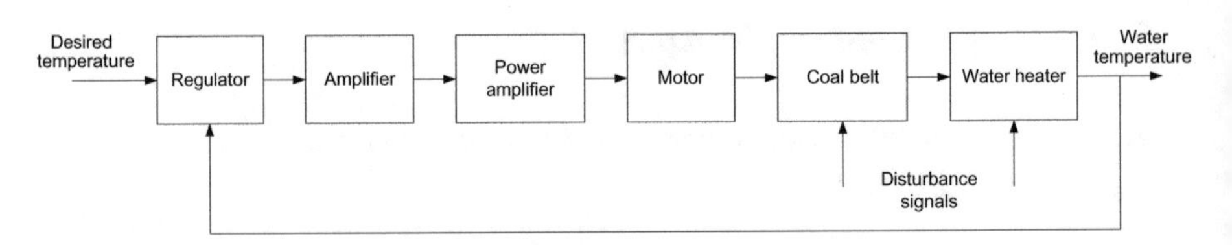

Fig. 1.22 Block diagram of temperature control

with thyristors. The speed of the motor is measured by a tachometer generator, which gives a voltage proportional to the speed (angular velocity). The error voltage is obtained by comparing this voltage with the reference signal voltage provided by the power supply. Its magnitude is amplified by the power amplifiers E1 and E2 and its shape is modified by a filter. Thus the manipulated variable is produced. The function of the manipulated variable is to change the firing angle of the thyristors. As a consequence, the terminal voltage, as the control signal, will be increased or decreased in order to reach the speed prescribed by the reference voltage of the power supply. A block diagram of the speed control is given in Fig. 1.24.

Level control, composition control, moisture control

Frequent tasks in industrial chemical processes are the following: level control in a tank, pressure control, temperature control, composition control of mixed materials, moisture control, etc. Figure 1.25 shows two solutions for liquid level control. In the upper figure the manipulation is executed by the control of the inflow. In the

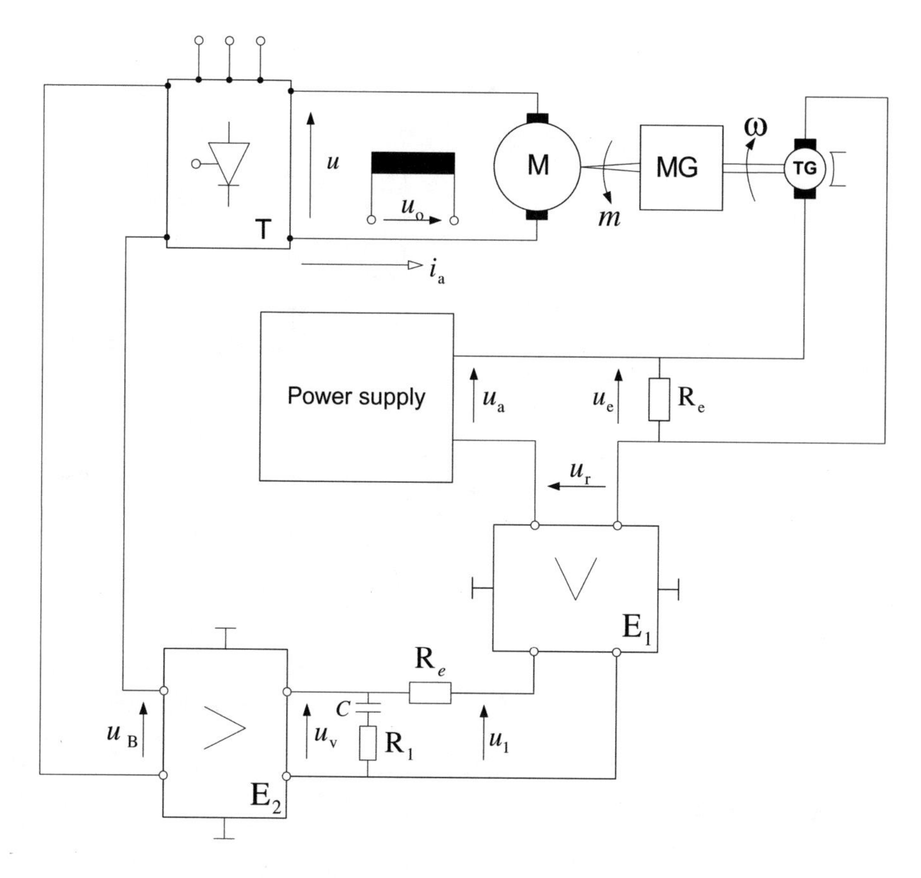

Fig. 1.23 Speed control of a DC motor

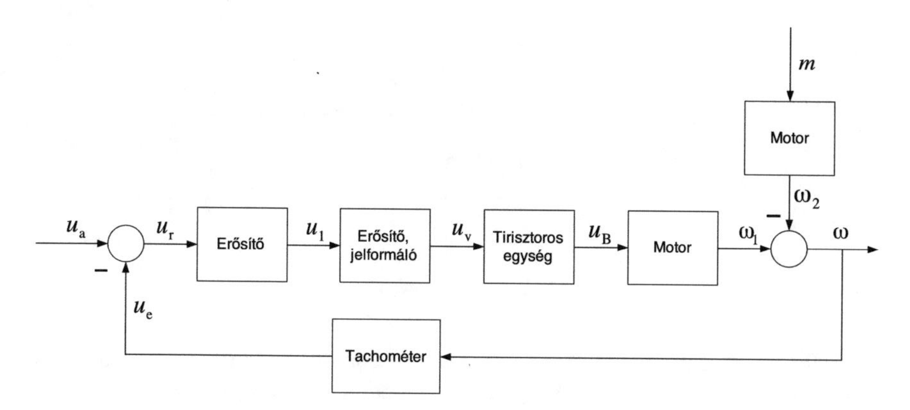

Fig. 1.24 Block diagram of speed control

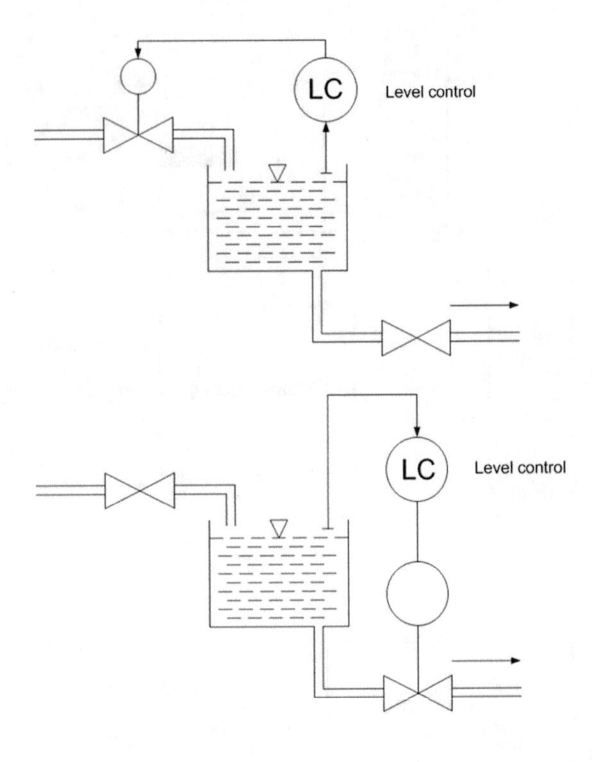

Fig. 1.25 Level control in a tank

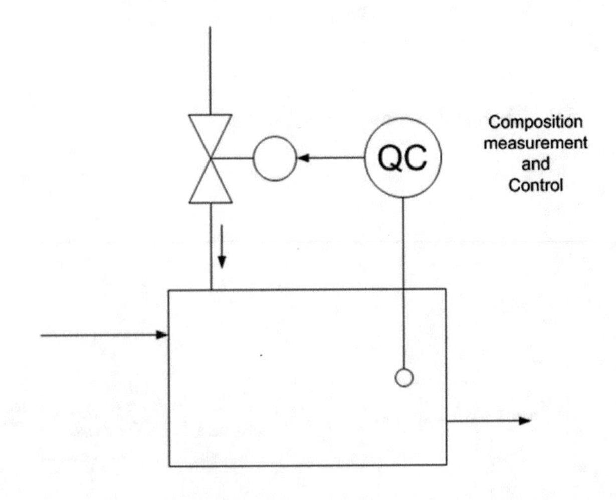

Fig. 1.26 pH control

lower figure the manipulation is executed by the control of the outflow. Figure 1.26 illustrates pH control. Figure 1.27 gives a schematic solution for the moisture control of a granular material in a drying process. The moisture content of the material is measured, and in case of its deviation from the desired value, the speed of the conveyor belt is modified or the inflow of the drying steam (or hot air) is changed.

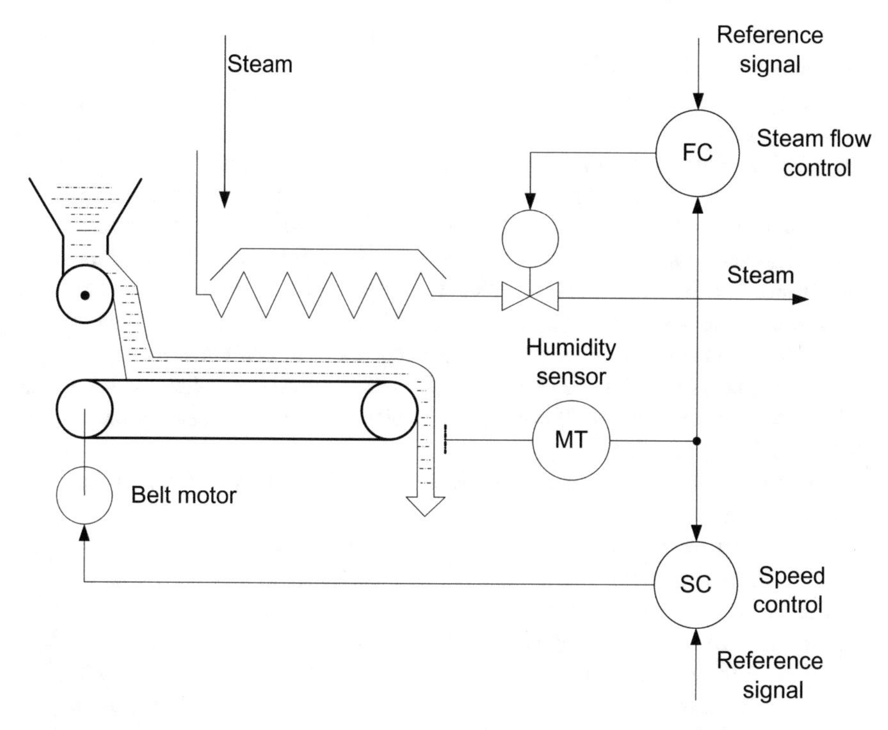

Fig. 1.27 Moisture control

1.2 On the History of Control

Control engineering even today is a developing discipline. New facilities and new techniques raise new theoretical questions, and open up the way to novel applications. Applying negative feedback is not a new principle, however: the ancient Greeks already used it. Looking back at the history of control engineering, some tendencies can be observed.

The application of negative feedback relates to the solution of engineering tasks. The development of control engineering is tightly connected to practical problems that waited for a solution in a stage of humanity's history. Some periods which had a significant influence at the development of control technique were

- the ancient Greek and Arab culture (~300 BC to ~1200 AD),
- the industrial revolution (18th century, but the beginnings already around 1600)
- the beginnings of telecommunication (1910–1945)
- the appearance of computers, the beginning of space research (1957–)

Considering these eras we may establish that humanity was looking first for their place in space and time, and then tried to shape the environment to make life more comfortable; industrial production contributed to this. Then, using also communication, humans found their place and position in society, and then tried to get connected to the universe.

Already the ancient Greeks used several automata. One of the first closed-loop control systems was the water clock of KTESIBIOS in Alexandria (270 BC). The equipment used a float to sense the level of a tank and to keep it at a constant value. If the water level in the tank decreased, a valve opened and refilled the tank. A constant level ensured a constant value of the outflow of the water. The outflowing water filled a second tank. The level of this tank changed proportionally to the time. The Byzantine PHILON (250 BC) also used a float controller to control the oil level of an oil lamp. HERON of Alexandria (first century AD) applied similar devices for level control, wine dosage, opening doors of churches, etc.

Arab engineers between 800 and 1200 AD used several controllers with floating balls. They initiated the on-off controllers, which operate by switching on and off the manipulating variable.

With the invention of mechanical clockwork, water clocks with floating balls were forgotten.

In the era of the industrial revolution many types of automatic equipment were invented. In these systems, the tasks of automatic level, temperature, pressure and speed control were carried out. Already from the beginning of the 17th century there were several control applications (speed control of windmills, temperature control of furnaces (Cornelis DREBBEL), pressure control (PAPIN), etc.). The discovery of the steam engine (SAVERY and NEWCOMEN, ~1700) indicates the beginning of the industrial revolution. The centrifugal controller of James WATT (Fig. 1.28) is considered the first industrial control system, which was applied to the speed control of a steam engine. The position of the centrifugal sensor depends on the speed of the steam engine. This sensor sets the position of the piston valve through the actuating lever, thus influencing the amount of steam inflowing to the

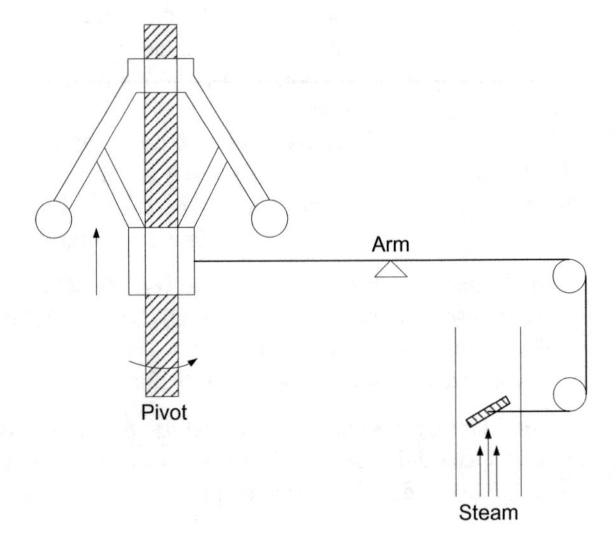

Fig. 1.28 Centrifugal controller

steam engine, changing its speed. (It is interesting to mention that almost another hundred years had to pass until Maxwell gave the exact mathematical description of the system with differential equations).

After the industrial revolution an essential step forward in the development of control engineering was the use of mathematical methods for the description of control circuits. This made possible a more rigorous and exact investigation of control systems.

A new era of control engineering started with the invention of the telephone, with the application of feedback operational amplifiers to compensate for the damping occurring in the transmission of the information.

During the Second World War a lot of high precision control systems were worked out, e.g., automatic flight control systems, radar antenna positioning systems, control equipment of submarines, etc. Then later on these techniques also gained applications in industrial production.

The general application of computers opened a new era in the development of control systems. The computer is no longer only an external device, to facilitate the control design, but becomes part of control systems in real time applications. The process and the process control computer are connected via peripherials, and the process control software calculates the control signal at every sampling time instant and forwards it to the process input. Thus the computer became a basic part of the control loop.

Industrial robots executing precision tasks appeared. The robot is a computer controlled automaton. Several times, human attributes have been imitated in robots, e.g. in robot manipulators the motion of the human hand is imitated. Mobile robots are aimed to be equipped with some intelligence, such as observing and avoiding obstacles moving in space.

Space research means a newer challenge for control systems. Tracking space-craft, placing artificial space objects in a given orbit requires extremely accurate, learning control systems which are able to adapt to changing circumstances. In these systems safe operation is extremely important.

Nowadays when realizing different control systems the control principles, the computer and communication systems and their interaction have to be considered together. The new technical possibilities facilitate new ways of control applications. The appearance of the new miniaturized sensors and manipulating elements opens new perspectives in control techniques. In industrial production processes, distributed control systems have appeared; a large number of control systems distributed in space are coordinated to ensure high quality production. These systems communicate, change information, forward commands and execute them in a coordinated way. Hardware and software elements (PLC-s, profibus, TCP/IP, industrial network standards, etc.) ensuring the operation at this level appeared.

Control theory deals with the construction and analysis and synthesis of closed-loop control systems. The classical period of control theory ($\sim$till 1960) gave the basic concepts of the operation, analysis and synthesis of closed-loop control systems based on negative feedback.

In the modern era of control theory ($\sim$1960–1980), the state space description of control systems and controller design methods based on this model have gained attention.

Nowadays design methods of robust reliable control systems which are less sensitive to parameter changes are in the forefront of interest. Control of non-linear systems, application of intelligent learning systems which are able to recognize environmental changes and adapt to them, application of distributed control systems using network connections and communication, open new perspectives in control theory and control engineering.

1.3 Systems and Models

Building a model is a significant part of analyzing a control system. The model describes the signal transfer properties of a system in mathematical form. With a model, the static and dynamic behavior of a system can be analyzed without performing experiments on the real system. Based on the model, calculations can be executed and the behavior of the system can be simulated numerically. A model of the system can also be used for controller design.

The choice of the elements of a control system is based on practical considerations. The operation of a control system can be followed in the structural diagram, which shows the connections and interactions of the individual units building the control system. The mathematical model of the elements of the control loop describes their signal transfer properties. In a control loop the signal transfer properties of all the elements are given by mathematical relationships. A block diagram can be considered as a mathematical model of the control loop. With a block diagram, the static and dynamic properties of the control system can be analyzed, and it can be determined whether the system satisfies the quality specifications.

The signal transfer properties of the individual elements can be given by mathematical relationships describing their physical operation. A deep understanding of the physical operation is required to derive its mathematical description. The parameters in the mathematical equations can be determined by calculations or by measurements.

The static and dynamic behavior of a system can also be obtained by analyzing the input signals and the output signals resulting from the effect of the input signals. For the execution of an experiment providing information for system analysis, it is important to choose the input signals appropriately. This procedure requires some form of a system model, and determines the parameters in such a way that the outputs of the system and that of the model be closest to each other in terms of a cost function. This procedure is called *identification*.

As the values of the parameters are generally determined by measurements, their values are not quite accurate, but usually the range of the parameter uncertainties can be given.

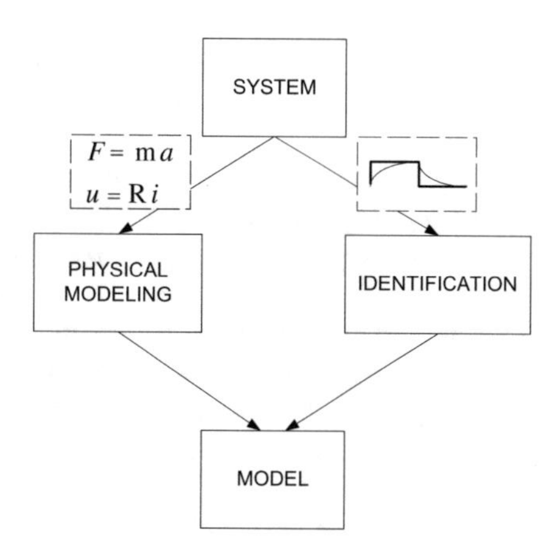

Fig. 1.29 Creation of a model of a system

To obtain a model of a system generally physical modeling and identification are used together (Fig. 1.29).

The model is reliable if its output for a given input approximates well the real output of the system. The domain of validity of the model can be obtained (e.g., in which range of the input signal it is valid).

1.3.1 Types of Models

A model is *static* if its output depends only on the actual value of its input signal. For example, a resistance where the input signal is the voltage and the output signal is the current is a static system. A model is *dynamic* if its output depends on previous signal values as well. An electrical circuit consisting of serially connected resistor and capacitor is a dynamical system, since the voltage drop on the capacitance depends on the charge, and thus on the previous values of the current.

A model can be *linear* or *non-linear*. The static characteristic plots the steady values of an output signal versus the steady values of an input signal. If the static characteristics are straight lines, the system is linear, otherwise it is non-linear.

A model can be *deterministic* or *stochastic*. The signals of a deterministic model can be described by analytical relationships. In a stochastic model, the signals can be given by probabilistic variables and contain uncertainties.

Spatially, a model can have either *lumped* or *distributed parameters*. Lumped parameter systems can be described by ordinary differential equations, while distributed parameter systems can be described by partial differential equations.

A model can be a *continuous-time* (CT) or a *discrete-time* (DT) model. A continuous-time model gives the relationship between its continuous input and output generally in the form of a differential equation. If the input and the output are sampled, the system is a discrete-time or sampled data system, where the relationship between the input and the output signals is described by a difference equation.

Considering the number of the input and the output signals, the model can be *Single Input Single Output* (*SISO*), *Multi Input Multi Output* (*MIMO*), *Single Input Multi Output* (*SIMO*) or *Multi Input Single Output* (*MISO*). Besides the input and the output signals *state variables* of the system can also be defined. The state variables are the internal variables of the system, whose current values have evolved through the previous changes of the signal in the system. Their values can not be changed abruptly when the input signals change abruptly. The current values of the input signals and that of the state variables determine the further motion of the system.

Our investigations will be restricted to the control of dynamic, linear, *SISO*, lumped parameter systems. The literature basically applies the following four methods to describe such systems:

- linear lumped parameter differential equations of order n
- state space equations
- the transfer function and frequency function
- time functions.

1.3.2 The Properties of a System

Some important system properties—which characterize the relationship between the input and the output—are linearity, causality and time invariance.

Linearity: A system is linear if the superposition and homogeneity principles are applicable to it. If for an input signal u_1 the output signal of the system is $y_1 = f(u_1)$, and for the input signal u_2 the output signal is $y_2 = f(u_2)$, then the superposition principle means that $y_1 + y_2 = f(u_1 + u_2)$; according to the homogeneity principle, a k-fold change in the input signal yields a k-fold change in the output signal: $k\,y = f(k\,u)$. It can also be stated that for the input signal $\alpha u_1 + \beta u_2$ the output signal is $\alpha y_1 + \beta y_2$.

Causality: at a given time instant the output depends on the past and the current input values, but it does not depend on future input values.

Time invariance: A system is time invariant if its response to the input signal does not depend on the time instant of applying the input signal: to an input signal shifted by a dead-time of τ, it gives the same response shifted by the dead-time τ (Fig. 1.30). In a time invariant system, for the delayed output the following relationship holds: $y_\tau(t) = y(t - \tau)$.

Linear time invariant systems generally are referred by the acronym *LTI*.

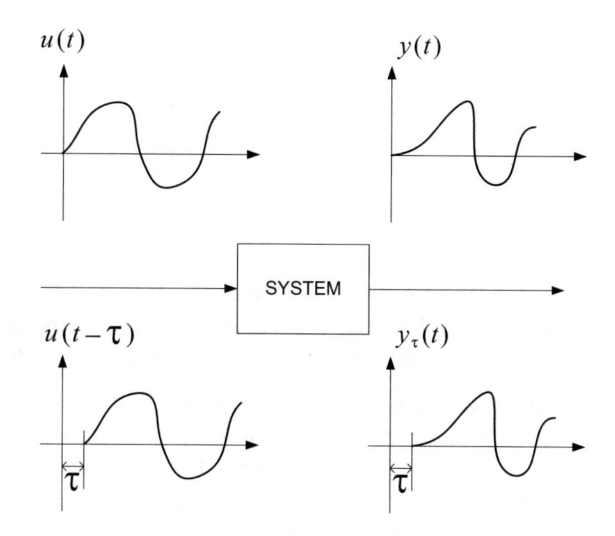

Fig. 1.30 Time invariant system

1.3.3 Examples of the Transfer Characteristics of Some Simple Systems

Next, some examples will demonstrate how to describe mathematically the signal transfer properties of physical systems, i.e., how to give the relationships between the input and the output signals. The description of the behavior of physical systems generally leads to differential equations.

Example 1.1 A mechanical system
Let us consider the mechanical system shown in Fig. 1.31, which can model a part of the chassis of a car. m denotes the mass, c_1 and c_2 are spring constants, and k is the damping coefficient of the oil brake. A concentrated mass is supposed. In the

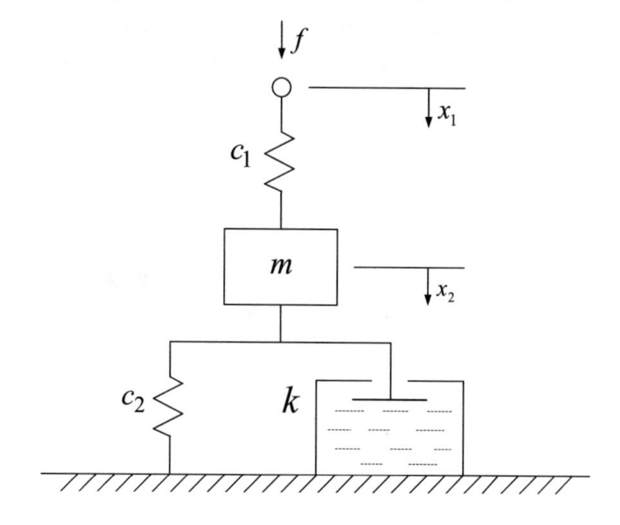

Fig. 1.31 Scheme of a mechanical system

springs, forces proportional to the position are created. The damping piston provides a braking force proportional to the velocity. The following force balance equations can be written. The force created by the upper spring is expressed as $c_1(x_1 - x_2) = f$. The equation expressing the balance of forces acting on the mass is

$$m\frac{\mathrm{d}^2x_2}{\mathrm{d}t^2} = c_1(x_1 - x_2) - c_2x_2 - k\frac{\mathrm{d}x_2}{\mathrm{d}t}.$$

It can be seen that the behavior of the system is described by a differential equation. By solving the differential equation, the motions x_1 and x_2 as function of time can be calculated as the responses to the given force. ■

Example 1.2 Direct current (DC) generator
Let us investigate the signal transfer of the externally excited DC generator shown in Fig. 1.32 between its input signal, the excitation voltage u_g, and its output signal, the armature voltage u_k. The resistance of the excitation coil is R_g and its inductance is L_g. The following differential equation can be written for the excitation circuit:

$$L_g\frac{\mathrm{d}i_g}{\mathrm{d}t} + R_g i_g = u_g$$

Assume that the machine works within the linear section of its magnetic characteristic, thus L_g can be considered constant. The generator is not loaded. The terminal voltage of the generator is proportional to the excitation flux, or supposing a linear magnetic characteristics the terminal voltage is proportional to the excitation current: $u_k = K_g i_g$, where K_g is a constant depending on the structural data of the machine, its units are [V/A]. ■

Example 1.3 A chemical process
Let us consider the mixing tank shown in Fig. 1.33. A solution of concentration c_o is mixed with water to obtain a solution of concentration c_k. The amount q_v of the inflow water is constant, the amount q_o of the inflow solution is controlled by a valve. The concentration is given by the amount of the dissolved material in one liter of the solution expressed in grams. The input signal of the system is the

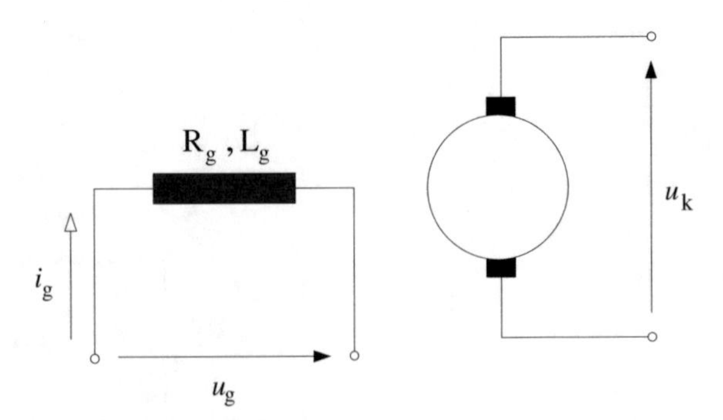

Fig. 1.32 Scheme of an externally excited direct current generator

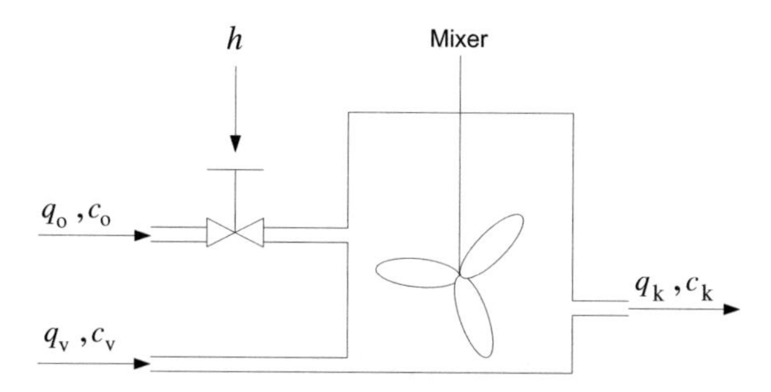

Fig. 1.33 Mixing tank

position h of the plunger, the output signal is the concentration c_k of the obtained solution. The amount of the inflow solution is proportional to the position of the plunger: $q_o = K\,h$. The amount of the outflow solution is the sum of the amount of the inflow solution and the inflow water: $q_k = q_o + q_v$. During time Δt the amount of the dissolved material getting into the tank of volume V is $q_o c_o \Delta t$, and at the same time dissolved material of amount $q_k c_k \Delta t$ leaves the tank. The change of the concentration is:

$$\Delta c_k = \frac{q_o c_o - q_k c_k}{V} \Delta t.$$

The differential equation of the system is obtained by taking the limit $\Delta t \to 0$:

$$\begin{aligned}\frac{dc_k}{dt} + \frac{q_k}{V} c_k &= \frac{dc_k}{dt} + \frac{q_o}{V} c_k + \frac{q_v}{V} c_k \\ &= \frac{dc_k}{dt} + \frac{K}{V} c_k h + \frac{q_v}{V} c_k = \frac{c_o K}{V} h\end{aligned}$$

The relationship is non-linear, as the product of the output signal c_k and the input signal h appears in the equation. But supposing $q_o \ll q_v$, then $q_k \approx q_v = \text{constant}$, and a constant q_k can be taken into consideration in the differential equation. Thus a linear differential equation is obtained.

$$\frac{dc_k}{dt} + \frac{q_v}{V} c_k = \frac{c_o K}{V} h$$

■

1.3.4 Linearization of Static Characteristics

Investigation of non-linear systems is a difficult task. The analysis can be simplified if the non-linear characteristics are linearized in a given vicinity of a working point. Thus in the surrounding of the working point the non-linear system is approximated by a linear model supposing only small changes in the input signals.

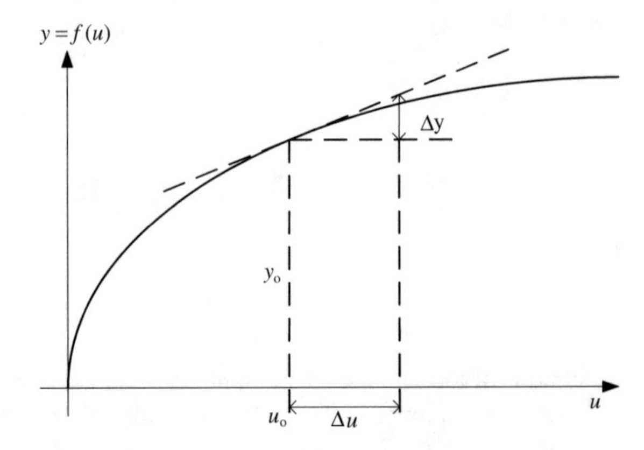

Fig. 1.34 A non-linear static characteristic with single input–single output

Let us consider the non-linear static characteristics $y = f(u)$ shown in Fig. 1.34. At the working point $u = u_o$; $y_o = f(u_o)$ the TAYLOR series of the function is:

$$y = y_o + \Delta y = f(u_o) + f'(u_o)\,(u - u_o) + \cdots$$

Neglecting the higher degree terms, the linearized model is given by

$$y - y_o = \Delta y = f'(u_o)(u - u_o) = f'(u_o)\Delta u$$

The linearized model replaces the static characteristics at the working point by the gradient. Of course the steepness depends on the working point.

Linearization in the case of several inputs

Let the output signal y be a function of the vector of the input variables $\mathbf{u} = [u_1, u_2, \ldots, u_n]^{\mathrm{T}}$. Thus y is a scalar-vector function. Let the vector $\mathbf{u}_o = [u_{1o}, u_{2o}, \ldots, u_{no}]^{\mathrm{T}}$ denote the working point. In a small vicinity of the working point the value of the output signal can be approximated by the TAYLOR expansion

$$y = y_o + \Delta y = f(\mathbf{u}_o) + \sum_{i=1}^{n} \left.\frac{\partial f(\mathbf{u})}{\partial u_i}\right|_{\mathbf{u}_o} (u_i - u_{io}) + \cdots$$
$$= f(\mathbf{u}_o) + \left[\left.\frac{d f(\mathbf{u})}{d\,\mathbf{u}}\right|_{\mathbf{u}_o} \right]^{\mathrm{T}} (\mathbf{u} - \mathbf{u}_o) + \cdots$$

Neglecting the second and higher order derivatives, the small change in the function $f(\boldsymbol{u})$ around the working point can be given by the following linear relationship:

$$\Delta y = \sum_{i=1}^{n} A_i\,\Delta u_i.$$

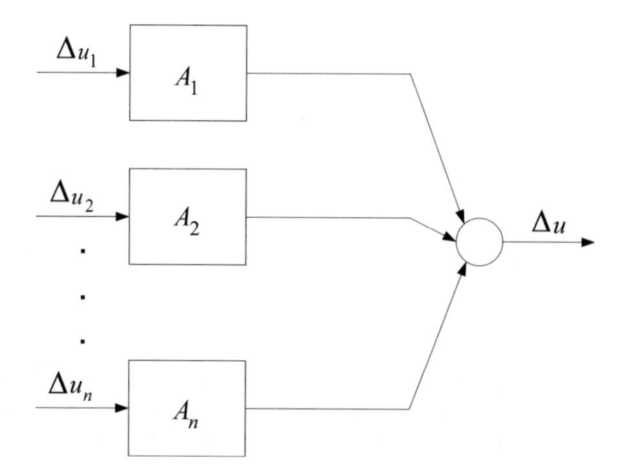

Fig. 1.35 Linearization of multi-input single-output static characteristics

The linearized block diagram is shown in Fig. 1.35. The A_i coefficients are the so called static transfer coefficients of the linearized model, whose values depend on the working point.

Example 1.4 Linearization of the moment equation of a DC motor

The moment m in a direct current (DC) motor is proportional to the product of the flux φ in the excitation coil and the armature current i (Fig. 1.36). The product of these two changing variables results in a non-linear relationship.

$$m = m_o + \Delta m = k\varphi i = k\varphi_o i_o + \left.\frac{\partial m}{\partial \varphi}\right|_{\varphi_o, i_o} \Delta\varphi + \left.\frac{\partial m}{\partial i}\right|_{\varphi_o, i_o} \Delta i$$

Determining the derivatives and considering that the value of the moment in the working point is $m_o = k\varphi_o i_o$, the change of the moment around the working point can be calculated according to the following relationship: $\Delta m = k i_o \Delta\varphi + k\varphi_o \Delta i$. ■

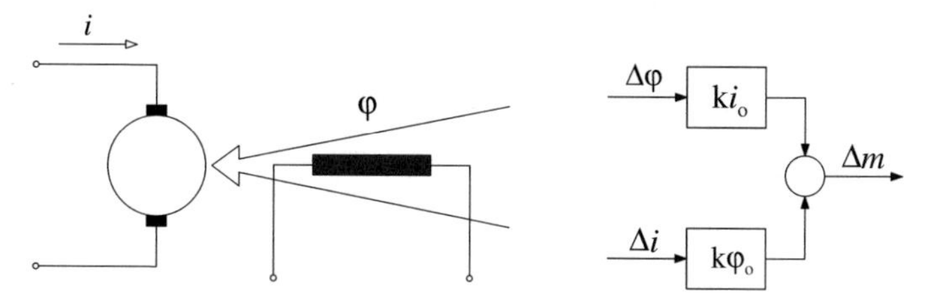

Fig. 1.36 The moment in the DC motor is proportional to the product of the excitation flux and the armature current

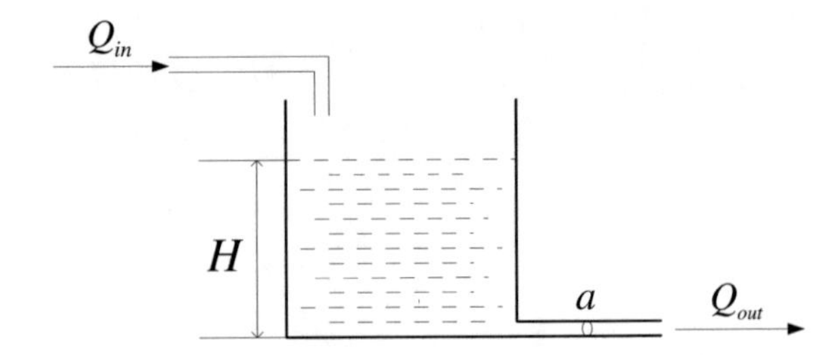

Fig. 1.37 Setting the liquid level in a tank

Example 1.5 Linearization of the tank equation
In a tank, the increase of the liquid level depends on the difference between the flow rate of the input liquid and that of the output liquid (Fig. 1.37). Let us denote the input flow by Q_{in}, and the output flow by Q_{out}, respectively. The cross section of the tank is denoted by A, and the cross section of the outflow tube is denoted by a. The liquid level is H. The change of the liquid level is described by the following differential equation:

$$A\frac{\mathrm{d}H}{\mathrm{d}t} = Q_{\text{in}} - Q_{\text{out}}$$

The output liquid flow depends on the velocity v of the outflow, which is proportional to the square root of the level.

$$Q_{\text{out}} = av = a\sqrt{2gH} = \beta\sqrt{H}$$

In steady-state, the level does not change, so the input and output flows are equal: $Q_{\text{in}} = Q_{\text{out}}$.

The steady-state value of the level will be $H = Q_{\text{in}}^2/\beta^2$. The static characteristic of the tank, viz., the relationship between the liquid level and the input flow, is non-linear (Fig. 1.38).

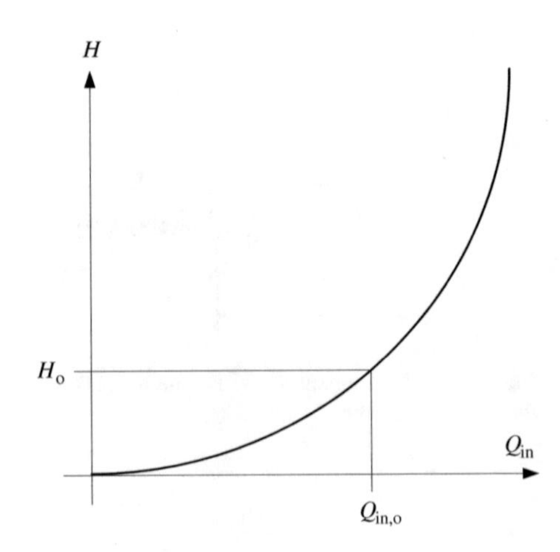

Fig. 1.38 Static characteristics of the tank

Let us denote the values of the working points by the index zero, and the changes around the working point with lower case letters

$$H = H_o + h$$
$$Q_{in} = Q_{in,o} + q_{in}$$

The outflow can be expressed with the first order TAYLOR approximation of the square root expression as

$$Q_{out} = \beta\sqrt{H} \approx \beta\sqrt{H_o} + \beta\frac{1}{2\sqrt{H_o}}h$$

The differential equation expressed with the working point values and the small changes around them is:

$$A\frac{d(H_o + h)}{dt} = Q_{in,o} + q_{in} - \beta\sqrt{H_o} - \frac{\beta}{2\sqrt{H_o}}h$$

As the derivative of a constant H_o working point value is zero, and $Q_{in,o} = \beta\sqrt{H_o}$, for the small changes around the working point the following differential equation can be given:

$$A\frac{dh}{dt} = q_{in} - \frac{\beta}{2\sqrt{H_o}}h$$

This is a linear differential equation whose parameters depend on the working point. ■

1.3.5 Relative Units

The transfer factors (gains) of the elements in a control system have dimensions. In the previous example of the liquid tank, the units of the working-point-dependent transfer gain resulting from the static characteristics is cm/(l/min). In the case of a motor, the output signal is the speed, the input signal is the voltage, thus the dimension of the transfer gain is (rad/s)/V. If the actual values of both the input and the output signals are related to their maximum values, the signals can be given with dimensionless relative values, which are between 0 and 1. The signals to be compared should be normalized identically. For example, the maximum values of the reference signal, the controlled signal and the error signal have to be the same.

Quantities with the dimension of time can also be given with relative values, if they are related to a maximum value chosen for the time variable.

As an example, let us consider the construction shown in Fig. 1.39. The DC motor M moves the rod R through transmission gears. The input signal of the motor

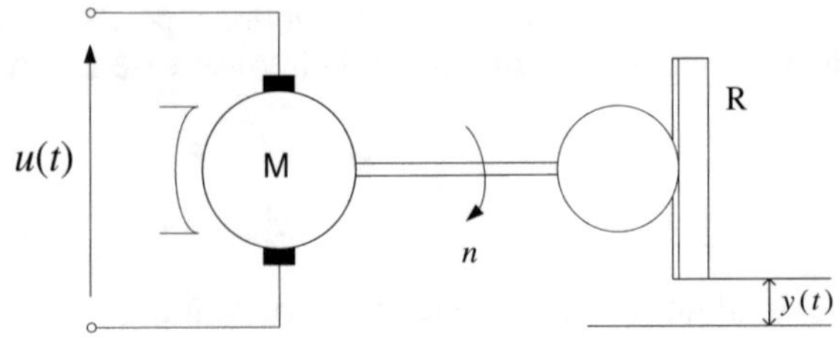

Fig. 1.39 Position control

is its terminal voltage $u(t)$, and its output signal is the position $y(t)$ of the rod (plunger). Neglecting the transients the displacement of the rod is proportional to the integral of the speed of the motor, and the speed is proportional to the terminal voltage. If the application of a terminal voltage of 200 V produces displacement of the rod by 5 cm within 10 s, then after time t the displacement is

$$y(t) = \frac{5\ \text{cm}}{10\ \text{s} \cdot 200\ \text{V}} \int_0^t u(t)\,\mathrm{d}t = 2.5 \cdot 10^{-3} \frac{\text{cm}}{\text{V s}} \int_0^t u(t)\,\mathrm{d}t$$

as the effect of the input voltage $u(t)$.

Let us take $t_{\max} = 50$ s as the unit of time, $y_{\max} = 20$ cm as the unit position and $u_{\max} = 200$ V as the unit of voltage. The relative units related to their basic units are:

$$t_{\text{rel}} = \frac{t}{t_{\max}} = \frac{t}{50\ \text{s}}; \quad y_{\text{rel}} = \frac{y}{y_{\max}} = \frac{y}{20\ \text{cm}}; \quad u_{\text{rel}} = \frac{u}{u_{\max}} = \frac{u}{200\ \text{V}}$$

With relative units the displacement of the rod can be given by the following relationship:

$$y_{\text{rel}}(t) = \frac{\frac{5\ \text{cm}}{20\ \text{cm}}}{\frac{10\ \text{s}}{50\ \text{s}} \cdot \frac{200\ \text{V}}{200\ \text{V}}} \int_0^{t_{\text{rel}}} u_{\text{rel}}(t)\,\mathrm{d}t_{\text{rel}} = 1.25 \int_0^{t_{\text{rel}}} u_{\text{rel}}(t)\,\mathrm{d}t_{\text{rel}}$$

1.4 Practical Aspects

The design and implementation of a control system is an iterative task. First the requirements set for the control system have to be formulated. Then based on the physical operation of the process, its mathematical model is established, whose parameters are determined by measurements and identification procedures. The controller is designed for the process model considering the given requirements. Then the operation of the control system is checked by simulation. If necessary, the controller is redesigned. During the implementation, the adjustment of the controller is refined.

In a control problem three basic tasks may occur.

It is necessary to create the model P of the process: the signal transfer properties of each element have to be determined based on the physical relationships describing the behavior of the element, or from its input and output measurement data by *identification* (Figs. 1.40 and 1.41).

If the input signal and the element P are known, the output signal can be determined and the behavior of the element can be analyzed (Fig. 1.42).

If the element P is given and the course of its required output signal is prescribed, then the task is to determine the input signal which ensures this behavior. The input of the plant is created by a control circuit. This is the synthesis or controller design task (Fig. 1.42).

Control engineering is an interdisciplinary area of science. The operation of the process is to be understood, to do this there is a need of knowledge of physical, chemical, biological, etc. phenomena. Mathematical knowledge is required for system modeling as well as the analysis and synthesis of control systems. To investigate the operation of control systems, knowledge is needed about signals, systems, and the behavior of systems with negative feedback. During the design, rational considerations and basic restrictions also have to be taken into account. The design has to cover economic, safety, environmental protection, etc. aspects as well. To fulfill a more complex control task, the coordinated work of different professionals is needed.

During the realization, the state of the system has to be observed—the considered output signal has to be measured by the appropriate measuring equipment, it is required to manipulate the process input—an actuator has to be selected. The measurement noise of the sensors, the signal ranges of the actuators, the limits of the produced actuating effects, all have to be taken into account. Several times the measured data have to be transferred across longer distances, thus data transfer has to be ensured. There are standards, so called protocols for data transfer which have

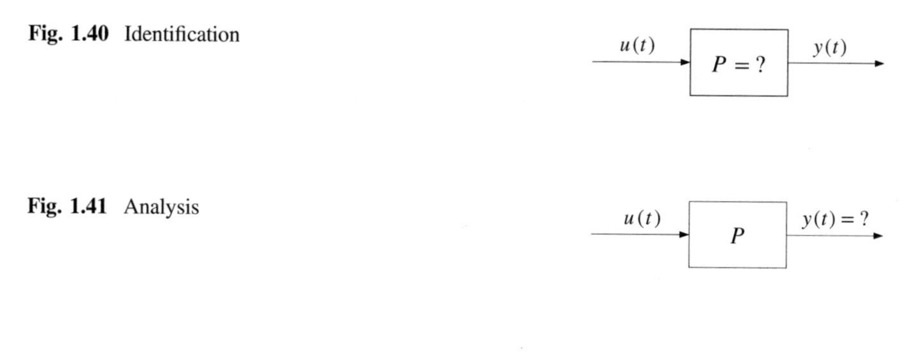

Fig. 1.40 Identification

Fig. 1.41 Analysis

Fig. 1.42 Synthesis

to be considered. The control signal has to be determined with an appropriate calculation algorithm, and has to be forwarded to the input of the process. In the design of the control algorithm the disturbances acting on the process, the uncertainties in the process parameters and also the restrictions due to practical realization have to be taken into account. During the control, real data are elaborated and real time signal transfer is realized. In signal transfer, non-deterministic signal delays do appear, which may distort the operation. The connection and exchange of information between the individual elements have to be addressed using appropriate interface elements.

Besides the continuous-time control systems computer control systems have gained more and more applications. The process and the process controller computer are connected via A/D (analog to digital) and D/A (digital to analog) converters. The computer executes the essential control functions in real time, repeatedly at the sampling instances. In industrial process control systems, distributed control systems are implemented, where spatially distributed control systems operate in an aligned fashion, communicating with each other.

Chapter 2
Description of Continuous Linear Systems in the Time, Operator and Frequency Domain

The aim of controlling a plant is to maintain the required value of the controlled (output) signal prescribed by the reference signal in spite of disturbances. The control system has to meet the quality specifications set for the control system. The quality specifications prescribe the static accuracy (the tolerable static error) of the control system and also the properties of its dynamic response (the settling time, the allowed value of the overshoot, etc.). The comparison of the factual and the prescribed behavior can be done based on the analysis of the static and dynamic response of the control system.

Various processes can be described mathematically by similar differential equations (or by a set of differential equations), which give the relationships between the individual variables and their changes. Mechanical motions, electrical and magnetic phenomena, heat processes, gas- and liquid flow, etc., can all be described by differential equations.

In a closed-loop control system different units executing specific control operations are connected to ensure the appropriate functioning of the process. The mathematical model of the closed-loop control system is a block diagram, which shows how the units are connected to each other and also represents the signal transfer properties of the individual units. Based on this model the operation of the closed-loop control system can also be given by a differential equation. In the sequel the behavior of systems described by lumped parameter, continuous linear differential equations will be investigated.

As the solution of the differential equation is sometimes cumbersome, several methods have been developed to simplify the calculations. Transforming the differential equation into the domain of the LAPLACE transform, an algebraic equation has to be solved instead of a differential equation. Examination of the process in the frequency domain provides fast approximate methods to evaluate the properties of the time response.

In the sequel, methods for analyzing lumped parameter, linear time invariant continuous-time systems in the time domain, the LAPLACE operator and the

L. Keviczky et al., *Control Engineering*, Advanced Textbooks in Control and Signal Processing, https://doi.org/10.1007/978-981-10-8297-9_2

frequency domain will be summarized. (These methods are known from the subject "Signals and Systems", here those relationships are considered which are important from control aspects.)

2.1 Description of Continuous Systems in the Time Domain

A continuous-time (CT) linear single-input single-output (*SISO*) time invariant system can be described in the time domain by a differential equation of order n or by a system constructed by a set of n first-order differential equations (the so-called state space equation), or it can be characterized by typical time responses given for typical input excitations.

2.1.1 Solution of an n-th Order Linear Differential Equations in the Time Domain

A linear CT time-invariant system can be described by the following n-th order differential equation:

$$\begin{aligned} &a_n y^{(n)}(t) + a_{n-1} y^{(n-1)}(t) + \cdots + a_1 \dot{y}(t) + a_o y(t) \\ &\quad = b_m u^{(m)}(t) + b_{m-1} u^{(m-1)}(t) + \cdots + b_1 \dot{u}(t) + b_o u(t) \end{aligned} \tag{2.1a}$$

where u denotes the input signal, y is the output signal, $\dot{y}$ is the first derivative of the output signal, $\dot{u}$ is the first derivative of the input signal, $y^{(n)}$ denotes the n-th derivative of the output signal, while $u^{(m)}$ denotes the m-th derivative of the input signal.

If the output responds with a delay (the so-called dead-time) to changes in the input signal, then the argument on the right side of the differential equation should be $t - T_d$, where T_d denotes the dead-time. Then the differential equation is given in the following form:

$$\begin{aligned} &a_n y^{(n)}(t) + a_{n-1} y^{(n-1)}(t) + \cdots + a_1 \dot{y}(t) + a_o y(t) \\ &\quad = b_m u^{(m)}(t - T_d) + b_{m-1} u^{(m-1)}(t - T_d) + \cdots + b_1 \dot{u}(t - T_d) + b_o u(t - T_d) \end{aligned} . \tag{2.1b}$$

Dead-time appears, e.g., in transport processes, where the change of the input signal can be measured with a delay in a farer measurement point. The necessary condition of physical realizability is

$$m \leq n \tag{2.2}$$

as only in the case of the fulfillment of this condition will the output signal remain finite for finite changes of the input signal.

From the theoretically infinite number of solutions of the differential equation that solution has to be chosen which satisfies the boundary conditions of the function y. The solution has to fulfill n conditions prescribed for $y(t)$ and its derivatives. The boundary conditions generally are initial conditions, i.e., they are given as $y(0), \dot{y}(0), \ldots, y^{(n-1)}(0)$.

The right side of the equation is the excitation $g(t)$

$$g(t) = b_m u^{(m)}(t) + b_{m-1} u^{(m-1)}(t) + \cdots + b_o u(t) \tag{2.3}$$

Equations (2.1a) and (2.1b) is an inhomogeneous differential equation, which if $g(t) = 0$ becomes a homogeneous equation.

In the following, different forms and solutions of the differential Eq. (2.1a) will be discussed, but the considerations can also be applied to Eq. (2.1b). Often the differential equation is written in the following, so called time constant form:

$$\begin{aligned} &T_n^n y^{(n)}(t) + T_{n-1}^{n-1} y^{(n-1)}(t) + \cdots + T_1 \dot{y}(t) + y(t) \\ &= A\left[\tau_m^m u^{(m)}(t) + \tau_{m-1}^{m-1} u^{(m-1)}(t) + \cdots + \tau_1 \dot{u}(t) + u(t)\right] \end{aligned} \tag{2.4}$$

where $A = b_o/a_o$ is the gain of the system, which gives the relation between the output and input signals in steady state. The gain is not a pure number, it has a physical dimension. $T_i = \sqrt[i]{a_i/a_o}$ and $\tau_j = \sqrt[j]{b_j/b_o}$ are time constants with the dimension of seconds.

The advantage of the time constant form is that even without solving the differential equation, on the basis of the parameters it is possible to approximately outline the course of the time responses for typical input signals.

The above definition of the system gain is valid only if a_o and b_o are different from zero. If e.g., $a_o = 0$, the gain is defined as $A = b_o/a_1$ and in this case the interpretation of the time constants is also changed.

The behavior of the system in the time domain can be obtained by solving the differential equation. The solution consists of two components, the general solution $y_h(t)$ of the homogeneous equation and one particular solution $y_i(t)$ of the inhomogeneous equation.

$$y(t) = y_h(t) + y_i(t) \tag{2.5}$$

The characteristic equation is obtained by substituting the derivatives of y multiplying y by the appropriate powers of s in the homogeneous equation. Thus the characteristic equation turns out to be

$$a_n s^n + a_{n-1} s^{n-1} + \cdots + a_1 s + a_o = 0. \tag{2.6}$$

The general solution of the homogeneous equation has the form

$$y_h(t) = k_1 e^{s_1 t} + k_2 e^{s_2 t} + \cdots + k_n e^{s_n t} \tag{2.7}$$

where $s_1, s_2, \ldots, s_n$ are the roots of the characteristic equation of the system (the roots of polynomials with real coefficients can only be real or complex conjugate pairs). The constants k_i have to be determined from the initial conditions.

If in the solution of the characteristic equation multiple roots show up, the corresponding exponential terms are multiplied by the powers of t. For example if there is a triple root, then the general solution of the homogeneous equation is given in the following form:

$$y_h(t) = \left(k_1 + k_2 t + k_3 t^2\right) e^{s_{1,2,3} t} + k_4 e^{s_4 t} + \cdots + k_n e^{s_n t} \tag{2.8}$$

Let $f(u)$ denote a particular solution of the inhomogeneous equation which depends on the input signal u. Supposing that $f(u)$ has been found by some procedure—e.g., by the method of variation of parameters or by simple considerations—the general solution of the differential Eqs. (2.1a) and (2.1b) becomes

$$y(t) = y_h(t) + f(u) = k_1 e^{s_1 t} + \cdots + k_n e^{s_n t} + f(u). \tag{2.9}$$

The constants k_i have to be determined by a knowledge of the initial conditions.

To solve the differential equation in the time domain often requires following a complicated and cumbersome procedure. The characteristic equation has an analytic solution only for $n \leq 4$. To find one particular solution of the inhomogeneous equation is a demanding computational task in the case of sophisticated input signals.

From the form of the differential equation some statements can be made concerning the initial and final values of the step response. Let us analyze the form (2.1a) of the differential equation. Let the input signal be a step given by $g(t) = b_o 1(t)$. At time point $t = 0$ only the highest derivative could jump. (I.e., the two sides of the differential equation have to be in balance at each time point. If there were a jump also in a lower order derivative of the output signal, this would result in a DIRAC impulse change in the higher order derivatives.)

$$a_n y^{(n)}(t = 0) = b_o,$$

So

$$y^{(n)}(t = 0) = b_o / a_n.$$

(Considering e.g., mechanical motion, when the force acting on the mass changes, first only the acceleration changes and this change will produce further changes in the velocity and the position.)

It has to be mentioned that if the excitation signal also contains the first derivative of the input signal, then at the initial point the n-th and also the $(n-1)$-th derivative of the output signal will jump. The general rule is that for a step-like excitation at time point $t = 0$ the $(n-m)$-th derivative of the output signal will jump. If the transients are decaying, all derivatives of the output signal will be zero, and the output signal will have settled at the value determined by the static gain: $y(t \to \infty) = b_o/a_o$.

The physical content behind the formal mathematical solution of the differential equation can be interpreted as follows.

The differential equation describes the motion of a system. The reason for the motion on the one hand is the input signal $u(t)$, and on the other hand, a component of the motion appears as a consequence of the past inputs, as before the appearance of the input signal at the time instant $t = 0$ the system was not in a steady state. The past history of the system is characterized unambiguously by its initial conditions. As a response to the excitation signal $g(t)$ a new steady state will be reached, which is determined by the solution of the inhomogeneous equation, which is independent of the initial conditions. This new steady state for time instant $t = 0$ would prescribe initial conditions which depend on the excitation. If the values of the actual initial conditions do not coincide with the initial values corresponding to the excitation, this indicates that the state of the system is different from the steady state prescribed by the excitation. This deviation can not disappear abruptly, as there are energy storing elements in the system which can only change their state gradually by energy conveyance or distraction. Changes in the state need a finite amount of time. The balancing movement is the transient motion which is described by the solution of the homogeneous differential equation.

The solution of the differential equation can be decomposed into a quasi-stationary and a transient component. The quasi-stationary component is the output signal of the system in steady state as a response to the input signal (see Appendix A.2). The transient component depends on the dynamics of the system, as determined by the roots of the characteristic equation.

As an example, let us analyze an electrical circuit consisting of a resistor and an inductor. A sinusoidal voltage gets switched on, as the input (Fig. 2.1a). The quasi-stationary steady state is represented by a sinusoidal alternating current $I(t)$ which is delayed, compared to the input alternating voltage by a given angle, determined by the parameters of the circuit. If the switching on of the voltage happens at time instant t_1 when the current is zero, then the state of the system coincides with the steady state corresponding to the input signal and in this case no transient motion occurs (Fig. 2.1b). But if the switching occurs at a time instant t_2 when the current has a non-zero value $I(t_2) \neq 0$, then the system is not in steady state. The deviation between the actual current $i(t_2) = 0$ and the steady state current $I(t_2)$ is compensated by the transient component $\Delta i(t)$, which is superposed onto

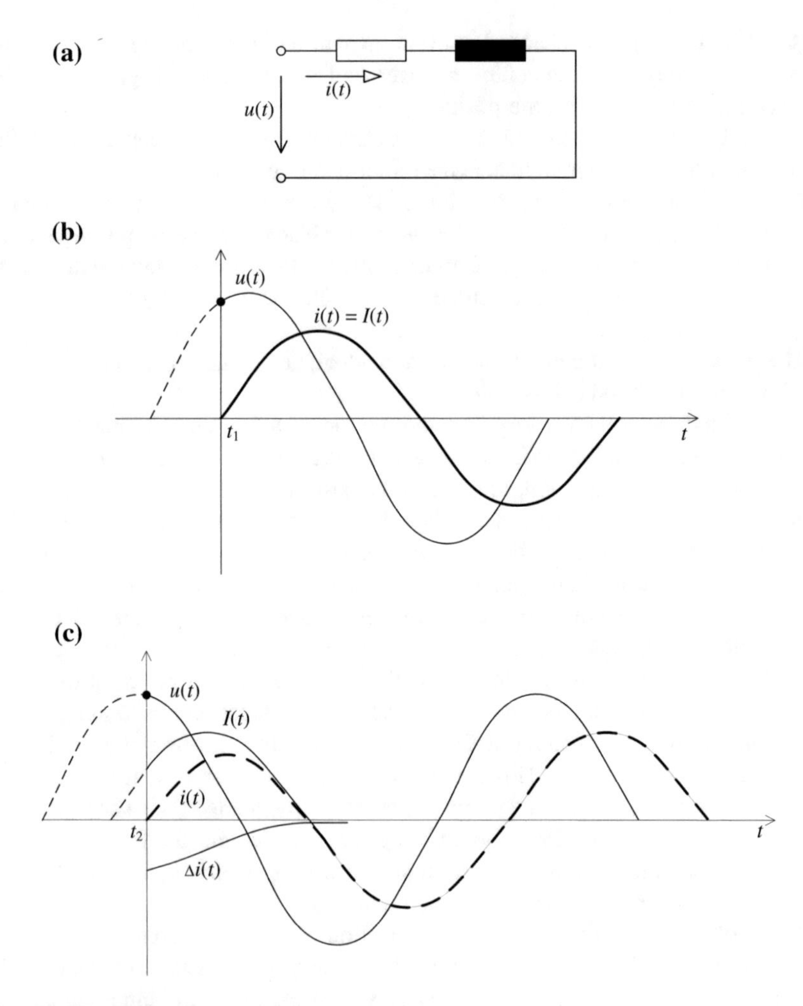

Fig. 2.1 RL circuit and its transients

$I(t)$. This transient component ensures the resulting zero value of the current at the switching time instant, and then it will decease exponentially (Fig. 2.1c).

The course of the motion of the transient shows the fundamental properties of the system. If the transient components are decreasing in time, then a new steady state corresponding to the excitation will be reached, i.e., the system is stable. But an increasing transient motion means unstable performance. In this case a new steady state will not be reached. Undamped oscillating periodic transient motion means a stability limit, when the system is resonant to sinusoidal input signals whose frequency is equal to the frequency of the transient oscillations. The stability of the system can be determined based on the roots of the characteristic equation.

To analyze the transient response it is enough to consider the solution of the homogeneous equation which provides the free motion of the system. The free response stems from the fact that the system is not in steady state at the time instant $t = 0$ (e.g., because the system previously had been moved away from its steady state). In this case a stable system tends to reach its steady state again through the transient motion. The transient phenomena of a system excited by an input signal are similar as a consequence of the superposition, but now the steady state value is replaced by the motion generated by the excitation input signal.

2.1.2 State Space Representation of Linear Differential Equations

The state of a system described by a differential equation at time instant $t = 0$ is unambiguously determined by the initial conditions. Besides the input and output signals inner signals can also be considered in the system, characterizing the state of the system at each time instant. These variables—the so called state variables—can be, e.g., the output signal and its derivatives. Their main property is that they can not respond abruptly to an abrupt change of the input signal: time is needed to gradually change their values. From the actual values of the state variables and the input signal, the value of the output signal at the next time instant can be determined.

Introducing the state variables the differential equation of order n can be transformed into a system of n first-order differential equations.

As an example let us consider the differential Eqs. (2.1a) and (2.1b) with excitation $g(t) = b_o\, u(t)$. Expressing $y^{(n)}$, the highest derivative, the differential equation can be represented by the block diagram shown in Fig. 2.2. On the basis of this block diagram, with the knowledge of the input signal and the initial

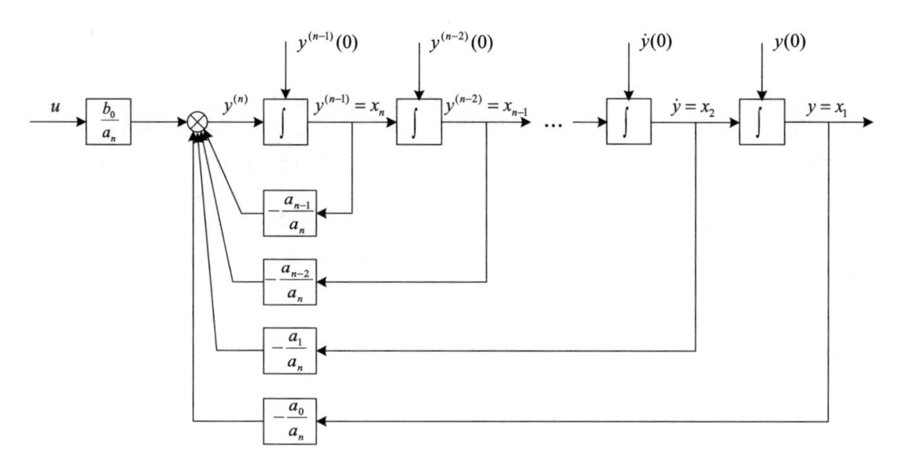

Fig. 2.2 State space form of the differential equation

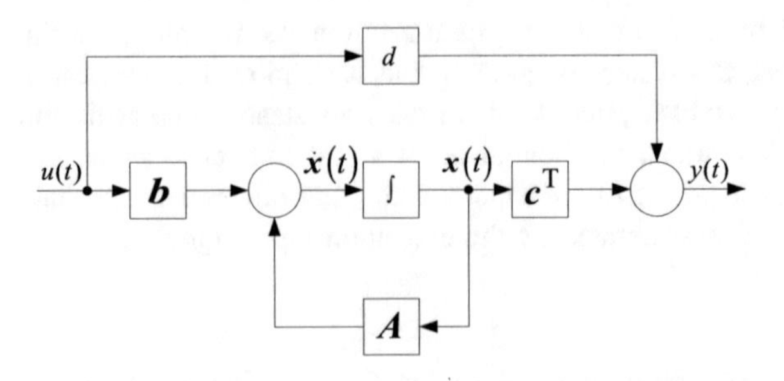

Fig. 2.3 State space representation of a dynamical system

conditions, the differential equation can be solved iteratively. In the block diagram the outputs of the integrators behave like state variables. Let us denote the state variables by $x_1, x_2, \ldots, x_n$. With these state variables the differential equation can be transformed to the following form.

$$\begin{aligned}
\dot{x}_1 &= x_2 \\
\dot{x}_2 &= x_3 \\
&\vdots \\
\dot{x}_n &= -\tfrac{a_0}{a_n} x_1 - \tfrac{a_1}{a_n} x_2 - \cdots - \tfrac{a_{n-1}}{a_n} x_n + \tfrac{b_0}{a_n} u \\
y &= x_1
\end{aligned} \tag{2.10}$$

In general, a system consisting of n first-order differential equations can be written in the following vector/matrix form.

$$\begin{aligned}
\dot{\boldsymbol{x}}(t) &= \boldsymbol{A}\,\boldsymbol{x}(t) + \boldsymbol{b}\,u(t) \\
y(t) &= \boldsymbol{c}^{\mathrm{T}}\boldsymbol{x}(t) + d\,u(t)
\end{aligned} \tag{2.11}$$

The elements of $\boldsymbol{x}$ are the state variables, $\boldsymbol{A}, \boldsymbol{b}, \boldsymbol{c}^{\mathrm{T}}$ are the matrices and vectors describing the system, and d is a scalar parameter. The output signal depends on the input signal generally through the state variables, but through the scalar gain by d a direct connection also exists between the input and the output signals. The state space representation of a dynamical system is shown in Fig. 2.3.

The state space form of a dynamical system also shows properties of the system which otherwise remain hidden when solving the differential equation describing the input/output relationship. Solving a set of first-order differential equations is generally simpler than solving the differential equation of order n.

Chapter 3 discusses the state space description of a control systems, the solution of the state equation and related topics.

2.1.3 Typical Input Excitations, Unit Impulse and Step Responses

The solution of the differential equation of the closed-loop control system gives the time evolution of the output signal for an arbitrary input signal. The calculation of one particular solution of the inhomogeneous equation is easier in the case of a simple input signal.

It is expedient to excite the system with a typical input signal which can generate a significant transient motion. Then the time evolution of the output signal will be characteristic for the signal transfer properties of the system, and consequences for the structure and the parameters of the system can be drawn from its shape.

When examining the behavior of a closed loop control system, it is expedient to choose an input signal resulting in a response which provides information about the reference signal tracking properties of the control system. If the system has to track and maintain a constant value, then a step-like input signal is appropriate. If it has to follow a changing reference signal, then a linearly changing ramp signal is to be chosen as input signal.

The most important typical input signals are the following:

- unit impulse function (Dirac delta): $\delta(t)$
- unit step function: $1(t)$,
- unit ramp function: $t\,1(t)$,
- unit parabolic function: $\frac{t^2}{2}1(t)$.

The responses obtained for the typical input signals are shown in Fig. 2.4.

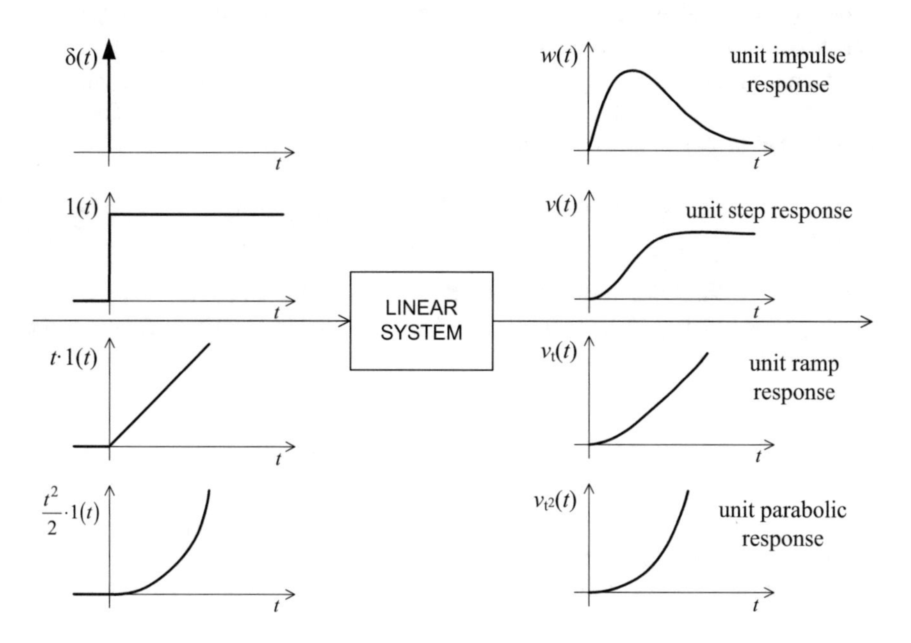

Fig. 2.4 Typical input signals and responses

The DIRAC delta is an impulse of unity area and infinite amplitude acting at the zero time instant. It is a mathematical abstraction, which can be derived as the limit of a rectangular impulse with width Δt and height $1/\Delta t$, when $\Delta t \to 0$. The *weighting function* denoted by $w(t)$ is the response of the system to a DIRAC delta input. The weighting function is characteristic for the system. From its evolution over time, one can draw conclusions about the structure and the parameters of the system, and even its stability. The weighting function characterizes the transient properties of the system. It behaves like the free response, as the exciting input signal acts for an infinitesimal time at time instant $t = 0$, but meanwhile, because of its finite energy content, it moves the output signal and its derivatives away from their steady position.

The *unit step* signal jumps at time instant $t = 0$ from 0 to 1. Its value is zero for $t < 0$, and is one for $t \geq 0$. The output of the system for a unit step input is called the *unit step response* and is denoted by $v(t)$.

The value of the *unit ramp* function for $t < 0$ is zero, and for $t \geq 0$ it is t. The response of the system to the ramp signal is called the *unit ramp response*.

The value of the *unit parabolic* function for $t < 0$ is zero, and for $t \geq 0$ it is $t^2/2$. The system response to this input is called the *unit parabolic response*.

The step, ramp and parabolic responses also characterize the system. The relationship between the typical input signals is the following:

$$\delta(t) = \frac{\mathrm{d}}{\mathrm{d}t} 1(t); \quad 1(t) = \frac{\mathrm{d}}{\mathrm{d}t} t1(t); \quad 1(t) = \frac{\mathrm{d}}{\mathrm{d}t^2} \frac{t^2}{2} 1(t). \tag{2.12}$$

(It has to be mentioned here that the unit step can not be differentiated according to the conventional definition of differentiation. In fact, the relationship between signals $\delta(t)$ and $1(t)$ can be interpreted using the theory of distributions.)

At the output of a linear system the relationship between the typical responses is the same as the relationship between the corresponding input signals. (This relationship can be derived by applying the linearity property.)

$$w(t) = \frac{\mathrm{d}v(t)}{\mathrm{d}t}; \quad v(t) = \frac{\mathrm{d}v_t(t)}{\mathrm{d}t}; \quad v_t(t) = \frac{\mathrm{d}v_{t^2}(t)}{\mathrm{d}t}. \tag{2.13}$$

Here $v_t(t)$ is the unit ramp response and $v_{t^2}(t)$ is the unit parabolic response (thus the weighting function is the derivative of the step response, the step response is the derivative of the ramp response, etc.).

2.1.4 System Response to an Arbitrary Input Signal

If the weighting function or the unit step response of the system is known, then with zero initial conditions the output can also be calculated for an arbitrary input signal. The response of the system will provide one particular solution of the inhomogeneous equation.

Let us determine the system response for an arbitrary input signal with the knowledge of the weighting function. The input signal $u(t)$ can be approximated by a series of shifted rectangular pulses (Fig. 2.5). Let the width of the pulses be $\Delta\tau$. The number of the pulses up to a given time point t is N. The area of a pulse is approximately $u(\tau)\Delta\tau$. The response of the system to a rectangular input pulse shifted by τ relative to time instant 0 is at time instant t approximately $w(t-\tau)u(\tau)\Delta\tau$. At a given time instant t the value of the output signal is influenced by all the pulses appearing as components of the input signal before the given time instant. In a linear system, the effect of the individual pulses on the output is superposed, thus the output signal can be approximately determined as

$$y(t) \approx \tilde{y}(t) = \sum_{i=1}^{N} w(t-\tau_i)\, u(\tau_i)\Delta\tau.$$

Taking the limit $\Delta\tau \to 0$ the output signal is expressed as

$$\begin{aligned} \tilde{y}(t) &= \sum_{i=1}^{N} w(t-\tau_i)\, u(\tau_i)\Delta\tau \to y(t) \\ &= \int_0^t w(t-\tau)\, u(\tau)\, d\tau, \quad \text{if } \Delta\tau \to 0. \end{aligned} \tag{2.14}$$

or substituting $t - \tau = \upsilon$

$$y(t) \approx \tilde{y}(t) = \sum_{i=1}^{N} w(\upsilon_i) u(t-\upsilon_i)\Delta\upsilon \tag{2.15}$$

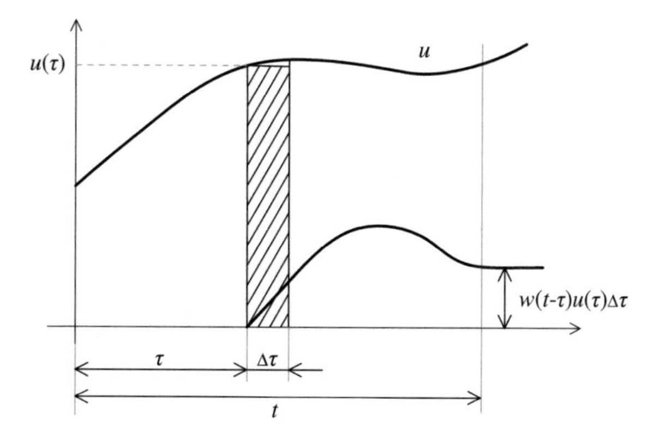

Fig. 2.5 Conceptual representation of the convolution integral

or taking the limit $\Delta\upsilon \to 0$

$$\begin{aligned} \tilde{y}(t) &= \sum_{i=1}^{N} w(\upsilon_i)\, u(t-\upsilon_i)\Delta\upsilon \to y(t) \\ &= \int_0^t w(\upsilon)\, u(t-\upsilon)\,\mathrm{d}\upsilon, \quad \text{if } \Delta\tau \to 0, \end{aligned} \tag{2.16}$$

Equations (2.14) and (2.16) give the *convolution integral* or the FALTUNG theorem. Applying the convolution integral instead of the solution of the differential equation a simpler expression is evaluated, but for a more complex input signal the calculation of this integral is also cumbersome.

Equation (2.15) provides a possibility for numerical evaluation in case the weighting function is decreasing. The values of the weighting function have to be given at sampling points $\upsilon_i = 0, \Delta\upsilon, 2\Delta\upsilon, \ldots, (N-1)\Delta\upsilon$. It is supposed that for the further course of the weighting function $w(i\,\Delta\upsilon) \approx 0$, if $i \geq N$. Besides the actual value of the input signal, $(N-1)$ previous values have to be stored.

The output signal can be approximately calculated as

$$\begin{aligned} \tilde{y}(t) \approx {} & [w(0)u(t) + w(\Delta\upsilon)u(t-\Delta\upsilon) + w(2\Delta\upsilon)u(t-2\Delta\upsilon) \\ & + \cdots + w((N-1)\Delta\upsilon)u(t-(N-1)\Delta\upsilon)]\Delta\upsilon \end{aligned}$$

(This form is also called the HANKEL form, or the weighting function model.)

The response of the system to an arbitrary input signal can also be calculated with the knowledge of the step response. The input signal can be approximated by a sum of shifted steps (Fig. 2.6). The output signal is obtained by superposing the responses to these shifted step inputs of given amplitudes.

The output signal can also be approximated by the following relationship:

$$\tilde{y}(t) = u(0)v(t) + \sum_{i=1}^{N} v(t-\tau_i)\Delta u(\tau_i) \tag{2.17}$$

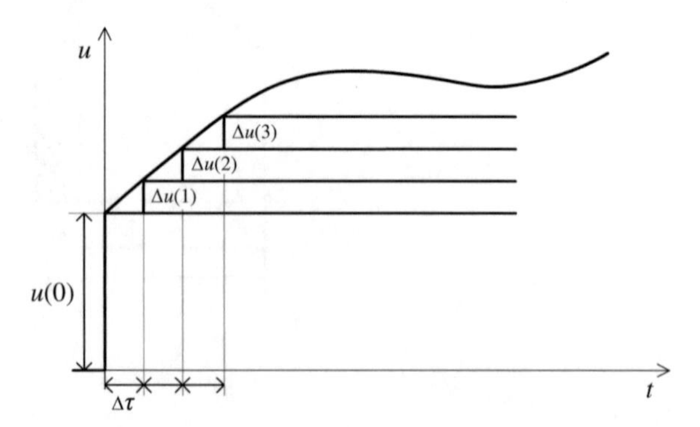

Fig. 2.6 The input signal can be built from superposed shifted step signals

or at the individual time points:

$$\begin{aligned}
&\tilde{y}(0) = u(0)v(0)\\
&\tilde{y}(\Delta\tau) = u(0)v(1) + \Delta u(1)v(0)\\
&\tilde{y}(2\Delta\tau) = u(0)v(2) + \Delta u(1)v(1) + \Delta u(2)v(0)\\
&\vdots
\end{aligned}$$

If $\Delta\tau$ is small, the output signal can be calculated with appropriate accuracy on the basis of the above relationship. If $\Delta\tau \to 0$ the output signal turns out to be

$$y(t) = u(0)\,v(t) + \int_0^t v(t-\tau_i)\,\frac{\mathrm{d}u(\tau)}{\mathrm{d}\tau}\mathrm{d}\tau \tag{2.18}$$

This expression is known as the DUHAMEL theorem.

2.1.5 Solution of a First-Order Differential Equation

A first-order differential equation is a special case of the n-th order differential equation given by Eq. (2.1a). Now $n = 1$, and let $m = 0$. Let us determine the weighting function and the step response of the system described by a first-order differential equation and derive the expression of the output signal for an arbitrary input excitation using the convolution integral. The differential equation takes the following form:

$$a_1\dot{y}(t) + a_o y(t) = b_o u(t) \tag{2.19}$$

Assume zero initial condition: $y(t = 0) = y(0)$. According to Eq. (2.4) the differential Eq. (2.19) gets normalized in the following time constant form:

$$T\dot{y}(t) + y(t) = Au(t) \tag{2.20}$$

where $T = a_1/a_o$ is the time constant and $A = b_o/a_o$ is the gain.

The behavior of the electrical circuit consisting of a resistor and an inductor shown in Fig. 2.1 can be described by a first-order differential equation. The KIRCHHOFF voltage law for this circuit is as follows:

$$L\frac{\mathrm{d}i(t)}{\mathrm{d}t} + Ri(t) = u(t).$$

The equation can be written in the form given by Eq. (2.20).

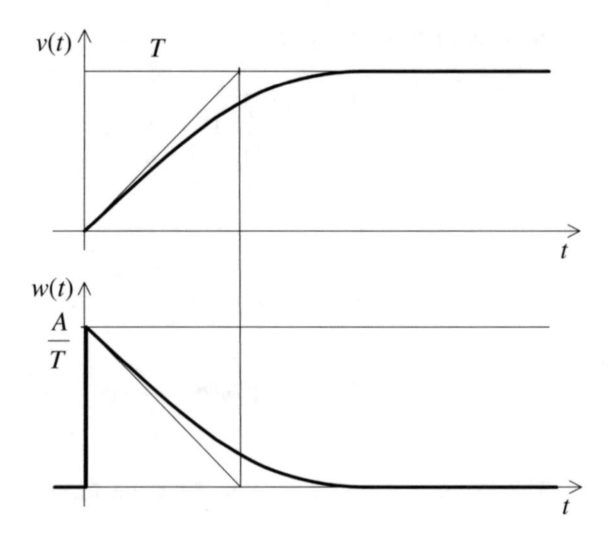

Fig. 2.7 Unit step response and weighting function of a system described by a first-order differential equation

Let us solve the differential equation applying a unit step input signal $u(t) = 1(t)$. The characteristic equation is $Ts + 1 = 0$. Its root is $s_1 = -1/T$. The general solution of the homogenous equation is $y_{\mathrm{h}}(t) = k_1 e^{-t/T}$. For unit step input in steady state the derivative of the output signal is zero, and $y_{\mathrm{ih}}(t) = y(t \to \infty) = A$.

The complete solution is $y(t) = y_{\mathrm{h}}(t) + y_{\mathrm{ih}}(t) = k_1 e^{-t/T} + A$. The value of the parameter k_1 can be determined from the knowledge of the initial condition: $y(0) = 0 = k_1 + A$. Thus, the complete solution, the analytical expression of the unit step response is

$$y(t) = v(t) = A\left(1 - e^{-t/T}\right), \quad t \geq 0 \tag{2.21}$$

which reaches its steady state value exponentially approximately within a time of $3T$ with an accuracy of 5%.

The derivative of the unit step response results in the weighting function

$$w(t) = \frac{\mathrm{d}v(t)}{\mathrm{d}t} = \frac{A}{T} e^{-t/T}. \tag{2.22}$$

The unit step response and the weighting function are shown in Fig. 2.7, where the time constant T can be indicated in the figure based on the relationship $\dot{v}(0) = w(0) = A/T$.

Knowing the weighting function the output signal can be calculated for an arbitrary input signal using the convolution integral. The complete solution considering also the effect of a non-zero initial condition is calculated according to the following relationship:

$$y(t) = \frac{A}{T}\left[e^{-\frac{1}{T}t}y(0) + \int_0^t e^{-\frac{1}{T}(t-\tau)}u(\tau)\mathrm{d}\tau\right]. \tag{2.23}$$

2.2 Transformation from the Time Domain to the Frequency and Operator Domains

An advantageous way to analyze lumped parameter differential equations is to use function transformations which transform the original functions of time to related functions. This transforms the original differential equation to an algebraic equation. Such transformations include the FOURIER and the LAPLACE transformations.

2.2.1 *FOURIER series, FOURIER integral, FOURIER transformation*

A periodic signal $y(t)$ can be expressed as the sum of harmonic (sinusoidal) components. This sum gives the FOURIER series, whose individual elements belong to discrete frequencies. Suppose the time period of the signal is T and its basic frequency $\omega_o = 2\pi/T$. The complex form of the FOURIER series is

$$y(t) = \sum_{n=-\infty}^{\infty} c_n e^{jn\omega_o t} \tag{2.24}$$

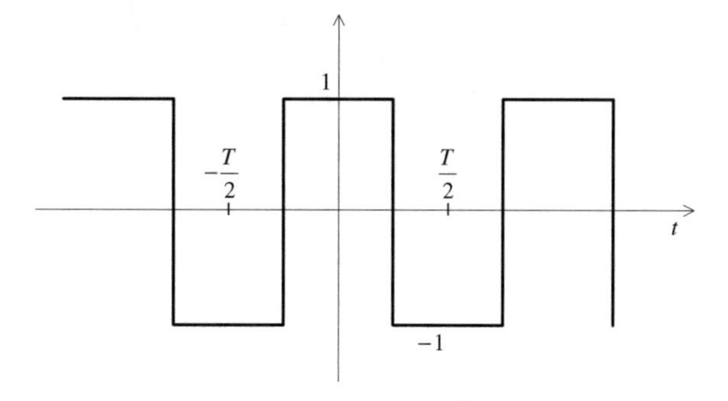

Fig. 2.8 Periodic signal

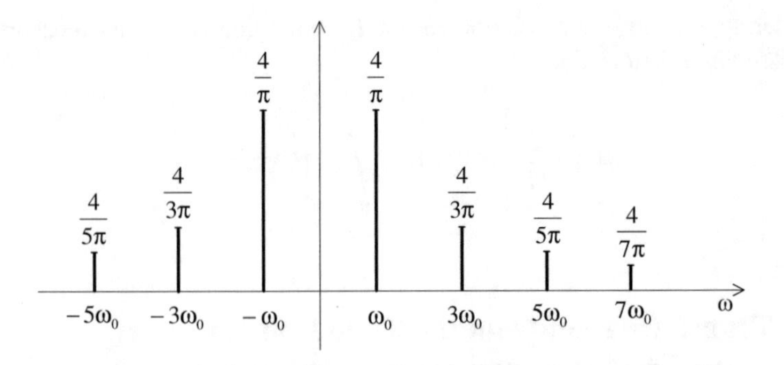

Fig. 2.9 Discrete frequency spectrum of a periodic signal

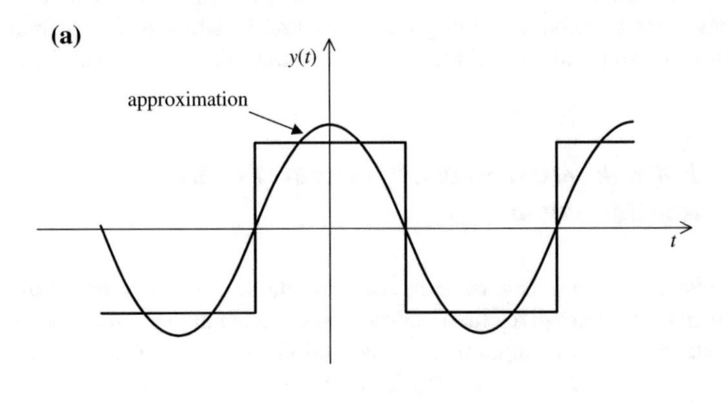

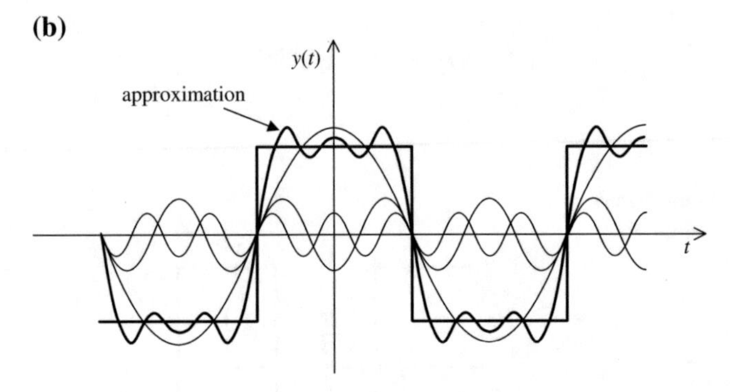

Fig. 2.10 Approximation of a periodic signal with harmonic components

where n is an integer and

$$c_n = \frac{1}{T} \int_{-T/2}^{T/2} y(t) e^{-jn\omega_o t} \mathrm{d}t \tag{2.25}$$

c_n is a complex number and further on $c_n = \bar{c}_{-n}$, where $\bar{c}$ denotes complex conjugate. The c_n are the amplitudes assigned to the discrete frequencies $\omega = n\omega_o$ and compose the amplitude spectrum of the periodic signal $y(t)$.

The FOURIER *series* can be given in real form as well, where the frequency components belonging to the same positive and negative frequency are closed up to sine and cosine functions.

Figure 2.8 shows a periodic function. Figure 2.9 gives the amplitude-frequency spectrum of the signal. Figure 2.10 illustrates the approximation of the function with the basic harmonic and with three FOURIER components, respectively. The more FOURIER components are considered, the better is the approximation of the periodic signal.

(It should be mentioned that the sine and cosine functions compose an orthogonal system. The FOURIER series is an orthogonal expansion of a periodic signal.)

In practice the input of a system generally is not periodic, but aperiodic (e.g., the unit step) in nature. An absolute integrable aperiodic function, where

$$\int_{-\infty}^{\infty} |y(t)| \mathrm{d}t = \text{finite}, \tag{2.26}$$

can be described in the form of a *FOURIER integral*, which is obtained by taking the limit $T \to \infty$ in the FOURIER series. That is, an aperiodic function can be considered as a periodic function whose time period tends to infinity. The derivation of an aperiodic function from a periodic function is illustrated in Fig. 2.11. By increasing the time period, the lines in the spectrum of the amplitude-frequency function are getting closer to each other, and in the limit the spectrum becomes continuous, every frequency appears in the signal with a certain weight. Instead of (2.24), the FOURIER integral is obtained by taking the limit $T \to \infty$:

$$y(t) = \frac{1}{2\pi} \int_{-\infty}^{\infty} Y(j\omega) e^{j\omega t} \mathrm{d}t \tag{2.27}$$

where $Y(j\omega)$ is the complex spectrum of the signal, the so called FOURIER transform of the signal $y(t)$, which is given by the following relationship:

$$Y(j\omega) = \int_{-\infty}^{\infty} y(t) e^{-j\omega t} \mathrm{d}t = \mathcal{F}\{y(t)\} \tag{2.28}$$

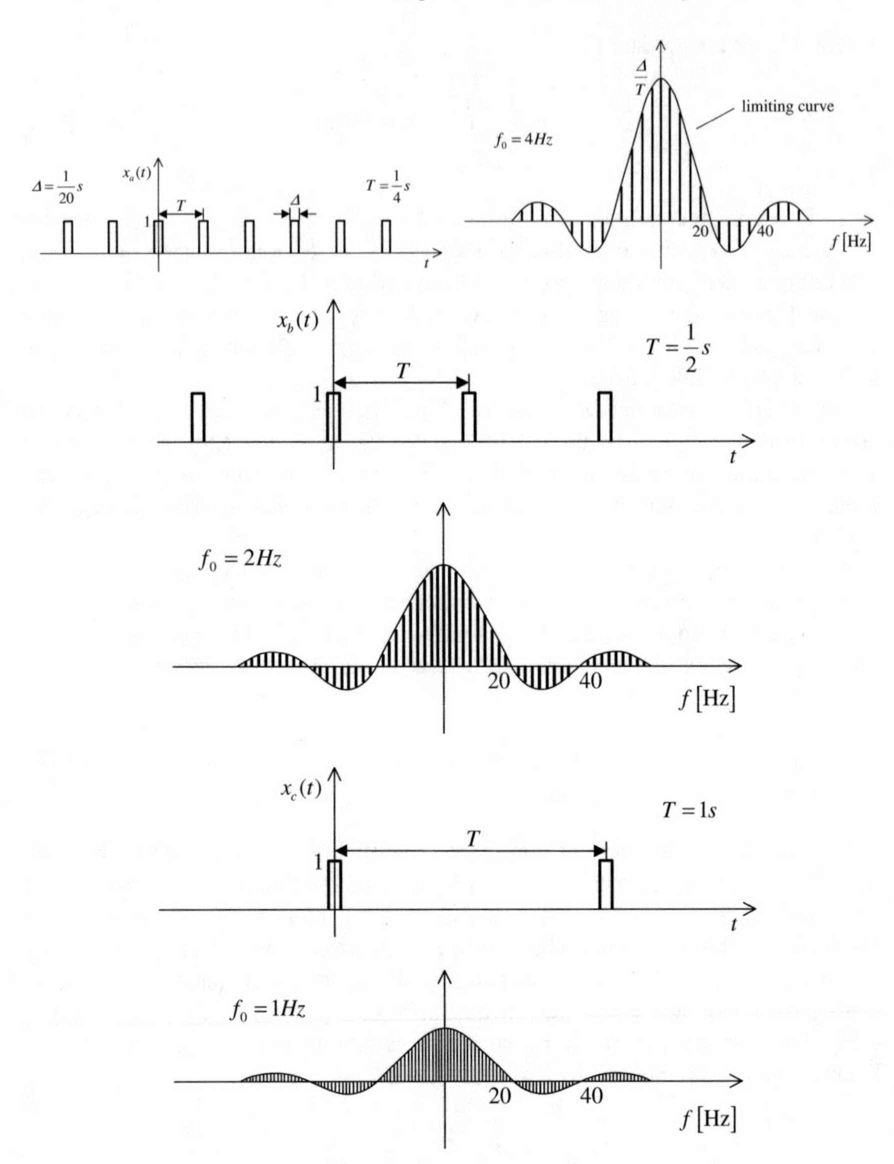

Fig. 2.11 Increasing the time period, the periodic function approximates an aperiodic function and the frequency spectrum becomes continuous

This is the basic expression of the FOURIER transform. The signal can be reconstructed from its FOURIER transform by the inverse FOURIER transformation, given by formula (2.27).

If $y(t)$ is different from zero only in the time domain $t \geq t_o$, then it is a one-sided time function and its FOURIER transform is also one-sided. Without restriction of generality, it can be supposed that $t_o = 0$. Then $y(t)$ is called a positive time function.

The FOURIER transform exists only if the signal is absolutely integrable, i.e., relationship (2.26) holds. This means that the square integral of the signal also exists, the signal has a finite energy content. Namely the energy can be expressed in the frequency domain by the PARSEVAL or the RAYLEIGH theorem as

$$\int_{-\infty}^{\infty} y^2(t)\mathrm{d}t = \frac{1}{2\pi} \int_{-\infty}^{\infty} Y(j\omega)Y(-j\omega)\mathrm{d}\omega. \tag{2.29}$$

Applying the FOURIER transformation to a differential equation an algebraic equation is obtained. Let us calculate the first time derivative of Eq. (2.27).

$$\dot{y}(t) = \frac{1}{2\pi} \int_{-\infty}^{\infty} j\omega Y(j\omega)e^{j\omega t}\mathrm{d}t.$$

It can be seen that the FOURIER transform of $\dot{y}(t)$ is $j\omega Y(j\omega)$, so in the frequency domain, differentiation by t is simplified to multiplication by $j\omega$.

It was seen that both the periodic and the aperiodic signals can be given by superposition of sinusoidal signals of different frequencies. Periodic signals can be approximated by the sum of sinusoidal signals of given discrete frequencies, where the higher frequency components appear with lower amplitude. Aperiodic signals contain all frequency components with a certain weighting. If a linear system is excited by a signal which is approximated by the sum of its sinusoidal components of different frequencies, using the superposition theorem the output signal can be approximated by the sum of the system responses for the individual components of the input signal. The approximation of the output signal is better if more frequency components are taken into account. Figure 2.12 shows the output of a system described by a second order differential equation in the case of a periodic rectangular input signal, and also illustrates the approximation of the input and the output signal with four and ten FOURIER components, respectively. It can be seen that both the input and the output signals are approximated well by ten components.

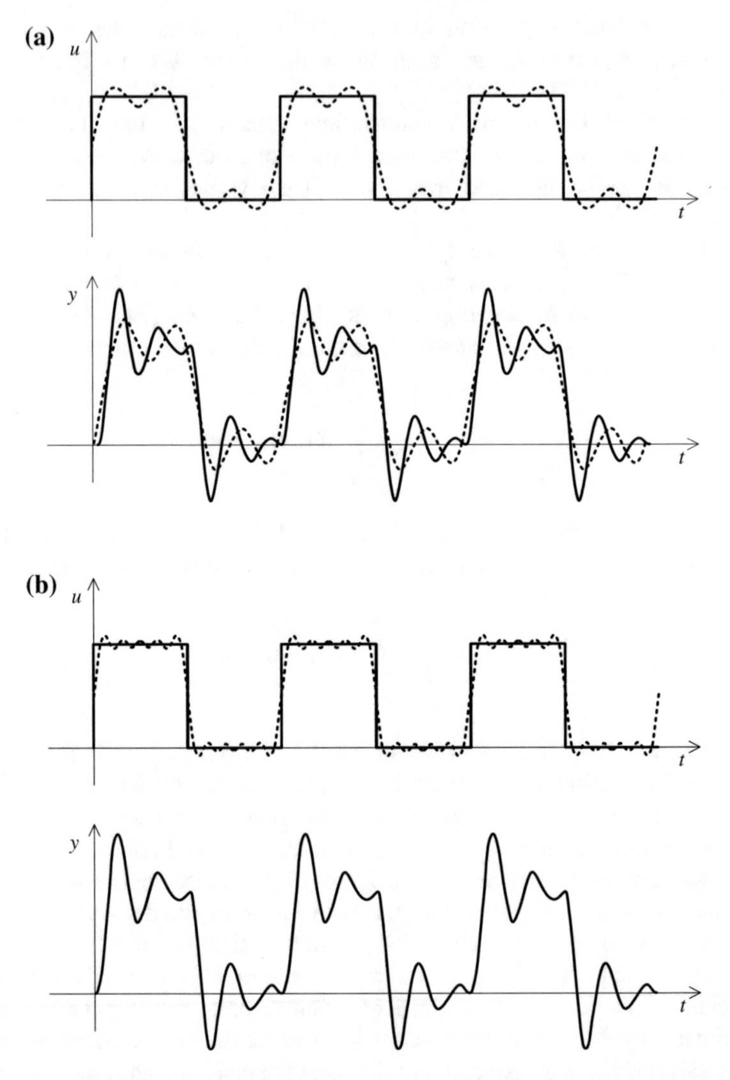

Fig. 2.12 Approximation of the periodic input and output signals of a second order system

Based on the above considerations, if the responses of a linear system are known for sinusoidal input signals, then theoretically its time response for an arbitrary input signal can also be given approximately.

2.2.2 The LAPLACE Transformation

Condition (2.26) of absolute integrability imposes a severe limit to the application of FOURIER transforms. This condition is not fulfilled for a number of practically applied signals (e.g., the unit step).

For practical applicability, the FOURIER transformation has to be modified to make it usable for non-integrable signals as well.

The scope of validity of the one-sided FOURIER transformation can be significantly extended if the function $y(t)$ to be transformed is first multiplied by the function $e^{-\sigma t}$, thus ensuring the condition of absolute integrability of the resulting function for a wide range of functions. Then the FOURIER transform of the resulting function is determined. Under the condition $\sigma > 0$, all the power functions, and under the condition $\sigma > \alpha$ also the exponential function $e^{\alpha t}$ with positive values of α, become absolutely integrable between $t = 0$ and ∞. The FOURIER transform of the function obtained by multiplying the original function with $e^{-\sigma t}$ is called the *LAPLACE transform* of the original function.

The LAPLACE transform for one-sided functions starting at $t = 0$:

$$\mathcal{L}\{y(t)\} = \int_{-\infty}^{\infty} y(t)e^{-\sigma t}e^{-j\omega t}\mathrm{d}t = \int_{0}^{\infty} y(t)e^{-st}\mathrm{d}t = Y(s),$$

where the transformation variable $s = \sigma + j\omega$ is a complex number with positive real part. Thus the LAPLACE transform of a function $y(t)$ is

$$Y(s) = \mathcal{L}\{y(t)\} = \int_{0}^{\infty} y(t)e^{-st}\mathrm{d}t \tag{2.30}$$

and the inverse LAPLACE transform is

$$y(t) = \mathcal{L}^{-1}\{Y(s)\} = \frac{1}{2\pi j}\int_{\sigma-j\infty}^{\sigma+j\infty} Y(s)e^{st}\mathrm{d}s. \tag{2.31}$$

The path of integration is to be chosen in such a way that $Y(s)$ be in its range of regularity, i.e., the singular places are to be the left of the path. In practical cases this general inversion formula can be replaced by methods which can be handled more easily, but with a narrower scope of validity (e.g., the expansion theorem). (Taking the limit $s \to j\omega$ the LAPLACE transform provides the FOURIER transform if it exists.) Table 2.1 gives the LAPLACE transforms of some important functions. All the functions are considered one-sided.

Table 2.1 LAPLACE transforms of some functions

$y(t)$	$Y(s)$
$\delta(t)$	1
$1(t)$	$\frac{1}{s}$
t	$\frac{1}{s^2}$
t^n	$\frac{n!}{s^{n+1}}$
e^{-at}	$\frac{1}{s+a}$
$1-e^{-at}$	$\frac{a}{s(s+a)}$
te^{-at}	$\frac{1}{(s+a)^2}$
$\frac{1}{(n-1)!}t^{n-1}e^{-at}$	$\frac{1}{(s+a)^n}$
$\sin(\omega t)$	$\frac{\omega}{s^2+\omega^2}$
$\cos(\omega t)$	$\frac{s}{s^2+\omega^2}$

Some important operational rules of the LAPLACE transformation follow.

Linearity

The LAPLACE transformation is a linear operation. If the individual time functions are multiplied by constants and summed, then the LAPLACE transform of the resulting function can be calculated in a similar way.

$$\mathcal{L}\{c_1y_1(t)+c_2y_2(t)\}=c_1Y_1(s)+c_2Y_2(s) \tag{2.32}$$

Differentiation

$$\begin{aligned}\mathcal{L}\{\dot{y}(t)\}&=sY(s)-y(-0)\\ \mathcal{L}\{\ddot{y}(t)\}&=s^2Y(s)-sy(-0)-\dot{y}(-0)\end{aligned} \tag{2.33}$$

If the function jumps at time instant $t=0$, in the LAPLACE transform of the derivative the initial value to be considered is the value of the function just before the jump ($t=-0$). If the initial values of the function and all of its derivatives are zeros, then differentiation with respect to time is reduced to multiplication by the appropriate power of s in the operator domain.

The differentiation of a LAPLACE transform with respect to s leads to multiplication in the time domain as follows:

$$\mathcal{L}\{ty(t)\}=-\frac{\mathrm{d}}{\mathrm{d}s}Y(s). \tag{2.34}$$

Integration

$$\mathcal{L}\left\{\int_0^t y(\tau)\mathrm{d}\tau\right\} = \frac{1}{s}Y(s) \tag{2.35}$$

Dead-time, shift in the s domain

$$\mathcal{L}\{y(t \pm \tau)\} = e^{\pm s\tau}Y(s); \quad y(t) = 0 \quad \text{if} \quad t < \tau \tag{2.36}$$

$$\mathcal{L}\{e^{-at}y(t)\} = Y(s+a). \tag{2.37}$$

Shifting the initial point of y to the right by τ in the operator domain means multiplication of the transformed function by $e^{-s\tau}$.

Initial and final value theorem

$$\begin{aligned} y(t = +0) &= \lim_{s\to\infty} sY(s) \\ y(t \to \infty) &= \lim_{s\to 0} sY(s) \end{aligned} \tag{2.38}$$

The relationship related to the steady state ($t \to \infty$) can only be applied if the poles of $Y(s)$ are on the left side of the complex plane, i.e., the transients are decaying, the steady state does exist (the relationship gives a false result, e.g., for a sinusoidal signal or for an exponentially increasing signal).

Convolution

$$\mathcal{L}\left\{\int_0^t y_1(\tau)y_2(t-\tau)\mathrm{d}\tau\right\} = Y_1(s)Y_2(s). \tag{2.39}$$

In the operator domain of the LAPLACE transformation, the convolution integral can be calculated by simply multiplying the LAPLACE transforms of the individual functions.

Inverse LAPLACE transform of a rational fraction

Calculation of the inverse LAPLACE transform by (2.31) is rarely applied. In general, analyzing linear systems with constant parameters, the LAPLACE transform of a signal is a rational fraction (i.e., a quotient of polynomials with real coefficients).

$$Y(s) = \frac{\mathcal{G}(s)}{\mathcal{H}(s)} = \frac{\beta_m s^m + \beta_{m-1}s^{m-1} + \cdots + \beta_o}{s^n + \alpha_{n-1}s^{n-1} + \cdots + \alpha_o}, \quad m \le n. \tag{2.40}$$

A rational function can be separated into partial fractions, and the inverse LAPLACE transform of the partial fraction can be calculated. This is the so called expansion theorem, which is simple if the denominator has single poles.

$$Y(s) = \sum_{i=1}^{n} \frac{r_i}{s - s_i} \quad \text{where} \quad r_i = \frac{\mathcal{G}(s_i)}{\mathcal{H}'(s_i)}, \tag{2.41}$$

$\mathcal{H}'$ is the derivative of $\mathcal{H}$ with respect to s. The time function is

$$y(t) = \sum_{i=1}^{n} r_i e^{s_i t}. \tag{2.42}$$

For a multiple pole, the number of terms in the partial fractional expansion of the rational fraction that must be employed is equal to the multiplicity of the pole. For example, if the i-th pole is a double pole, then the partial fraction terms are

$$\frac{r_{i1}}{s - s_i} + \frac{r_{i2}}{(s - s_i)^2}, \tag{2.43}$$

whose inverse transform according to Table 2.1 is $(r_{i1} + t r_{i2}) e^{s_i t}$.

2.2.3 *The Transfer Function*

Applying the LAPLACE transformation to the differential equation, an algebraic equation is obtained. With zero initial conditions the derivatives are simply replaced by multiplications by the appropriate powers of the variable s. The solution of the algebraic equation gives the LAPLACE transform of the output signal. By the inverse LAPLACE transformation, we get the output signal in the time domain.

Applying the LAPLACE transformation to the differential Eq. (2.1a) supposing zero initial conditions the following equation is obtained:

$$\begin{aligned} &a_n s^n Y(s) + a_{n-1} s^{n-1} Y(s) + \cdots + a_1 s Y(s) + a_o Y(s) \\ &= b_m s^m U(s) + b_{m-1} s^{m-1} U(s) + \cdots + b_1 s U(s) + b_o U(s) \end{aligned}$$

or

$$Y(s) = \frac{b_m s^m + b_{m-1} s^{m-1} + \cdots + b_1 s + b_o}{a_n s^n + a_{n-1} s^{n-1} + \cdots + a_1 s + a_o} U(s) = H(s) U(s) \tag{2.44a}$$

where $Y(s) = \mathcal{L}\{y(t)\}$, $U(s) = \mathcal{L}\{u(t)\}$, and $H(s)$ is the so called *transfer function*. For physically realizable systems, $m \leq n$. In this case the transfer function is called *proper*. If the stricter condition $m<n$ is also fulfilled, the $H(s)$ transfer function is *strictly proper*. The difference in the degrees, $n-m$, is called the pole-excess.

For systems containing also series dead-time the LAPLACE transform of differential equation (2.1b) is

$$a_n s^n Y(s) + a_{n-1} s^{n-1} Y(s) + \cdots + a_1 s Y(s) + a_o Y(s)$$
$$= \left[b_m s^m U(s) + b_{m-1} s^{m-1} U(s) + \cdots + b_1 s U(s) + b_o U(s) \right] e^{-sT_d}$$

or

$$Y(s) = \frac{b_m s^m + b_{m-1} s^{m-1} + \cdots + b_1 s + b_o}{a_n s^n + a_{n-1} s^{n-1} + \cdots + a_1 s + a_o} e^{-sT_d} U(s) = H(s)U(s) \tag{2.44b}$$

The transfer function of a system is the ratio of the LAPLACE transforms of its output and input signals (Fig. 2.13).

$$H(s) = \frac{Y(s)}{U(s)} \tag{2.45}$$

Different forms of the transfer function

In the sequel, systems without dead-time will be considered. The transfer function can be given in *polynomial/polynomial form* as

$$H(s) = \frac{b_m s^m + b_{m-1} s^{m-1} + \cdots + b_1 s + b_o}{a_n s^n + a_{n-1} s^{n-1} + \cdots + a_1 s + a_o} \tag{2.46}$$

The numerator and the denominator have real or complex conjugate poles. Let us denote the roots of the numerator—the zeros of the transfer function—by $z_1, z_2, \ldots, z_m$, and the roots of the denominator—the poles of the transfer function—by $p_1, p_2, \ldots, p_n$. The *zero-pole-gain form* of the transfer function is

$$H(s) = k \frac{(s-z_1)(s-z_2)\cdots(s-z_m)}{(s-p_1)(s-p_2)\cdots(s-p_n)}, \tag{2.47}$$

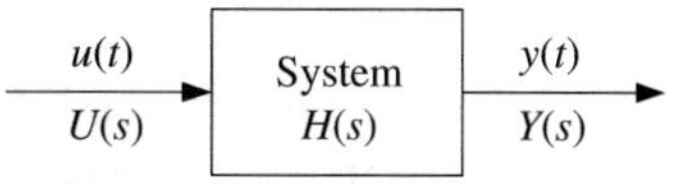

Fig. 2.13 A linear system can be described by its transfer function

u(t) U(s) → System H(s) → y(t) Y(s)

where the value of the gain factor is $k = b_m/a_n$. The transfer function can also be given in *partial fractional form* as (supposing single roots):

$$H(s) = \sum_{i=1}^{n} \frac{r_i}{s - p_i}, \tag{2.48}$$

where p_i denotes the poles while r_i denotes the residues of the transfer function. Both the poles and the residues can take real or complex conjugate values.

In control applications many times it is advantageous to feature the reciprocals of the roots, the so called time constants in the transfer function. Introducing the notations $\tau_i = -1/z_i$ and $T_i = -1/p_i$ the *time constant form* of the transfer function is obtained as:

$$H(s) = A\frac{(1 + s\tau_1)(1 + s\tau_2)\ldots(1 + s\tau_m)}{(1 + sT_1)(1 + sT_2)\ldots(1 + sT_n)}, \tag{2.49}$$

where τ_i and T_i are real or complex numbers and A is the gain whose value is expressed as

$$A = \frac{b_o}{a_o} = k\frac{(-z_1)\ldots(-z_m)}{(-p_1)\ldots(-p_n)}.$$

For zeros and poles that have the values of zero, the conversion is not performed.

It is reasonable to combine the complex conjugate pairs of root both in the numerator and the denominator into second order terms with real coefficients. Let, e.g., $p_1 = \alpha + j\beta$ and $p_2 = \bar{p}_1 = \alpha - j\beta$ be complex conjugate poles. Multiplying together the root factors, the following relationship is obtained:

$$\begin{aligned}(s - p_1)(s - p_2) &= (s - \alpha - j\beta)(s - \alpha + j\beta)\\ &= s^2 - 2\alpha s + \alpha^2 + \beta^2\\ &= s^2 + 2\xi\omega_o s + \omega_o^2\end{aligned}$$

where $\omega_o^2 = \alpha^2 + \beta^2$ and $\xi = -\alpha/\omega_o$.

In time constant form,

$$s^2 + 2\xi\omega_o s + \omega_o^2 = \omega_o^2\left(1 + \frac{2\xi}{\omega_o}s + \frac{1}{\omega_o^2}s^2\right).$$

Introducing the time constant $T_o = 1/\omega_o$ the part of the right hand side of the equation in brackets can be written in a form like $1 + 2\xi T_o s + T_o^2 s^2$. The frequency ω_o is called the *natural frequency* and ξ is the *damping factor* of the second degree term.

Combining the terms with complex conjugate roots the transfer function can be written in the following form.

$$H(s) = \frac{A}{s^i} \frac{\prod_1^c (1 + s\tau_j) \prod_1^c \left(1 + 2\zeta_j \tau_{oj} s + s^2 \tau_{oj}^2\right)}{\prod_1^e (1 + s\tau_j) \prod_1^f \left(1 + 2\xi_j T_{oj} s + s^2 T_{oj}^2\right)} \tag{2.50}$$

If $i < 0$, the element contains the effect of a differentiation.

If $i = 0$, the element is proportional.

If $i > 0$, the element contains the effect of an integration.

These effects clearly appear when the transients generated by the input signal have already decayed.

In the case of dead-time the above transfer functions have to be multiplied by e^{-sT_d}.

The relation of the transfer function to the weighting function and the unit step response

With the transfer function the output signal can be determined as the response to a given input excitation.

$$\begin{aligned} Y(s) &= H(s)U(s) \\ y(t) &= \mathcal{L}^{-1}\{Y(s)\} = \mathcal{L}^{-1}\{H(s)U(s)\} \end{aligned} \tag{2.51}$$

Knowing the transfer function the weighting function and the unit step response can easily be calculated.

The *weighting function* is the system response to a DIRAC delta impulse in the time domain. As $\mathcal{L}\{\delta(t)\} = 1$, the LAPLACE transform of the weighting function is the transfer function

$$Y(s) = U(s)H(s) = \mathcal{L}\{\delta(t)\}H(s) = H(s)$$

Hence the weighting function is

$$w(t) = \mathcal{L}^{-1}\{H(s)\} \quad \text{and vice versa} \quad H(s) = \mathcal{L}\{w(t)\} \tag{2.52}$$

Thus the LAPLACE transform of the weighting function of a system is the transfer function of the system. With the knowledge of the weighting function the system response to an arbitrary input signal can be determined with the convolution integral. In the domain of the LAPLACE transformation, convolution is transformed to multiplication:

$$y(t) = \mathcal{L}^{-1}\{Y(s)\} = \mathcal{L}^{-1}\{H(s)U(s)\} = \int_0^t w(t - \tau)u(\tau)d\tau. \tag{2.53}$$

The *unit step response* is the system response in the time domain to a unit step input signal. The LAPLACE transform of the unit step response is

$$Y(s) = U(s)H(s) = \mathcal{L}\{1(t)\}H(s) = \frac{1}{s}H(s).$$

The unit step response can be obtained by an inverse LAPLACE transformation of the above expression.

$$v(t) = \mathcal{L}^{-1}\left\{\frac{H(s)}{s}\right\}. \tag{2.54}$$

The unit step response of a proportional element is illustrated in Fig. 2.14.

The initial and final value of the unit step response and also the initial values of its derivatives can be determined on the basis of the transfer function. Using the initial value theorem the initial value of the unit step response is:

$$v(0) = \lim_{s\to\infty} s\frac{H(s)}{s} = \lim_{s\to\infty} H(s) \tag{2.55}$$

The initial value of the r-th order derivative of the unit step response is:

$$v^{(r)}(0) = \lim_{s\to\infty} s\frac{s^r H(0)}{s} = \lim_{s\to\infty} s^r H(0) \tag{2.56}$$

Letting $s \to \infty$ in the transfer function (2.44a) and (2.44b), the highest degree terms dominate in the numerator and the denominator:

$$v^{(r)}(0) = \lim_{s\to\infty} s^r \frac{b_m s^m}{a_n s^n} = \lim_{s\to\infty} \frac{b_m}{a_n}\frac{s^r}{s^{n-m}}. \tag{2.57}$$

If the degrees of the numerator and the denominator are identical, the unit step response jumps at time instant $t = 0$. If there is a difference between the degrees of the denominator and the numerator, there is a jump at $t = 0$ in the derivative of order $r = n - m$, and the value of the lower order derivatives is zero at $t = 0$. If the

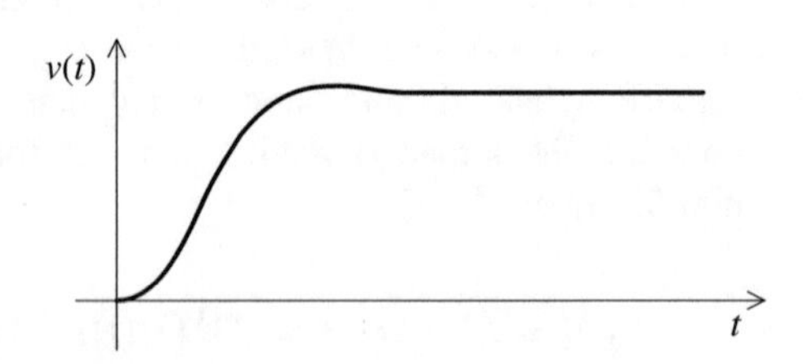

Fig. 2.14 Unit step response

difference in the degree is 1, the value of the unit step response at $t = 0$ is zero, but the value of the first derivative is different from zero: the unit step response starts with a finite slope. If the difference of the degrees of the denominator and the numerator is 2, then the initial value of the unit step response and also of its first derivative (initial slope) is zero, and the initial value of the second derivative is non-zero. The greater the degree difference is, the better the step response fits to the time axis at the initial point $t = 0$.

The steady state value of the step response (supposing that a steady state is reached at all, i.e., in (2.50) the real value of all of the poles is negative) is

$$v(t \to \infty) = \lim_{s \to 0} s \frac{H(s)}{s} = \lim_{s \to 0} H(s) \tag{2.58}$$

In this case in expression (2.50) the terms containing the variable s can be neglected.

If $i = 0$, the steady state value of the system for a unit step input will settle down to the value of the static gain A. Elements with this property are called proportional elements. If $i > 0$, the element has the effect of integration and the output signal tends to infinity if $t \to \infty$, linearly if $i = 1$, and quadratically if $i = 2$. If $i < 0$, the element has the effect of a differentiation and the steady value of the step response is zero (see Fig. 2.15).

The poles of the transfer function characterize the transient response. Real poles result in aperiodic transients, while complex conjugate poles provide oscillating transients. A pole at the origin means an integration. Poles on the left side of the complex plane give decreasing transients, while poles on the right side lead to increasing transients. Figure 2.16 shows poles of systems located in different areas of the complex plane as well as the shapes of the corresponding weighting functions.

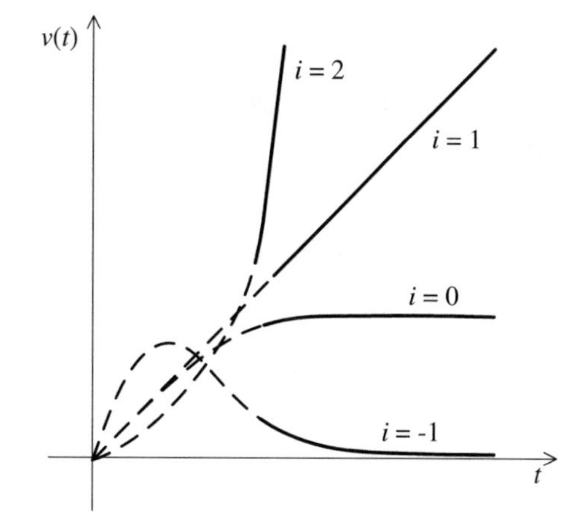

Fig. 2.15 Stationary behavior of the unit step response

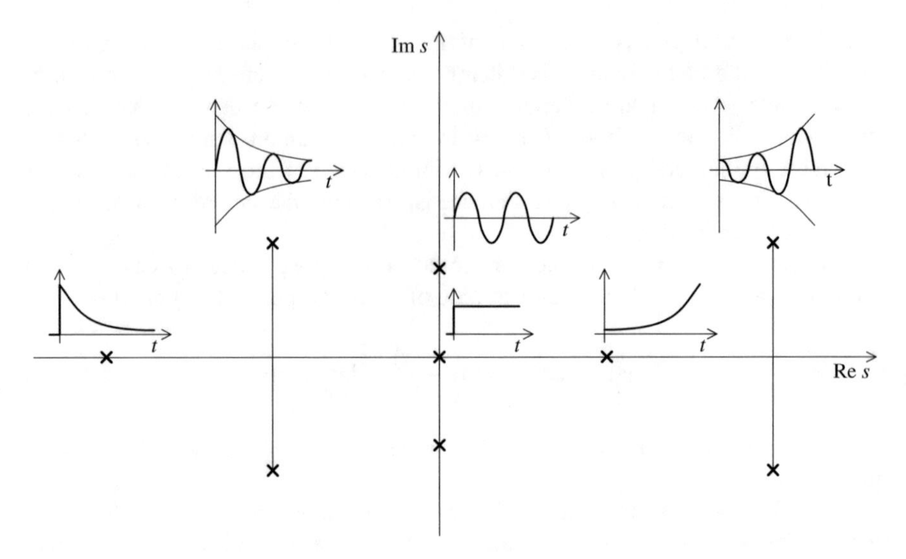

Fig. 2.16 Poles of the transfer function and types of the related weighting functions

2.2.4 *Basic Connections of Elementary Blocks, Block-Scheme Algebra, Equivalent Block Manipulations*

In a closed-loop control system the elements are connected to each other. The way they are connected and interact get defined in block diagrams.

The three basic ways of connecting elements are

- serial connection,
- parallel connection, and
- feedback connection.

Let us determine the resulting transfer functions of the different basic connection types.

Serial connection

In serial connection the output of the considered element is the input of the next one (Fig. 2.17).

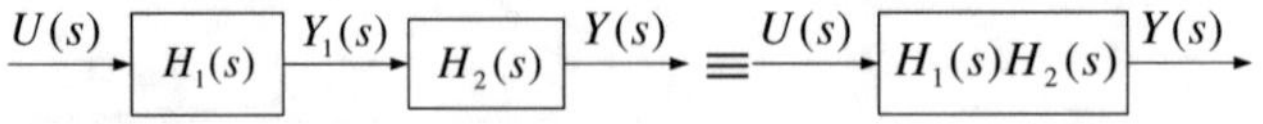

$U(s) \rightarrow H_1(s) \xrightarrow{Y_1(s)} H_2(s) \rightarrow Y(s) \equiv U(s) \rightarrow H_1(s)H_2(s) \rightarrow Y(s)$

Fig. 2.17 Serial connection

The LAPLACE transform of the output signal is

$$Y(s) = Y_1(s)H_2(s) = U(s)H_1(s)H_2(s)$$

Thus the resulting transfer function is obtained by multiplying the transfer functions of the individual elements.

$$H(s) = H_1(s)H_2(s)$$

Parallel connection

In parallel connection the input of the individual elements is the same, and the outputs are summarized (Fig. 2.18). The LAPLACE transform of the output signal is

$$Y(s) = Y_1(s) + Y_2(s) = U(s)[H_1(s) + H_2(s)]$$

Thus the resulting transfer function is the sum of the transfer functions of the individual elements.

$$H(s) = H_1(s) + H_2(s)$$

It is to be emphasized that above contraction of H_1 and H_2 is only possible if their inputs are the same and their outputs are summarized exclusively in one common point.

Feedback connection

We talk about feedback if the output of an element—passing through another element—is added or subtracted from its input. Addition realizes positive feedback, while subtraction means negative feedback. The basic connection of a closed loop control system is the negative feedback. Based on Fig. 2.19 let us determine the resulting transfer function of a feedback circuit.

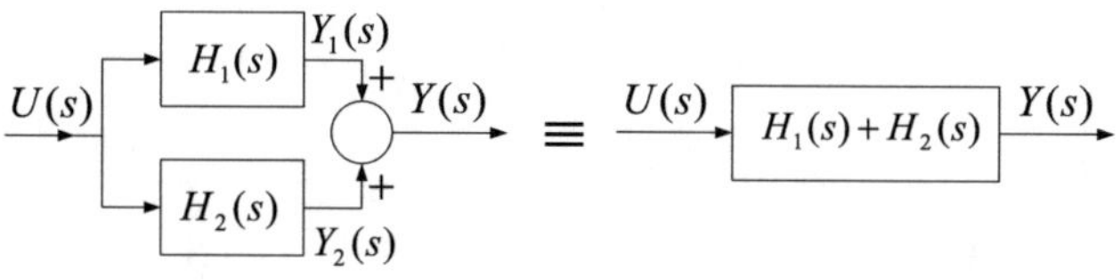

Fig. 2.18 Parallel connection

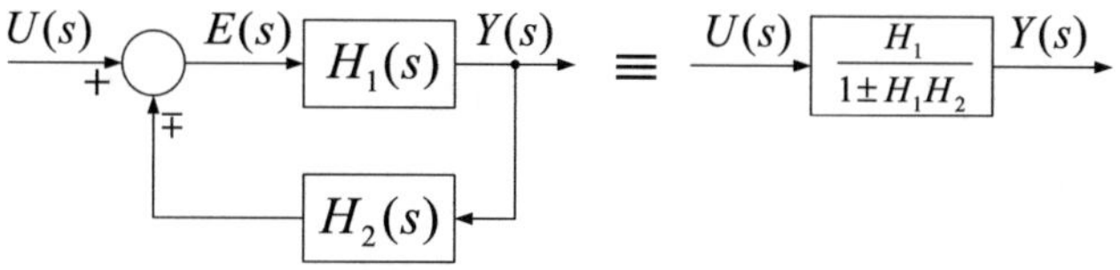

Fig. 2.19 Feedback scheme

The LAPLACE transform of the output signal is

$$Y(s) = E(s)H_1(s) = [U(s) - H_2(s)Y(s)]H_1(s)$$

Rearranging $Y(s) = [H_1(s)/(1 + H_1(s)H_2(s))]U(s)$, the resulting transfer function is

$$H(s) = \frac{H_1(s)}{1 \pm H_1(s)H_2(s)}.$$

In the denominator the negative sign stands for positive feedback. The transfer function $L(s) = H_1(s)H_2(s)$ is called the *loop transfer function*.

The mathematical analysis of a closed-loop control system is greatly facilitated by block diagrams. The analysis can be simplified in many cases if the block diagram is converted to another, equivalent form using conversion rules. With a conversion, a simpler form or a more advantageous structure for the calculations can be obtained. Blocks and signals can be relocated with the conversion, but the effects of the individual input signals on the output signals have to remain unchanged. In the sequel some rules for equivalent conversions will be presented.

The junction points from the same signal can be interchanged (Fig. 2.20). The location of the summation points can be interchanged (Fig. 2.21). Figure 2.22 shows the equivalent relocation of the summation points. Relocation of a junction point is shown in Fig. 2.23.

Example 2.1 In the block diagram of Fig. 2.24 the system is given by two serially connected elements characterized by their transfer functions. The disturbance acts between the two elements. Let us transform the disturbance to the output or to the input. Figure 2.25 shows the converted block diagrams. ■

Example 2.2 Let us determine the resulting transfer function of the complex control scheme shown in Fig. 2.26 between the output signal y and the reference signal r.

The steps of the conversion of the block diagram and the calculation of the resulting transfer function are shown in Fig. 2.27. ■

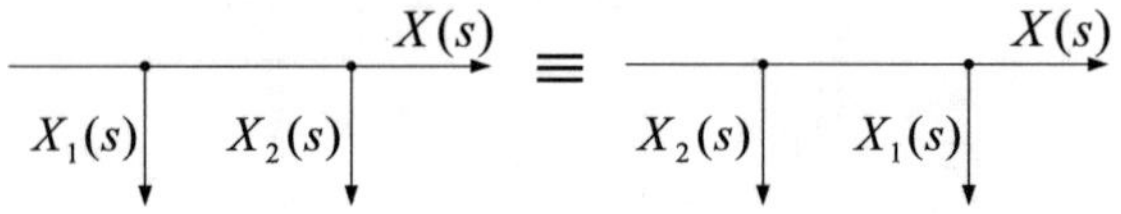

Fig. 2.20 The junction points are interchangeable

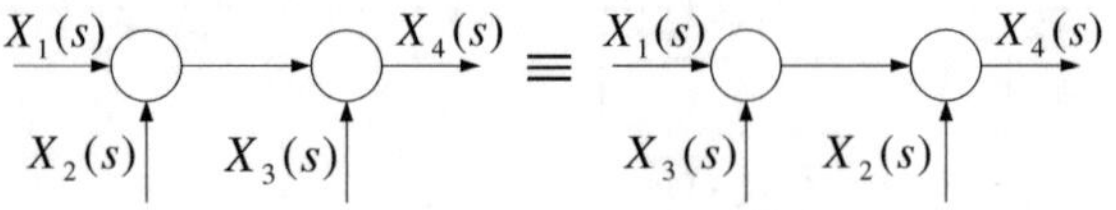

Fig. 2.21 The summation points are interchangeable

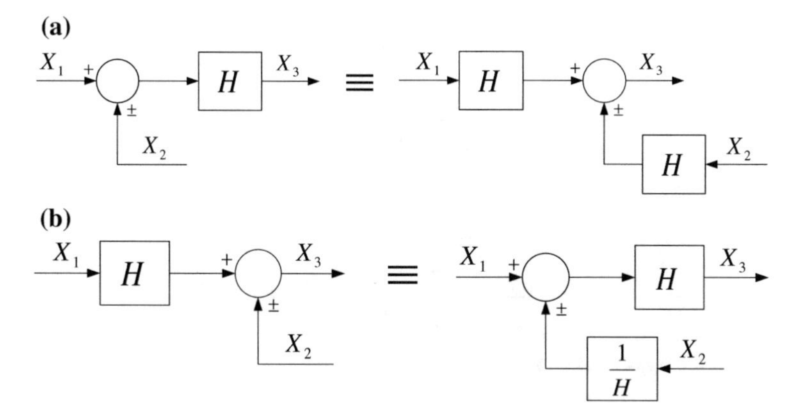

Fig. 2.22 Equivalent relocation of the summation points

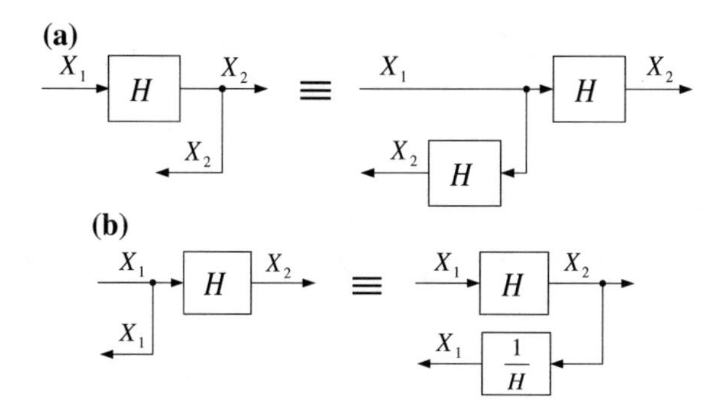

Fig. 2.23 Relocation of a junction point

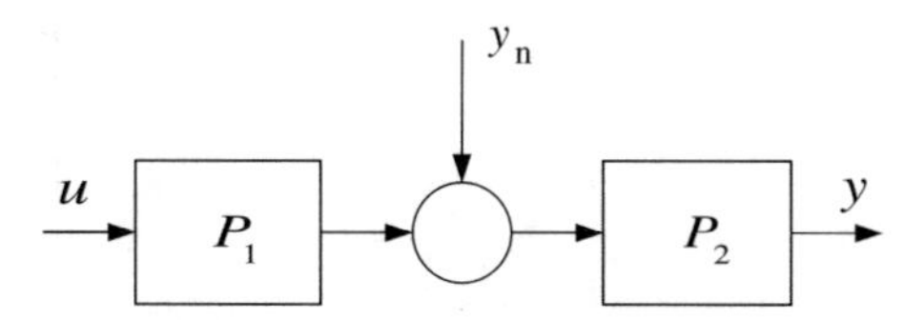

Fig. 2.24 The disturbance acts between the two serially connected elements

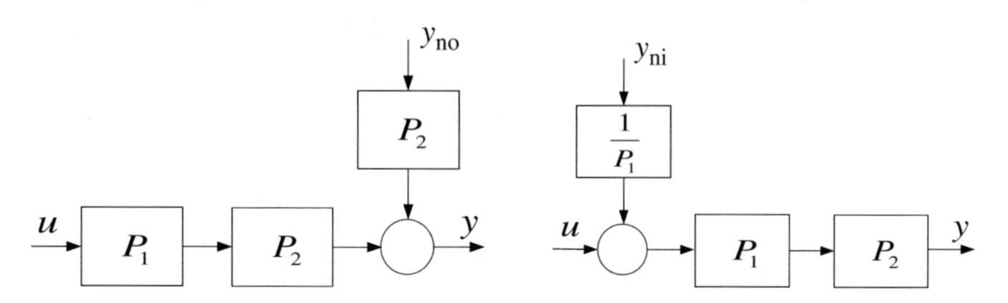

Fig. 2.25 The disturbance can be relocated to the output or to the input of the process

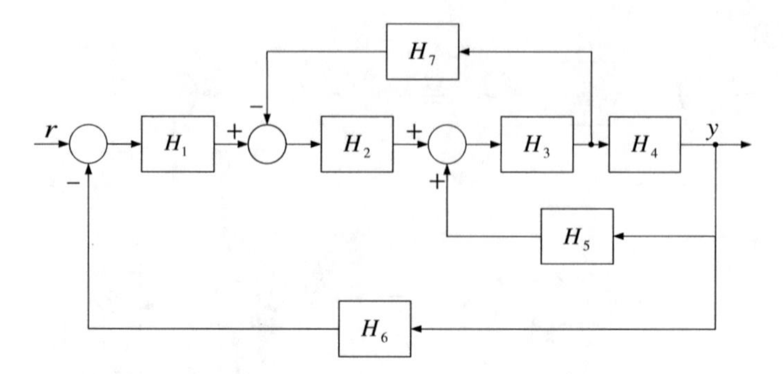

Fig. 2.26 Multiloop control system

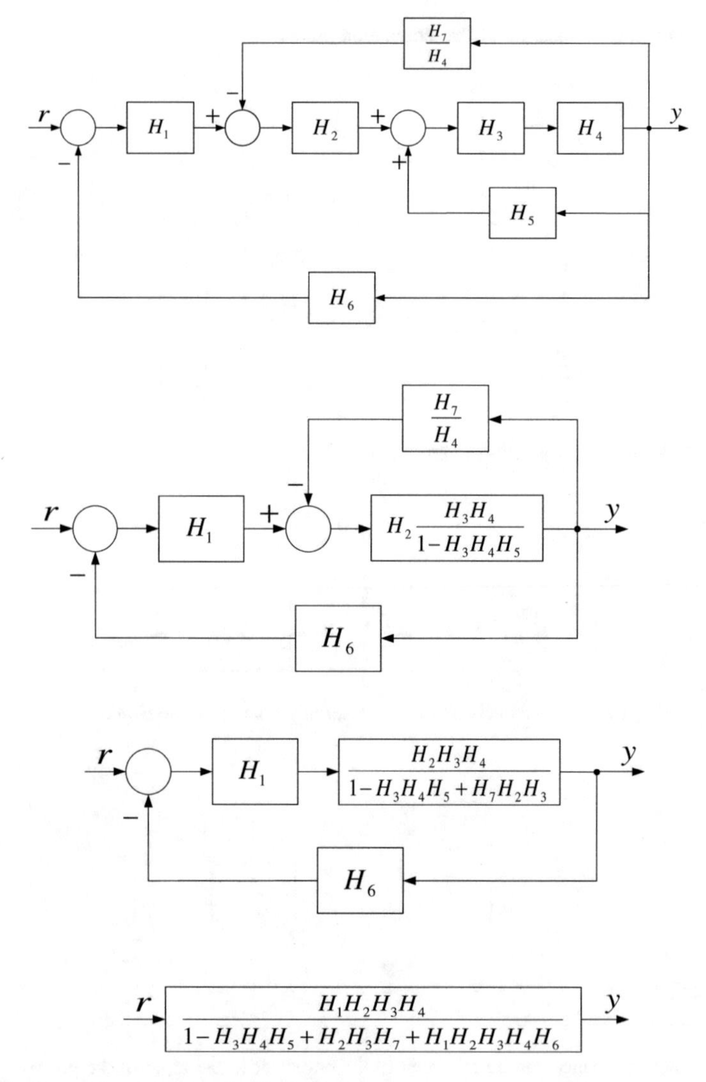

Fig. 2.27 Steps of conversion of a block diagram

Example 2.3 Let us give the resulting transfer function of the circuit shown in Fig. 2.28.

The steps of the conversion of the block diagram and the determination of the resulting transfer function are given in Fig. 2.29. ■

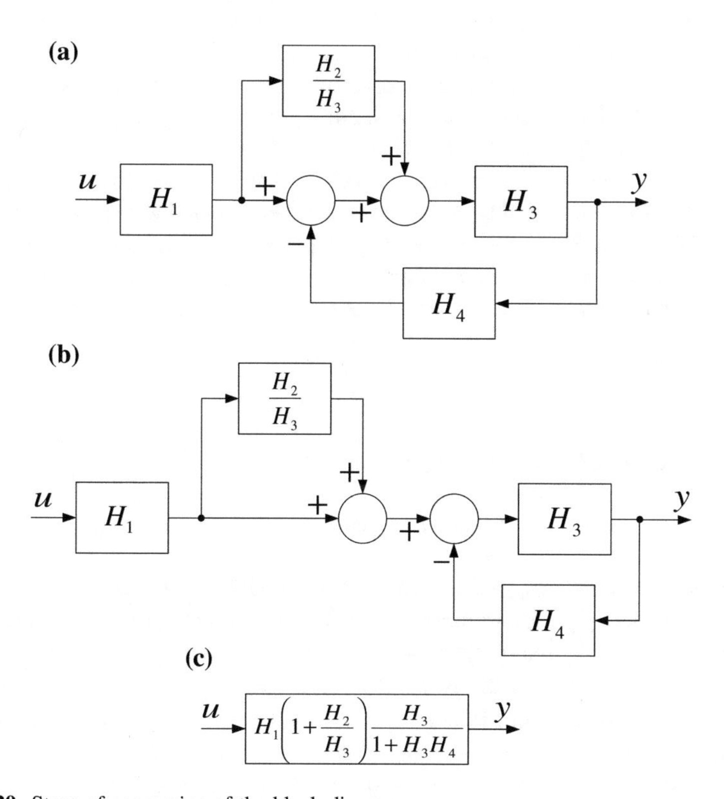

Fig. 2.29 Steps of conversion of the block diagram

Fig. 2.28 Control system with forward path

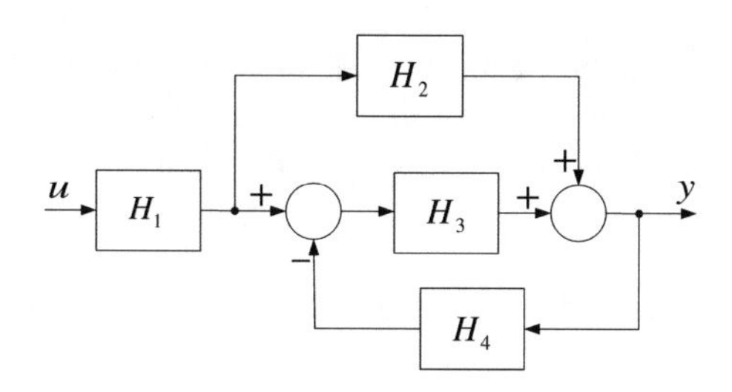

2.3 Investigation of Linear Dynamical Systems in the Frequency Domain

In the sequel, the outputs of a linear system for sinusoidal input signals will be investigated. The system responses for sinusoidal input signals—as we have seen in Sect. 2.2 in relation to FOURIER analysis—contain basic information about system responses to other, non-sinusoidal inputs, as well, as a given input signal can be expanded to a sum of sinusoidal components. In a linear system, summing the responses for the individual sinusoidal input signal components yields an approximation of the output signal for the given input signal.

The basic property of stable linear systems is that for sinusoidal input signals in *steady state*, after the decaying of the transients they respond with sinusoidal output signals of the same frequency as that of the input signal (Fig. 2.30). The amplitude and the phase angle of the output signal, however, depend on the frequency.

Let the input signal of the system be $u(t) = A_\mathrm{u}\sin(\omega t + \varphi_\mathrm{u})$, $t \geq 0$. The output signal is

$$y(t) = y_\mathrm{steady}(t) + y_\mathrm{transient}(t).$$

The output signal in steady (quasi-stationary) state is

$$y_\mathrm{steady}(t) = A_\mathrm{y}\sin(\omega t + \varphi_\mathrm{y}).$$

(Let us remark that $y_\mathrm{steady}(t)$ is generally not equal to the steady state final value y_ss of the transient signal introduced earlier.) The *frequency function* is a complex function representing the frequency dependence of two system properties, the *amplitude ratio* $A_\mathrm{y}/A_\mathrm{u}$ and the *phase difference* $(\varphi_\mathrm{y} - \varphi_\mathrm{u})$. It can be proven that *formally* the frequency function can be derived from the transfer function by substituting $s = j\omega$, which gives the direct relationship between the operator domain of the LAPLACE transformation and the frequency domain.

$$H(j\omega) = H(s)|_{s=j\omega} = |H(j\omega)|e^{j\varphi(\omega)} = a(\omega)e^{j\varphi(\omega)} \tag{2.59}$$

In the frequency function, the expressions for the *amplitude function* $a(\omega)$ (the absolute value of the frequency function) and the *phase function* $\varphi(\omega)$ (the phase angle of the frequency function) are

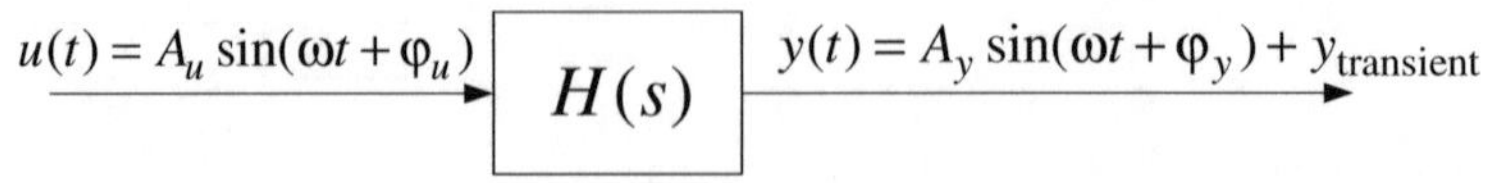

$$u(t) = A_u \sin(\omega t + \varphi_u) \longrightarrow \boxed{H(s)} \longrightarrow y(t) = A_y \sin(\omega t + \varphi_y) + y_\mathrm{transient}$$

Fig. 2.30 Response of a linear system to a sinusoidal input signal

$$a(\omega) = |H(j\omega)| = \frac{A_y(\omega)}{A_u(\omega)} \quad \text{and}$$
$$\varphi(\omega) = \arg\{H(j\omega)\} = \varphi_y(\omega) - \varphi_u(\omega)$$

Proof Suppose the transfer function of the system is

$$H(s) = k\frac{(s-z_1)(s-z_2)\ldots(s-z_m)}{(s-p_1)(s-p_2)\ldots(s-p_n)}$$

For the sake of simplicity, let us suppose there are only single poles and zeros. The LAPLACE transform of the sinusoidal input signal is

$$U(s) = \frac{A_u\omega}{s^2+\omega^2}.$$

The output signal can be written in partial fractional form as

$$Y(s) = U(s)H(s) = \frac{A_u\omega}{s^2+\omega^2}k\frac{(s-z_1)(s-z_2)\cdots(s-z_m)}{(s-p_1)(s-p_2)\cdots(s-p_n)}$$
$$= \frac{\alpha}{s+j\omega} + \frac{\bar{\alpha}}{s-j\omega} + \frac{\beta_1}{s-p_1} + \frac{\beta_2}{s-p_2} + \cdots + \frac{\beta_n}{s-p_n}$$

where α and $\bar{\alpha}$ are complex conjugate residues. By means of the inverse LAPLACE transformation the output signal in the time domain is calculated to be

$$y(t) = \mathcal{L}^{-1}\{Y(s)\} = \alpha e^{-j\omega t} + \bar{\alpha}e^{j\omega t} + \beta_1 e^{p_1 t} + \beta_2 e^{p_2 t} + \cdots + \beta_n e^{p_n t}$$

For a stable system, the transients resulting from those partial fractions which contain the poles of the system are decreasing and the quasi-stationary response—as seen from the above formula—is a sinusoidal signal with the same frequency as that of the input signal.

Let us determine the values of the residues α and $\bar{\alpha}$ based on the above partial fractional description given for $Y(s)$.

$$\alpha = \frac{A_u\omega}{s^2+\omega^2}H(s)(s+j\omega)\bigg|_{s=-j\omega} = -\frac{A_u}{2j}H(-j\omega)$$
$$\text{and} \quad \bar{\alpha} = \frac{A_u}{2j}H(j\omega)$$

So

$$Y_{\text{steady}}(s) = -\frac{A_\text{u}}{2j}H(-j\omega)\frac{1}{s+j\omega} + \frac{A_\text{u}}{2j}H(j\omega)\frac{1}{s-j\omega}$$
$$= -\frac{A_\text{u}}{2j}|H(j\omega)|e^{-j\varphi(\omega)}\frac{1}{s+j\omega} + \frac{A_\text{u}}{2j}|H(j\omega)|e^{j\varphi(\omega)}\frac{1}{s-j\omega}$$

and the steady state, quasi-stationary component of the output signal is

$$y_{\text{steady}}(t) = \frac{A_\text{u}}{2j}|H(j\omega)|\left[e^{j(\omega t+\varphi)} - e^{-j(\omega t+\varphi)}\right] = A_\text{u}|H(j\omega)|\sin(\omega t+\varphi)$$

Thus it has been proved that the frequency function can be obtained from the transfer function by substituting $s = j\omega$, i.e., $H(j\omega) = H(s)|_{s=j\omega}$.

Note that the frequency function is the Fourier transform of the weighting function, if it exists.

When a system is excited by an input signal which produces transients, initially the high frequencies (faster in time) are dominant, and subsequently the low frequency properties are dominant. Taking the limit $j\omega \to 0$ gives the steady state, i.e., the steady state value of the unit step response ($t \to \infty$) is equal to the amplitude of the frequency function at $\omega = 0$. The initial value of the unit step response is equal to the value of the frequency function as $\omega \to \infty$.

2.3.1 Graphical Representations of the Frequency Functions

The frequency function can be plotted in several forms. The *Nyquist diagram* draws the frequency function in the complex plane as a polar diagram. For each value of the frequency function in the selected frequency range a point can be given in the complex plane corresponding to the pair of values $a(\omega)$ and $\varphi(\omega)$. Connecting these points by a contour forms the Nyquist diagram. When plotting the Nyquist diagram, generally the frequency is taken between zero and infinity (Fig. 2.31). The arrow shows the direction of increasing frequency parameter. Often the curve is supplemented by values calculated for negative frequencies. In this case the diagram is called the complete Nyquist diagram. The part of the diagram given for the frequency range $-\infty < \omega < 0$ (indicated by the dashed line in the figure) is the mirror image of the curve plotted for positive frequencies related to the real axis. The Nyquist diagram can also be considered as the conformal mapping of the straight line $s = j\omega$, $-\infty < \omega < \infty$ according to the function $H(s)$.

Fig. 2.31 Nyquist diagram

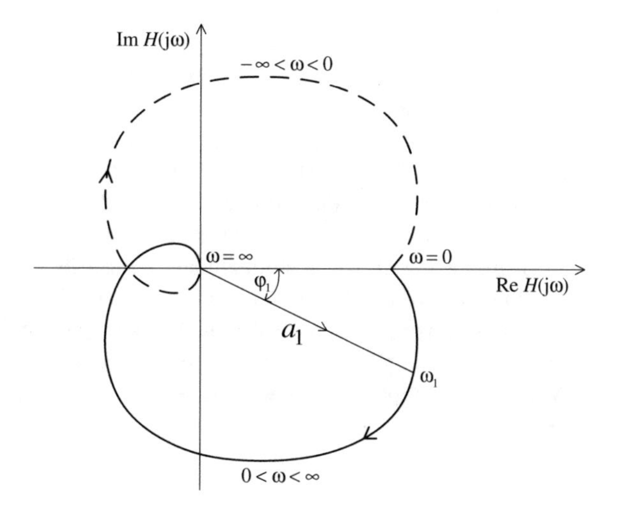

The shape of the Nyquist diagram characterizes the system. Analyzing the Nyquist diagram a qualitative picture can be obtained of important system properties (e.g., stability).

The *Bode diagram* simultaneously plots the absolute value $a(\omega)$ and the phase angle $\varphi(\omega)$ of the frequency function versus the frequency in a given frequency range (Fig. 2.32). Generally the frequency scale is logarithmic in order to cover a

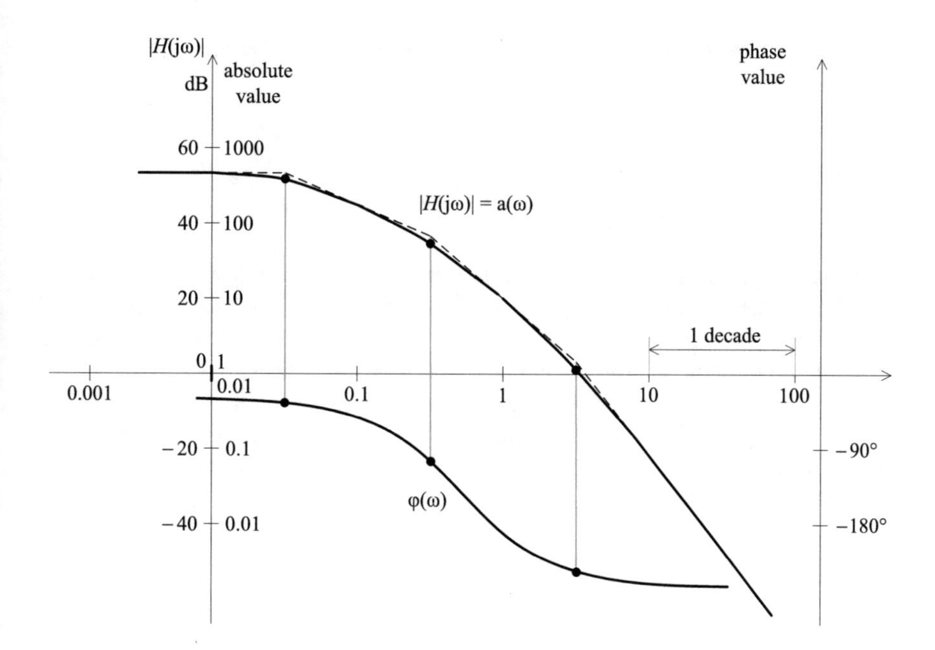

Fig. 2.32 Bode diagram

wide frequency range. The frequency range when the frequency is changed by a factor 10 is called a decade. (In music, the word octave is used, which gives a frequency band when the frequency is changed by a factor of two.) The absolute value—following telecommunication tradition—is scaled in decibels. The decibel (dB) employs the base 10 (decimal) logarithm of a number, then this value is multiplied by 20. The phase angle is drawn in a linear scale.

The advantage of the Bode diagram on the one hand is that multiplying individual frequency function components—because of the logarithmic scale—the Bode diagrams of the individual components are simply added. On the other hand a further advantage is that generally the Bode amplitude-frequency diagram can well be approximated by its asymptotes. From the course and from the breakpoints of the asymptotic amplitude-frequency curve, a quick evaluation can be made about fundamental system properties.

2.4 Transfer Characteristics of Typical Basic Blocks

As was seen, a linear time-invariant (*LTI*) system can be described by the differential Eq. (2.1b),

$$\begin{aligned} & a_n y^{(n)}(t) + a_{n-1} y^{(n-1)}(t) + \cdots + a_1 \dot{y}(t) + a_o y(t) \\ & = b_m u^{(m)}(t - T_d) + b_{m-1} u^{(m-1)}(t - T_d) + \cdots + b_1 \dot{u}(t - T_d) + b_o u(t - T_d) \end{aligned}$$

or in time constant form it can be given by the transfer function in the following form:

$$H(s) = \frac{A}{s^i} \frac{\prod_1^c (1 + s\tau_j) \prod_1^d \left(1 + 2\zeta_j \tau_{oj} s + s^2 \tau_{oj}^2\right)}{\prod_1^e (1 + sT_j) \prod_1^f \left(1 + 2\xi_j T_{oj} s + s^2 T_{oj}^2\right)} e^{-sT_d} \tag{2.60}$$

In particular cases the order of the differential equation is given, and possibly some terms are missing, and the transfer function contains only some elements of the general form. The general linear element described by $H(s)$ can be built as the combination of some appropriately chosen simple basic elements.

In the sequel, the time and frequency characteristics of the most important transfer elements will be investigated. These elements are the proportional, integrating, differentiating, dead-time and lag elements, and elements obtained by their series and parallel connection.

2.4.1 Ideal Basic Blocks

The ideal basic elements are the pure proportional, integrating, differentiating elements and the dead-time element.

Proportional (P) element

Its differential equation is $a_o y(t) = b_o u(t)$, which actually is an algebraic equation. A proportional element is for instance an amplifier in its linearity range. The transfer function is a constant, also called the gain factor.

$$H(s) = H_P(s) = A = b_o/a_o. \tag{2.61}$$

Its weighting function is a Dirac delta of area A, its unit step response is a step function of amplitude A. Its Nyquist diagram is a single point at the real axis. Its Bode amplitude diagram is a straight line parallel to the frequency axis, its phase angle is zero at all frequencies. The characteristics are shown in Table 2.2.

Integrating (I) element

Its differential equation is

$$a_1 \frac{\mathrm{d}y(t)}{\mathrm{d}t} = b_o u(t),$$

or in time constant form

$$T_I \frac{\mathrm{d}y(t)}{\mathrm{d}t} = u(t), \quad \text{or} \quad \frac{\mathrm{d}y(t)}{\mathrm{d}t} = K_I u(t),$$

where $K_I = 1/T_I$ and T_I is the integrating time constant. The solution of the differential equation is:

$$y(t) = \frac{1}{T_I}\int_0^t u(t)\mathrm{d}t + c$$

With zero input signal the system maintains the output signal corresponding to its previous state. The integrating element has a memory property. The signal at its output can be constant only if the value of its input signal is zero. The actual value of its output signal depends on the past values of the input signal.

An example for a physical realization of an integrating element is a liquid tank if its input signal is the rate of inflow of the liquid and its output signal is the level in the tank, or the relationship between the terminal voltage of a capacitor and its

Table 2.2 Ideal basic elements

Name	$H(s)$	$w(t)$	$v(t)$	Nyquist	Bode
Proportional element 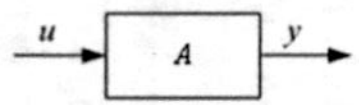 (u → A → y)	A	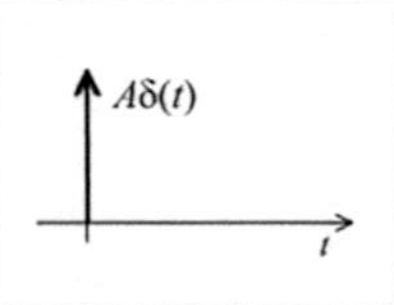 $A\delta(t)$; t	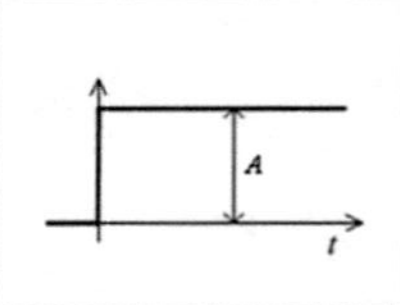 A; t	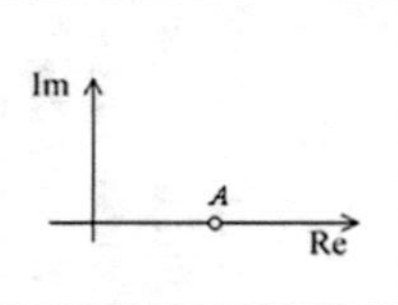 Im; A; Re	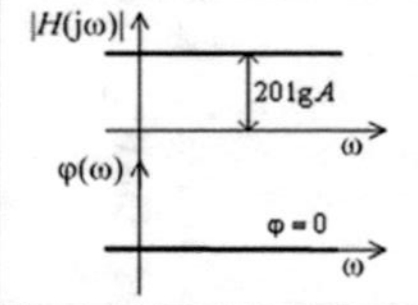 $\|H(j\omega)\|$; $20\lg A$; ω; $\varphi(\omega)$; $\varphi = 0$; ω
Integrating element (u → $\frac{1}{T_I}\int \ldots dt$ → y) 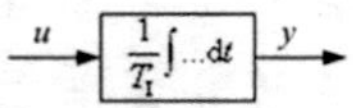	$\frac{A_I}{s} = \frac{1}{sT_I}$	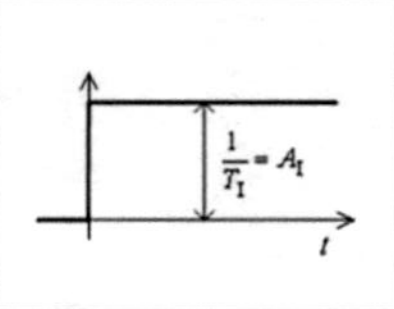 $\frac{1}{T_I} = A_I$; t	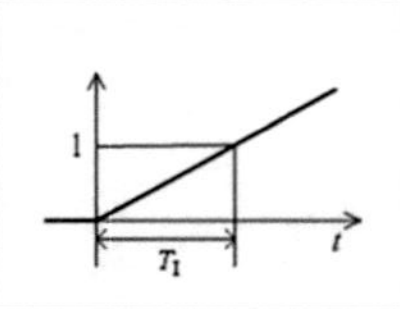 1; T_I; t	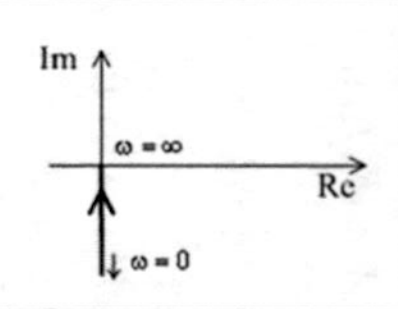 Im; $\omega = \infty$; $\omega = 0$; Re	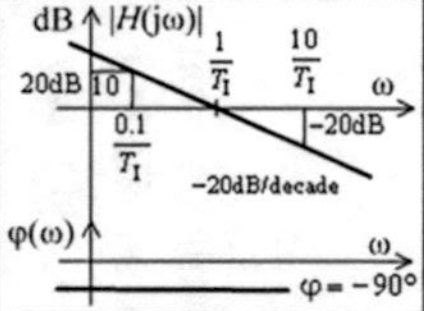 dB; $\|H(j\omega)\|$; 20dB; 10; $\frac{0.1}{T_I}$; $\frac{1}{T_I}$; $\frac{10}{T_I}$; −20dB; −20dB/decade; ω; $\varphi(\omega)$; $\varphi = -90°$
Double integrator (u → $A\int \ldots dt$ → $\int \ldots dt$ → y) 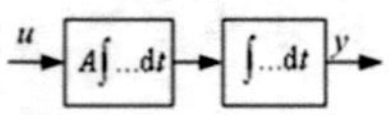	$\frac{A}{s^2} = \frac{1}{s^2 T_I^{\,2}}$	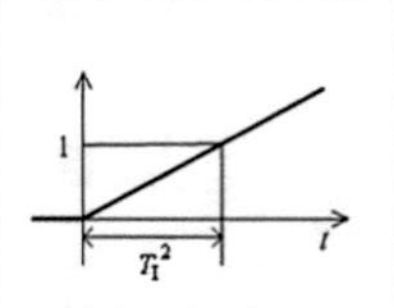 1; T_I^2; t	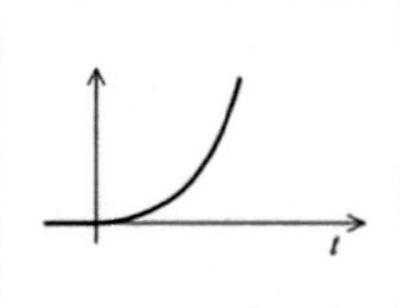 t	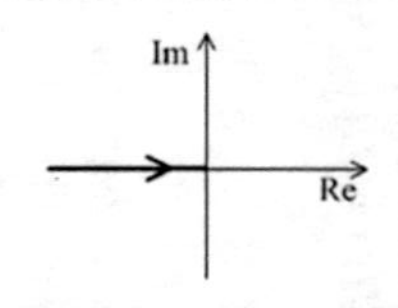 Im; Re	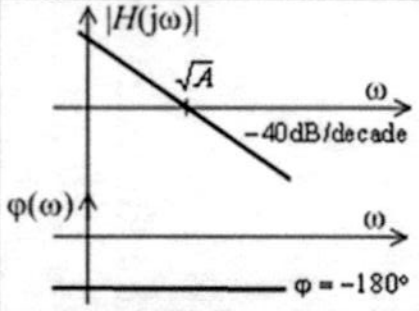 $\|H(j\omega)\|$; $\sqrt{A}$; −40dB/decade; ω; $\varphi(\omega)$; $\varphi = -180°$
Differentiating element (u → $\tau_D \frac{du(t)}{dt}$ → y) 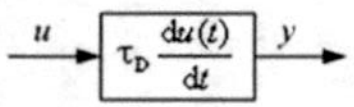	$s\tau_D$	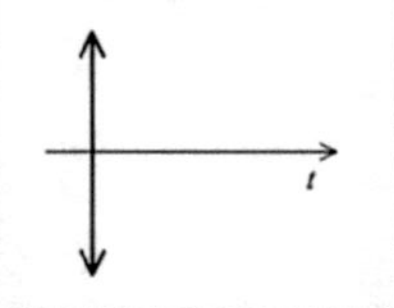 t	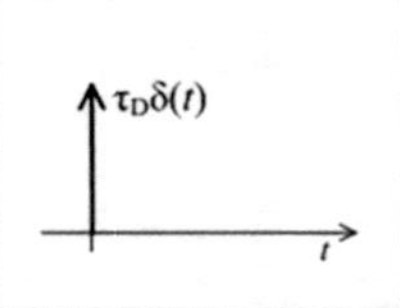 $\tau_D\delta(t)$; t	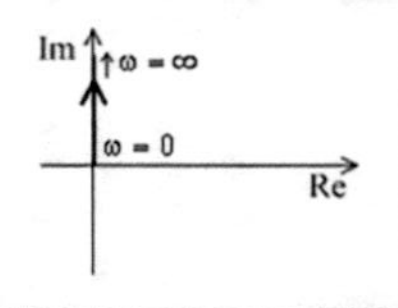 Im; $\omega = \infty$; $\omega = 0$; Re	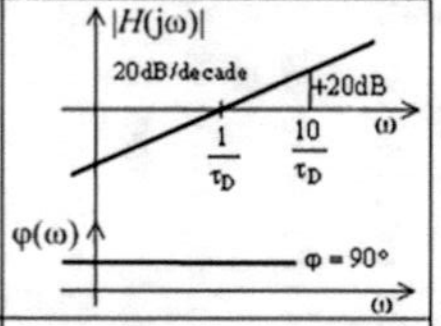 $\|H(j\omega)\|$; 20dB/decade; +20dB; $\frac{1}{\tau_D}$; $\frac{10}{\tau_D}$; ω; $\varphi(\omega)$; $\varphi = 90°$
Dead time (u → [u, y] → y) 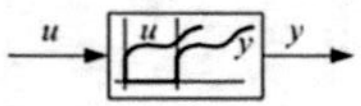	Ae^{-sT_d}	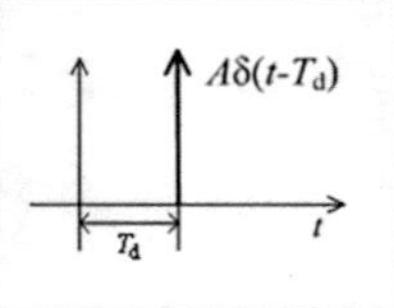 $A\delta(t-T_d)$; T_d; t	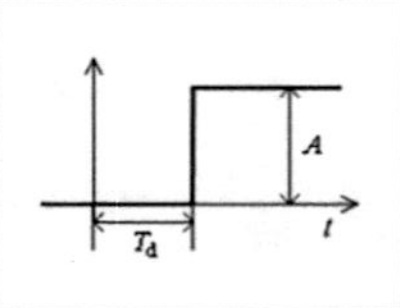 A; T_d; t	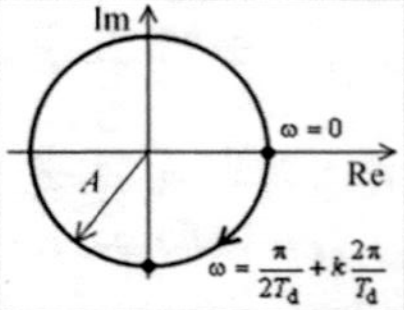 Im; A; $\omega = 0$; Re; $\omega = \frac{\pi}{2T_d} + k\frac{2\pi}{T_d}$	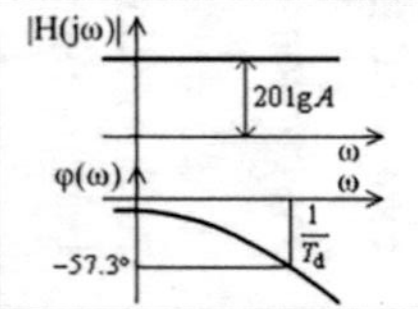 $\|H(j\omega)\|$; $20\lg A$; ω; $\varphi(\omega)$; $\frac{1}{T_d}$; −57.3°

charging current, or the relationship between the change of the angular position of a motor as the function of the angular speed. Its transfer function is:

$$H(s) = H_{\mathrm{I}}(s) = \frac{1}{sT_{\mathrm{I}}} = \frac{K_{\mathrm{I}}}{s} \tag{2.62}$$

and its frequency function is

$$H_{\mathrm{I}}(j\omega) = \frac{1}{j\omega T_{\mathrm{I}}} = \frac{K_{\mathrm{I}}}{j\omega}.$$

Its weighting function is a step, its unit step response is a ramp, which reaches unity after an elapsed time equal to the integrating time constant. Its NYQUIST diagram for positive ω values is a straight line going through the negative imaginary axis. The amplitude of the frequency function is $20\lg|H(j\omega)| = -20\lg \omega T_{\mathrm{I}}$, so the BODE amplitude-frequency diagram is a straight line of slope −20 dB/decade, crossing the 0 dB axis at $1/T_{\mathrm{I}}$. When the frequency is increased by a factor of ten, the amplitude is decreased by a factor of ten (it decreases by −20 dB). The value of the phase angle is −90° at all frequencies. The characteristic functions are shown in Table 2.2.

Note that if the element contains two integrating effects, its transfer function is $H(s) = K_{\mathrm{I}}/s^2 = 1/s^2T_{\mathrm{I}}^2$, its NYQUIST diagram goes through the negative real axis, the slope of the straight line of its BODE amplitude diagram is −40 dB/decade, which crosses the zero dB axis at frequency $\sqrt{K_{\mathrm{I}}}$, and its phase angle is −180° at all frequencies.

Differentiating (D) element

Its differential equation in time constant form is

$$y(t) = \tau_{\mathrm{D}} \frac{\mathrm{d}u(t)}{\mathrm{d}t}$$

The corresponding transfer function is

$$H(s) = H_{\mathrm{D}}(s) = s\tau_{\mathrm{D}}, \tag{2.63}$$

and the frequency function:

$$H_{\mathrm{D}}(j\omega) = j\omega\tau_{\mathrm{D}}.$$

Its weighting function consists of two DIRAC delta signals of the same area but of opposite sign. Its unit step response is a DIRAC delta of area τ_{D}. Its ramp response is a step of amplitude τ_{D}. Differentiating elements in reality appear only in systems where impulses and step-like inputs are excluded and can not be applied. The ideal differentiating element can not be realized, as a real physical device is not able to produce a DIRAC delta pulse as a response to a step input. It can be seen that the differentiating element gives zero output for a constant input signal. Therefore a *D* element is never connected serially in a closed control loop, as it would break off the loop in steady state.

The Nyquist diagram of the differentiating element for positive ω values is a straight line going through the positive imaginary axis. Its Bode diagram is a straight line of slope +20 dB/decade crossing the zero dB axis at $1/\tau_D$. The phase angle is +90° at all frequencies. The characteristic curves are given in Table 2.2.

An example for the physical realization of the ideal differentiating element is a transformer with open secondary circuit, where the input signal is the primary current, and the output signal is the induced voltage in the secondary coil. But in the primary circuit because of known physical laws, the primary current can not be changed in a step-like fashion.

Dead-time (H) element

Real processes often contain dead-time. If in a technological process a material (solid, liquid or gaseous) is transported from one place to another, then in the model of the process a transportation delay, the so called dead-time has to be considered.

In the dead-time element a delay T_d appears between the output and the input signals, which can be described by the following time function:

$$y(t) = \begin{cases} 0, & \text{if} \quad t < T_d \\ u(t - T_d), & \text{if} \quad t \geq T_d \end{cases}$$

Its differential equation is an algebraic equation:

$$a_o y(t) = b_o u(t - T_d), \quad \text{or} \quad y(t) = Au(t - T_d).$$

Its transfer function is a transcendental function:

$$H(s) = H_H(s) = Ae^{-sT_d}. \tag{2.64}$$

The frequency function is

$$H_H(j\omega) = Ae^{-j\omega T_d},$$

where the absolute value and the phase angle are expressed by

$$a(\omega) = \left| e^{-j\omega T_d} \right| = A \quad \text{and} \quad \varphi(\omega) = \arg\left\{ e^{-j\omega T_d} \right\} = -\omega T_d.$$

The characteristic functions are shown in Table 2.2.

The weighting function is a Dirac delta signal of area *A* shifted by the delay T_d, the unit step response is a step of amplitude *A* shifted by T_d. The Nyquist diagram consists of overlapping circles of radius *A* with their centers in the origin, where the endpoint of vector $H_H(j\omega)$ turns by an angle of $-\omega T_d$ with increasing ω. Vectors shifted by angle 2π are coincident. The Bode amplitude diagram is a straight line parallel to the frequency axis (it is the same as the amplitude diagram of the ideal P element), and the phase angle changes with the frequency in a linear way. At frequency $\omega = 1/T_{\rm d}$ the phase angle is -1 rad $= -57.3°$.

Dead-time exists in every real system, but its effect is significant only if the time of the transients in the system is comparable to the dead-time. In describing mass and energy transfer phenomena, the dead-time can not be neglected (mass transfer on conveyor or pipeline, convection, etc.).

2.4.2 Lag Blocks

Operations described by the ideal basic elements are influenced by energy storage elements which always show up in real devices. Their effect is taken into consideration by the so called lag elements. The basic types are the first and the second order lag element.

First order lag element

This element can be described by the following differential equation given in time constant form:

$$T\frac{{\rm d}y(t)}{{\rm d}t} + y(t) = Au(t)$$

Solving for the derivative of the output signal:

$$\frac{{\rm d}y(t)}{{\rm d}t} = \frac{A}{T}u(t) - \frac{1}{T}y(t).$$

The output signal can be obtained by integrating its derivative. According to the above expression the element can be represented as an integrator fed back by a constant (Fig. 2.33).

The transfer function of this element is

$$H(s) = H_{\rm T}(s) = \frac{A}{1+sT} \tag{2.65}$$

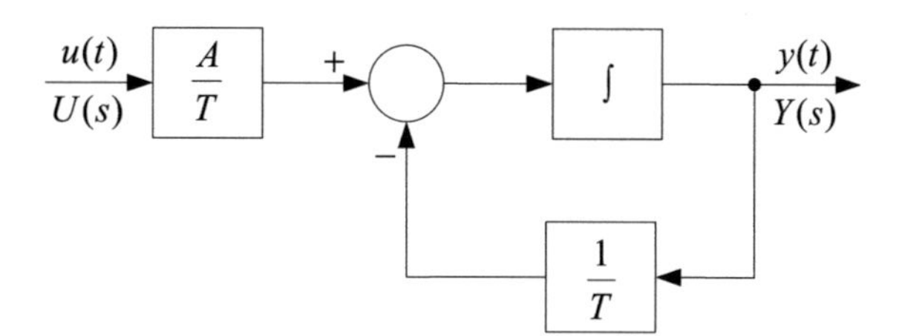

Fig. 2.33 The first order lag element can be interpreted as an integrator fed back by a constant gain

and its frequency function is

$$H(j\omega) = \frac{A}{1 + j\omega T}.$$

By the inverse LAPLACE transformation, expressions for the weighting function and the unit step response can be obtained:

$$w(t) = \frac{A}{T} e^{-t/T} \quad \text{and} \quad v(t) = A\left(1 - e^{-t/T}\right), \quad t \geq 0.$$

The functions are shown in Fig. 2.34. Let us observe in the figure the excised sections cut by the initial slopes of the unit step response and the weighting function, respectively.

For positive ω values the NYQUIST diagram is a half circle, which starts at $\omega = 0$ from point A of the real axis of the complex plane, and goes to the origin as $\omega \to \infty$ (Fig. 2.35).

To determine the BODE diagram let us express the absolute value of the frequency function.

$$20\lg|H(j\omega)| = 20\lg A - 20\lg\sqrt{1 + \omega^2 T^2}$$

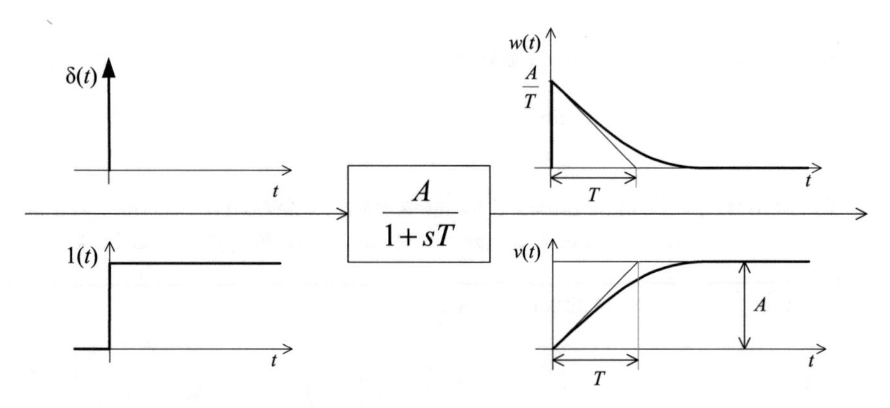

Fig. 2.34 Weighting function and unit step response of a first order lag element

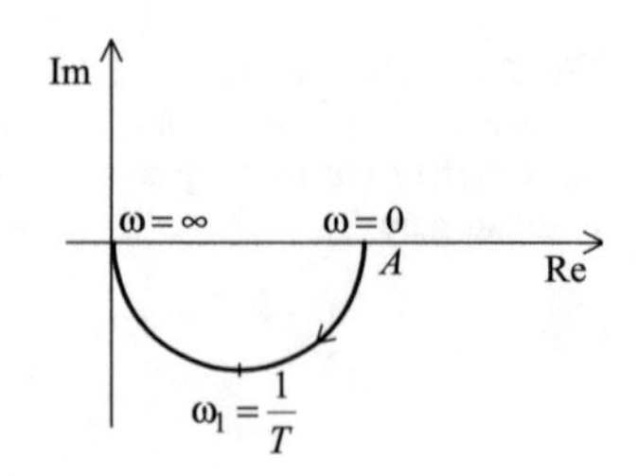

Fig. 2.35 NYQUIST diagram of a first order lag element

Supposing $A = 1$ the first term is zero. Let us apply the following approximations:

If $\omega T << 1$, $20\lg|H(j\omega)| \approx 0$

If $\omega T >> 1$, $20\lg|H(j\omega)| \approx -20\lg\omega T$

If $\omega T = 1$, $20\lg|H(j\omega)| = -20\lg\sqrt{2} \approx -3\text{dB}$

The approximate Bode diagram goes along the 0 dB axis till the so called corner frequency $\omega_1 = 1/T$, then it continues with a straight line of slope −20 dB/decade. At the corner frequency the accurate value of the amplitude is $-20\lg\sqrt{2} \approx -3$ dB (Fig. 2.36). In the figure the accurate diagram is denoted by a thin line. If the gain is not unity, then the Bode diagram is shifted parallel up or down by ($20\lg A$). The system can be considered as a low-pass filter which passes the low frequency signals, and attenuates the high frequency signals.

In the low frequency domain, the first order lag element can be approximated by a proportional element and in the high frequency domain by an integrating element. In the time domain this means that as $t \to \infty$ the element shows proportional properties, its output signal settles down to a constant value corresponding to the unit step input, while for $t = 0$, when the input signal is switched on, it shows an integrating effect.

The expression for the phase angle is $\varphi(\omega) = -\text{arctg}\,\omega T$. The phase function is shown in Fig. 2.36. At the corner frequency $\omega_1 = 1/T$ the phase shift is −45°, the slope of the curve is −66°/decade (see A.2.1 of Appendix A.5 for the element $H(s) = 1 + sT$).

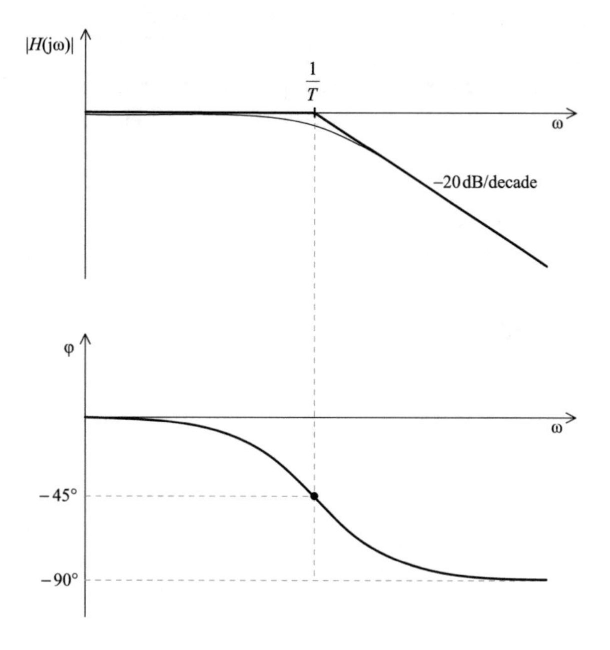

Fig. 2.36 Bode diagram of a first order lag element

An example of a first order lag element is an electrical circuit consisting of a serially connected resistor and inductor, where the input signal is the terminal voltage and the output is the current.

Second order oscillating (ξ) *element*

The second order proportional element can be described by the following differential equation:

$$a_2 \frac{\mathrm{d}^2 y(t)}{\mathrm{d}t^2} + a_1 \frac{\mathrm{d}y(t)}{\mathrm{d}t} + a_o y(t) = b_o u(t).$$

This differential equation describes, for instance, the behavior of an electrical circuit consisting of a resistor R, an inductor L and a capacitor C (Fig. 2.37), or a mechanical system consisting of a mass m, a spring with spring constant c and a fluid friction element with friction coefficient k (Fig. 2.38).

For the electrical circuit the following KIRCHHOFF voltage law holds:

$$u = iR + L\frac{\mathrm{d}i}{\mathrm{d}t} + \frac{1}{C}\int i\mathrm{d}t$$

The input signal is the terminal voltage u, the output signal is either the current i or the charge $q = \int i\mathrm{d}t$.

$$L\frac{\mathrm{d}^2 q}{\mathrm{d}t^2} + R\frac{\mathrm{d}q}{\mathrm{d}t} + \frac{1}{C}q = u$$

In the mechanical example the following differential equation gives the relationship between the external force F applied to the mass and the position h.

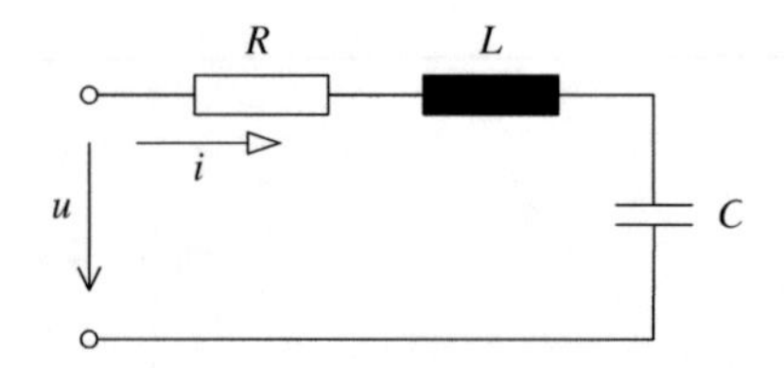

Fig. 2.37 The behavior of an electrical circuit consisting of a resistor, an inductor and a capacitor can be described by a second order differential equation

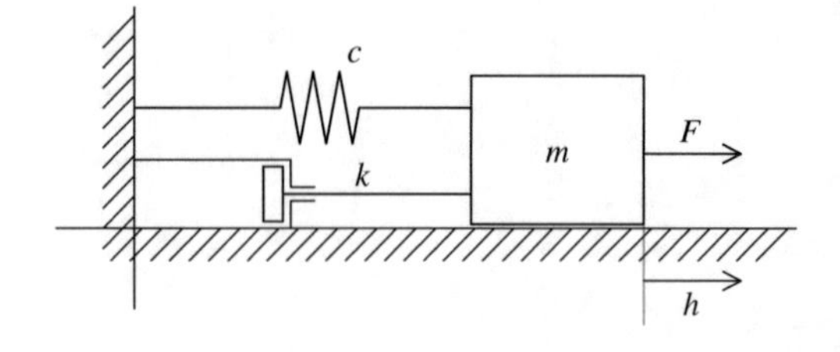

Fig. 2.38 The behavior of a mechanical system consisting of a mass, a spring and a fluid friction element can be described by a second order differential equation

$$m\frac{\mathrm{d}^2h}{\mathrm{d}t^2}+k\frac{\mathrm{d}h}{\mathrm{d}t}+ch=F$$

It is worth mentioning that there is a close analogy between the electrical circuit and the mechanical system.

Dividing both sides of the differential equation by the coefficient a_o, the differential equation can be transformed to the so called time constant form:

$$T^2\frac{\mathrm{d}^2y(t)}{\mathrm{d}t^2}+2\xi T\frac{\mathrm{d}y(t)}{\mathrm{d}t}+y(t)=Au(t)$$

where $A=b_o/a_o$ is the gain, which gives the steady state value of the output signal for unit step input, $T=\sqrt{a_2/a_o}$ is the time constant, $\xi=a_1/2\sqrt{a_oa_2}$ is the damping factor, all of which influence the dynamic behavior of the system. The transfer function of this element is

$$H(s)=H_\xi(s)=\frac{A}{1+2\xi Ts+T^2s^2}. \tag{2.66}$$

The poles of the second order oscillating element (the roots of the denominator) are:

$$s_{1,2}=-\frac{\xi}{T}\pm\frac{1}{T}\sqrt{\xi^2-1}. \tag{2.67}$$

The poles are real negative values if $\xi>1$; we get two coinciding negative real values if $\xi=1$; and we get two complex conjugate values if $\xi<1$. Let us determine the step responses for each of the three cases.

(a) Aperiodic case, $\xi>1$. The transfer function can be interpreted as two, serially connected first order lag elements:

$$H(s)=\frac{A/T_1T_2}{(s+1/T_1)(s+1/T_2)},\quad \text{where}$$
$$T_1=-1/s_1\quad\text{and}\quad T_2=-1/s_2.$$

The unit step response is

$$\begin{aligned}v(t)&=\mathcal{L}^{-1}\left\{\frac{1}{s}H(s)\right\}=\mathcal{L}^{-1}\left\{\frac{A/T_1T_2}{s(s+1/T_1)(s+1/T_2)}\right\}\\&=A\left(1-\frac{T_1}{T_1-T_2}e^{-t/T_1}+\frac{T_2}{T_1-T_2}e^{-t/T_2}\right)\end{aligned} \tag{2.68}$$

and the weighting function is

$$w(t) = \mathcal{L}^{-1}\{H(s)\} = A\left(\frac{1}{T_1 - T_2}e^{-t/T_1} - \frac{1}{T_1 - T_2}e^{-t/T_2}\right) \tag{2.69}$$

The weighting function is the derivative of the unit step response. The initial value and the initial derivative of the unit step response are zero. The initial value of the weighting function is zero, and the value of its first derivative is A/T_1T_2.

(b) Aperiodic boundary case, $\xi = 1$. The transfer function is:

$$H(s) = \frac{A}{(1+sT)^2} = \frac{A/T^2}{(s+1/T)^2}.$$

The weighting function can be calculated by the inverse LAPLACE transformation of the transfer function:

$$w(t) = \frac{A}{T^2}te^{-t/T}, \quad t \geq 0. \tag{2.70}$$

The unit step response is:

$$v(t) = \mathcal{L}^{-1}\left\{\frac{1}{s}\frac{A/T^2}{(s+1/T)^2}\right\} = \mathcal{L}^{-1}\left\{\frac{\alpha}{s} + \frac{\beta}{s+1/T} + \frac{\gamma}{(s+1/T)^2}\right\}$$

where $\alpha = A$, $\beta = -A$ and $\gamma = -A/T$. Thus

$$v(t) = A\left(1 - e^{-t/T} - \frac{1}{T}te^{-t/T}\right), \quad t \geq 0. \tag{2.71}$$

(c) Oscillating case, $\xi < 1$. The transfer function is:

$$H(s) = \frac{A}{1+2\xi Ts + T^2s^2} = \frac{A/T^2}{(s-s_1)(s-s_2)},$$

where

$$s_{1,2} = -\frac{\xi}{T} \pm j\frac{1}{T}\sqrt{1-\xi^2} = -\xi\omega_o \pm j\omega_p = a \pm jb$$

are complex conjugate poles.

Fig. 2.39 Poles of a second order oscillating element

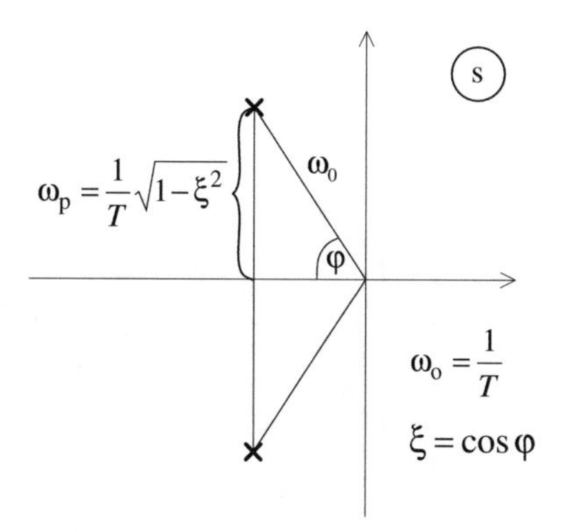

Here $\omega_o = 1/T$ is the so called *natural frequency*, which is the absolute value of the vector starting from the origin and pointing to one of the complex poles. The poles of the oscillating element are depicted in Fig. 2.39.

Here $\omega_p = b = \sqrt{1-\xi^2}/T$ is the oscillation frequency of the periodic component of the unit step response and the weighting function, respectively; it is the imaginary part of the pole. Thus $\omega_o^2 = a^2 + b^2$ and $\cos\varphi = \xi$, where φ is the angle of the vector representing the pole formed with the negative real axis. If T changes and ξ is constant, the poles move on a straight line forming an angle φ with the negative real axis.

The weighting function is obtained by the inverse LAPLACE transformation of the transfer function:

$$w(t) = \frac{A\omega_o}{\sqrt{1-\xi^2}} e^{-\xi\omega_o t}\sin\omega_p, \quad t \geq 0. \tag{2.72}$$

and the unit step response is:

$$v(t) = A\left[1 - \frac{e^{-\xi\cdot\omega_o t}}{\sqrt{1-\xi^2}}\left(\sqrt{1-\xi^2}\cos\omega_p t + \sin\omega_p t\right)\right], \quad t \geq 0. \tag{2.73}$$

Figure 2.40 shows the unit step responses for different damping factors.

The overshoot of the unit step response expressed in percentages in case of $\xi < 1$ (obtained by differentiating the unit step response) is

$$\sigma = \frac{v_{max} - v_{ss}}{v_{ss}} 100\% = e^{\frac{-\xi\pi}{\sqrt{1-\xi^2}}} 100\%. \tag{2.74}$$

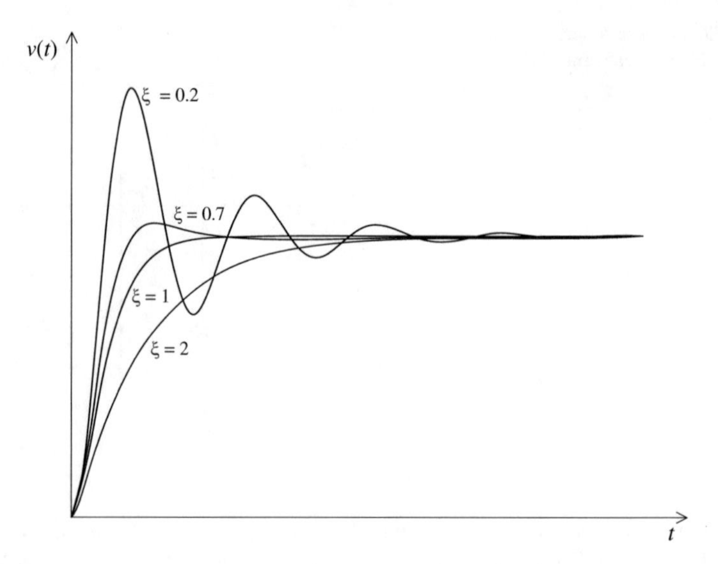

Fig. 2.40 Unit step responses of a second order oscillating element with damping factors $\xi = 0.2, 0.7, 1, 2$

The location of the first maximum of the unit step response (peak time) is

$t_c = \frac{\pi}{\omega_p} = \frac{\pi}{\omega_o\sqrt{1-\xi^2}}$ and the corresponding peak frequency is

$$\omega_p = \omega_o\sqrt{1-\xi^2}. \tag{2.75}$$

The settling time is defined as the time instant t_α, such that for $t > t_\alpha$ the unit step response function remains within a given $\Delta\%$ band around its steady state value. The following condition can be given for the envelope of the unit step response:

$$e^{-\xi\omega_o t_\alpha} = \frac{\Delta}{100},$$

whence the settling time is $t_\alpha = \ln(100/\Delta)/\xi\omega_o$. For $\Delta = 2\%$ or $\Delta = 5\%$ the settling time is approximately $4/\xi\omega_o$ or $3/\xi\omega_o$, respectively.

Let us now determine the frequency function of the oscillating element. The expression of the absolute value of the frequency function is:

$$|H(j\omega)| = \frac{A}{\sqrt{(1-\omega^2T^2)^2 + 4\xi^2T^2\omega^2}}. \tag{2.76}$$

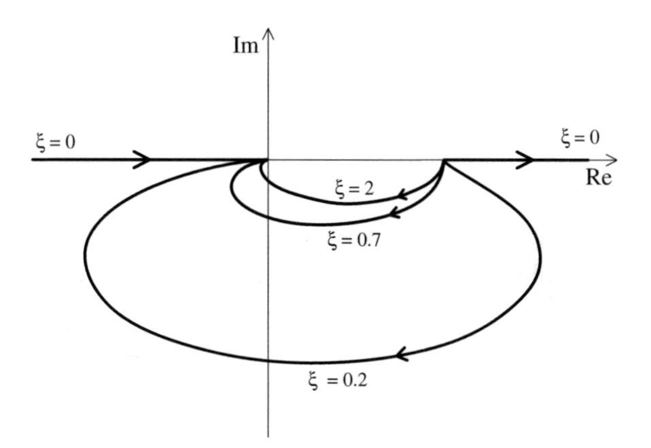

Fig. 2.41 NYQUIST diagram of a second order lag element

and its phase angle is

$$\varphi(\omega) = -\mathrm{arctg}\frac{2\xi T\omega}{1 - \omega^2 T^2} \tag{2.77}$$

The NYQUIST diagram (Fig. 2.41) at $\omega = 0$ starts at point A of the real axis of the complex plane. As $\omega \to \infty$ it goes to the origin. In between it passes through two quarters of the complex plane. If $\xi < 0.5$, in a given frequency range the curve shows amplification, the values of the amplitudes exceed the value taken at $\omega = 0$. If $\xi = 0$, the curve runs on the real axis, and at $\omega = \omega_o = 1/T$ it has a discontinuity.

Let us determine the BODE amplitude-frequency curve and its asymptotic approximation. The expression of the absolute value in decibels is

$$20\lg|H(j\omega)| = 20\lg A - 20\lg\sqrt{(1 - \omega^2T^2)^2 + 4\xi^2T^2\omega^2}$$

with $A = 1$ the first term is zero.

Let us apply the following approximations:

if $\omega T \ll 1$, $20\lg|H(j\omega)| \approx 0$
if $\omega T \gg 1$, $20\lg|H(j\omega)| \approx -40\lg\omega T$, as besides the fourth degree term all the other terms can be neglected,
if $\omega T = 1$, $20\lg|H(j\omega)| = -20\lg 2\xi$. At the corner frequency (also called the breakpoint frequency) the absolute value depends only on the damping factor.

Supposing $A = 1$ at the natural frequency the absolute value of the frequency function is $|H(j\omega_o)| = 1/2\xi$, and the phase angle is $-90°$. (For the slope of the phase-frequency curve $-132°/\xi$/decade is obtained according to the calculations in A.2.1 of Appendix A.5) Another characteristic point of the frequency function is the *resonance frequency* (ω_r), where the amplitude takes its maximum value.

Calculating the derivative of the expression of the absolute value and setting it equal to zero, the following relationship is obtained:

$$\omega_r = \frac{1}{T}\sqrt{1-2\xi^2}. \tag{2.78}$$

A resonance frequency exists if $\xi < \sqrt{0.5} \approx 0.707$. At this frequency the absolute value is

$$|H(j\omega_r)| = \frac{1}{2\xi\sqrt{1-\xi^2}}. \tag{2.79}$$

The *cut-off frequency* (ω_c) is defined as the frequency where the absolute value of the frequency function is unity. Supposing the gain $A = 1$, this condition is fulfilled if

$$\left(1-\omega_c^2T^2\right)^2 + 4\xi^2T^2\omega_c^2 = 1,$$

whence

$$\omega_c = \frac{1}{T}\sqrt{2\left(1-2\xi^2\right)}. \tag{2.80}$$

The relation between the characteristic frequencies (in case of $\xi < 0.5$) is

$$\omega_r < \omega_p < \omega_o < \omega_c. \tag{2.81}$$

If $\xi < 1$, the first three frequencies are very close to each other (Fig. 2.42).

Often when the asymptotic Bode amplitude-frequency curve is plotted, the accurate amplitude values are calculated only at the resonance and the natural frequencies, where significant amplification may occur.

The Bode diagrams for damping factors $\xi = 0.2, 0.7, 1, 2$ are given in Fig. 2.43. In the low frequency domain the asymptote of the amplitude-frequency curve is a horizontal line, and in the high frequency domain when $\omega \gg 1/T$ the asymptote is a straight line of slope −40 dB/decade. (If $\xi > 1$, it is expedient to decompose the transfer function into a product of two first order lag elements, and in between the two breakpoints to put in an additional asymptote of slope −20 dB/decade.) The phase-frequency curve starts at 0°, then it tends to reach −180°, while its value at the natural frequency is −90°. Its steepness is bigger if the damping factor is smaller. The smaller the damping factor is, the higher the tendency towards oscillations and overshoot in the time domain and high amplification in the frequency domain. For damping factors higher than 0.6 the overshoot is within 10%, and in the amplitude-frequency function there is no significant amplification.

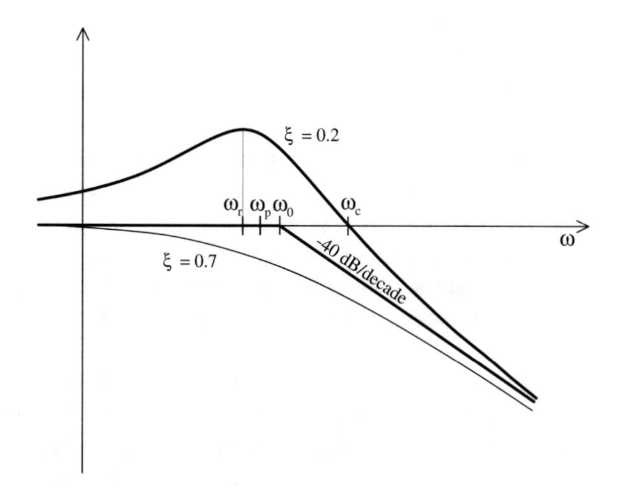

Fig. 2.42 Amplitude-frequency diagram of the oscillating element

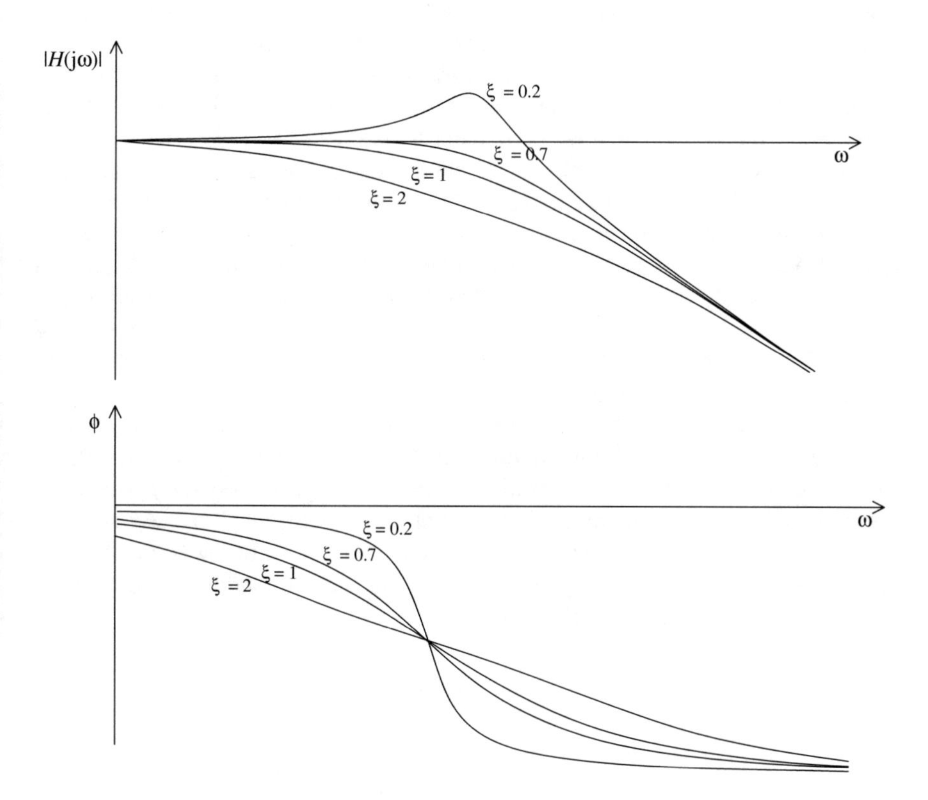

Fig. 2.43 Bode diagram of the oscillating element for $\xi = 0.2, 0.7, 1, 2$

2.4.3 Proportional, Integrating and Differentiating Lag Blocks

Complex elements can be assembled by the serial or parallel connection of the basic elements.

By series connection of the pure proportional, integrating or differentiating elements with lag elements first order, second order, *n*th order proportional (*PT1*, *PT2*,…), first order, second order, *n*th order integrating (*IT1*, *IT2*,…), as well as first order, second order, *n*th order differentiating elements (*DT1*, *DT2*,…) can be derived.

The transfer functions, unit step responses, NYQUIST and BODE amplitude-frequency diagrams of these elements are given in Table 2.3. The NYQUIST diagrams are obtained by multiplying the NYQUIST diagrams of the series component elements. In the considered frequency values the vectors of the individual components have to be multiplied (the phase angles are added, the absolute values are multiplied). The multiplication has to be executed for all the considered frequencies. The approximate BODE amplitude diagrams can easily be composed by adding the asymptotic BODE diagrams of the serially connected components.

In the unit step responses, the proportional, integrating or differentiating characteristics of an element are represented by the steady state performance. The lag elements influence the initial response and the transients.

In the low frequency domain the NYQUIST diagram shows a performance corresponding to the NYQUIST diagram of the proportional, integrating or differentiating element, then with increasing frequency it passes through as many quarters of the plane as the number of the lag elements suggests. The asymptotic BODE diagram in the low frequency domain starts according to the proportional, integrating or differentiating effect. Each lag element produces a change of the slope with −20 dB/decade at the reciprocal of the corresponding time constant. Thus the parameters of the element can be read from the approximate BODE diagram.

Often it is sufficient to plot the approximate BODE amplitude-frequency diagram. In some systems, besides the approximate diagram, it is also necessary to accurately determine the amplitude at some critical points or in a given frequency range. For example, in the case of an integrating element serially connected to a second order lag element containing complex conjugate poles, the course of the NYQUIST or the BODE diagram has also to be given accurately in the surroundings of the corner point. The transfer function of the element is

$$H(s) = \frac{K_{\mathrm{I}}}{s(1 + 2\xi Ts + s^2T^2)}.$$

Table 2.3 Basic elements

Transfer function	Unit step response	Nyquist diagram	Bode diagram
Proportional elements $A; \frac{A}{1+sT_1}; \frac{A}{(1+sT_1)(1+sT_2)\ldots}$	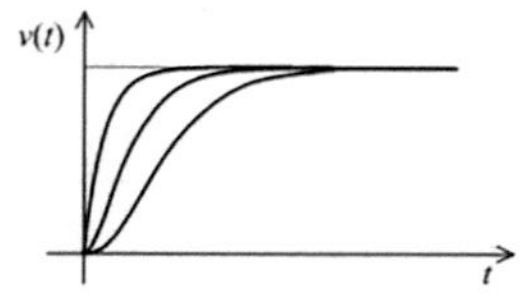 $v(t)$, t	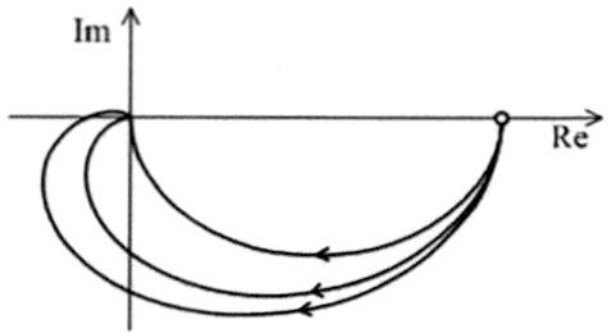 Im, Re	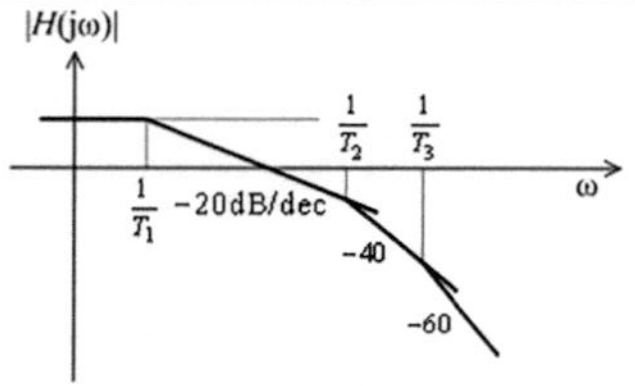 $\lvert H(j\omega)\rvert$, $\frac{1}{T_1}$, $\frac{1}{T_2}$, $\frac{1}{T_3}$, −20dB/dec, −40, −60, ω
Integrating elements $\frac{K_I}{s}; \frac{K_I}{s(1+sT_1)}; \frac{K_I}{s(1+sT_1)(1+sT_2)\ldots}$	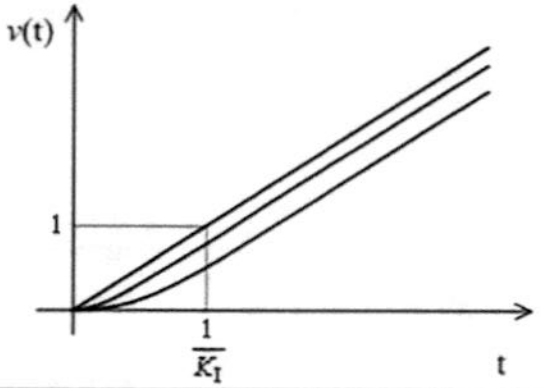 $v(t)$, 1, $\frac{1}{K_I}$, t	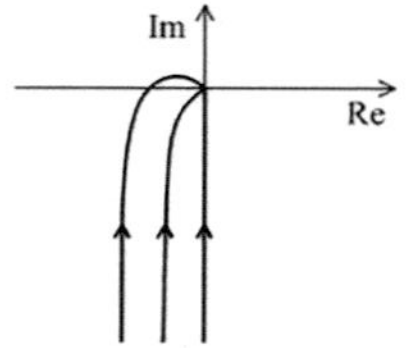 Im, Re	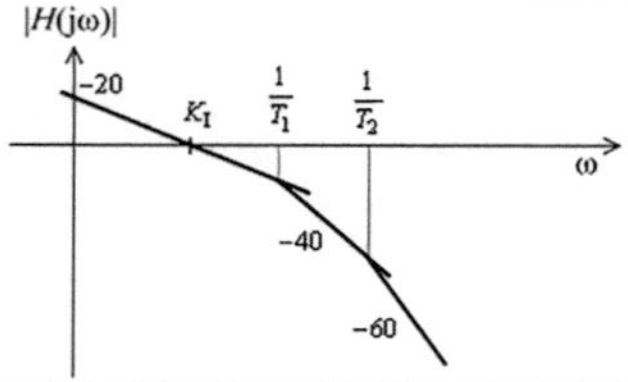 $\lvert H(j\omega)\rvert$, −20, K_I, $\frac{1}{T_1}$, $\frac{1}{T_2}$, −40, −60, ω
Double integrators $\frac{K_I}{s^2}; \frac{K_I}{s^2(1+sT_1)}; \frac{K_I}{s^2(1+sT_1)(1+sT_2)\ldots}$	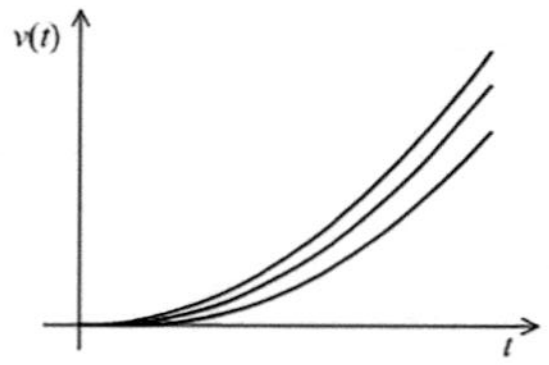 $v(t)$, t	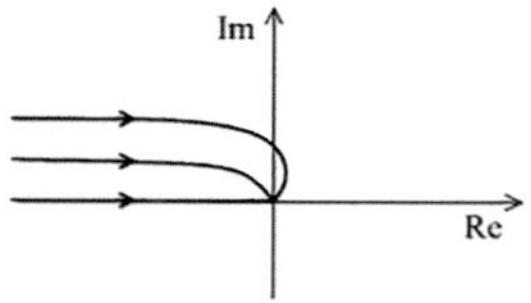 Im, Re	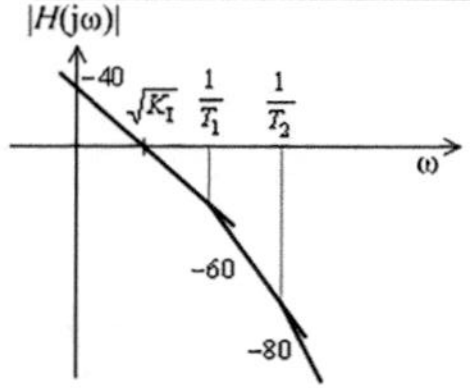 $\lvert H(j\omega)\rvert$, −40, $\sqrt{K_I}$, $\frac{1}{T_1}$, $\frac{1}{T_2}$, −60, −80, ω
Differentiating elements $s\tau; \frac{s\tau}{(1+sT_1)}; \frac{s\tau}{(1+sT_1)(1+sT_2)\ldots}$	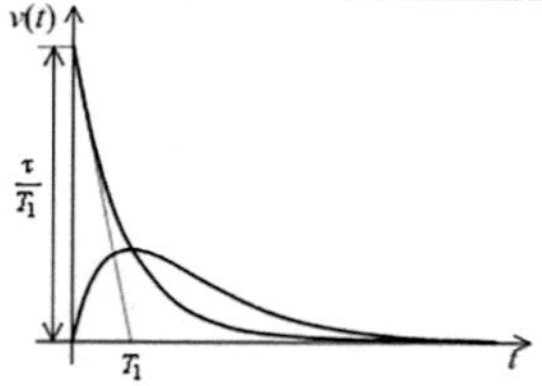 $v(t)$, $\frac{\tau}{T_1}$, T_1, t	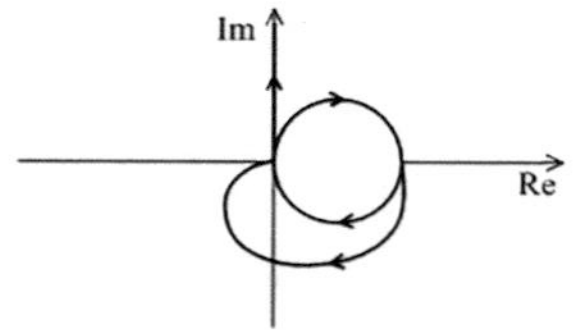 Im, Re	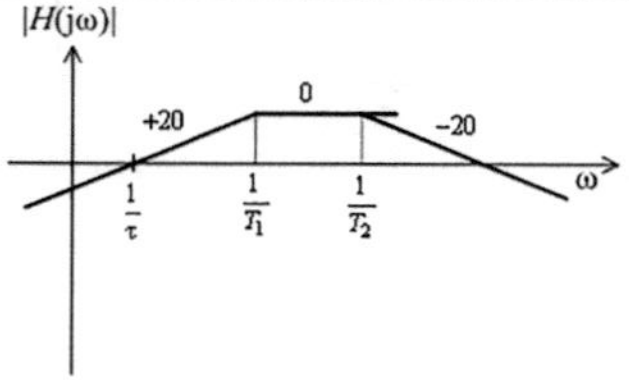 $\lvert H(j\omega)\rvert$, +20, 0, −20, $\frac{1}{\tau}$, $\frac{1}{T_1}$, $\frac{1}{T_2}$, ω

The Bode and the Nyquist diagrams with different damping factors ξ are shown in Fig. 2.44. With a small damping factor, the frequency function may have very high amplifications around the natural frequency of the second order oscillating element.

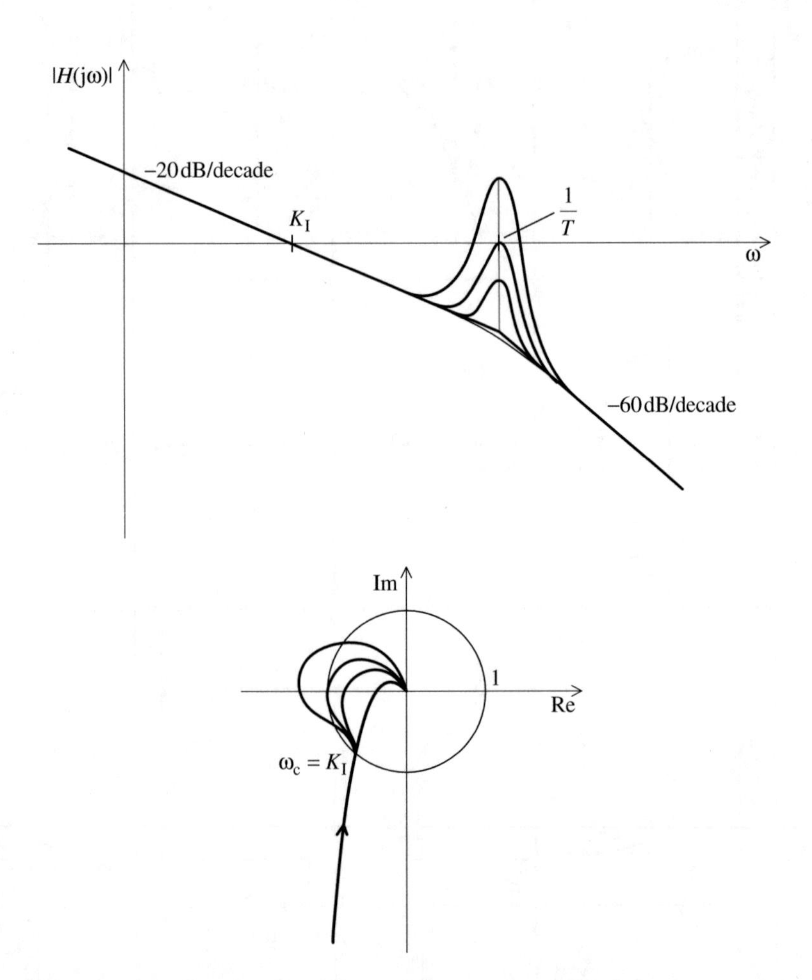

Fig. 2.44 Bode and Nyquist diagrams of a second order oscillating block serially connected to an integrating element

2.4.4 Influence of the Zeros of the Transfer Function

The zeros are the roots of the numerator of the transfer function given in Eq. (2.47).

$$H(s) = \frac{k(s - z_1)(s - z_2)\ldots(s - z_m)}{\mathcal{D}(s)}$$

Here $\mathcal{D}(s)$ denotes the denominator of the transfer function. The zeros are $z_1, z_2, \ldots, z_m$. The system can be accelerated by inserting zeros located at the left side of the complex plane. Let us analyze the characteristic functions of the so called ideal *PD* element

$$1 + s\tau = \tau(s + 1/\tau)$$

appearing in the numerator. Here the zero is $z_1 = -1/\tau$. The unit step response, the NYQUIST and the BODE diagrams are shown in Fig. 2.45.

This element in itself is unrealizable, for a DIRAC delta appears in its unit step response. The approximate BODE diagram is the mirror image relative to the frequency axis of the BODE diagram of the first order lag element. The amplitude-frequency curve can be approximated by 0 dB till the corner point $1/\tau$, and by a straight line of slope +20 dB/decade beyond it. Its phase angle is positive, $\varphi(\omega) = +\mathrm{arctg}\omega\tau$.

By inserting a zero, the system can be accelerated. To demonstrate this effect let us consider the circuit in Fig. 2.46, where a phase-lead element is connected serially to a first order lag element. For a unit step input signal, at the first instant a signal of ten times amplitude appears at the output of the *PLead* element, which is the input of the first order lag element. At the beginning the first order lag element acts as if it should reach this value according to its time constant, thus its output starts with a remarkable slope, and when its input signal is decreased, its output has almost reached the required steady value. The cost of this acceleration is the so called *overexcitation*, which is the ratio of the initial and final values of the signal at the input of the element. Acceleration can be reached if the overexcitation is larger than 1. Often it is expedient to apply mathematical pole cancellation, when a pole causing the undesirable slow behavior is cancelled by a zero, and a pole leading to a more favorable behavior is inserted into the system.

In a realizable way a zero can be inserted into a system only together with a pole. Let us determine the characteristic functions of the realizable

$$H(s) = A\frac{1 + s\tau}{1 + sT}$$

element for $\tau < T$ (phase-lag: *PLag*) and for $\tau > T$ (phase-lead: *PLead*). The unit step response, the NYQUIST and the BODE diagrams are shown in Table 2.4.

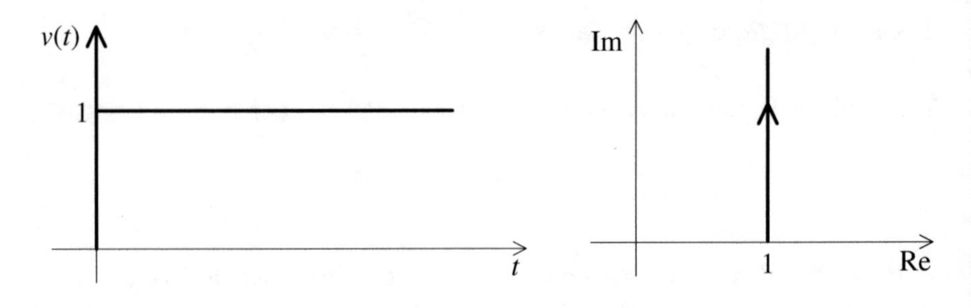

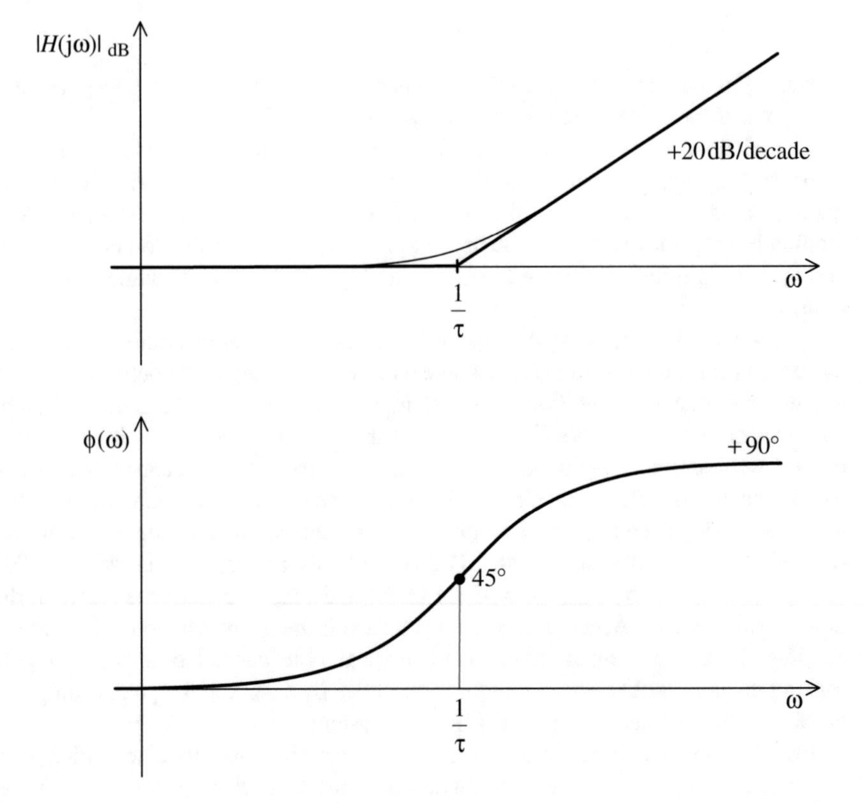

Fig. 2.45 Unit step response, NYQUIST and BODE diagrams of the ideal *PD* element with transfer function $1 + s\tau$

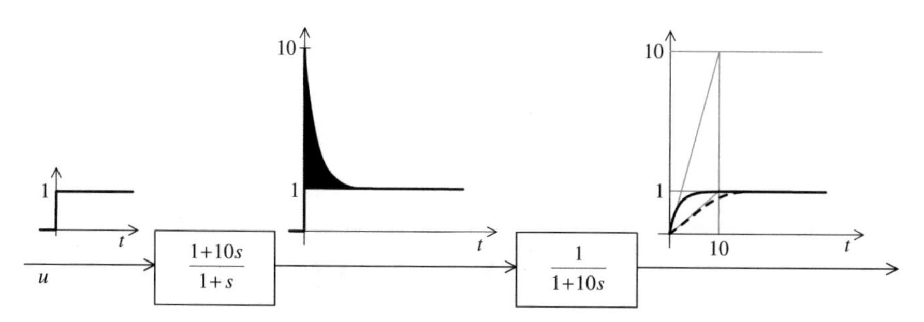

Fig. 2.46 Inserting a zero may accelerate the system at the cost of certain overexcitation

To demonstrate the accelerating effect of the zeros let us consider the transfer function below:

$$H(s) = \frac{1 + s\tau}{(1 + s)(1 + 10s)}.$$

Let the value of the time constant τ in the numerator be 0, 1, 5 and 10. The step responses are shown in Fig. 2.47.

If there are only poles (which are stable) in the transfer function, the phase angle curve versus the frequency changes monotonically. Phase angles belonging to the poles are negative. Inserting zeros adds some positive phase angles to the original phase function, and the monotonity of the phase function is impaired. "Buckling" appears in a given frequency range of the Nyquist diagram. (Later on it will be shown that with the appropriate choice of zeros the Nyquist diagram can be modified expediently to evade regions of the complex plane which are undesirable considering the transient behavior.) The slope of the asymptotic Bode diagram is changed by +20 dB/decades at the breakpoint corresponding to the zeros, and the phase angle is modified by positive values. Figure 2.48 shows the change of the Nyquist diagram, while Fig. 2.49 illustrates the change of the Bode diagram when a zero is inserted into the system.

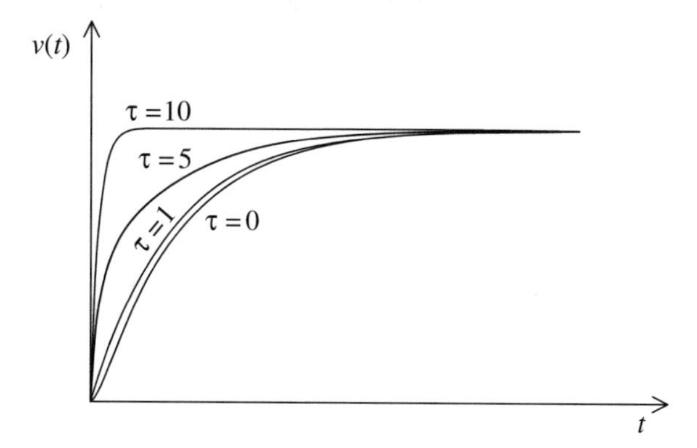

Fig. 2.47 Unit step responses with different values of the zero

Table 2.4 Characteristic functions of the phase-lag and phase-lead elements

Transfer function	Unit step response	Nyquist diagram	Bode diagram
$H(s) = A\dfrac{1+s\tau}{1+sT}$ $\tau < T$ Phase-lag element	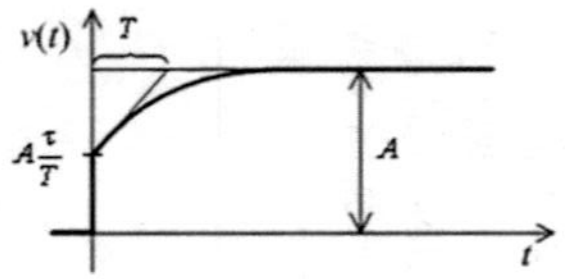 $v(t)$, T, $A\frac{\tau}{T}$, A, t	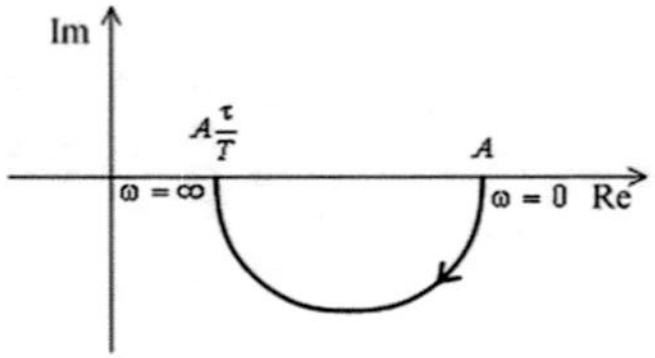 Im, $A\frac{\tau}{T}$, $\omega = \infty$, A, $\omega = 0$, Re	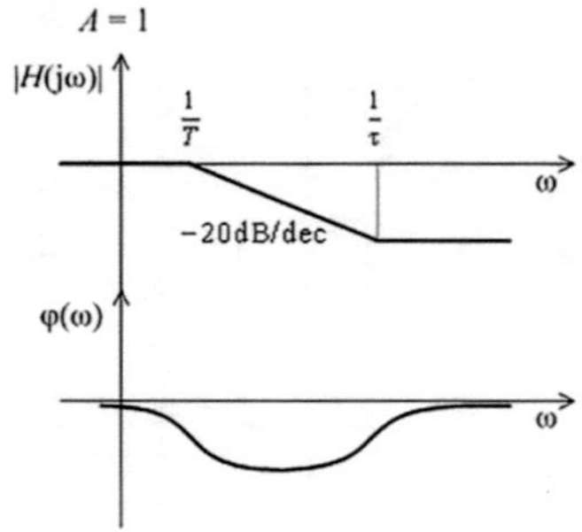 $A = 1$, $\lvert H(j\omega)\rvert$, $\frac{1}{T}$, $\frac{1}{\tau}$, -20dB/dec, ω, $\varphi(\omega)$, ω
$H(s) = A\dfrac{1+s\tau}{1+sT}$ $\tau > T$ Phase-lead element	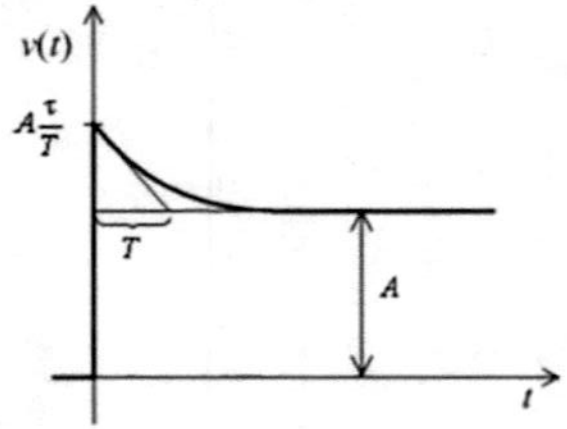 $v(t)$, $A\frac{\tau}{T}$, T, A, t	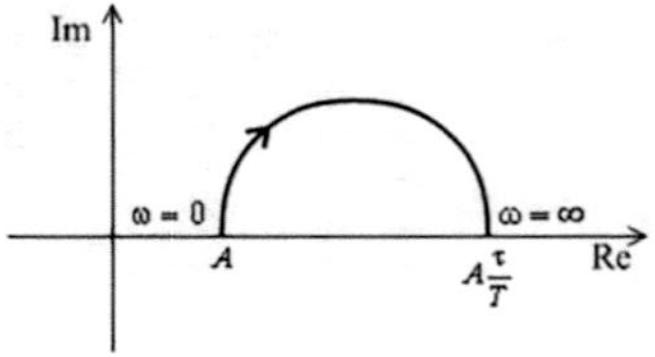 Im, $\omega = 0$, A, $\omega = \infty$, $A\frac{\tau}{T}$, Re	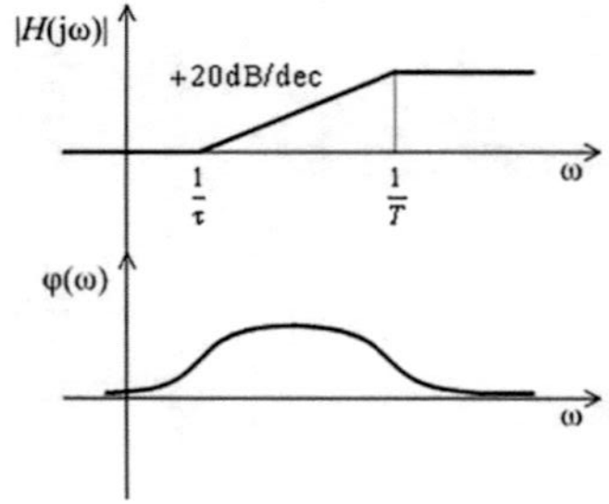 $\lvert H(j\omega)\rvert$, $+20\text{dB/dec}$, $\frac{1}{\tau}$, $\frac{1}{T}$, ω, $\varphi(\omega)$, ω

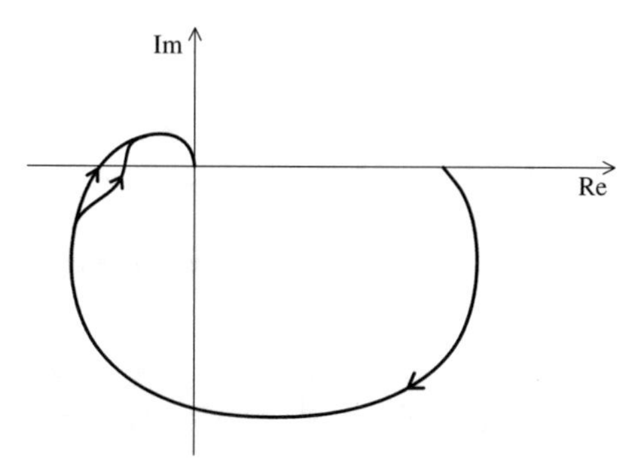

Fig. 2.48 Insertion of a zero changes the monotonicity of the phase diagram: a "buckling" appears in the NYQUIST diagram

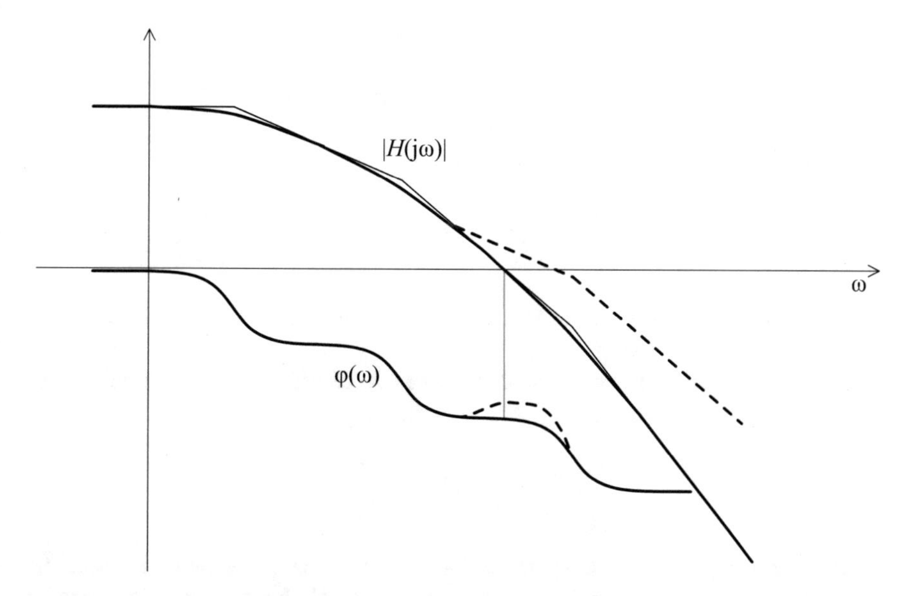

Fig. 2.49 The effect of insertion of a zero in the BODE diagram

2.4.5 Non-minimum Phase Systems

Non-minimum phase systems are systems, whose zeros are located on the right hand side of the complex plane.

If a system is of minimum phase, i.e., the zeros of its transfer function are on the left hand side of the complex plane, then the phase angle belonging to the poles is negative, and the phase angle belonging to the zeros is positive. Thus the phase angle curve can be unambiguously assigned to the asymptotic amplitude curve.

A pole in the right (unstable) half-plane modifies the BODE phase-frequency diagram by positive phase angle. A zero in the right half plane modifies it by negative phase angles, thus a zero does not decrease a negative phase angle, but increases it. (This property motivated the denomination of non-minimum phase systems.)

To illustrate the non-minimum phase property let us consider the following two transfer functions:

$$H_a(s) = \frac{1+sT}{1+sT_1} \quad \text{and} \quad H_b(s) = \frac{1-sT}{1+sT_1}.$$

For positive values of T_1 and T both systems are stable. H_a has a left hand, while H_b has a right hand side zero. The amplitude-frequency functions of the two systems are the same:

$$a(\omega) = \sqrt{\frac{1+(\omega T)^2}{1+(\omega T_1)^2}},$$

while their phase angles differ from each other:

$$\varphi_a(\omega) = -\arctan\frac{\omega(T_1-T)}{1+\omega^2 T_1 T} \quad \text{and} \quad \varphi_b(\omega) = -\arctan\frac{\omega(T_1+T)}{1+\omega^2 T_1 T}.$$

Comparing the two curves in Fig. 2.50a it is clearly seen that in each frequency range $|\varphi_a(\omega)|$ is less than $|\varphi_b(\omega)|$.

Every non-minimum phase system H_{nmp} can be converted to the product of a so called all-pass phase element H_{ap} and a minimum phase element H_{mp}.

$$H_{nmp}(s) = \frac{1-sT}{1+sT_1} = \frac{1-sT}{1+sT}\frac{1+sT}{1+sT_1} = H_{ap}(s)H_{mp}(s). \tag{2.82}$$

The property of an all-pass non-minimum phase element is that its absolute-frequency function is a unity constant at all frequencies. The transfer function of the n-th order all-pass element in case of real poles is

$$H_{ap}(s) = \prod_{i=1}^{n}\frac{1-sT_i}{1+sT_i} = \prod_{i=1}^{n}\frac{s-s_i}{s+s_i}. \tag{2.83}$$

Non-minimum phase systems have an unusual behavior in the time domain. For example, in the case of one right side zero, the unit step response starts in the direction opposite to its steady state value, then changing direction it finally reaches its steady state. Figure 2.50b shows the unit step response of the system given by

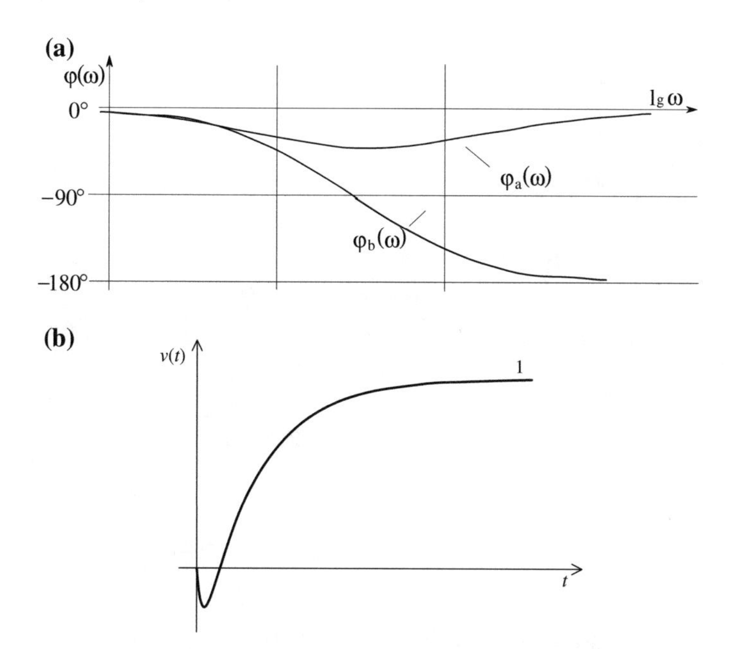

Fig. 2.50 Frequency function and unit step response of a non-minimum phase system

the transfer function $H(s) = (1 - 4s)/(1 + s)(1 + 10s)$. Chemical processes and furnaces often have the non-minimum phase property.

2.4.6 *Quick Drawing of Asymptotic Bode Diagrams*

The asymptotic Bode diagram can be easily drawn based on the previous considerations. In the case of proportional elements the Bode amplitude diagram starts parallel to the frequency axis with zero phase. The Bode amplitude diagram of a system containing one integrator starts with slope of -20 dB/decade and with phase $-90°$, while in the case of two integrators it starts with slope -40 dB/decade and with phase $-180°$. Lag elements change the slope of the asymptotic Bode amplitude diagram at the breakpoints by -20 dB/decade. Zeros change the slope by $+20$ dB/decade at the corresponding breakpoints. Dead-time does not modify the amplitude diagram, but significantly changes the phase angle.

The corner frequencies of the asymptotic amplitude diagram are the reciprocals of the time constants. As an example the asymptotic Bode amplitude-frequency diagram belonging to the transfer function

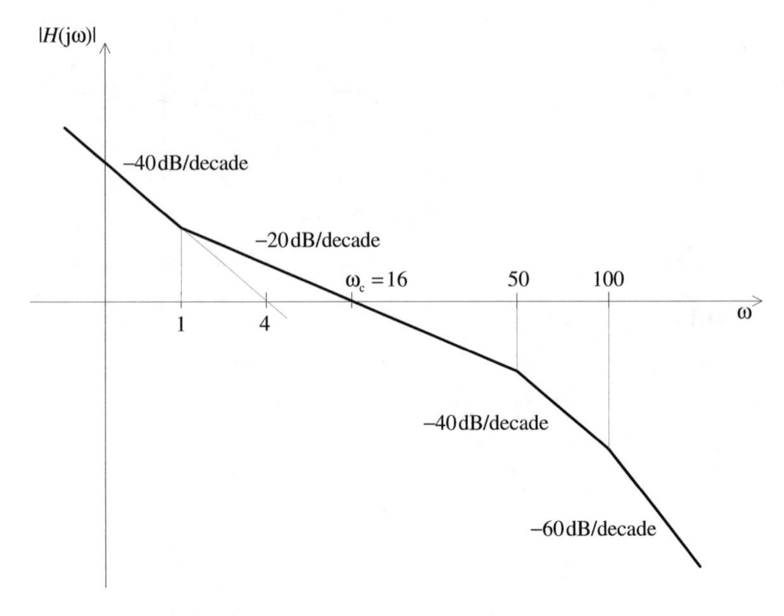

Fig. 2.51 Rapid plotting of the asymptotic Bode amplitude-frequency diagram

$$H(s) = \frac{16(1+s)}{s^2(0.02s)(1+0.01s)}$$

is shown in Fig. 2.51. Around the crossing point with the 0 dB axis the absolute value of the frequency function can be approximated by the expression $16/\omega$ (the absolute value of the lag elements here can be approximated by 1). At the crossing point the absolute value is 1, thus $\omega_c \approx 16$. With similar approximations the characteristic values of the Bode amplitude diagram can easily be determined.

2.4.7 *Influence of Parameter Changes*

When modeling a real system, generally the parameters of the differential equation or the transfer function describing the system are determined by measurements considering a modeling procedure. Thus their values are non accurate, and they may vary within a given range around their nominal values. When analyzing the system or designing the controller, it is important to take into consideration the effects of changes in the parameters.

From the time constant form of the transfer function, it is easy to determine the effect of changes in the parameters on the characteristic functions of the system. Table 2.5 illustrates the effect of parameter changes on the unit step responses and the Bode diagrams for some typical elements.

Table 2.5 Effects of parameter changes

System	Parameter	Unit step response	Bode amplitude	Bode phase
$\frac{K}{1+sT}$	K	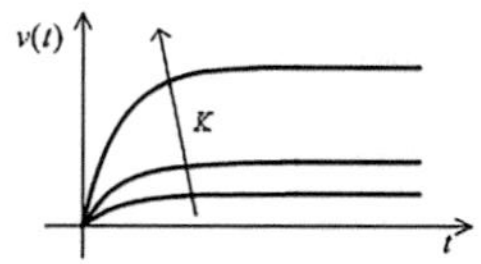	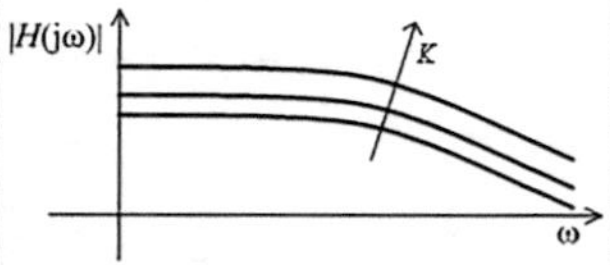	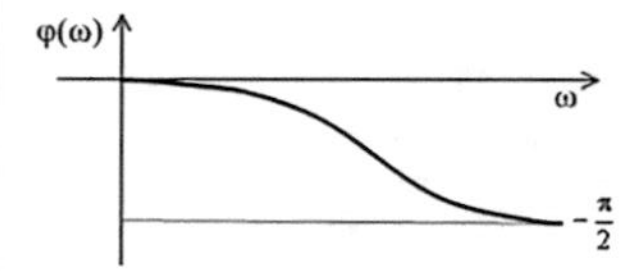
	T	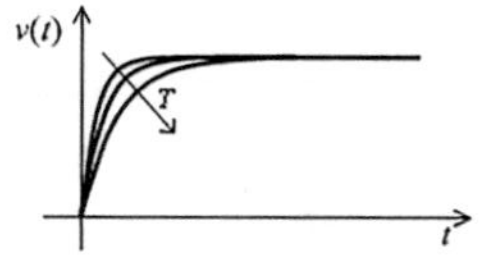	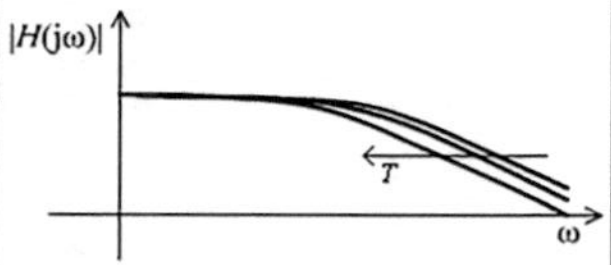	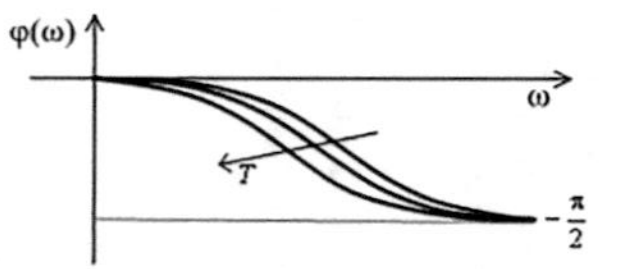
$\frac{1}{1+2\xi Ts+T^2s^2}$	ξ	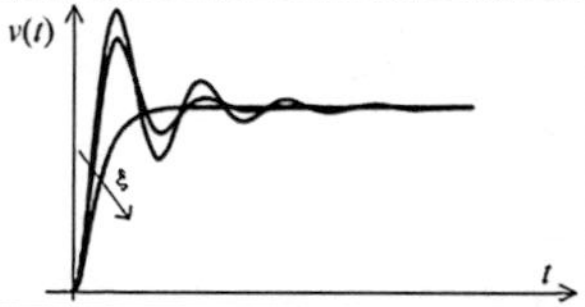	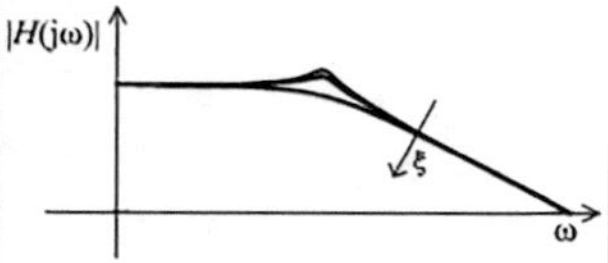	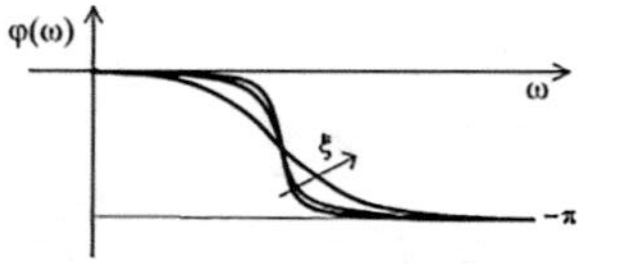
$\frac{1+\tau s}{(1+s)^2}$	τ	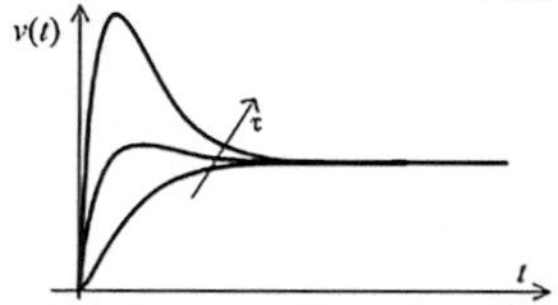	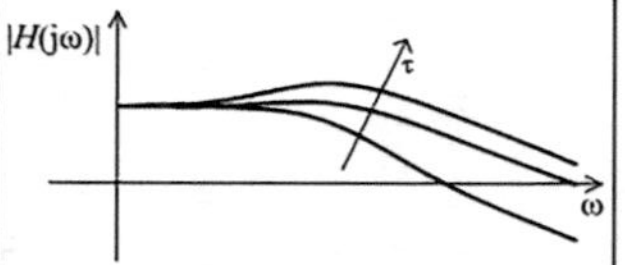	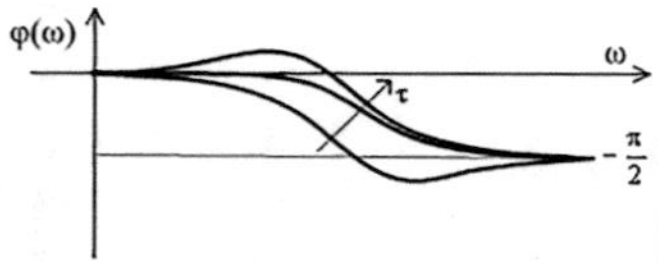
$\frac{1-\tau s}{(1+s)^2}$	τ	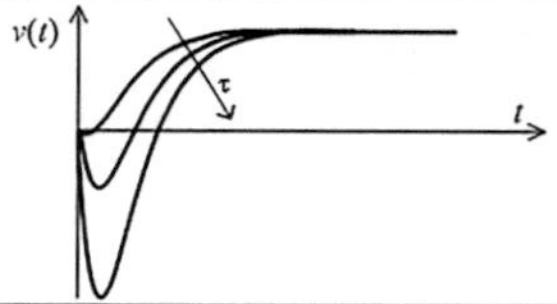	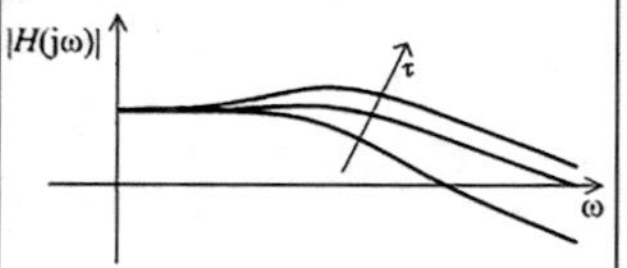	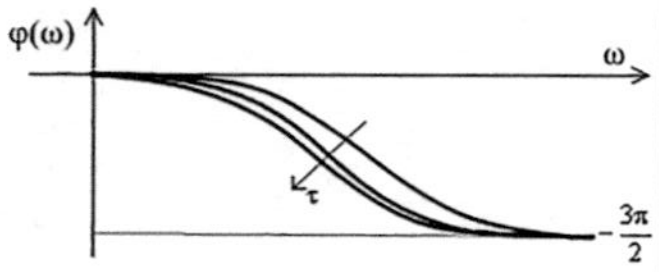

2.5 Approximate Descriptions

In practice it often makes sense to approximate a higher order model of a system by a simpler model which can be treated more easily. Besides the larger time constants, the smaller ones generally can be neglected, or their effect can be interpreted as a dead-time effect. In the sequel some frequently applied approximations will be given. It is an important remark that if the controller is designed for the lower order approximate model of the system, the performance of the control system has to be checked always with the higher order original model of the system!

2.5.1 Dominant Pole Pair

A closed-loop control system is often characterized by its so called dominant pole or dominant pole pair. The *dominant pole* is defined as the pole of the transfer function which is the closest to the imaginary axis. The pair of complex conjugate poles which is the closest to the imaginary axis is called the *dominant pole pair* (Fig. 2.52). If the other poles to the left are far away from the dominant poles (their real part is at least three times the value of the real part of the dominant poles), then the transients due to these poles have practically decayed by the time the effect of the dominant poles prevails. Thus the effect of the poles far to the left can be neglected and the behavior of the complete system can be approximated well by the behavior of a second order oscillating element formed by the dominant pole pair

$$H(s) = \frac{1}{1 + 2\xi Ts + T^2 s^2} = \frac{\omega_o^2}{\omega_o^2 + 2\xi\omega_o s + s^2}.$$

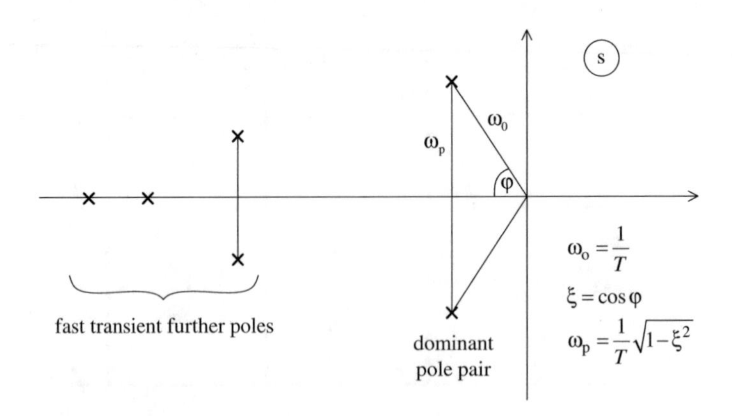

Fig. 2.52 Dominant pole pair

(See the description of the properties of the second order oscillating (ξ) element discussed previously in this chapter.)

2.5.2 Approximation of Higher Order Plants by First- and Second-Order Time Lag Models with Dead-Time

The unit step response of a proportional element containing several lags starts from zero if the degree of the denominator of its transfer function is higher than the degree of its numerator. The pole excess (the number of the poles minus the number of the zeros) indicates which derivative has non zero value at the initial point.

Aperiodic proportional elements with several time lags can be well approximated by a first order lag serially connected to a latent dead-time (T_L) (Fig. 2.53). If the system shows oscillating behavior, it can be well approximated by a second order oscillating element with latent dead-time.

Sometimes the aperiodic proportional element with several lags is taken into account with an equivalent pure dead-time (Fig. 2.54).

2.5.3 Approximation of a Dead-Time by Rational Transfer Functions

The transfer function of the pure dead-time element is the transcendental function $H_H(s) = e^{-sT_d}$, which can be approximated by a rational fraction.

The pure dead-time element can be approximated by the series connection of an infinite number of first order lag elements with the same time constant. As known from mathematics e^{-x} can be expressed as the limit of the series

$$e^{-x} = \lim_{n\to\infty}\left(1+\frac{x}{n}\right)^{-n}.$$

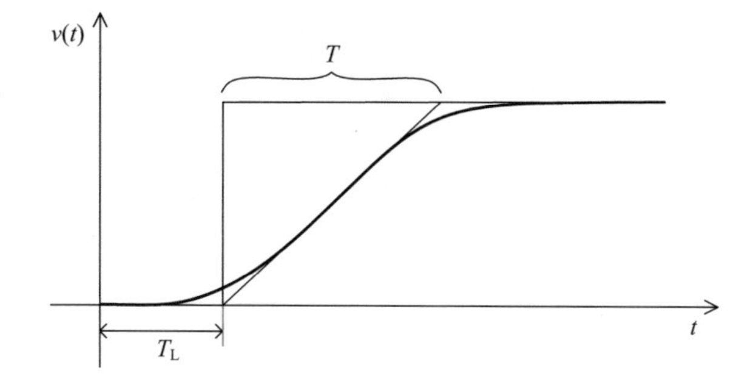

Fig. 2.53 A proportional element containing several lags can be approximated by a first-order lag element serially connected to a latent dead-time

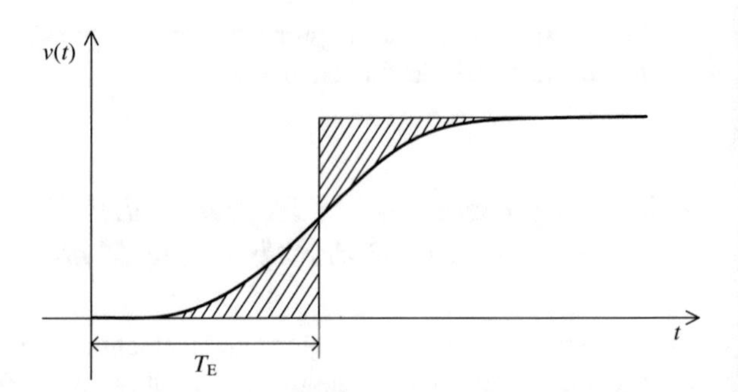

Fig. 2.54 Approximation of a proportional element containing more lags with an equivalent T_{E} pure dead-time

Applying the formula for the transfer function of a dead-time element the STREJC *approximation* is obtained:

$$H_{\mathrm{H}}(s) = e^{-sT_{\mathrm{d}}} = \lim_{n\to\infty}\left(1 + s\frac{T_{\mathrm{d}}}{n}\right)^{-n} \approx \frac{1}{\left(1 + \frac{T_{\mathrm{d}}}{n}s\right)^{n}} \tag{2.84}$$

By this relationship the dead-time element is approximated by a lag element containing a pole with multiplicity n, whose equivalent dead-time is $nT_{\mathrm{d}}/n = T_{\mathrm{d}}$. The more lags are serially connected, the better the approximation is. Figure 2.55 shows the unit step response of the STREJC approximation of the dead-time element for $n = 2, 5, 10$.

Another approximation of the dead-time element is the PADE approximation, which approximates the transfer function of the dead-time by non-minimum phase rational fractions, where the first elements of the TAYLOR expansion are equal to the first elements of the TAYLOR expansion of the exponential expression of the transfer function of the dead-time element.

The n-th order PADE approximation gives a rational fraction where there are n zeros and n poles, differing only in their sign.

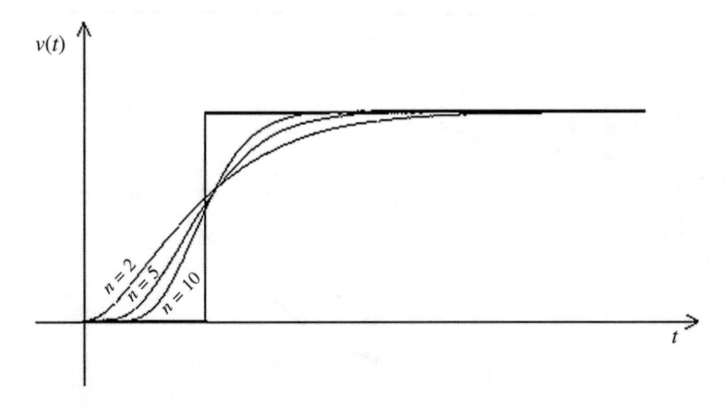

Fig. 2.55 Unit step responses of the dead-time element with STREJC approximation for n = 2, 5, 10

$$H_{\mathrm{H}}(s) = e^{-sT_{\mathrm{d}}} \approx \frac{(s-s_1)\ldots(s-s_n)}{(s+s_1)\ldots(s+s_n)} = H_{\mathrm{PADE}}(s). \tag{2.85}$$

The absolute value of the frequency function of the rational fraction for any value of $s = j\omega$ is 1, thus, similarly to the frequency function of the dead-time element, only its phase angle changes with the frequency. Such elements are called all-pass filters. The PADE approximation is a non-minimum phase rational function. The poles s_i and the coefficients of the numerator and the denominator can be determined by taking the first $N+M+1$ terms of the TAYLOR expansion of the transfer function $H_{\mathrm{H}}(s)$ as the first $N+M+1$ terms of the rational fraction transfer function $H_{\mathrm{PADE}}(s)$. Here M is the degree of the numerator and N is the degree of the denominator of the rational function ($M \leq N$).

$$e^{-x} = \sum_{i=0}^{\infty} b_i x^i \approx \frac{\sum_{k=0}^{M} d_k x^k}{\sum_{j=0}^{N} c_j x^j}.$$

In the equation there are $N+M+2$ unknown coefficients, therefore choosing $c_o = 1$, $N+M+1$ linear equations can be written for the remaining $N+M+1$ parameters, assuming the condition mentioned above. With this method the following form is obtained for the case $N = M = 3$:

$$e^{-x} \approx \frac{1 - \frac{1}{2}x + \frac{1}{10}x^2 - \frac{1}{120}x^3}{1 + \frac{1}{2}x + \frac{1}{10}x^2 + \frac{1}{120}x^3}.$$

Expressions of the first- and second order PADE approximations according to similar calculations are the following:

$$e^{-x} \approx \frac{1 - \frac{1}{2}x}{1 + \frac{1}{2}x}, \quad e^{-x} \approx \frac{1 - \frac{1}{2}x + \frac{1}{12}x^2}{1 + \frac{1}{2}x + \frac{1}{12}x^2}.$$

Table 2.6 First, second and third order PADE approximations of the dead-time element $e^{-sT_{\mathrm{d}}}$

Dead-time element	$H(s) = e^{-sT_{\mathrm{d}}}$
First order PADE approximation	$H(s) \approx \frac{2 - sT_{\mathrm{d}}}{2 + sT_{\mathrm{d}}}$
Second order PADE approximation	$H(s) \approx \frac{12 - 6sT_{\mathrm{d}} + (sT_{\mathrm{d}})^2}{12 + 6sT_{\mathrm{d}} + (sT_{\mathrm{d}})^2}$
Third order PADE approximation	$H(s) \approx \frac{120 - 60sT_{\mathrm{d}} + 12(sT_{\mathrm{d}})^2 - (sT_{\mathrm{d}})^3}{120 + 60sT_{\mathrm{d}} + 12(sT_{\mathrm{d}})^2 + (sT_{\mathrm{d}})^3}$

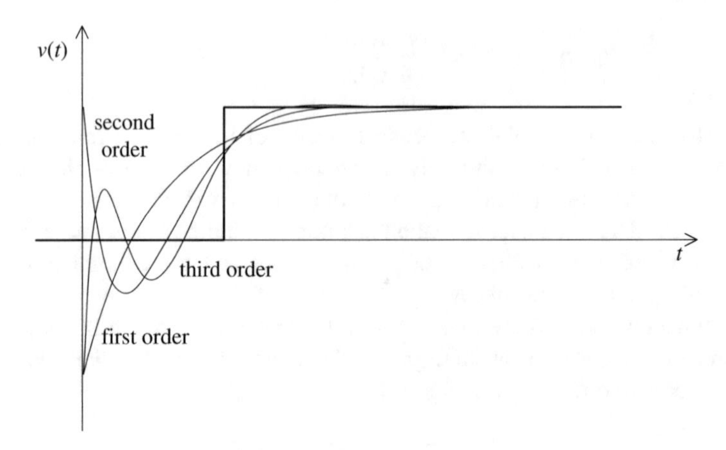

Fig. 2.56 Unit step responses of first-, second- and third-order PADE approximations of a dead-time element

The approximations are summarized in Table 2.6. The higher the degree of the rational fraction, the more terms are identical in the TAYLOR approximations of the two functions.

Figure 2.56 shows the unit step responses of the PADE approximation for the first-, second- and third order cases. It can be seen that the approximations do not fit well at the initial point, but approximate well the steady state. With higher order approximations the step responses fit better the step response of the dead-time element.

2.6 Examples of the Description of Continuous-Time Systems

When creating the model of a process, we wish to characterize the transfer properties between the input and the output signals. The process can be described by its differential equation, state equation or by its transfer functions. To determine the process model the physical operation of the process should be known as precisely as possible. Then the physical operation has to be described by mathematical relationships. The parameters in the equations can be determined based on a priori knowledge or by measurements.

Generally, physical processes are non-linear. To approximate non-linear systems by linear models which can be handled more easily, the most commonly applied technique is the linearization of the equations describing the system in the vicinity of the operating point. In this case the range of the changes where the linearization is valid has to be specified.

Often it is expedient to introduce relative units which give the actual values of the signals relative to their maximum values. Thus the values of the variables lie between 0 and 1.

For controlling a system, a model of the process is required. The control system is designed to meet the quality specifications taking the model of the process into consideration. In the sequel, the determination of mathematical models of some processes will be shown. The aim of the discussion is to show how we can reach the model giving the relationship between the input and the output signals starting from the physical description of the operation. (Based on the obtained model often further, deeper considerations may become necessary to describe more accurately the properties of the process.)

2.6.1 Direct Current (DC) Motor

Let us analyze the signal transfer properties of an armature controlled DC motor with constant external excitation. The scheme of the motor is given in Fig. 2.57.

The output signal of the motor is the angular velocity ω, the input signals are the terminal voltage (armature voltage) u_a and the load torque m_t acting on the shaft (disturbance). The excitation voltage u_e is constant.

One control task might be to keep the angular velocity of the motor at a given constant value despite the changing load. The manipulating variables could be the armature voltage and the excitation voltage (which at first is considered constant). The resistance and the inductance of the armature are denoted by R_a and L_a, respectively. The load torque reduced to the motor shaft is denoted by m_t and the armature current is denoted by i_a.

First let us think about the physical operation of the motor. The motor converts electrical energy to mechanical energy. In the excitation coil, a constant field current is created, which creates a constant magnetic flux in the air gap (it is supposed that this flux is proportional to the excitation current). Switching on the

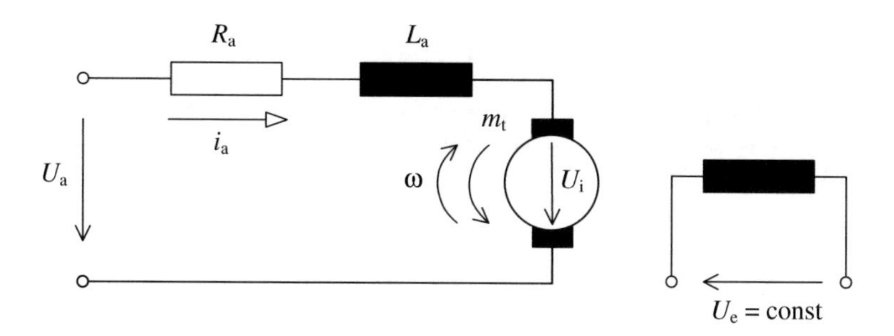

Fig. 2.57 Scheme of a DC motor with external excitation

armature voltage u_a, an armature current i_a is created in the armature circuit, which generates flux in the armature. The interaction of the excitation flux and the armature flux generates a torque which—working against the load—rotates the rotor. When the rotor rotates, a back emf u_i is induced according to the LENZ law.

The inputs of the plant are the armature voltage u_a and the load torque m_t (disturbance). The output signal is the angular velocity ω. The armature current i_a can also be considered as an output signal, which could take high values on starting, breaking or loading the motor.

The behavior of the system can be characterized by the KIRCHHOFF equation written for the armature circuit and the equation describing the mechanical motion. In the armature circuit the armature voltage keeps in balance with the sum of the resistive, inductive and the induced voltages. The induced voltage is proportional to the product of the constant excitation voltage and the angular speed by a factor of k_1. The difference of the motor torque and the disturbance load torque provides the accelerating torque, which can be expressed by the product of the inertia load and the angular acceleration. The motor torque is proportional to the product of the excitation flux and the armature current by a factor of k_2. The mathematical equations describing the behavior of the motor are:

$$\begin{aligned} u_a &= i_a R_a + L_a \frac{\mathrm{d}i_a}{\mathrm{d}t} + u_i \quad & u_i &= k_1 \varphi \omega \\ m - m_t &= \Theta \frac{\mathrm{d}\omega}{\mathrm{d}t} & m &= k_2 \varphi\, i_a \end{aligned} \tag{2.86}$$

(It should be mentioned that k_1 and k_2 are the same constants.) Let us find the LAPLACE transforms of the equations above:

$$\begin{aligned} u_a(s) &= i_a(s)(R_a + sL_a) + k_1 \varphi \omega(s) \\ k_2 \varphi\, i_a(s) - m_t(s) &= \Theta s \omega(s) \end{aligned}$$

Based on these equations—following the cause-effect relations—the block diagram shown in Fig. 2.58 can be derived. The difference between the armature

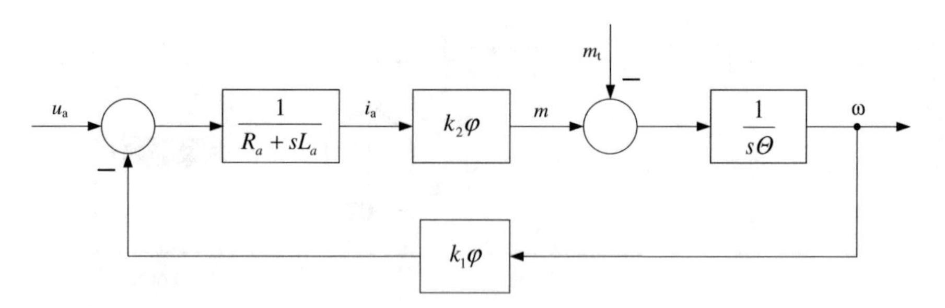

Fig. 2.58 Block diagram of a DC motor

voltage and the induced voltage creates the armature current. The interaction of the armature current and the excitation flux generates the rotation torque of the machine. The difference of the motor torque and the load torque provides the acceleration torque, which finally determines the angular velocity.

Two state variables can be given in the system: these are the angular velocity ω and the armature current i_a, whose instantaneous values are determined by the past motions in the system, and their values can not be changed abruptly by an abrupt change of the input signals. In the block diagram they appear at the output of an integrator (let us observe that a lag element can always be built as an integrator fed back by a constant). Let us describe the state equation of the motor. The general form of the state equation is:

$$\dot{\boldsymbol{x}} = \boldsymbol{A}\boldsymbol{x} + \boldsymbol{B}\boldsymbol{u}$$
$$y = \boldsymbol{c}^{\mathrm{T}}\boldsymbol{x} + \boldsymbol{d}^{\mathrm{T}}\boldsymbol{u}$$

where $\boldsymbol{x}$ is the state vector, $\boldsymbol{u}$ denotes the vector of the input signals and y is the output signal (*MISO* system). Here

$$\boldsymbol{x} = \begin{bmatrix} x_1 \\ x_2 \end{bmatrix} = \begin{bmatrix} i_a \\ \omega \end{bmatrix}; \quad \boldsymbol{u} = \begin{bmatrix} u_a \\ m_t \end{bmatrix}. \tag{2.87}$$

The state equation is obtained by rearranging the differential equation.

$$\begin{aligned} \frac{\mathrm{d}i_a}{\mathrm{d}t} &= \frac{1}{L_a}u_a - \frac{R_a}{L_a}i_a - \frac{1}{L_a}k_1\varphi\omega \\ \frac{\mathrm{d}\omega}{\mathrm{d}t} &= \frac{1}{\Theta}k_1\varphi i_a - \frac{1}{\Theta}m_t \end{aligned} \tag{2.88}$$

In vector-matrix form it becomes the following.

$$\begin{aligned} \begin{bmatrix} \dot{x}_1 \\ \dot{x}_2 \end{bmatrix} &= \begin{bmatrix} \frac{\mathrm{d}i_a}{\mathrm{d}t} \\ \frac{\mathrm{d}\omega}{\mathrm{d}t} \end{bmatrix} = \overbrace{\begin{bmatrix} -\frac{R_a}{L_a} & \frac{-k_1\varphi}{L_a} \\ \frac{k_2\varphi}{\Theta} & 0 \end{bmatrix}}^{\boldsymbol{A}} \begin{bmatrix} i_a \\ \omega \end{bmatrix} + \overbrace{\begin{bmatrix} \frac{1}{L_a} & 0 \\ 0 & -\frac{1}{\Theta} \end{bmatrix}}^{\boldsymbol{B}} \begin{bmatrix} u_a \\ m_t \end{bmatrix} \\ y = \omega &= \overbrace{\begin{bmatrix} 0 & 1 \end{bmatrix}}^{\boldsymbol{c}^{\mathrm{T}}} \begin{bmatrix} i_a \\ \omega \end{bmatrix} + \overbrace{\begin{bmatrix} 0 & 0 \end{bmatrix}}^{\boldsymbol{d}^{\mathrm{T}}} \begin{bmatrix} u_a \\ m_t \end{bmatrix} \end{aligned} \tag{2.89}$$

The overall transfer functions between the output and the input signals in the feedback system can be obtained based on the block diagram. The same result is obtained if the superposition principle is applied to a linear system. In this case, first the LAPLACE transforms stemming from the original differential equations are considered supposing zero load disturbance torque, then the ratio of the LAPLACE transforms of the angular velocity and the armature voltage is expressed, and finally, supposing zero armature voltage, the relation between the angular velocity and the load torque is determined. (Of course the transfer relations can be obtained from the state space representation form, as well.)

The overall transfer functions are:

$$\left.\frac{\omega(s)}{u_{\mathrm{a}}(s)}\right|_{m_{\mathrm{t}}=0}=\frac{\frac{k_2\varphi}{\Theta L_{\mathrm{a}}}}{s^2+s\frac{R_{\mathrm{a}}}{L_{\mathrm{a}}}+\frac{k_1k_2\varphi^2}{\Theta L_{\mathrm{a}}}}=\frac{\frac{1}{k_1\varphi}}{s^2\frac{\Theta L_{\mathrm{a}}}{k_1k_2\varphi^2}+s\frac{\Theta R_{\mathrm{a}}}{k_1k_2\varphi^2}+1}=\frac{A_{\mathrm{m}}}{s^2T_{\mathrm{m}}T_{\mathrm{e}}+sT_{\mathrm{m}}+1}, \tag{2.90}$$

where $A_{\mathrm{m}}=1/k_1\varphi$ is the transfer gain of the motor. Multiplying it with the steady value of the armature voltage yields the steady value of the angular velocity. $T_{\mathrm{m}}=\Theta R_{\mathrm{a}}/k_1k_2\varphi^2$ is the so called electromechanical time constant, whose value depends both on the electrical and mechanical parameters. $T_{\mathrm{e}}=L_{\mathrm{a}}/R_{\mathrm{a}}$ is the electrical time constant. The relationship between the angular velocity and the load torque is

$$\left.\frac{\omega(s)}{m_{\mathrm{t}}(s)}\right|_{u_{\mathrm{a}}=0}=\frac{-\frac{1}{\Theta}\left(s+\frac{R_{\mathrm{a}}}{L_{\mathrm{a}}}\right)}{s^2+s\frac{R_{\mathrm{a}}}{L_{\mathrm{a}}}+\frac{k_1k_2\varphi^2}{\Theta L_{\mathrm{a}}}}=\frac{-\frac{R_{\mathrm{a}}}{k_1k_2\varphi^2}\left(1+s\frac{L_{\mathrm{a}}}{R_{\mathrm{a}}}\right)}{s^2\frac{\Theta L_{\mathrm{a}}}{k_1k_2\varphi^2}+s\frac{\Theta R_{\mathrm{a}}}{k_1k_2\varphi^2}+1}=-\frac{A_{\mathrm{t}}(1+sT_{\mathrm{e}})}{s^2T_{\mathrm{m}}T_{\mathrm{e}}+sT_{\mathrm{m}}+1}. \tag{2.91}$$

Here the gain factor A_{t} gives the static effect of the load torque on the angular velocity. The negative sign indicates that by increasing the load, the steady value of the angular velocity is decreased.

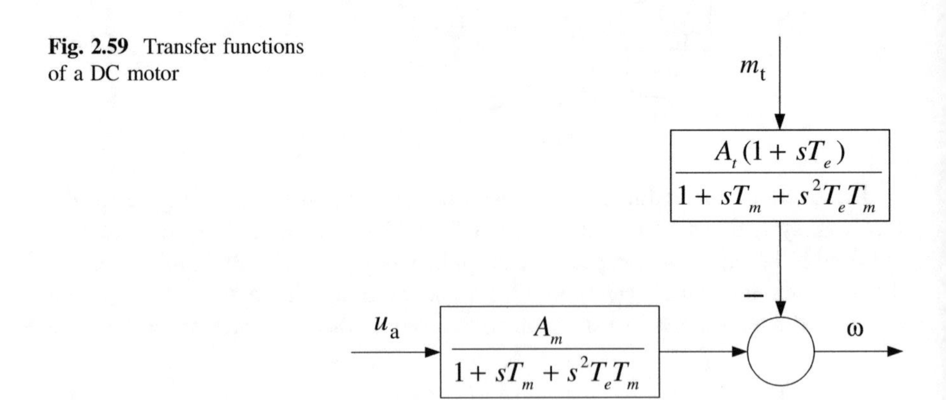

Fig. 2.59 Transfer functions of a DC motor

With the overall transfer functions, the model of the motor can also be described by the block scheme shown in Fig. 2.59.

The relationship between the angular velocity and the armature voltage can be characterized by a proportional element with two time lags. In the relationship between the angular velocity and the load torque, besides the second order proportional effect, there is also a parallel second order differentiating effect. Whether the evolution of the transients is aperiodic or oscillating depends on the ratio of the electrical and the electromechanical time constants. If $T_m < 4T_e$, oscillations appear in the transients.

Let us derive the model of the motor also for the case when the excitation voltage is varying. The magnetic flux in the air gap is proportional to the excitation current i_e. The resistance of the excitation coil is R_e, its inductance is L_e. Thus the inputs of the motor are: the armature voltage u_a, the excitation voltage u_e and the disturbance load torque m_t. Its output signals are the angular velocity ω and the armature current i_a. The operation of the motor is described by the following equations:

$$\begin{aligned} u_a &= i_a R_a + L_a \frac{\mathrm{d}i_a}{\mathrm{d}t} + u_i \qquad u_i = k_1 \varphi \omega \\ m - m_t &= \Theta \frac{\mathrm{d}\omega}{\mathrm{d}t} \qquad\qquad m = k_3 \varphi\, i_a = k_2 k_3 i_e i_a \\ \varphi &= k_2 i_e \end{aligned} \tag{2.92}$$

The system is non-linear, as the induced voltage and the torque of the motor can be given as the product of two changing variables. Let us express from the equations above the derivatives of the state variables i_a and ω. The state equation of the system is

$$\begin{aligned} \frac{\mathrm{d}i_a}{\mathrm{d}t} &= -\frac{R_a}{L_a} i_a - \frac{k_1 k_2}{L_a} i_e \omega + \frac{1}{L_a} u_a = f_1(i_a, i_e, \omega, u_a) \\ \frac{\mathrm{d}\omega}{\mathrm{d}t} &= \frac{1}{\Theta} k_2 k_3 i_e i_a - \frac{1}{\Theta} m_t = f_2(i_a, i_e, m_t) \end{aligned} \tag{2.93}$$

Let us formulate a linearized model of the system valid for small changes around the operating point. In the state equation, the variables are replaced by the sum of their operating points and the small variations around it. The operating points of the individual variables are denoted by $u_{ao}, i_{ao}, i_{eo}, m_{to}, \omega_o$. Around the operating point the variables are given as

$$\begin{aligned} u_a &= u_{ao} + \Delta u_a \\ i_a &= i_{ao} + \Delta i_a \\ i_e &= i_{eo} + \Delta i_e \\ m_t &= m_{to} + \Delta m_t \\ \omega &= \omega_o + \Delta\omega \end{aligned} \tag{2.94}$$

In a small vicinity of the operating point, a multivariable function can be expressed as the sum of the operating point and the small variations around it:

$$f \approx f_o + \Delta f = f_o + \sum_i \left.\frac{\partial f}{\partial x_i}\right|_{x_{1o},x_{2o}} \Delta x_i.$$

For the non-linear expressions in our state equation the following relationships are valid:

$$\begin{aligned} i_e\omega &= i_{eo}\omega_o + \Delta(i_e\omega) = i_{eo}\omega_o + i_{eo}\Delta\omega + \omega_o\Delta i_e \\ i_e i_a &= i_{eo} i_{ao} + i_{eo}\Delta i_a + i_{ao}\Delta i_e \end{aligned} \tag{2.95}$$

The linearized state equations valid around the operating point are:

$$\begin{aligned} \frac{d(i_{ao} + \Delta i_a)}{dt} &= -\frac{R_a}{L_a}(i_{ao} + \Delta i_a) + \frac{1}{L_a}(u_{ao} + \Delta u_a) \\ &\quad - \frac{k_1 k_2}{L_a} i_{eo}\omega_o - \frac{k_1 k_2}{L_a} i_{eo}\Delta\omega - \frac{k_1 k_2}{L_a}\omega_o \Delta i_e \\ \frac{d(\omega_o + \Delta\omega)}{dt} &= \frac{k_2 k_3}{\Theta} i_{eo} i_{ao} + \frac{k_2 k_3}{\Theta} i_{eo}\Delta i_a + \frac{k_2 k_3}{\Theta} i_{ao}\Delta i_e \\ &\quad - \frac{1}{\Theta} m_{to} - \frac{1}{\Theta}\Delta m_t \end{aligned} \tag{2.96}$$

From the equations above the following relationships are valid among the operating points:

$$\begin{aligned} \frac{di_{ao}}{dt} &= 0 = -\frac{R_a}{L_a} i_{ao} + \frac{1}{L_a} U_{ao} - \frac{k_1 k_2}{L_a} i_{eo}\omega_o \\ \frac{d\omega_o}{dt} &= 0 = -\frac{k_2 k_3}{\Theta} i_{eo} i_{ao} - \frac{1}{\Theta} m_{to} \end{aligned} \tag{2.97}$$

It can be seen that the values of the operating points are not independent. The operating point values of the load torque and the excitation current determine the operating point value of the armature current. The operating point values of the armature current, the armature voltage and the excitation current determine the operating point value of the angular velocity.

For small changes around the operating point, the following equations can be written:

$$\begin{aligned} \frac{d\Delta i_a}{dt} &= -\frac{R_a}{L_a}\Delta i_a - \frac{k_1 k_2}{L_a} i_{eo}\Delta\omega + \frac{1}{L_a}\Delta u_a - \frac{k_1 k_2}{L_a}\omega_o \Delta i_e \\ \frac{d\Delta\omega}{dt} &= \frac{k_2 k_3}{\Theta} i_{eo}\Delta i_a + \frac{k_2 k_3}{\Theta} i_{ao}\Delta i_e - \frac{1}{\Theta}\Delta m_t \end{aligned} \tag{2.98}$$

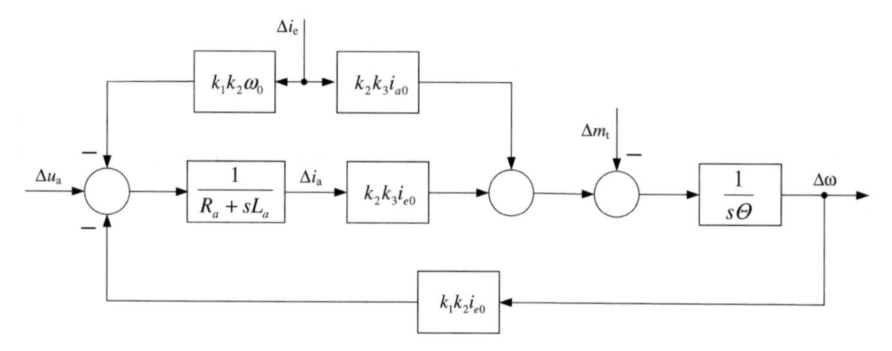

Fig. 2.60 Block diagram of a DC motor with variable external excitation considering linearization

Let us describe the linearized equations in vector-matrix form:

$$\begin{bmatrix} \frac{d\Delta i_a}{dt} \\ \frac{d\Delta\omega}{dt} \end{bmatrix} = \overbrace{\begin{bmatrix} -\frac{R_a}{L_a} & -\frac{k_1k_2}{L_a} \\ \frac{k_2k_3}{\Theta} & 0 \end{bmatrix}}^{A} \begin{bmatrix} \Delta i_a \\ \Delta\omega \end{bmatrix} + \overbrace{\begin{bmatrix} \frac{1}{L_a} & -\frac{k_1k_2}{L_a}\omega_o & 0 \\ 0 & \frac{k_2k_3}{\Theta}i_{ao} & -\frac{1}{\Theta} \end{bmatrix}}^{B} \begin{bmatrix} \Delta u_a \\ \Delta i_a \\ \Delta m_t \end{bmatrix}. \tag{2.99}$$

Using the LAPLACE transform of the equations, the block diagram of the linearized system is depicted in Fig. 2.60.

The overall transfer functions are

$$\left.\frac{\Delta\omega}{\Delta u_a}\right|_{\substack{\Delta i_e=0\\ \Delta m_t=0}} = \frac{\frac{k_2k_3i_{eo}}{s\Theta(R_a+sL_a)}}{1+\frac{k_1k_2^2k_3i_{eo}^2}{s\Theta(R_a+sL_a)}} = \frac{\frac{1}{k_1k_2i_{eo}}}{1+s\frac{\Theta R_a}{k_1k_2^2k_3i_{eo}^2}+s^2\frac{\Theta L_a}{k_1k_2^2k_3i_{eo}^2}}$$

$$\left.\frac{\Delta\omega}{\Delta u_a}\right|_{\substack{\Delta u_a=0\\ \Delta m_t=0}} = \frac{k_2k_3i_{ao}\frac{1}{s\Theta}-\frac{k_1k_2^2k_3\omega_oi_{eo}}{s\Theta R_a+sL_a}}{1+\frac{k_1k_2^2k_3i_{eo}^2}{s\Theta(R_a+sL_a)}} = \frac{\frac{i_{ao}}{k_1k_2i_{eo}^2}(R_a+sL_a)-\frac{\omega_o}{i_{eo}}}{1+s\frac{\Theta R_a}{k_1k_2^2k_3i_{eo}^2}+s^2\frac{\Theta L_a}{k_1k_2^2k_3i_{eo}^2}} \tag{2.100}$$

$$\left.\frac{\Delta i_a}{\Delta m_t}\right|_{\substack{\Delta u_a=0\\ \Delta i_e=0}} = \frac{\frac{k_1k_2i_{eo}}{s\Theta(R_a+sL_a)}}{1+\frac{k_1k_2^2k_3i_{eo}^2}{s\Theta(R_a+sL_a)}} = \frac{\frac{1}{k_2k_3i_{eo}}}{1+s\frac{\Theta R_a}{k_1k_2^2k_3i_{eo}^2}+s^2\frac{\Theta L_a}{k_1k_2^2k_3i_{eo}^2}}$$

The transfer functions have a second order proportional character. It can be seen that the individual transfer functions have the same denominator. The values of the parameters (transfer gains and time constants) also depend on the values of the operating point. The excitation current can be changed around its nominal value only within a restricted range to avoid a "runaway" of the motor. Increasing the excitation current—depending on the value of the operating point—in steady state may increase or decrease the value of the angular velocity. If $i_{ao}R_a/k_1k_2i_{eo} > \omega_o$, an increase of the excitation current increases the angular velocity, otherwise it decreases it.

2.6.2 Modeling of a Simple Liquid Tank System

Determining the relationship between the inflow and outflow of a liquid and the level in the tank is a basic task not only in technological processes, but also in logistical and general economic systems and also in biological processes. Let us consider first the simplest formulation of this task for the case seen in Fig. 2.61.

Here h is the level of the fluid, $A(h)$ is the cross section of the tank, a is the cross section of the tube of the outflow fluid, $u = q_{in}$ is the inflow fluid stream, while $y = q_{out}$ is the outflow fluid stream. Supposing a constant fluid density the following relationship can be written for the change of the amount V of fluid in the tank:

$$\frac{dV}{dt} = q_{in} - q_{out} = u - y \tag{2.101}$$

For the outflow fluid, the following relationship holds:

$$q_{out} = a\sqrt{2gh}. \tag{2.102}$$

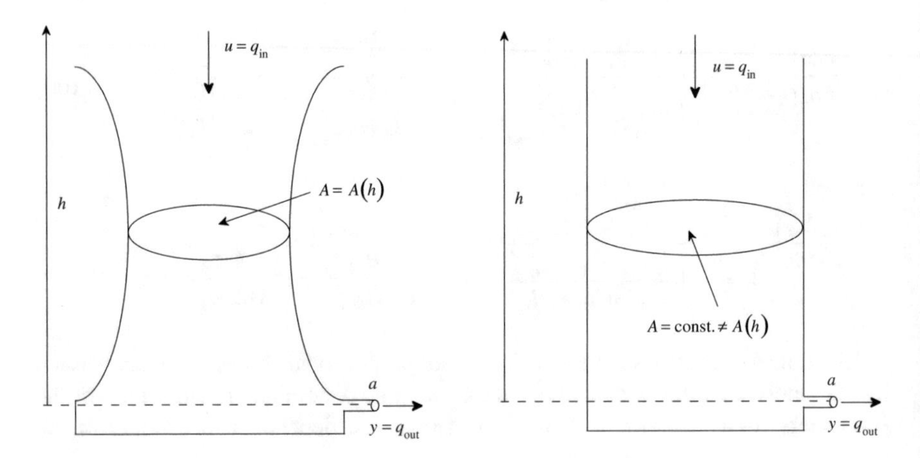

Fig. 2.61 Fluid tanks with varying and constant cross sections

According to the principle of conservation of energy the potential energy of the liquid upside is transformed to motion energy of the outflow liquid, i.e., $mgh = mv^2/2$. Hence the velocity v of the outflow fluid is expressed as $v = \sqrt{2gh}$, and the volumetric velocity is $q_{out} = av$.

The state variable can be chosen in different ways. One possible method is that the accumulation in the tank is expressed with the fluid level. In case of varying cross section

$$V = \int_0^h A(x)\mathrm{d}x. \tag{2.103}$$

The previous equations can be written in the following form:

$$\begin{aligned} \frac{\mathrm{d}h}{\mathrm{d}t} &= \frac{1}{A(h)}(q_{in} - q_{out}) = \frac{1}{A(h)}\left(q_{in} - a\sqrt{2gh}\right) = \frac{1}{A(h)}\left(u - a\sqrt{2gh}\right) \\ y &= q_{out} = a\sqrt{2gh} \end{aligned} \tag{2.104}$$

Thus the behavior of the tank can be described by a first order non-linear differential equation. The state equation is linearized around the equilibrium operating point $q_{out} = q_{in} = u = y$:

$$q_{out} = q_{in} = q_o = a\sqrt{2gh_o}$$

whence

$$h_o = \frac{q_o^2}{2ga^2}.$$

The cross section at a height of h_o is denoted by A_o. The linearized equation written for the signal changes Δh and Δq_{out} is

$$\begin{aligned} \frac{\mathrm{d}\Delta h}{\mathrm{d}t} &= \frac{a\sqrt{2gh_o}}{2A_o h_o}\Delta h + \frac{1}{A_o}\Delta q_{in} = -\frac{q_o}{2A_o h_o}\Delta h \\ &\quad + \frac{1}{A_o}\Delta q_{in} = -\frac{1}{T_o}\Delta h + \frac{1}{A_o}\Delta q_{in} \\ \Delta q_{out} &= \frac{a\sqrt{2gh}}{h_o}\Delta h = \frac{q_o}{h_o}\Delta h \end{aligned} \tag{2.105}$$

The quantity

$$T_o = \frac{2A_o h_o}{q_o} = 2 \times \frac{\text{total fluid amount}\left[\mathrm{m}^3\right]}{\text{fluid stream}\left[\mathrm{m}^3/\mathrm{s}\right]} \tag{2.106}$$

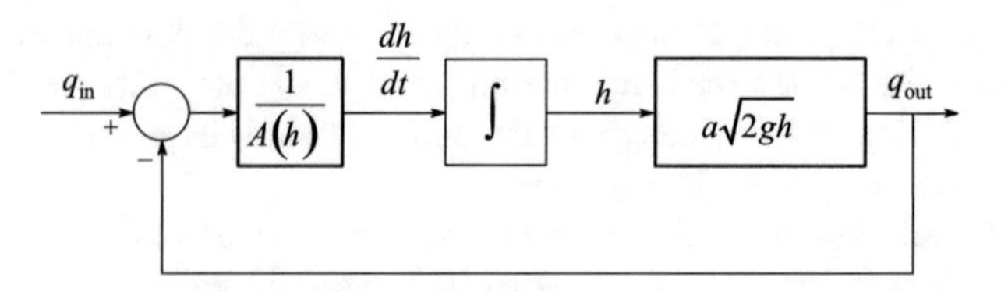

Fig. 2.62 Equivalent block diagram of a fluid tank

can be considered as the time constant of the system. A time of $T_o/2$ is needed to fill the volume $A_o h_o$ with fluid stream q_o. An equivalent block diagram is shown in Fig. 2.62. From the figure the transfer function can be given formally as

$$H(s) = \frac{1}{1 + \dfrac{A(h)}{a\sqrt{2gh}}s} \tag{2.107}$$

where the virtual time constant is $T_o = A(h)/a\sqrt{2gh}$. It is worthwhile to compare it with the time constant obtained by linearization around the operating point

$$T_o = \frac{2A_o h_o}{a\sqrt{2gh_o}} = 2h_o \frac{A_o}{a\sqrt{2gh_o}} = 2h_o T(h_o). \tag{2.108}$$

(The transfer function is formal as it is valid only in the linear range.)

Fig. 2.63 Scheme of a two tank system

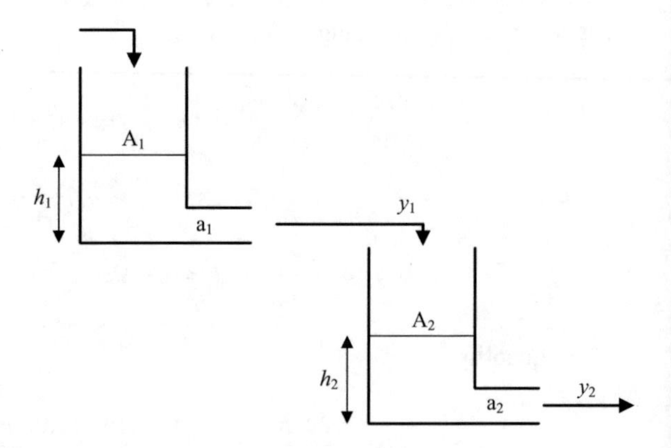

2.6.3 A Simple Two Tank System

Let us now consider the two tank system shown in Fig. 2.63. The liquid flow into the upper tank is controlled by a valve. The liquid flows from the upper tank to the lower tank and then it flows out. One task could be to keep the levels of the liquid in the tanks constant.

The output signals are the levels in the tanks, the input signal is the amount of the inflow liquid. To build a model of the system let us think over the physical operation of the system. The velocity of the liquid flowing out of a tank depends on the level of the liquid in that tank. The balance of the potential and the kinetic energy is given by the following equation:

$$mgh = \frac{1}{2}mv^2.$$

This relationship is valid for both tanks. Hence the velocity of the outflow liquid can be expressed as $v = \sqrt{2gh}$.

The process can be characterized with the following signals:

- u the amount of the inflow liquid into the upper tank,
- h_1 the level in the upper tank,
- A_1 and A_2 cross sections of the tanks,
- a_1 and a_2 cross sections of the tubes of the outflow liquid from the two tanks,
- y_1 the outflow liquid stream from the upper tank,
- h_2 the level in the lower tank,
- y_2 the outflow liquid stream from the lower tank.

The outflow depends on the velocity of the liquid, the cross section of the outflow pipe, and the viscosity factor μ of the liquid: $y = av\mu = a\sqrt{2g}\sqrt{h}\mu = k\sqrt{h}$.

It can be seen that the relationship is non-linear. The rise in the level in the tanks depends on the difference between the amount of inflow and outflow. The smaller the cross section of the tank, the faster the rise. The following differential equations can be written for the two tanks:

$$\begin{aligned}\frac{\mathrm{d}h_1}{\mathrm{d}t} &= \frac{1}{A_1}\left(u - k_1\sqrt{h_1}\right) = \beta_1 u - \alpha_1\sqrt{h_1} \\ \frac{\mathrm{d}h_2}{\mathrm{d}t} &= \frac{1}{A_2}\left(k_1\sqrt{h_1} - k_2\sqrt{h_2}\right) = \beta_2\sqrt{h_1} - \alpha_2\sqrt{h_2}\end{aligned} \tag{2.109}$$

where

$$\beta_1 = \frac{1}{A_1};\quad \alpha_1 = \frac{k_1}{A_1};\quad \beta_2 = \frac{k_1}{A_2};\quad \alpha_2 = \frac{k_2}{A_2}.$$

As the levels can be considered as state variables, the two first order differential equations above give the non-linear state equation of the system. It is expedient to give the individual variables in relative units, relating their actual values to their maximum values.

$$h_{\mathrm{rel}} = \frac{h}{h_{\max}} \quad \text{and} \quad u_{\mathrm{rel}} = \frac{u}{u_{\max}}. \tag{2.110}$$

Thus the differential equations can be given in the following form:

$$\begin{aligned} \frac{\mathrm{d}h_{1,\mathrm{rel}}}{\mathrm{d}t} &= \beta_{1,\mathrm{rel}} u_{\mathrm{rel}} - \alpha_{1,\mathrm{rel}} \sqrt{h_{1,\mathrm{rel}}} \\ \frac{\mathrm{d}h_{2,\mathrm{rel}}}{\mathrm{d}t} &= \beta_{2,\mathrm{rel}} \sqrt{h_{1,\mathrm{rel}}} - \alpha_{2,\mathrm{rel}} \sqrt{h_{2,\mathrm{rel}}} \end{aligned} \tag{2.111}$$

where

$$\begin{aligned} \beta_{1,\mathrm{rel}} &= \frac{1}{A_1} \frac{u_{\max}}{h_{1,\max}}; \quad \alpha_{1,\mathrm{rel}} = \frac{k_1}{A_1} \frac{\sqrt{h_{1,\max}}}{h_{1,\max}} \\ \beta_{2,\mathrm{rel}} &= \frac{k_1}{A_2} \frac{\sqrt{h_{1,\max}}}{h_{2,\max}}; \quad \alpha_{2,\mathrm{rel}} = \frac{k_2}{A_2} \frac{\sqrt{h_{2,\max}}}{h_{2,\max}}. \end{aligned}$$

For easier handling, often a non-linear system is linearized around given operating points. Thus for small variations the system can be considered linear. Let us give the TAYLOR expansion of the non-linear function $\sqrt{h}$ around the operating point h_o.

$$f(h) = \sqrt{h} \approx \sqrt{h_o} + \frac{1}{2\sqrt{h_o}} (h - h_o)$$

The higher order terms of the TAYLOR expansion are neglected. Let us denote the small variations by Δ, thus

$$h - h_o = \Delta h, \quad u_{\mathrm{rel}} = u_o + \Delta u, \quad h_{1,\mathrm{rel}} = h_{1o,\mathrm{rel}} + \Delta h_1, \quad h_{2,\mathrm{rel}} = h_{2o,\mathrm{rel}} + \Delta h_2$$

The linearized equations are:

$$\begin{aligned} \frac{\mathrm{d}\left(h_{1o,\mathrm{rel}} + \Delta h_1\right)}{\mathrm{d}t} &= \beta_{1,\mathrm{rel}} (u_o + \Delta u) - \alpha_{1,\mathrm{rel}} \left(\sqrt{h_{1o,\mathrm{rel}}} + \frac{1}{2\sqrt{h_{1o,\mathrm{rel}}}} \Delta h_1 \right) \\ \frac{\mathrm{d}\left(h_{2o,\mathrm{rel}} + \Delta h_1\right)}{\mathrm{d}t} &= \beta_{2,\mathrm{rel}} \left(\sqrt{h_{1o,\mathrm{rel}}} + \frac{1}{2\sqrt{h_{1o,\mathrm{rel}}}} \Delta h_1 \right) \\ &\quad - \alpha_{2,\mathrm{rel}} \left(\sqrt{h_{2o,\mathrm{rel}}} + \frac{1}{2\sqrt{h_{2o,\mathrm{rel}}}} \Delta h_2 \right) \end{aligned} \tag{2.112}$$

The relationships between the operating points are

$$\begin{aligned}\frac{\mathrm{d}h_{1o,rel}}{\mathrm{d}t} &= 0 = \beta_{1,rel}u_o - \alpha_{1,rel}\sqrt{h_{1o,rel}} \\ \frac{\mathrm{d}h_{2o,rel}}{\mathrm{d}t} &= 0 = \beta_{2,rel}\sqrt{h_{1o,rel}} - \alpha_{2,rel}\sqrt{h_{2o,rel}}\end{aligned} \tag{2.113}$$

For small changes around the operating point, the following equations are obtained.

$$\begin{aligned}\frac{\mathrm{d}\Delta h_1}{\mathrm{d}t} &= \beta_{1,rel}\Delta u - \frac{\alpha_{1,rel}}{2\sqrt{h_{1o,rel}}}\Delta h_1 \\ \frac{\mathrm{d}\Delta h_2}{\mathrm{d}t} &= \frac{\beta_{2,rel}}{2\sqrt{h_{1o,rel}}}\Delta h_1 - \frac{\alpha_{2,rel}}{2\sqrt{h_{2o,rel}}}\Delta h_2\end{aligned} \tag{2.114}$$

It can be seen that the parameters in the equations depend on the operating points. The parameters can be determined from measurements (filling and emptying the tanks), and also from geometrical data. Let us give the transfer function of the linearized process when the output signal is Δh_2, the change of the level in the lower tank, and the input signal is Δu, the change of the inflow liquid. Let us introduce the following notations:

$$\delta_1 = \beta_{1,rel}; \quad \gamma_1 = \frac{\alpha_{1,rel}}{2\sqrt{h_{1o,rel}}},$$

$$\delta_2 = \frac{\beta_{2,rel}}{2\sqrt{h_{1o,rel}}}; \quad \gamma_2 = \frac{\alpha_{2,rel}}{2\sqrt{h_{2o,rel}}}.$$

With this notation, the state equations are

$$\begin{aligned}\frac{\mathrm{d}\Delta h_1}{\mathrm{d}t} &= \delta_1\Delta u - \gamma_1\Delta h_1 \\ \frac{\mathrm{d}\Delta h_2}{\mathrm{d}t} &= \delta_2\Delta h_1 - \gamma_2\Delta h_2\end{aligned}$$

and hence the transfer function is

$$H(s) = \delta_1\frac{1/s}{1+\gamma_1/s}\delta_2\frac{1/s}{1+\gamma_2/s} = \frac{K}{(1+sT_2)(1+sT_1)}. \tag{2.115}$$

The process can be considered as a proportional two lag element where the time constants and the gain are as follows.

$$T_1 = \frac{1}{\gamma_1} = \frac{2\sqrt{h_{1o,\mathrm{rel}}}}{\alpha_{1,\mathrm{rel}}}; \quad T_2 = \frac{1}{\gamma_2} = \frac{2\sqrt{h_{2o,\mathrm{rel}}}}{\alpha_{2,\mathrm{rel}}};$$
$$K = \frac{2\beta_{1,\mathrm{rel}}\sqrt{h_{2o,\mathrm{rel}}}}{\alpha_{2,\mathrm{rel}}}. \tag{2.116}$$

It can be seen that the parameters depend on the operating point.

2.6.4 A Simple Heat Process

Let us analyze the heat process of a system consisting of two heat sources. The arrangement is shown in Fig. 2.64. The temperature changes in several connected pieces in electrical equipment can be modeled by analyzing the heat transfer processes in two embedded bodies. For example the heat transfer processes in slots of electrical machines, where the copper winding is placed in the iron slots can be analyzed on the basis of this model.

A body of mass m_2 and specific heat g_2 where power p_2 is converted to heat closes around a body of mass m_1 and specific heat g_1 where the heat power is p_1. The surface of the bodies contacting each other is denoted f_1 and has heat transfer coefficient h_1. That in contact with the external environment is denoted by f_2, and has the heat transfer coefficient h_2.

Let us determine the change of temperatures υ_1 and υ_2 in the two bodies after switching on the heat generation, supposing that earlier the temperature of the system was equal to the environmental temperature. The input signals of the system are the heating powers p_1 and p_2, respectively, the disturbance is the environmental temperature υ_o, the output signals are the υ_1 and υ_2 temperatures of the bodies.

The behavior of the system can be described as follows. The temperature of the two bodies starts growing after switching on the heat.

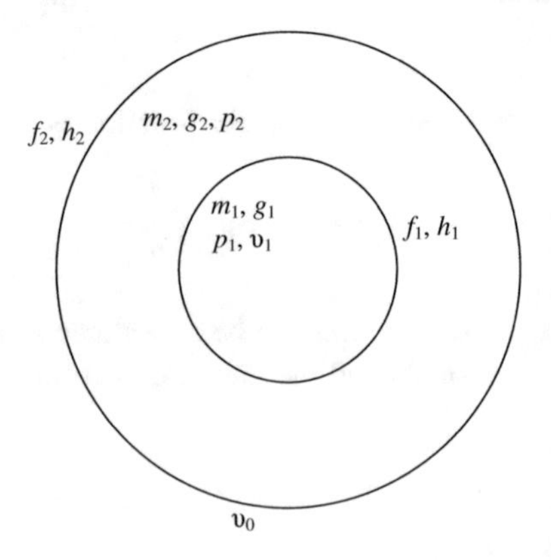

Fig. 2.64 A system consisting of two heat sources

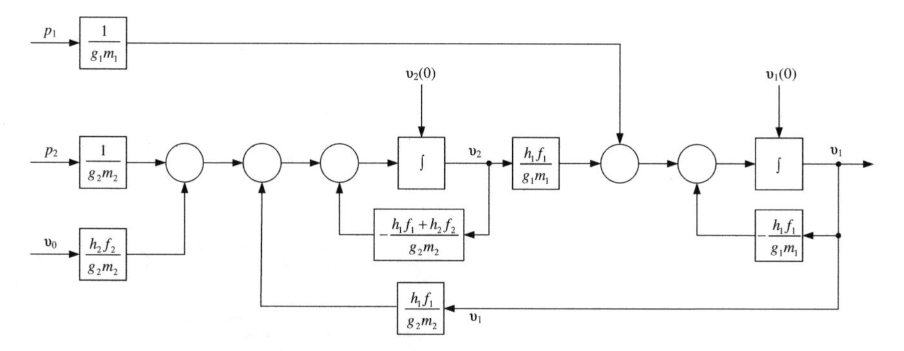

Fig. 2.65 Block diagram of the heat process

One part of the generated heat energy is stored in the heat capacity of the bodies, increasing their temperatures, while the second part—as the effect of the temperature difference—leaves, entering the environment through the interfacial surface. It is supposed that the bodies are homogeneous, because of their good heat transfer properties a temperature difference does not take shape inside the bodies.

In the inner body the heat generated through time Δt partly increases the temperature of the body by $\Delta\upsilon_1$ degrees, and partly leaves into the outer body. The heat transfer depends on the difference of the temperatures in the two bodies, on the size of the interfacing surface and on the heat transfer coefficient. In the outer body the heat generated by the heat power p_2 is added to the amount of heat coming from the inner body. This resulting heat partly increases the temperature of the outer body by $\Delta\upsilon_2$ degrees, and partly goes into the environment.

The temperatures υ_1 and υ_2 can be taken as the state variables of the system. Describing the heat balance equations for the two bodies yields:

$$\begin{aligned} p_1\Delta t &= g_1m_1\Delta\upsilon_1 + h_1f_1(\upsilon_1-\upsilon_2)\Delta t \\ p_2\Delta t + h_1f_1(\upsilon_1-\upsilon_2)\Delta t &= g_2m_2\Delta\upsilon_2 + h_2f_2(\upsilon_2-\upsilon_o)\Delta t \end{aligned} \tag{2.117}$$

Dividing the equations by $g_1m_1\Delta t$ and $g_2m_2\Delta t$, respectively, then taking the limit $\Delta t \to 0$ and rearranging the equations, expressing the derivatives of the temperature changes the following equations are obtained.

$$\begin{aligned} \frac{\mathrm{d}\upsilon_1}{\mathrm{d}t} &= -\frac{h_1f_1}{g_1m_1}\upsilon_1 + \frac{h_1f_1}{g_1m_1}\upsilon_2 + \frac{1}{g_1m_1}p_1 \\ \frac{\mathrm{d}\upsilon_2}{\mathrm{d}t} &= \frac{h_1f_1}{g_2m_2}\upsilon_1 - \frac{h_1f_1+h_2f_2}{g_2m_2}\upsilon_2 + \frac{1}{g_2m_2}p_2 + \frac{h_2f_2}{g_2m_2}\upsilon_o \end{aligned} \tag{2.118}$$

These equations give the state equation of the system. Based on these equations, a block diagram of the system can be built (Fig. 2.65). From the block diagram, the resulting transfer functions between the individual output and input signals can be calculated and the time responses of the output signals for given input signal changes can also be derived.

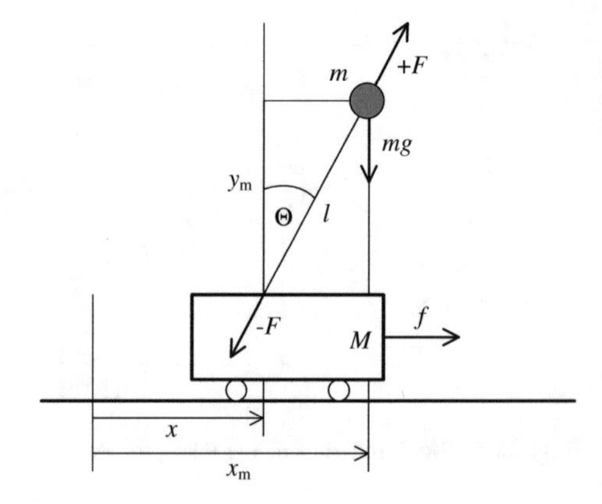

Fig. 2.66 Mechanical model of the inverted pendulum

2.6.5 *The Moving Inverted Pendulum*

The diagram of the investigated mechanical system is shown in Fig. 2.66. A rod of length l mounted on a cart of mass M is considered weightless. It can be moved around a joint. At the end of the rod, a ball of mass m is mounted. A force f acts on the cart. This device is the so called inverted pendulum, which has an unstable mechanical behavior. One control task could be to keep the mass m at the upper balance point by appropriate movements of the cart. A juggler in the circus realizes such a hand control, balancing the rod. A similar construction can work as part of a mobile robot. The stable control of an unstable process is not an easy task. A control algorithm is to be designed considering the model of the plant.

Let us consider the NEWTON equations describing the mechanical behavior of the system to move the masses M and m. In the rigid rod compulsion forces $+F$ and $-F$ arise. The force-acceleration relations for the rigid rod in various directions are:

$$\begin{aligned} M\ddot{x} &= f - F\sin\Theta \\ m\ddot{x}_m &= F\sin\Theta \\ m\ddot{y}_m &= F\cos\Theta - mg \end{aligned} \tag{2.119}$$

Adding the first two equations, and then multiplying the second equation by $\cos\Theta$ and the third equation by $\sin\Theta$ and subtracting them from each other the expression of the F compulsion force can be eliminated from the equations.

With these manipulations the following equations are obtained:

$$\begin{aligned} M\ddot{x} + m\ddot{x}_\mathrm{m} &= f \\ M\ddot{x}_\mathrm{m}\cos\Theta - M\ddot{y}_\mathrm{m}\sin\Theta &= mg\sin\Theta \end{aligned} \tag{2.120}$$

Let us calculate the derivatives of x_m and y_m.

$$
\begin{aligned}
x_{\mathrm{m}} &= x + l\sin\Theta & y_{\mathrm{m}} &= l\cos\Theta \\
\dot{x}_{\mathrm{m}} &= \dot{x} + l\dot{\Theta}\cos\Theta & \dot{y}_{\mathrm{m}} &= -l\Theta\sin\Theta \\
\ddot{x}_{\mathrm{m}} &= \dot{x} - l\dot{\Theta}^2\sin\Theta + l\ddot{\Theta}\cos\Theta & \ddot{y}_{\mathrm{m}} &= -l\dot{\Theta}^2\cos\Theta - l\ddot{\Theta}\sin\Theta
\end{aligned}
\tag{2.121}
$$

Substituting the expressions for $\ddot{x}_{\mathrm{m}}$ and $\ddot{y}_{\mathrm{m}}$ into the previous equations, the following non-linear differential equations are obtained to describe the behavior of the inverted pendulum.

$$
\begin{aligned}
&(M+m)\ddot{x} - ml\dot{\Theta}^2\sin\Theta + ml\ddot{\Theta}\cos\Theta = f \\
&m\ddot{x}\cos\Theta + ml\ddot{\Theta} = mg\sin\Theta
\end{aligned}
$$

Expressing the second order derivatives from both equations yields the following

$$
\begin{aligned}
\ddot{x} &= \frac{1}{M+m}\left(f + ml\dot{\Theta}^2\sin\Theta - ml\ddot{\Theta}\cos\Theta\right) \\
\ddot{\Theta} &= \frac{g}{l}\sin\Theta - \frac{1}{l}\ddot{x}\cos\Theta
\end{aligned}
\tag{2.122}
$$

For small movements around the perpendicular position and for small changes in the angular position, a linearized model of the system can now be given. It is assumed that the approximations $\Theta \approx 0$ and $\dot{\Theta} \approx 0$, $\sin\Theta \approx \Theta$ and $\cos\Theta \approx 1$ can be employed. With these simplifying assumptions the equations can be transformed to the following form:

$$
\begin{aligned}
\ddot{x} &= -\frac{mg}{M}\Theta + \frac{1}{M}f \\
\ddot{\Theta} &= \frac{g}{l}\frac{M+m}{M}\Theta - \frac{1}{Ml}f
\end{aligned}
\tag{2.123}
$$

Here, Θ, $\dot{\Theta}$, x and $\dot{x}$ can be chosen as the state variables. The state equation can be written in the form

$$
\begin{bmatrix} \dot{\Theta} \\ \ddot{\Theta} \\ \dot{x} \\ \ddot{x} \end{bmatrix} = \begin{bmatrix} 0 & 1 & 0 & 0 \\ \frac{g}{l}\frac{M+m}{M} & 0 & 0 & 0 \\ 0 & 0 & 0 & 1 \\ -\frac{mg}{M} & 0 & 0 & 0 \end{bmatrix} \begin{bmatrix} \Theta \\ \dot{\Theta} \\ x \\ \dot{x} \end{bmatrix} + \begin{bmatrix} 0 \\ -\frac{1}{Ml} \\ 0 \\ \frac{1}{M} \end{bmatrix} f
\tag{2.124}
$$

The output signal can be Θ or the state variables in x. From these equations the transfer functions between the outputs Θ or x and the input signal f can be determined:

$$H_1(s) = \frac{\Theta(s)}{f(s)} = \frac{-1/Ml}{s^2 - \alpha^2} = -\frac{1/Ml}{(s+\alpha)(s-\alpha)} \tag{2.125}$$

and

$$H_2(s) = \frac{x(s)}{f(s)} = \frac{\frac{1}{M}(s^2 - \beta^2)}{s^2(s^2 - \alpha^2)} = \frac{\frac{1}{M}(s+\beta)(s-\beta)}{s^2(s+\alpha)(s-\alpha)} \tag{2.126}$$

where $\alpha^2 = \frac{g}{l}\frac{M+m}{M}$ and $\beta^2 = \frac{g}{l}$. It can be seen that the transfer function of the linearized system contains an unstable pole, and also a non-minimum phase zero does appear. The control of this system is not a simple task.

Chapter 3
Description of Continuous-Time Systems in State-Space

The so-called state-equations are widely used in the scientific and engineering fields for the description of dynamical systems. The necessity for this kind of description is explained in different ways. Perhaps the easiest way is the recognition that the operation of a wide class of complex dynamical systems can be modeled with relatively high precision by the first order vector differential equations

$$\begin{aligned}\frac{\mathrm{d}\boldsymbol{x}(t)}{\mathrm{d}t} &= \dot{\boldsymbol{x}}(t) = \boldsymbol{f}[\boldsymbol{x}(t), u(t)] \\ y(t) &= g[\boldsymbol{x}(t), u(t)]\end{aligned} \tag{3.1}$$

The state variables of the system as scalar components are collected in a vector $\boldsymbol{x}$ called the state vector. The system input is u, and the output is y. The dimension of $\boldsymbol{x}$ is called the degree or the order of the system. The function $\boldsymbol{f}(\boldsymbol{x}, u)$ represents the varying "speed" of the state vector as the function of the states and the input signal. The function $g(\boldsymbol{x}, u)$ is called sensor or measurement function since it provides the output of the system. Let's call attention here to the fact that $\boldsymbol{f}(\boldsymbol{x}, u)$ and $g(\boldsymbol{x}, u)$ do not depend on time in an explicit way. (But here we emphasize that nevertheless the signals of the state-equations obviously depend on time!) This kind of system is called a time-invariant system. The state variables contain the information about the past of the system, and the future values of the signals can be predicted, therefore the state vector behaves like the memory of the system.

In engineering systems the state vectors are often related to the basic physical processes, where the relations necessary to describe the storage of the mass, flow, impulse, and power, have to be determined. (It has to be noted, however, that in certain fields, e.g. in chemistry, the definition of the state vector is different from the above general system-theoretical concept: it mostly reflects the variables—like pressure, temperature, composition, etc.—representing the physico-chemical state of the investigated material, mixture, compound, etc.)

L. Keviczky et al., *Control Engineering*, Advanced Textbooks in Control and Signal Processing, https://doi.org/10.1007/978-981-10-8297-9_3

The state variables as coordinates define a space (*state-space*). The state vector $\boldsymbol{x}(t)$ is interpreted in this space. The motion of the end point of the vector represents the motion of the system. The curve described by the motion of the end point of the state vector gives the state-trajectory.

A special class of the non-linear dynamical systems is given by the Eq. (3.1), whose possible equilibrium state $(\boldsymbol{x}_o, u_o)$ (where $\dot{\boldsymbol{x}} = \boldsymbol{0}$) is obtained from the equation

$$\boldsymbol{f}(\boldsymbol{x}_o, u_o) = \boldsymbol{0}. \tag{3.2}$$

(Remark: In general, several equilibrium states can be obtained. These equilibrium states can provide different stable states. The performance of these states requires the investigation of the second order derivatives of $\boldsymbol{f}(\boldsymbol{x}, u)$.)

The static systems can be described by degenerate state-equations, since they do not have memory, or the corresponding states, so they can be described by the second equation of (3.1) by itself

$$y = g(u) \tag{3.3}$$

Taking the TAYLOR-expansion at the point u_o, we get

$$y = g(u_o) + \frac{dg(u_o)}{du}(u - u_o) + \cdots = g(u_o) + g'(u_o)(u - u_o) + \cdots \tag{3.4}$$

and the linearized model

$$y - y_o = \Delta y = y - g(u_o) = g'(u_o)(u - u_o) = g'(u_o)\Delta u \tag{3.5}$$

can be obtained from the first order term of (3.4).

The linearized model replaces the original curve with its tangent at the operating point u_o and establishes a static linear connection between the changes $(\Delta y, \Delta u)$ around the operating point.

Actually the linearization of the state-space Eq. (3.1) can also be given in a very similar way.

With the following notation, valid for changes around the equilibrium state $(\boldsymbol{x}_o, u_o)$, i.e.

$$\boldsymbol{x} = \boldsymbol{x}_o + \Delta\boldsymbol{x}; \quad u = u_o + \Delta u; \quad y = y_o + \Delta y \tag{3.6}$$

let us calculate the first order linearized approach of (3.1)

$$\begin{aligned} \frac{d\boldsymbol{x}}{dt} &= \boldsymbol{f}(\boldsymbol{x}_o + \Delta\boldsymbol{x}, u_o + \Delta u) \approx \boldsymbol{f}(\boldsymbol{x}_o, u_o) + \frac{d\boldsymbol{f}(\boldsymbol{x}_o, u_o)}{d\boldsymbol{x}^T}\Delta\boldsymbol{x} + \frac{d\boldsymbol{f}(\boldsymbol{x}_o, u_o)}{du}\Delta u \\ y &= g(\boldsymbol{x}_o + \Delta\boldsymbol{x}, u_o + \Delta u) \approx g(\boldsymbol{x}_o, u_o) + \frac{dg(\boldsymbol{x}_o, u_o)}{d\boldsymbol{x}^T}\Delta\boldsymbol{x} + \frac{dg(\boldsymbol{x}_o, u_o)}{du}\Delta u \end{aligned} \tag{3.7}$$

Let us use the fact that at the equilibrium point $\boldsymbol{f}(\boldsymbol{x}_\mathrm{o}, u_\mathrm{o}) = 0$ and let us introduce the notation $y_\mathrm{o} = g(\boldsymbol{x}_\mathrm{o}, u_\mathrm{o})$, so the linearized model valid for small changes takes the form

$$\begin{aligned} \frac{\mathrm{d}(\boldsymbol{x}-\boldsymbol{x}_\mathrm{o})}{\mathrm{d}t} = \frac{\mathrm{d}\Delta\boldsymbol{x}}{\mathrm{d}t} &= \boldsymbol{A}(\boldsymbol{x}-\boldsymbol{x}_\mathrm{o}) + \boldsymbol{b}(u-u_\mathrm{o}) = \boldsymbol{A}\Delta\boldsymbol{x} + \boldsymbol{b}\Delta u \\ y - y_\mathrm{o} = \Delta y &= \boldsymbol{c}^\mathrm{T}(\boldsymbol{x}-\boldsymbol{x}_\mathrm{o}) + d(u-u_\mathrm{o}) = \boldsymbol{c}^\mathrm{T}\Delta\boldsymbol{x} + d\Delta u \end{aligned} \tag{3.8}$$

where the following notations are employed

$$\begin{aligned} \boldsymbol{A} &= \frac{\mathrm{d}\boldsymbol{f}(\boldsymbol{x}_\mathrm{o}, u_\mathrm{o})}{\mathrm{d}\boldsymbol{x}^\mathrm{T}}; \quad & \boldsymbol{b} &= \frac{\mathrm{d}\boldsymbol{f}(\boldsymbol{x}_\mathrm{o}, u_\mathrm{o})}{\mathrm{d}u} \\ \boldsymbol{c}^\mathrm{T} &= \frac{\mathrm{d}g(\boldsymbol{x}_\mathrm{o}, u_\mathrm{o})}{\mathrm{d}\boldsymbol{x}^\mathrm{T}}; \quad & d &= \frac{\mathrm{d}g(\boldsymbol{x}_\mathrm{o}, u_\mathrm{o})}{\mathrm{d}u} \end{aligned} \tag{3.9}$$

The obtained model is a linear time-invariant (*LTI*) system, i.e. it does not change in time. It is a widely used practice that the original variables $\boldsymbol{x}, u, y$ are used instead of the small changes $(\Delta\boldsymbol{x}, \Delta u, \Delta y)$ for simplicity, but they are considered as the changes around the operating point. In this way we arrive at the linear, constant parameter (*LTI*) state-space equation of the system generally applied in the theory of systems and control,

$$\begin{aligned} \frac{\mathrm{d}x(t)}{\mathrm{d}t} &= \boldsymbol{A}\boldsymbol{x}(t) + \boldsymbol{b}u(t) & \qquad & \frac{\mathrm{d}x}{\mathrm{d}t} = \boldsymbol{A}\boldsymbol{x} + \boldsymbol{b}u \\ & & \text{or simply} & \\ y(t) &= \boldsymbol{c}^\mathrm{T}\boldsymbol{x}(t) + du(t) & & y = \boldsymbol{c}^\mathrm{T}\boldsymbol{x} + du \end{aligned} \tag{3.10}$$

Here the parameter matrices of the system are $\boldsymbol{A}, \boldsymbol{b}, \boldsymbol{c}^\mathrm{T}, d$. Since in this book single-input single-output (*SISO*) systems are considered, in the n-order case, $\boldsymbol{A}$ is an $(n \times n)$ square matrix, which is the so called state matrix, $\boldsymbol{b}$ is an $(n \times 1)$ column vector, $\boldsymbol{c}^\mathrm{T}$ is a row vector of dimensions $(1 \times n)$, and d is a scalar. The block diagram of the state-Eq. (3.10) can be seen in Fig. 3.1.

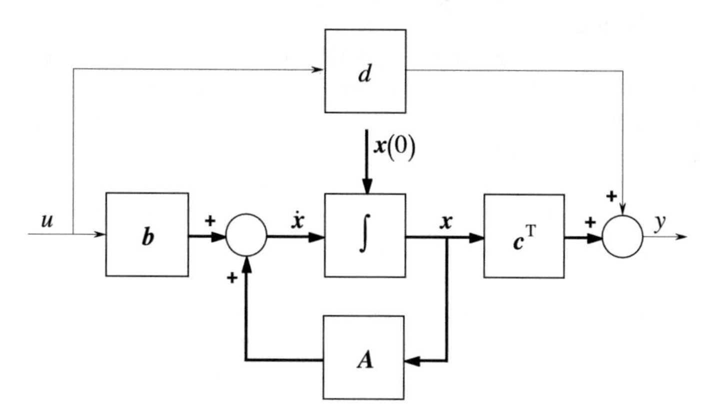

Fig. 3.1 Block diagram of the linear time invariant system

3.1 Solution of the State-Equations in the Complex Frequency Domain

The state-equations can be transferred to the complex frequency domain by the LAPLACE transformation of (3.10). Let us denote the transformed time functions $\boldsymbol{x}, u, y$ by $\boldsymbol{X}(s), U(s), Y(s)$. Taking the rules of transformation of derivatives into account we get

$$\begin{aligned} s\boldsymbol{X}(s) &= \boldsymbol{AX}(s) + \boldsymbol{b}U(s) + \boldsymbol{x}(0) = \boldsymbol{AX}(s) + \boldsymbol{b}U(s) + \boldsymbol{Ix}(0) \\ Y(s) &= \boldsymbol{c}^{\mathrm{T}}\boldsymbol{X}(s) + dU(s) \end{aligned} \tag{3.11}$$

In the first equation, the vector of the initial conditions $\boldsymbol{x}(0)$ can be considered as an input which has an effect on the system via the identity matrix $\boldsymbol{I}$. From the first equation we get

$$\boldsymbol{X}(s) = (s\boldsymbol{I} - \boldsymbol{A})^{-1}[\boldsymbol{b}U(s) + \boldsymbol{x}(0)] = (s\boldsymbol{I} - \boldsymbol{A})^{-1}\boldsymbol{b}U(s) + (s\boldsymbol{I} - \boldsymbol{A})^{-1}\boldsymbol{x}(0). \tag{3.12}$$

According to the rules of matrix inversion

$$(s\boldsymbol{I} - \boldsymbol{A})^{-1} = \frac{\mathbf{adj}(s\boldsymbol{I} - \boldsymbol{A})}{\det(s\boldsymbol{I} - \boldsymbol{A})} = \frac{\mathbf{adj}(s\boldsymbol{I} - \boldsymbol{A})}{\mathcal{A}(s)} = \frac{\boldsymbol{\Psi}(s)}{\mathcal{A}(s)} = \boldsymbol{\Phi}(s). \tag{3.13}$$

Here $\boldsymbol{\Psi}(s) = \mathbf{adj}(s\boldsymbol{I} - \boldsymbol{A})$ is the transpose of a matrix whose elements are the signed sub-determinants belonging to the corresponding elements of the matrix $(s\boldsymbol{I} - \boldsymbol{A})$. The determinant of that matrix, $\det(s\boldsymbol{I} - \boldsymbol{A})$ is the denominator of the transfer function, and is an n-degree polynomial in s:

$$\mathcal{A}(s) = s^n + k_1 s^{n-1} + \cdots + k_{n-1}s + k_n = \prod_{i-1}^{n}(s - \lambda_i) = \det(s\boldsymbol{I} - \boldsymbol{A}). \tag{3.14}$$

$\mathcal{A}(s)$ is the so-called characteristic polynomial of the matrix $\boldsymbol{A}$. The roots $\lambda_1, \ldots \lambda_n$ of the characteristic equation $\mathcal{A}(s) = 0$ are the eigenvalues of $\boldsymbol{A}$, called the poles of the system.

The elements of the matrix in the numerator of (3.13) are also polynomials in s, but since they come from an $(n-1)$ order sub-determinant, they can have at most order $(n-1)$, consequently the quotients of each element and $\mathcal{A}(s)$ represent strictly proper transfer functions.

According to (3.12) the motion of the state vector is determined by the initial condition $\boldsymbol{x}(0)$ and the input signal $U(s)$. Since the characteristic polynomial is in the denominators of all the elements depending on $\boldsymbol{x}(0)$, as a consequence of the expansion theorem, their time functions are exclusively determined by the poles of the system. This part of the solution describes the motion of a un-excited system from any initial position to its equilibrium point and it exclusively depends on one of the parameters, i.e., on the state matrix $\boldsymbol{A}$ both in the frequency and time domains.

In the case of excitation, in each element of the solution depending on $U(s)$, the denominator contains not only $\mathcal{A}(s)$ but also the denominator polynomial of $U(s)$, so the time functions depend not only on the poles of the system but also on the poles of the input. This part of the solution gives the motion of the excited system.

From Eqs. (3.11) and (3.12), the output is

$$Y(s) = \boldsymbol{c}^{\mathrm{T}}(s\boldsymbol{I} - \boldsymbol{A})^{-1}[\boldsymbol{b}U(s) + \boldsymbol{x}(0)] + dU(s). \tag{3.15}$$

The output of the excited motion, when $\boldsymbol{x}(0) = \boldsymbol{0}$ is

$$Y(s) = \left[\boldsymbol{c}^{\mathrm{T}}(s\boldsymbol{I} - \boldsymbol{A})^{-1}\boldsymbol{b} + d\right]U(s). \tag{3.16}$$

Thus the transfer function of the system is

$$P(s) = \frac{Y(s)}{U(s)} = \boldsymbol{c}^{\mathrm{T}}(s\boldsymbol{I} - \boldsymbol{A})^{-1}\boldsymbol{b} + d\big|_{d=0} = \boldsymbol{c}^{\mathrm{T}}(s\boldsymbol{I} - \boldsymbol{A})^{-1}\boldsymbol{b} = \frac{\mathcal{B}(s)}{\mathcal{A}(s)}. \tag{3.17}$$

The first term of $P(s)$ is strictly proper since it consists of a linear combination of only proper elements (see (3.13), i.e., the order of the adjoint is always lower than that of the determinant). Thus if $d = 0$, then $P(s)$ is strictly proper, the order of the numerator being lower by at least one than that of the denominator. If $d \neq 0$, then $P(s)$ is proper, i.e. the order of the numerator is equal to that of the denominator. The physical meaning of d is how the input directly influences the output without any dynamics. Note, that this effect does not disappear even for very high frequencies, thus $P(j\omega \to \infty) = d$. This means, at the same time, that the jump of the transfer function at time $t = 0$ is $v(t = 0) = d$. In practice the case $d \neq 0$ is usually traced back to the case $d = 0$ by introducing a new output $\tilde{y} = y - du$. The case $d \neq 0$ can also be considered as an imperfect linearization which needs a certain correction.

Example 3.1 Let the parameter matrices of the system be

$$\boldsymbol{A} = \begin{bmatrix} -3 & -2 \\ 1 & 0 \end{bmatrix}; \quad \boldsymbol{b} = \begin{bmatrix} 1 \\ 0 \end{bmatrix}; \quad \boldsymbol{c}^{\mathrm{T}} = [2 \quad 2],$$

and compute the transfer function using (3.17).

$$P(s) = \frac{B(s)}{A(s)} = \boldsymbol{c}^{\mathrm{T}}(s\boldsymbol{I} - \boldsymbol{A})^{-1}\boldsymbol{b} = \frac{1}{s^2 + 3s + 2}[2 \quad 2]\begin{bmatrix} s & -2 \\ 1 & s+3 \end{bmatrix}\begin{bmatrix} 1 \\ 0 \end{bmatrix}$$

$$= \frac{1}{s^2 + 3s + 2}[2s + 2 \quad -4 + 2s + 6]\begin{bmatrix} 1 \\ 0 \end{bmatrix} = \frac{2s + 2}{s^2 + 3s + 2} = \frac{s + 1}{0.5s^2 + 1.5s + 1}$$

■

3.2 Solution of the State-Equations in the Time Domain

The solution of the state-Eq. (3.10) in time domain can also be given in closed form

$$\boldsymbol{x}(t) = e^{\boldsymbol{A}t}\boldsymbol{x}(0) + \int_0^t e^{\boldsymbol{A}(t-\tau)}\boldsymbol{b}u(\tau)\mathrm{d}\tau = e^{\boldsymbol{A}t}\boldsymbol{x}(0) + \left[\int_0^t e^{\boldsymbol{A}(t-\tau)}u(\tau)\mathrm{d}\tau\right]\boldsymbol{b}. \quad (3.18)$$

The first term represents the motion of the un-excited system starting from the initial point $\boldsymbol{x}(0)$, the second term is the convolution integral, i.e., the excited motion starting from the initial point $\boldsymbol{x}(0) = \boldsymbol{0}$.

To check (3.18), let us differentiate the above equation with respect to time:

$$\frac{\mathrm{d}\boldsymbol{x}(t)}{\mathrm{d}t} = \boldsymbol{A}e^{\boldsymbol{A}t}\boldsymbol{x}(0) + \int_0^t \boldsymbol{A}e^{\boldsymbol{A}(t-\tau)}\boldsymbol{b}u(\tau)\mathrm{d}\tau + \boldsymbol{b}u(t) = \boldsymbol{A}\boldsymbol{x}(t) + \boldsymbol{b}u(t), \quad (3.19)$$

which proves the correctness of (3.18). (See the detailed derivation in Chap. A.3.1 of Appendix A.5.) Here, $e^{\boldsymbol{A}t}$ is the fundamental matrix of the system, which is defined by its TAYLOR-series, convergent for all t, as is valid for matrix functions in general.

$$e^{\boldsymbol{A}t} = \boldsymbol{I} + \boldsymbol{A}t + \frac{1}{2}(\boldsymbol{A}t)^2 + \cdots + \frac{1}{n!}(\boldsymbol{A}t)^n + \cdots. \quad (3.20)$$

By differentiating the equation, a very interesting and important feature of the fundamental matrix can be obtained.

$$\begin{aligned}\frac{\mathrm{d}e^{\boldsymbol{A}t}}{\mathrm{d}t} &= \boldsymbol{A} + \boldsymbol{A}^2 t + \frac{1}{2}\boldsymbol{A}^3 t^2 + \cdots + \frac{1}{(n-1)!}\boldsymbol{A}^n t^{n-1} + \cdots \\ &= \boldsymbol{A}\left(\boldsymbol{I} + \boldsymbol{A}t + \frac{1}{2}(\boldsymbol{A}t)^2 + \cdots + \frac{1}{n!}(\boldsymbol{A}t)^n + \cdots\right) = \boldsymbol{A}e^{\boldsymbol{A}t} = e^{\boldsymbol{A}t}\boldsymbol{A}\end{aligned} \quad (3.21)$$

Comparing (3.12) and (3.18), the LAPLACE-transform of the fundamental matrix for $U(s) = 0$ is

$$\mathcal{L}\{e^{\boldsymbol{A}t}\} = (s\boldsymbol{I} - \boldsymbol{A})^{-1} = \boldsymbol{\Phi}(s), \quad (3.22)$$

which provides a new relationship for the computation of the fundamental matrix:

$$e^{\boldsymbol{A}t} = \mathcal{L}^{-1}\left\{(s\boldsymbol{I} - \boldsymbol{A})^{-1}\right\} = \mathcal{L}^{-1}\{\boldsymbol{\Phi}(s)\} \quad (3.23)$$

Combining (3.10) and (3.18) shows that the output of the system is

$$y(t) = \boldsymbol{c}^{\mathrm{T}} e^{\boldsymbol{A}t}\boldsymbol{x}(0) + \boldsymbol{c}^{\mathrm{T}}\left[\int_0^t e^{\boldsymbol{A}(t-\tau)}u(\tau)\mathrm{d}\tau\right]\boldsymbol{b} + du(t) \tag{3.24}$$

In the case of zero initial conditions ($\boldsymbol{x}(0) = \boldsymbol{0}$) and $d = 0$, the weighting function of the system for the excitation $u(t) = \delta(t)$ can be easily obtained from the last equation

$$\begin{aligned} w(t) &= \boldsymbol{c}^{\mathrm{T}} e^{\boldsymbol{A}t}\boldsymbol{b} = \boldsymbol{c}^{\mathrm{T}}\mathcal{L}^{-1}\left\{(s\boldsymbol{I}-\boldsymbol{A})^{-1}\right\}\boldsymbol{b} = \mathcal{L}^{-1}\left\{\boldsymbol{c}^{\mathrm{T}}(s\boldsymbol{I}-\boldsymbol{A})^{-1}\boldsymbol{b}\right\} \\ &= \mathcal{L}^{-1}\{\boldsymbol{c}^{\mathrm{T}}\boldsymbol{\Phi}(s)\boldsymbol{b}\} = \mathcal{L}^{-1}\{P(s)|_{d=0}\} \end{aligned} \tag{3.25}$$

See the details for the weighting function computation in Chap. A.3.2 of Appendix A.5.

As a consequence of matrix function operations and the CAYLEY-HAMILTON theorem the fundamental matrix can also be computed in the form of finite sum:

$$e^{\boldsymbol{A}\tau} = \alpha_{\mathrm{o}}(\tau)\boldsymbol{I} + \alpha_1(\tau)\boldsymbol{A} + \cdots + \alpha_{\mathrm{n}-1}(\tau)\boldsymbol{A}^{n-1} \tag{3.26}$$

since the state matrix $\boldsymbol{A}$ satisfies its characteristic equation, i.e.

$$\mathcal{A}(\boldsymbol{A}) = \boldsymbol{0}. \tag{3.27}$$

(See the proofs in A.3.3 of Appendix A.5.)

3.3 Transformation of the State-Equations, Canonical Forms

The input and output signals of a system are usually certain physical variables. The state variables, however, depend on the chosen coordinate system. The parameter matrices $\boldsymbol{A}, \boldsymbol{b}, \boldsymbol{c}^{\mathrm{T}}$ also depend on the coordinate system. Introduce the new state vector z, which can be obtained from $\boldsymbol{x}$ by a linear transformation $z = \boldsymbol{T}\boldsymbol{x}$ where $\boldsymbol{T}$ is regular. Using (3.10) the new state-equations are

$$\begin{aligned} \frac{\mathrm{d}z}{\mathrm{d}t} &= \boldsymbol{T}(\boldsymbol{A}\boldsymbol{x} + \boldsymbol{b}u) = \boldsymbol{T}\boldsymbol{A}\boldsymbol{T}^{-1}z + \boldsymbol{T}\boldsymbol{b}u = \tilde{\boldsymbol{A}}z + \tilde{\boldsymbol{b}}u \\ y &= \boldsymbol{c}^{\mathrm{T}}\boldsymbol{x} + du = \boldsymbol{c}^{\mathrm{T}}\boldsymbol{T}^{-1}z + du = \tilde{\boldsymbol{c}}^{\mathrm{T}}z + \tilde{d}u \end{aligned} \tag{3.28}$$

where

$$\tilde{A} = TAT^{-1}, \quad \tilde{b} = Tb, \quad \tilde{c}^{\mathrm{T}} = c^{\mathrm{T}}T^{-1}, \quad \tilde{d} = d. \tag{3.29}$$

It is easy to check that the weighting function and the transfer function of the system are invariant under linear transformations:

$$w(t) = \tilde{c}^{\mathrm{T}} e^{\tilde{A}t}\tilde{b} = c^{\mathrm{T}}T^{-1}e^{TAT^{-1}t}Tb = c^{\mathrm{T}}e^{At}b \tag{3.30}$$

$$H(s) = \tilde{c}^{\mathrm{T}}\left(sI - \tilde{A}\right)^{-1}\tilde{b} = c^{\mathrm{T}}T^{-1}\left(sI - TAT^{-1}\right)^{-1}Tb = c^{\mathrm{T}}(sI - A)^{-1}b \tag{3.31}$$

In (3.30) the following simple identity (obtained by TAYLOR series of e^x) was employed

$$e^{\left(TAT^{-1}\right)t} = Te^{At}T^{-1}. \tag{3.32}$$

It is well known, that a linear transformation has certain special directions in which the vectors keep their directions, only their lengths change by a factor of λ_i, i.e.

$$Av_i = \lambda_i v_i, \quad i = 1,\ldots,n. \tag{3.33}$$

Here v_i is the eigenvector of A, and λ_i is its eigenvalue. The eigenvalue problem can also be formulated in a different way, i.e. as a homogeneous system equations in n unknown variables

$$(\lambda_i I - A)v = \mathbf{0}, \tag{3.34}$$

where the variables are the components of v. This system of equations has a solution different from the trivial one ($v = \mathbf{0}$) if the condition

$$\det(\lambda_i I - A) = \mathcal{A}(\lambda_i) = 0 \tag{3.35}$$

is satisfied, i.e. the eigenvalues λ_i are the roots of the characteristic polynomial. If the roots are single, then the total number of the eigenvalues is n, and each has only one eigenvector of unit length.

3.3.1 Diagonal Canonical Form

In the case of single eigenvalues, by choosing a special transformation matrix T_{d}, one can make $T_{\mathrm{d}}A(T_{\mathrm{d}})^{-1}$ diagonal:

$$\tilde{A}_{\mathrm{d}} = T_{\mathrm{d}} A (T_{\mathrm{d}})^{-1} = \Lambda = \begin{bmatrix} \lambda_1 & 0 & \dots & 0 \\ 0 & \lambda_2 & \dots & 0 \\ \vdots & \vdots & \ddots & \vdots \\ 0 & 0 & \dots & \lambda_n \end{bmatrix} = A_{\mathrm{d}} = \mathbf{diag}[\lambda_1, \lambda_2, \dots, \lambda_n]. \tag{3.36}$$

The necessary transformation matrix T_{d} is the inverse of the matrix of the eigenvectors

$$T_{\mathrm{d}} = [v_1, v_2, \dots, v_n]^{-1}. \tag{3.37}$$

The canonical state-equation (canonical form) obtained by the diagonal transformation is

$$\frac{\mathrm{d}z}{\mathrm{d}t} = \begin{bmatrix} \lambda_1 & 0 & \dots & 0 \\ 0 & \lambda_2 & \dots & 0 \\ \vdots & \vdots & \ddots & \vdots \\ 0 & 0 & \dots & \lambda_n \end{bmatrix} z + \begin{bmatrix} \beta_1 \\ \beta_2 \\ \vdots \\ \beta_n \end{bmatrix} u = \Lambda z + \beta u \tag{3.38}$$

$$y = [\gamma_1 \quad \gamma_2 \quad \dots \quad \gamma_n] z + du = \gamma^{\mathrm{T}} z + du$$

The transfer function of the transformed system is

$$P(s) = \sum_{i=1}^{n} \frac{\beta_i \gamma_i}{s - \lambda_i} + d. \tag{3.39}$$

Thus the transfer function can be obtained in partial fraction form from the canonical one. Note that the eigenvalues of A appear in the denominator. The transfer function remains unchanged if the product of β_i and γ_i remains constant. Thus there are a great many canonical forms which are different in the matrices β and γ^{T} but have the same transfer function.

In the case of single poles, the state-equation system in the canonical coordinates consists of n independent first order differential equations. Each individual state variable can be assigned to an individual pole of the system.

If the characteristic equation has multiple roots, the matrix A_{d} can be diagonalized only in exceptional cases, but its JORDAN-form, in general, can be given by

$$J = \begin{bmatrix} J_1 & \mathbf{0} & \dots & \mathbf{0} \\ \mathbf{0} & J_2 & \dots & \mathbf{0} \\ \vdots & \vdots & \ddots & \vdots \\ \mathbf{0} & \mathbf{0} & \dots & J_m \end{bmatrix}. \tag{3.40}$$

Here, each $\boldsymbol{J}_i$ is a square matrix (a JORDAN block) of dimension equal to the multiplicity of the eigenvalue λ_i, whose main diagonal contains the eigenvalues and there are ones in the first off-diagonal right from the main diagonal, all the other elements are zeros. If, e.g., λ_1 is a triple eigenvalue, the sub-matrix $\boldsymbol{J}_1$ has the form

$$\boldsymbol{J}_1 = \begin{bmatrix} \lambda_1 & 1 & 0 \\ 0 & \lambda_1 & 1 \\ 0 & 0 & \lambda_1 \end{bmatrix}. \tag{3.41}$$

(The number of ones depends on how many linearly independent eigenvectors can be found for the multiple eigenvalue λ_1. If only one such an eigenvector exists —which is the normal case corresponding to (3.38)—then all elements of the off-diagonal are ones. If the number of the independent eigenvectors has increased by one compared to the previous case, then the number of ones decreases by one. If there exists the same number of independent eigenvectors as the multiplicity, the JORDAN block is diagonal. In other cases, finding the transformation matrix needs special considerations, which are not discussed here.)

3.3.2 Controllable Canonical Form

It is the most common practice in modeling to directly derive the state-equations from the differential equations formulated for the physical variables. In many cases, however, the initial information is a transfer function, or a linear differential equation of order n. This procedure is often called the description or construction (reconstruction) of the state-equations. Suppose that the operation of the system is described by the differential equation

$$\frac{\mathrm{d}^n y}{\mathrm{d}t^n} + a_1 \frac{\mathrm{d}^{n-1} y}{\mathrm{d}t^{n-1}} + \cdots + a_n y = b_1 \frac{\mathrm{d}^{n-1} u}{\mathrm{d}t^{n-1}} + \cdots + b_n u. \tag{3.42}$$

The equation valid for the LAPLACE-transforms is

$$Y(s) = \frac{b_1 s^{n-1} + \cdots + b_{n-1} s + b_n}{s^n + a_1 s^{n-1} + \cdots + a_{n-1} s + a_n} U(s) = \frac{\mathcal{B}(s)}{\mathcal{A}(s)} U(s) = P(s) U(s). \tag{3.43}$$

Introduce the following state variables with their LAPLACE-transforms

$$\begin{aligned} X_1(s) &= \frac{s^{n-1}}{\mathcal{A}(s)} U(s) \\ X_2(s) &= \frac{s^{n-2}}{\mathcal{A}(s)} U(s) = \frac{1}{s} X_1(s) \qquad \frac{\mathrm{d}x_2}{\mathrm{d}t} = x_1 \\ &\vdots \qquad\qquad\qquad\qquad\qquad\qquad\qquad \vdots \\ X_n(s) &= \frac{1}{\mathcal{A}(s)} U(s) = \frac{1}{s} X_{n-1}(s) \qquad \frac{\mathrm{d}x_n}{\mathrm{d}t} = x_{n-1} \end{aligned} \tag{3.44}$$

On this basis,

$$sX_1(s) = -a_1X_1(s) - \cdots - a_nX_n(s) + U(s) \qquad \frac{\mathrm{d}x_1}{\mathrm{d}t} = -a_1x_1 - \cdots - a_nx_n + u$$
$$Y(s) = b_1X_1(s) + \cdots + b_nX_n(s) \qquad y = b_1x_1 + \cdots + b_nx_n \tag{3.45}$$

Thus the resulting state-equations are

$$\frac{\mathrm{d}\boldsymbol{x}}{\mathrm{d}t} = \begin{bmatrix} -a_1 & -a_2 & \ldots & -a_{n-1} & -a_n \\ 1 & 0 & \ldots & 0 & 0 \\ 0 & 1 & \ldots & 0 & 0 \\ \vdots & \vdots & \ddots & \vdots & \vdots \\ 0 & 0 & \ldots & 1 & 0 \end{bmatrix} \boldsymbol{x} + \begin{bmatrix} 1 \\ 0 \\ 0 \\ \vdots \\ 0 \end{bmatrix} u \tag{3.46}$$

$$y = \begin{bmatrix} b_1 & b_2 & \ldots & b_{n-1} & b_n \end{bmatrix} \boldsymbol{x}$$

This form with its special system matrices is called the *controllable canonical form* or *phase-variable form*

$$\boldsymbol{A}_{\mathrm{c}} = \begin{bmatrix} -a_1 & -a_2 & \ldots & -a_{n-1} & -a_n \\ 1 & 0 & \ldots & 0 & 0 \\ 0 & 1 & \ldots & 0 & 0 \\ \vdots & \vdots & \ddots & \vdots & \vdots \\ 0 & 0 & \ldots & 1 & 0 \end{bmatrix}; \boldsymbol{b}_{\mathrm{c}} = \begin{bmatrix} 1 \\ 0 \\ 0 \\ \vdots \\ 0 \end{bmatrix}; \boldsymbol{c}_{\mathrm{c}}^{\mathrm{T}} = \begin{bmatrix} b_1 & b_2 & \ldots & b_{n-1} & b_n \end{bmatrix}. \tag{3.47}$$

The special feature of this form is that every state variable, except the last one, is the derivative of the next state variable in the action direction, and all state variables are fed back to the first one. The feedback factors are the negative coefficients of the characteristic equation which appear in the first row of matrix $\boldsymbol{A}$. The input has effect only on x_1. The feedforward factors representing the output are the coefficients of the numerator of the transfer function.

If $P(s)$ is not strictly proper, i.e., $\mathcal{B}(s) = b'_{\mathrm{o}}s^n + b'_1s^{n-1} + \cdots + b'_{n-1}s + b'_n$, the $d = b'_{\mathrm{o}}$ also occurs in the state-equation. In this case new coefficients b_i must be computed from the original coefficients b'_i by the following decomposition

$$P(s) = \frac{\mathcal{B}(s)}{\mathcal{A}(s)} = \frac{b'_{\mathrm{o}}s^n + b'_1s^{n-1} + \cdots + b'_{n-1}s + b'_n}{s^n + a_1s^{n-1} + \cdots + a_{n-1}s + a_n}$$
$$= b_{\mathrm{o}} + \frac{b_1s^{n-1} + \cdots + b_{n-1}s + b_n}{s^n + a_1s^{n-1} + \cdots + a_{n-1}s + a_n}. \tag{3.48}$$

The second term is already strictly proper and the coefficients of the numerator can be computed by the relationships $b_i = b'_i - b'_o a_i = b'_i - b_o a_i \, (b_o = b'_o)$.

The characteristic polynomial of the controllable canonical form is

$$\mathcal{A}(s) = \det \begin{bmatrix} s+a_1 & a_2 & \dots & a_{n-1} & a_n \\ -1 & s & \dots & 0 & 0 \\ 0 & -1 & \dots & 0 & 0 \\ \vdots & \vdots & \ddots & \vdots & \vdots \\ 0 & 0 & \dots & -1 & s \end{bmatrix} = \mathcal{A}_n(s) = s\mathcal{A}_{n-1}(s) + a_n, \quad (3.49)$$

where a recursive relationship is obtained by decomposing the last row. It is obvious that

$$\mathcal{A}_n(s) = s^n + a_1 s^{n-1} + \ldots + a_{n-1}s + a_n = \mathcal{A}(s) \quad (3.50)$$

i.e. the characteristic polynomial is the denominator of the transfer function. Therefore the special matrix $\boldsymbol{A}_c$ is called the accompanying (complementary) matrix of $\mathcal{A}(s)$.

Note that the parameter matrix selection

$$\bar{\boldsymbol{A}}_c = \begin{bmatrix} 0 & 1 & \dots & 0 & 0 \\ \vdots & \vdots & \ddots & \vdots & \vdots \\ 0 & 0 & \dots & 1 & 0 \\ 0 & 0 & \dots & 0 & 1 \\ -a_n & -a_{n-1} & \dots & -a_2 & -a_1 \end{bmatrix}; \; \bar{\boldsymbol{b}}_c = \begin{bmatrix} 0 \\ 0 \\ 0 \\ \vdots \\ 1 \end{bmatrix}; \; \bar{\boldsymbol{c}}_c^T = [b_n \quad b_{n-1} \quad \dots \quad b_2 \quad b_1] \quad (3.51)$$

also provides the controllable canonical form, where the serial number of the state variables is the opposite of what appeared in the form (3.47).

3.3.3 Observable Canonical Form

To create this form, let us introduce the state variables with their Laplace-transforms according to the recursions

$$\begin{aligned} &X_1(s) = Y(s) \\ &sX_1(s) = -a_1X_1(s) + X_2(s) + b_1U(s) && \frac{dx_1}{dt} = -a_1x_1 + x_2 + b_1u \\ &sX_2(s) = -a_2X_1(s) + X_3(s) + b_2U(s) && \frac{dx_2}{dt} = -a_2x_1 + x_3 + b_2u \\ &\vdots && \vdots \\ &sX_n(s) = -a_nX_1(s) + b_nU(s) && \frac{dx_n}{dt} = -a_nx_1 + b_nu \end{aligned} \quad (3.52)$$

where $Y(s)$ corresponds to (3.43). Based on the relationships above the following state-equations can be written

$$\frac{\mathrm{d}\boldsymbol{x}}{\mathrm{d}t} = \begin{bmatrix} -a_1 & 1 & 0 & \dots & 0 \\ -a_2 & 0 & 1 & \dots & 0 \\ \vdots & \vdots & \vdots & \ddots & 0 \\ -a_{n-1} & 0 & 0 & \dots & 1 \\ -a_n & 0 & 0 & \dots & 0 \end{bmatrix} \boldsymbol{x} + \begin{bmatrix} b_1 \\ b_2 \\ \vdots \\ b_{n-1} \\ b_n \end{bmatrix} u \tag{3.53}$$

$$y = [1 \quad 0 \quad \dots \quad 0 \quad 0]\boldsymbol{x}$$

This form with its special system matrices

$$\boldsymbol{A}_\mathrm{o} = \begin{bmatrix} -a_1 & 1 & 0 & \dots & 0 \\ -a_2 & 0 & 1 & \dots & 0 \\ \vdots & \vdots & \vdots & \ddots & \vdots \\ -a_{n-1} & 0 & 0 & \dots & 1 \\ -a_n & 0 & 0 & \dots & 0 \end{bmatrix}, \quad \boldsymbol{b}_\mathrm{o} = \begin{bmatrix} b_1 \\ b_2 \\ \vdots \\ b_{n-1} \\ b_n \end{bmatrix}, \quad \boldsymbol{c}_\mathrm{o}^\mathrm{T} = [1 \quad 0 \quad \dots \quad 0 \quad 0] \tag{3.54}$$

is called observable canonical form. The special feature of this form is that its output is the state variable x_1 itself, which is fed back to the inputs of all the state variables. The feedback factors are the negative coefficients of the characteristic equation, and thus they appear in the first column of $\boldsymbol{A}_\mathrm{o}$. Note that the parameter matrix selection

$$\bar{\boldsymbol{A}}_\mathrm{o} = \begin{bmatrix} 0 & \dots & 0 & 0 & -a_n \\ 1 & \dots & 0 & 0 & -a_{n-1} \\ \vdots & \ddots & \vdots & \vdots & \vdots \\ 0 & \dots & 1 & 0 & -a_2 \\ 0 & \dots & 0 & 1 & -a_1 \end{bmatrix}, \quad \bar{\boldsymbol{b}}_\mathrm{o} = \begin{bmatrix} b_n \\ b_{n-1} \\ \vdots \\ b_2 \\ b_1 \end{bmatrix}, \quad \bar{\boldsymbol{c}}_\mathrm{o}^\mathrm{T} = [0 \quad 0 \quad \dots \quad 0 \quad 1] \tag{3.55}$$

also provides the observable canonical form, where the serial number of the state variables is the opposite to what appeared in (3.53).

If $P(s)$ is not strictly proper, $d = b_\mathrm{o}$ also appears in the state-equation and the statements made in connection with (3.48) are also valid.

(If the poles and the partial fractional forms of the transfer function are known, then further canonical forms can be constructed.)

3.4 The Concepts of Controllability and Observability

A very important question of control is how to influence arbitrarily all state variables by the input. This question can be answered by the controllability theorem introduced by KALMAN.

A system is *state controllable* if its state vector can be driven from an initial state $\boldsymbol{x}(t_o)$ to an arbitrary final state $\boldsymbol{x}(t_v)$ in a finite time $(t_v - t_o)$ by a control signal u. If this definition is fulfilled only for the output, then the system is *output controllable*. In the case of linear time invariant systems, the starting time is chosen $(t_o = 0)$, and the initial state can be given as $\boldsymbol{x}(0)$. By this definition, the controllability is connected to the system. If the controllability exists for a certain initial state, then it remains for any initial state, since from any $\boldsymbol{x}(0)$ the system can be driven to $\boldsymbol{x}(t_v)$ by an appropriate control signal.

Controllability can best be explained in canonical coordinates. If in the canonical form (3.38) β_i is zero for a state variable, then this state can not be controlled. This means, that there is no parallel component, but only a perpendicular component, of any control to the eigenvector belonging to the eigenvalue λ_i, thus the effect of the control always remains in the plane perpendicular to the eigenvector. (As a consequence of the canonical form the system can only be controlled if the poles of the canonical coordinates are different.)

If the system is not state controllable, the output, however, may be controllable if at least one state variable is controllable and the γ_i belonging to it is not zero [see (3.39)].

In coordinates different from the canonical ones, the above conditions can not be directly recognized because of the relationships between the state variables, therefore they have to be replaced by more general criteria.

For simplicity, choose the initial condition $\boldsymbol{x}(0) = \boldsymbol{0}$. Then the solution of the state-Eq. (3.18) has the form

$$\boldsymbol{x}(t) = \left[\int_0^t e^{\boldsymbol{A}(t-\tau)} u(\tau) \mathrm{d}\tau\right] \boldsymbol{b} = \left[\int_0^t e^{\boldsymbol{A}\tau} u(t-\tau) \mathrm{d}\tau\right] \boldsymbol{b} \tag{3.56}$$

and using the finite sum form of the fundamental matrix [see (3.26)], it can be written as

$$e^{\boldsymbol{A}\tau} = \alpha_o(\tau)\boldsymbol{I} + \alpha_1(\tau)\boldsymbol{A} + \cdots + \alpha_{n-1}(\tau)\boldsymbol{A}^{n-1} \tag{3.57}$$

The solution of the state-equation is obtained in closed form as

$$\boldsymbol{x}(t) = \boldsymbol{b}\int_0^t \alpha_o(\tau) u(\tau) \mathrm{d}\tau + \boldsymbol{A}\boldsymbol{b}\int_0^t \alpha_1(\tau) u(\tau) \mathrm{d}\tau + \cdots + \boldsymbol{A}^{n-1}\boldsymbol{b}\int_0^t \alpha_{n-1}(\tau) u(\tau) \mathrm{d}\tau. \tag{3.58}$$

Thus the right-hand side of the equation is a linear combination of the columns of the *controllability matrix*

$$\boldsymbol{M}_{\mathrm{c}} = \begin{bmatrix} \boldsymbol{b} & \boldsymbol{A}\boldsymbol{b} & \dots & \boldsymbol{A}^{n-1}\boldsymbol{b} \end{bmatrix} \tag{3.59}$$

Thus the condition of that each point of the state-space be reachable means that $\boldsymbol{M}_{\mathrm{c}}$ must have n linearly independent columns, i.e. $\boldsymbol{M}_{\mathrm{c}}$ must be invertible and regular. Since $\boldsymbol{M}_{\mathrm{c}}$ depends on $\boldsymbol{A}$ and $\boldsymbol{b}$, the controllability of the pair $\boldsymbol{A};\boldsymbol{b}$ is a quite accepted convention.

If the above statements are referred to the output, then the condition of *output controllability* is that at least one element of

$$\boldsymbol{m}_{\mathrm{c}}^{\mathrm{T}} = \begin{bmatrix} \boldsymbol{c}^{\mathrm{T}}\boldsymbol{b} & \boldsymbol{c}^{\mathrm{T}}\boldsymbol{A}\boldsymbol{b} \dots \boldsymbol{c}^{\mathrm{T}}\boldsymbol{A}^{n-1}\boldsymbol{b} \end{bmatrix} \tag{3.60}$$

must be non zero.

The controllability matrix of the controllable form (3.46) has the special form

$$\boldsymbol{M}_{\mathrm{c}}^{\mathrm{c}} = \begin{bmatrix} 1 & a_1 & a_2 & \dots & a_{n-1} \\ 0 & 1 & a_1 & \dots & a_{n-2} \\ \vdots & \vdots & \vdots & \ddots & \vdots \\ 0 & 0 & 0 & \dots & a_1 \\ 0 & 0 & 0 & \dots & 1 \end{bmatrix}^{-1}, \tag{3.61}$$

which can be seen very easily by taking the product $\boldsymbol{M}_{\mathrm{c}}^{\mathrm{c}}\left(\boldsymbol{M}_{\mathrm{c}}^{\mathrm{c}}\right)^{-1}$

$$\begin{bmatrix} \boldsymbol{b}_{\mathrm{c}} & \boldsymbol{A}_{\mathrm{c}}\boldsymbol{b}_{\mathrm{c}} & \dots & (\boldsymbol{A}_{\mathrm{c}})^{n-1}\boldsymbol{b}_{\mathrm{c}} \end{bmatrix} \begin{bmatrix} 1 & a_1 & a_2 & \dots & a_{n-1} \\ 0 & 1 & a_1 & \dots & a_{n-2} \\ \vdots & \vdots & \vdots & \ddots & \vdots \\ 0 & 0 & 0 & \dots & a_1 \\ 0 & 0 & 0 & \dots & 1 \end{bmatrix} \tag{3.62}$$

$$= \begin{bmatrix} \boldsymbol{w}_{\mathrm{o}} & \boldsymbol{w}_1 & \boldsymbol{w}_2 & \dots & \boldsymbol{w}_{n-1} \end{bmatrix}$$

Based on the special construction of the state matrices $\boldsymbol{A}_{\mathrm{c}}$ and $\boldsymbol{b}_{\mathrm{c}}$ [see (3.47)], it can be seen that

$$\begin{aligned} \boldsymbol{w}_{\mathrm{o}} &= \boldsymbol{b}_{\mathrm{c}} \\ \boldsymbol{w}_1 &= a_1\boldsymbol{b}_{\mathrm{c}} + \boldsymbol{A}_{\mathrm{c}}\boldsymbol{b}_{\mathrm{c}} \\ &\vdots \\ \boldsymbol{w}_{n-1} &= a_{n-1}\boldsymbol{b}_{\mathrm{c}} + a_{n-2}\boldsymbol{A}_{\mathrm{c}}\boldsymbol{b}_{\mathrm{c}} + \cdots + (\boldsymbol{A}_{\mathrm{c}})^{n-1}\boldsymbol{b}_{\mathrm{c}} \end{aligned} \tag{3.63}$$

where the following recursive relationship holds:

$$\boldsymbol{w}_k = a_k \boldsymbol{b}_{\mathrm{c}} + \boldsymbol{A}_{\mathrm{c}} \boldsymbol{w}_{k-1}. \tag{3.64}$$

The use of the recursive relationship

$$\begin{bmatrix} \boldsymbol{w}_{\mathrm{o}} & \boldsymbol{w}_1 & \boldsymbol{w}_2 & \dots & \boldsymbol{w}_{n-1} \end{bmatrix} = \begin{bmatrix} 1 & 0 & 0 & \dots & 0 \\ 0 & 1 & 0 & \dots & 0 \\ 0 & 0 & 1 & \dots & 0 \\ \vdots & \vdots & \vdots & \ddots & \vdots \\ 0 & 0 & 0 & \dots & 1 \end{bmatrix} = \boldsymbol{I} \tag{3.65}$$

proves the validity of (3.61).

The special matrix $\boldsymbol{M}_{\mathrm{c}}^{\mathrm{c}}$ obtained by the controllable canonical form—which is always derived from the transfer function—is regular, since it is the inverse of a regular matrix. (The determinant of a triangular matrix is the product of its diagonal elements, which is now equal to one.) The name of this canonical form comes from the above features, where only the observability (see later) can be investigated by the pair $\boldsymbol{A}_{\mathrm{c}}; \boldsymbol{c}_{\mathrm{c}}^{\mathrm{T}}$.

It is an interesting question how the linear transformation $\boldsymbol{z} = \boldsymbol{T}\boldsymbol{x}$ influences the controllability matrix. Based on (3.29), one can write

$$\begin{aligned} \tilde{\boldsymbol{b}} &= \boldsymbol{T}\boldsymbol{b} \\ \tilde{\boldsymbol{A}}\tilde{\boldsymbol{b}} &= \boldsymbol{T}\boldsymbol{A}\boldsymbol{T}^{-1}\boldsymbol{T}\boldsymbol{b} = \boldsymbol{T}\boldsymbol{A}\boldsymbol{b} \\ &\vdots \\ \tilde{\boldsymbol{A}}^{n-1}\tilde{\boldsymbol{b}} &= \boldsymbol{T}\boldsymbol{A}^{n-1}\boldsymbol{b} \end{aligned} \tag{3.66}$$

on the basis of which it follows that

$$\tilde{\boldsymbol{M}}_{\mathrm{c}} = \begin{bmatrix} \tilde{\boldsymbol{b}} & \tilde{\boldsymbol{A}}\tilde{\boldsymbol{b}} & \dots & \tilde{\boldsymbol{A}}^{n-1}\tilde{\boldsymbol{b}} \end{bmatrix} = \boldsymbol{T}\begin{bmatrix} \boldsymbol{b} & \boldsymbol{A}\boldsymbol{b} & \dots & \boldsymbol{A}^{n-1}\boldsymbol{b} \end{bmatrix} = \boldsymbol{T}\boldsymbol{M}_{\mathrm{c}} \tag{3.67}$$

Based on the above form of the controllability matrix, any controllable systems can be rewritten into controllable canonical form by using the transformation matrix $\boldsymbol{T}_{\mathrm{c}} = \boldsymbol{M}_{\mathrm{c}}^{\mathrm{c}}(\boldsymbol{M}_{\mathrm{c}})^{-1}$.

The controllability matrix, however, is not always derived from the transfer function. In this case, of course, the direct investigation of the controllability matrix $\boldsymbol{M}_{\mathrm{c}}$ is required.

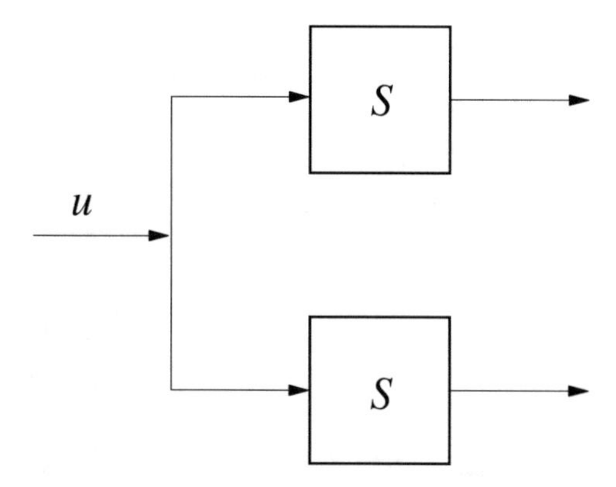

Fig. 3.2 A non-controllable system

Example 3.2 The complete state-equation of a system (see the block diagram in Fig. 3.2) consisting of identical first order sub-systems is

$$\frac{\mathrm{d}\boldsymbol{x}}{\mathrm{d}t} = \begin{bmatrix} -1 & 0 \\ 0 & -1 \end{bmatrix} \boldsymbol{x} + \begin{bmatrix} 1 \\ 1 \end{bmatrix} u = \boldsymbol{A}\boldsymbol{x} + \boldsymbol{b}u. \tag{3.68}$$

The controllability matrix is

$$\boldsymbol{M}_{\mathrm{c}} = [\boldsymbol{b} \quad \boldsymbol{A}\boldsymbol{b}] = \begin{bmatrix} 1 & -1 \\ 1 & -1 \end{bmatrix} \tag{3.69}$$

which is singular, so the system is not controllable. ■

An other essential question of control is, whether each state variable can be observed by measuring the output. This question can be answered by the observability theorem introduced by KALMAN.

Observability—being related to the controllability—gives an answer to the question, whether the initial state at the starting point of the measurements can be reconstructed by measuring the input and output signals of a system of unknown state during a certain time. The system is observable if $\boldsymbol{x}(t_o)$ can be determined from the signals $y(t)$ and $u(t)$ observed in the interval $t_o < t < t_v$.

It is enough to perform the investigation only for $u(t) \equiv 0$, i.e., for the motion generated by the initial values. Observability can be diagnosed in the most easiest way in canonical coordinates. Two criteria have to be fulfilled: the signal y must depend on all canonical state variables; and the poles of the systems must be different. Thus if any γ_i in (3.38) is zero, the output does not have any information concerning the given canonical state variable, so it cannot be reconstructed from the measurements. This means that there is no observation which would have parallel component, to the eigenvector belonging to the eigenvalue λ_i, only perpendicular component, so the effect of the observation always remains in the plane perpendicular to the eigenvector.

In coordinates different from the canonical ones, the above conditions can not directly be recognized due to the interrelationships between the state variables, therefore they have to be replaced by more general criteria.

In the discussion of controllability, the controllability of the state variables was investigated and the output was disregarded. Here in the discussion of observability, the input is disregarded, as was mentioned earlier. Consider the following system.

$$\begin{aligned} \frac{\mathrm{d}\boldsymbol{x}}{\mathrm{d}t} &= \boldsymbol{A}\boldsymbol{x} \\ y &= \boldsymbol{c}^{\mathrm{T}}\boldsymbol{x} \end{aligned} \tag{3.70}$$

By consecutive differentiations of the output the equation

$$\left[y, \frac{\mathrm{d}y}{\mathrm{d}t}, \ldots, \frac{\mathrm{d}^{n-1}y}{\mathrm{d}t^{n-1}}\right]^{\mathrm{T}} = \begin{bmatrix} \boldsymbol{c}^{\mathrm{T}} \\ \boldsymbol{c}^{\mathrm{T}}\boldsymbol{A} \\ \vdots \\ \boldsymbol{c}^{\mathrm{T}}\boldsymbol{A}^{n-1} \end{bmatrix} \boldsymbol{x} \tag{3.71}$$

is obtained and the state vector can be unambiguously determined from the output and its derivatives, if the *observability matrix*

$$\boldsymbol{M}_{\mathrm{o}} = \begin{bmatrix} \boldsymbol{c}^{\mathrm{T}} \\ \boldsymbol{c}^{\mathrm{T}}\boldsymbol{A} \\ \vdots \\ \boldsymbol{c}^{\mathrm{T}}\boldsymbol{A}^{n-1} \end{bmatrix} \tag{3.72}$$

has n linearly independent rows. Thus $\boldsymbol{M}_{\mathrm{o}}$ must be invertible and regular. Since $\boldsymbol{M}_{\mathrm{o}}$ depends on $\boldsymbol{A}$ and $\boldsymbol{c}^{\mathrm{T}}$, this problem is used to be cited as the observability of the pair $\boldsymbol{A};\boldsymbol{c}^{\mathrm{T}}$. (In (3.71)—due to the CAYLEY-HAMILTON theorem—there is no need to compute derivates of higher order than $(n-1)$, see A.3.3 of Appendix 5.).

The observability matrix of the observable canonical form is very special

$$\boldsymbol{M}_{\mathrm{o}}^{\mathrm{o}} = \begin{bmatrix} 1 & 0 & \ldots & 0 & 0 \\ a_1 & 1 & \ldots & 0 & 0 \\ a_2 & a_1 & \ddots & 0 & 0 \\ \vdots & \vdots & \ldots & 1 & 0 \\ a_{n-1} & a_{n-2} & \ldots & a_1 & 1 \end{bmatrix}^{-1} \tag{3.73}$$

which can be seen very easily, if the product $\left(\boldsymbol{M}_{\mathrm{o}}^{\mathrm{o}}\right)^{-1}\boldsymbol{M}_{\mathrm{o}}$ is computed, i.e.

$$\begin{bmatrix} 1 & 0 & \dots & 0 & 0 \\ a_1 & 1 & \dots & 0 & 0 \\ a_2 & a_1 & \ddots & 0 & 0 \\ \vdots & \vdots & \dots & 1 & 0 \\ a_{n-1} & a_{n-2} & \dots & a_1 & 1 \end{bmatrix} \begin{bmatrix} \mathbf{c}_{\mathrm{o}}^{\mathrm{T}} \\ \mathbf{c}_{\mathrm{o}}^{\mathrm{T}}\mathbf{A}_{\mathrm{o}} \\ \vdots \\ \mathbf{c}_{\mathrm{o}}^{\mathrm{T}}(\mathbf{A}_{\mathrm{o}})^{n-1} \end{bmatrix} = \begin{bmatrix} \mathbf{w}_{\mathrm{o}}^{\mathrm{T}} \\ \mathbf{w}_1^{\mathrm{T}} \\ \vdots \\ \mathbf{w}_{n-1}^{\mathrm{T}} \end{bmatrix} \tag{3.74}$$

Based on the special construction [see (3.54)] of the system matrices, one has

$$\begin{aligned} \mathbf{w}_{\mathrm{o}}^{\mathrm{T}} &= \mathbf{c}_{\mathrm{o}}^{\mathrm{T}} \\ \mathbf{w}_1^{\mathrm{T}} &= a_1\mathbf{c}_{\mathrm{o}}^{\mathrm{T}} + \mathbf{c}_{\mathrm{o}}^{\mathrm{T}}\mathbf{A}_{\mathrm{o}} \\ &\vdots \\ \mathbf{w}_{n-1}^{\mathrm{T}} &= a_{n-1}\mathbf{c}_{\mathrm{o}}^{\mathrm{T}} + a_{n-2}\mathbf{c}_{\mathrm{o}}^{\mathrm{T}}\mathbf{A}_{\mathrm{o}} + \ldots + \mathbf{c}_{\mathrm{o}}^{\mathrm{T}}(\mathbf{A}_{\mathrm{o}})^{n-1} \end{aligned} \tag{3.75}$$

where there exists the following recursive relationship

$$\mathbf{w}_k^{\mathrm{T}} = a_k\mathbf{c}_{\mathrm{o}}^{\mathrm{T}} + \mathbf{w}_{k-1}^{\mathrm{T}}\mathbf{A}_{\mathrm{o}} \tag{3.76}$$

Using this recursive relationship

$$\begin{bmatrix} \mathbf{w}_{\mathrm{o}}^{\mathrm{T}} \\ \mathbf{w}_1^{\mathrm{T}} \\ \vdots \\ \mathbf{w}_{n-1}^{\mathrm{T}} \end{bmatrix} = \begin{bmatrix} 1 & 0 & 0 & \dots & 0 \\ 0 & 1 & 0 & \dots & 0 \\ 0 & 0 & 1 & \dots & 0 \\ \vdots & \vdots & \vdots & \ddots & \vdots \\ 0 & 0 & 0 & \dots & 1 \end{bmatrix} = \mathbf{I} \tag{3.77}$$

which proves the validity of (3.73).

The special $\mathbf{M}_{\mathrm{o}}^{\mathrm{o}}$ obtained by the observable canonical form—which is always derived from the transfer function—is always regular, since it is the inverse of a regular matrix. (The determinant of a triangle matrix is the product of the diagonal elements, which is now equal to one.) The name of this canonical form comes from the above features, where only the controllability can be investigated by the pair $\mathbf{A}_{\mathrm{o}}; \mathbf{b}_{\mathrm{o}}$.

It is an interesting question how the linear transformation $\mathbf{z} = \mathbf{T}\mathbf{x}$ influences the observability matrix. Based on (3.29), one has that

$$\begin{aligned} \tilde{\mathbf{c}}^{\mathrm{T}} &= \mathbf{c}^{\mathrm{T}}\mathbf{T}^{-1} \\ \tilde{\mathbf{c}}^{\mathrm{T}}\tilde{\mathbf{A}} &= \mathbf{c}^{\mathrm{T}}\mathbf{T}^{-1}\mathbf{T}\mathbf{A}\mathbf{T}^{-1} = \mathbf{c}^{\mathrm{T}}\mathbf{A}\mathbf{T}^{-1} \\ &\vdots \\ \tilde{\mathbf{c}}^{\mathrm{T}}\tilde{\mathbf{A}}^{n-1} &= \mathbf{c}^{\mathrm{T}}\mathbf{A}^{n-1}\mathbf{T}^{-1} \end{aligned} \tag{3.78}$$

In matrix form, this is

$$\tilde{\boldsymbol{M}}_{\mathrm{o}} = \begin{bmatrix} \tilde{\boldsymbol{c}}^{\mathrm{T}} \\ \tilde{\boldsymbol{c}}^{\mathrm{T}}\tilde{\boldsymbol{A}} \\ \vdots \\ \tilde{\boldsymbol{c}}^{\mathrm{T}}\tilde{\boldsymbol{A}}^{n-1} \end{bmatrix} = \begin{bmatrix} \boldsymbol{c}^{\mathrm{T}} \\ \boldsymbol{c}^{\mathrm{T}}\boldsymbol{A} \\ \vdots \\ \boldsymbol{c}^{\mathrm{T}}\boldsymbol{A}^{n-1} \end{bmatrix} \boldsymbol{T}^{-1} = \boldsymbol{M}_{\mathrm{o}}\boldsymbol{T}^{-1} \tag{3.79}$$

Based on the above form of the observability matrix, any observable system can be rewritten in observable canonical form by using the transformation matrix $\boldsymbol{T}_{\mathrm{o}}^{-1} = (\boldsymbol{M}_{\mathrm{o}})^{-1}\boldsymbol{M}_{\mathrm{o}}^{\mathrm{o}}$ (i.e., $\boldsymbol{T}_{\mathrm{o}} = \left(\boldsymbol{M}_{\mathrm{o}}^{\mathrm{o}}\right)^{-1}\boldsymbol{M}_{\mathrm{o}}$).

The observability matrix, however, is not always derived from the transfer function. In this case, of course, the direct investigation of the observability matrix $\boldsymbol{M}_{\mathrm{o}}$ is required.

Example 3.3 The complete state-equation of a system consisting of identical first order subsystems (see the block diagram in Fig. 3.3) is

$$\frac{\mathrm{d}\boldsymbol{x}}{\mathrm{d}t} = \begin{bmatrix} -1 & 0 \\ 0 & -1 \end{bmatrix} \boldsymbol{x} = \boldsymbol{A}\boldsymbol{x}$$
$$y = \begin{bmatrix} 1 & 1 \end{bmatrix} = \boldsymbol{c}^{\mathrm{T}}\boldsymbol{x} \tag{3.80}$$

The observability matrix is

$$\boldsymbol{M}_{\mathrm{o}} = \begin{bmatrix} 1 & 1 \\ -1 & -1 \end{bmatrix} \tag{3.81}$$

which is singular, thus the system is not observable. ■

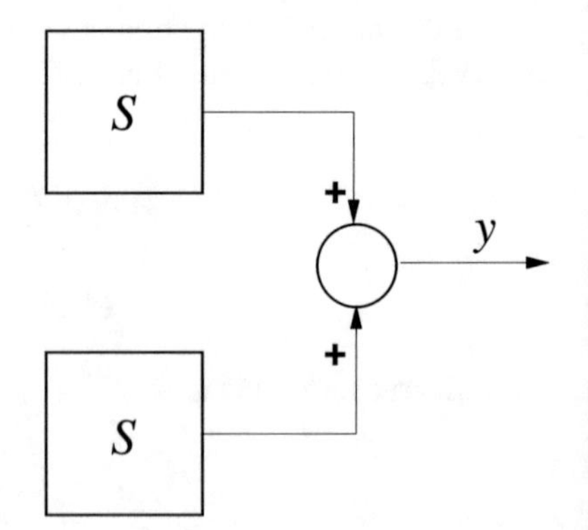

Fig. 3.3 A non-observable system

3.4.1 The Kalman Decomposition

The concepts of controllability and observability make it possible to understand the structure of a linear system. Remember that the space of controllable states is the sub-space defined by the columns of the controllability matrix. If its dimension is n, then the whole space is controllable. Let us introduce the notation $\boldsymbol{x}_\mathrm{c}$ for the controllable states, and $\boldsymbol{x}_{\bar{\mathrm{c}}}$ for the non-controllable states. In this case the state-equation is

$$\frac{\mathrm{d}}{\mathrm{d}t}\begin{bmatrix}\boldsymbol{x}_\mathrm{c}\\ \boldsymbol{x}_{\bar{\mathrm{c}}}\end{bmatrix}=\begin{bmatrix}\boldsymbol{A}_{11} & \boldsymbol{A}_{12}\\ \boldsymbol{0} & \boldsymbol{A}_{22}\end{bmatrix}\begin{bmatrix}\boldsymbol{x}_\mathrm{c}\\ \boldsymbol{x}_{\bar{\mathrm{c}}}\end{bmatrix}+\begin{bmatrix}\boldsymbol{b}_1\\ \boldsymbol{0}\end{bmatrix}u \tag{3.82}$$

where it can be clearly seen from the structure that the states $\boldsymbol{x}_{\bar{\mathrm{c}}}$ cannot be influenced by u. Similarly, let us introduce the notation $\boldsymbol{x}_\mathrm{o}$ for the observable states and $\boldsymbol{x}_{\bar{\mathrm{o}}}$ for the non-observable states. Then the state-equation

$$\begin{aligned}\frac{\mathrm{d}}{\mathrm{d}t}\begin{bmatrix}\boldsymbol{x}_\mathrm{o}\\ \boldsymbol{x}_{\bar{\mathrm{o}}}\end{bmatrix}&=\begin{bmatrix}\boldsymbol{A}_{11} & \boldsymbol{0}\\ \boldsymbol{A}_{21} & \boldsymbol{A}_{22}\end{bmatrix}\begin{bmatrix}\boldsymbol{x}_\mathrm{o}\\ \boldsymbol{x}_{\bar{\mathrm{o}}}\end{bmatrix}\\ y&=\begin{bmatrix}\boldsymbol{c}_1^\mathrm{T} & \boldsymbol{0}^\mathrm{T}\end{bmatrix}\begin{bmatrix}\boldsymbol{x}_\mathrm{o}\\ \boldsymbol{x}_{\bar{\mathrm{o}}}\end{bmatrix}\end{aligned} \tag{3.83}$$

is obtained, where it can be well seen, that there is no component in the output for the states $\boldsymbol{x}_{\bar{\mathrm{o}}}$.

A linear system can be decomposed into four sub-systems:

- S_co controllable and observable $\boldsymbol{x}_\mathrm{co}$
- $S_{\mathrm{c}\bar{\mathrm{o}}}$ controllable and non-observable $\boldsymbol{x}_{\mathrm{c}\bar{\mathrm{o}}}$
- $S_{\bar{\mathrm{c}}\mathrm{o}}$ non-controllable and observable $\boldsymbol{x}_{\bar{\mathrm{c}}\mathrm{o}}$
- $S_{\overline{\mathrm{co}}}$ non-controllable and non-observable $\boldsymbol{x}_{\overline{\mathrm{co}}}$

where the corresponding state variables are also presented (the entering arrows mean the effect of the input and the regarding state sub-system). The complete Kalman decomposition of the linear system is

$$\begin{aligned}\frac{\mathrm{d}}{\mathrm{d}t}\begin{bmatrix}\boldsymbol{x}_\mathrm{co}\\ \boldsymbol{x}_{\mathrm{c}\bar{\mathrm{o}}}\\ \boldsymbol{x}_{\bar{\mathrm{c}}\mathrm{o}}\\ \boldsymbol{x}_{\overline{\mathrm{co}}}\end{bmatrix}&=\begin{bmatrix}\boldsymbol{A}_{11} & \boldsymbol{0} & \boldsymbol{A}_{13} & \boldsymbol{0}\\ \boldsymbol{A}_{21} & \boldsymbol{A}_{22} & \boldsymbol{A}_{23} & \boldsymbol{A}_{24}\\ \boldsymbol{0} & \boldsymbol{0} & \boldsymbol{A}_{33} & \boldsymbol{0}\\ \boldsymbol{0} & \boldsymbol{0} & \boldsymbol{A}_{43} & \boldsymbol{A}_{44}\end{bmatrix}\begin{bmatrix}\boldsymbol{x}_\mathrm{co}\\ \boldsymbol{x}_{\mathrm{c}\bar{\mathrm{o}}}\\ \boldsymbol{x}_{\bar{\mathrm{c}}\mathrm{o}}\\ \boldsymbol{x}_{\overline{\mathrm{co}}}\end{bmatrix}+\begin{bmatrix}\boldsymbol{b}_1\\ \boldsymbol{b}_2\\ \boldsymbol{0}\\ \boldsymbol{0}\end{bmatrix}u=\boldsymbol{A}\boldsymbol{x}+\boldsymbol{b}u\\ y&=\begin{bmatrix}\boldsymbol{c}_1^\mathrm{T} & \boldsymbol{0}^\mathrm{T} & \boldsymbol{c}_2^\mathrm{T} & \boldsymbol{0}^\mathrm{T}\end{bmatrix}\boldsymbol{x}\end{aligned} \tag{3.84}$$

The block diagram representing each sub-system is shown in Fig. 3.4. Following the arrows of the block diagram it can be seen that the input influences the

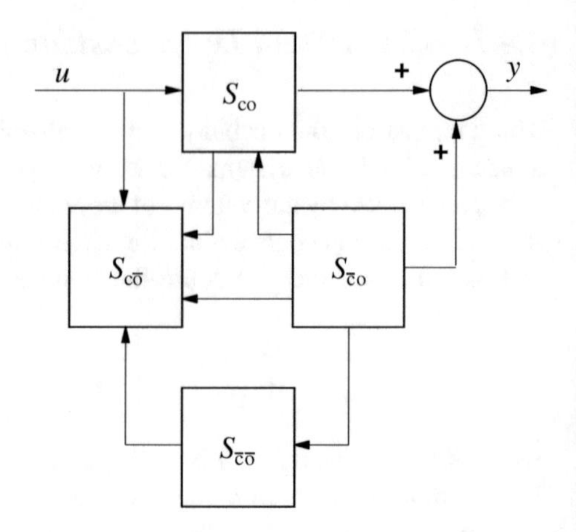

Fig. 3.4 KALMAN decomposition of the linear system

sub-systems S_{co} and $S_{c\bar{o}}$, but the output depends only on the sub-systems S_{co} and $S_{\bar{c}o}$. The sub-system $S_{\overline{co}}$ does not belong either to the input or to the output.

The transfer function of the entire system can be obtained by simple computation

$$P(s) = \boldsymbol{c}_1^{\mathrm{T}}(s\boldsymbol{I} - \boldsymbol{A}_{11})^{-1}\boldsymbol{b}_1, \tag{3.85}$$

i.e., it is completely determined by the sub-system S_{co}. In contrast it can be stated that only the controllable and observable sub-system of the whole system can be determined from the transfer function.

3.4.2 *The Effect of Common Poles and Zeros*

A very old problem of control, namely the canceling of poles and zeros, can be explained by the KALMAN decomposition. To illustrate it let us consider the following example.

Example 3.4 Let the transfer function of the process be

$$P(s) = \frac{Y(s)}{U(s)} = \frac{s-1}{s-1} = 1, \tag{3.86}$$

i.e., the numerator and the denominator have common roots, thus one zero and one pole are equal. In this case a common root, at the same time, means also an unstable pole. It can be seen easily that the following differential equation corresponds formally to the transfer function (3.86):

$$\frac{\mathrm{d}y}{\mathrm{d}t} - y = \frac{\mathrm{d}u}{\mathrm{d}t} - u. \tag{3.87}$$

The solution obtained by integration of the differential equation is

$$y(t) = u(t) + ce^t \tag{3.88}$$

where c is a constant. In the method of canceling, it must never be forgotten that the complete solution of the state-equation is performed according to (3.18), which contains also the initial condition, whose dynamics (an un-excited system) depends on the poles of the whole system, even also on the possibly cancelled pole. If this pole is unstable, then its non-disappearing effect occurs unpleasantly in the solution.

The trivial system $y(t) = u(t)$ obtained from (3.86) after the pole cancellation is obviously not equal to (3.88). The Eq. (3.86) can be brought to the following form

$$P(s) = b_\mathrm{o} + \frac{b_1}{s-1} = d + \frac{b_1}{s-1} = 1 + \frac{0}{s-1} \tag{3.89}$$

Based on this the controllable canonical form can be easily written as

$$\frac{\mathrm{d}x_1}{\mathrm{d}t} = x_1 + u; \quad y = u \tag{3.90}$$

which is not observable, and an observable canonical form can also be defined as

$$\frac{\mathrm{d}x_2}{\mathrm{d}t} = x_2; \quad y = x_2 + u \tag{3.91}$$

which is not controllable. ■

The KALMAN-form of the whole system corresponding to Eqs. (3.90) and (3.91) is shown in Fig. 3.5, which consists of the sub-systems $S_{\mathrm{c}\bar{\mathrm{o}}}$, $S_{\bar{\mathrm{c}}\mathrm{o}}$ and S_{co}. The S_{co} is a static system with transfer function $P(s) = 1$. The $S_{\mathrm{c}\bar{\mathrm{o}}}$ is a non-observable but controllable subsystem, while $S_{\bar{\mathrm{c}}\mathrm{o}}$ is not controllable, but is an observable sub-system.

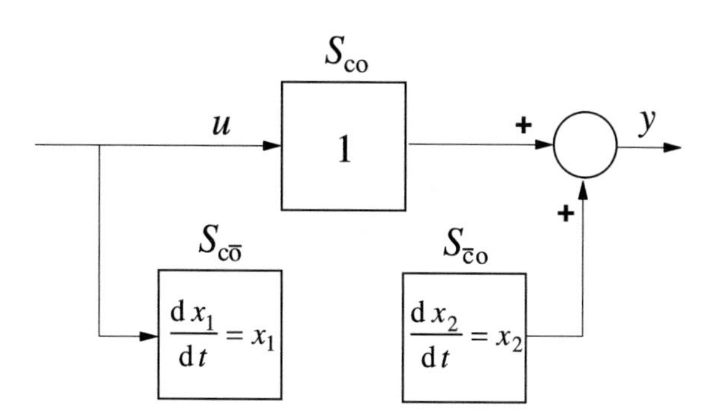

Fig. 3.5 The complete KALMAN-form of the system of transfer function (3.86)

Note that if the transfer function of the system is given, then first the common divisors of the numerator and denominator have to be investigated. The common factor can only be a common root. It is reasonable to continue the simplification until there are no more common divisors. Such polynomials are called relatively prime. A transfer function $P(s) = \mathcal{B}(s)/\mathcal{A}(s)$ is called irreducible (i.e., can not be simplified) if the polynomials $\mathcal{A}(s)$ and $\mathcal{B}(s)$ are relatively prime, which is an algebraic condition for the special DIOPHANTINE (or BEZOUT) equation

$$\mathcal{A}(s)\mathcal{X}(s) + \mathcal{B}(s)\mathcal{Y}(s) = 1 \tag{3.92}$$

to have a solution, i.e., the corresponding SILVESTER matrix must be regular (see more details in Chap. 9).

If a transfer function is not reducible, the related state-equation corresponds to the controllable and observable sub-system S_{co} of the KALMAN-form and the other sub-systems do not exist.

Equations (3.90) and (3.91) can be generalized to the case when the controllable and observable system $S_{co}\{\mathbf{A}; \mathbf{b}; \mathbf{c}^T; d\}$ is irreducible, and the numerator and also the denominator of the transfer function $P(s)$ are extended by a common factor $(s - p)$ referring to a real pole. For this general case the state-equation of the redundant non-controllable and non-observable system can be given as

$$\dot{\mathbf{x}}_r = \begin{bmatrix} \dot{\mathbf{x}} \\ \dot{x}_1 \\ \dot{x}_2 \end{bmatrix} = \begin{bmatrix} \mathbf{A} & \mathbf{0} & \mathbf{0} \\ \mathbf{0}^T & p & 0 \\ \mathbf{0}^T & 0 & p \end{bmatrix} \begin{bmatrix} \mathbf{x} \\ x_1 \\ x_2 \end{bmatrix} + \begin{bmatrix} \mathbf{b} \\ 1 \\ 0 \end{bmatrix} u = \mathbf{A}_r \mathbf{x}_r + \mathbf{b}_r u \quad . \tag{3.93}$$

$$y = \begin{bmatrix} \mathbf{c}^T & 0 & 1 \end{bmatrix} \mathbf{x}_r + du = \mathbf{c}_r^T \mathbf{x}_r + du$$

Example 3.5 Assume that the transfer function of the process is

$$P(s) = \frac{2(s+1)}{(s+1)(s+2)} = \frac{2s+2}{s^2+3s+2} = \frac{b_1 s + 2}{s^2+3s+2}.$$

It is easy to form the parameter matrices of the controllable canonical form, which are

$$\mathbf{A}_c = \begin{bmatrix} -3 & -2 \\ 1 & 0 \end{bmatrix}; \quad \mathbf{b}_c = \begin{bmatrix} 1 \\ 0 \end{bmatrix} \quad \text{and} \quad \mathbf{c}_c^T = [2 \quad 2].$$

The controllability matrix of this canonical form is

$$\mathbf{M}_c^c = [\mathbf{b}_c \quad \mathbf{A}_c \mathbf{b}_c] = \begin{bmatrix} 1 & -3 \\ 0 & 1 \end{bmatrix}.$$

It is easy to check that the determinant of this matrix is $\det\left(\boldsymbol{M}_{\mathrm{c}}^{\mathrm{c}}\right)=1$, as a consequence the process is controllable. The observability matrix of this canonical form is

$$\boldsymbol{M}_{\mathrm{o}}^{\mathrm{c}}=\begin{bmatrix}\boldsymbol{c}_{\mathrm{c}}^{\mathrm{T}}\\ \boldsymbol{c}_{\mathrm{c}}^{\mathrm{T}}\boldsymbol{A}_{\mathrm{c}}\end{bmatrix}=\begin{bmatrix}2 & 2\\ -4 & -4\end{bmatrix}.$$

Note that the determinant of this matrix is $\det\left(\boldsymbol{M}_{\mathrm{o}}^{\mathrm{c}}\right)=0$; as a consequence the process is not observable.

Now form the parameter matrices of the observable canonical form, which are

$$\boldsymbol{A}_{\mathrm{o}}=\begin{bmatrix}-3 & 1\\ -2 & 0\end{bmatrix};\quad \boldsymbol{b}_{\mathrm{o}}=\begin{bmatrix}2\\ 2\end{bmatrix}\quad \text{and}\quad \boldsymbol{c}_{\mathrm{o}}^{\mathrm{T}}=[1\quad 0].$$

The controllability matrix of this canonical form is

$$\boldsymbol{M}_{\mathrm{c}}^{\mathrm{o}}=[\boldsymbol{b}_{\mathrm{o}}\quad \boldsymbol{A}_{\mathrm{o}}\boldsymbol{b}_{\mathrm{o}}]=\begin{bmatrix}2 & -4\\ 2 & -4\end{bmatrix}.$$

It is easy to check that the determinant of this matrix is $\det\left(\boldsymbol{M}_{\mathrm{c}}^{\mathrm{o}}\right)=0$; as a consequence the process is not controllable. The observability matrix of this canonical form is

$$\boldsymbol{M}_{\mathrm{o}}^{\mathrm{o}}=\begin{bmatrix}\boldsymbol{c}_{\mathrm{o}}^{\mathrm{T}}\\ \boldsymbol{c}_{\mathrm{o}}^{\mathrm{T}}\boldsymbol{A}_{\mathrm{o}}\end{bmatrix}=\begin{bmatrix}1 & 0\\ -3 & 1\end{bmatrix}.$$

Note that the determinant of this matrix is $\det\left(\boldsymbol{M}_{\mathrm{o}}^{\mathrm{o}}\right)=1$; as a consequence the process is observable. This example shows and explains very nicely the meaning of the above matrices.

Observe and check that the irreducible equivalent transfer function

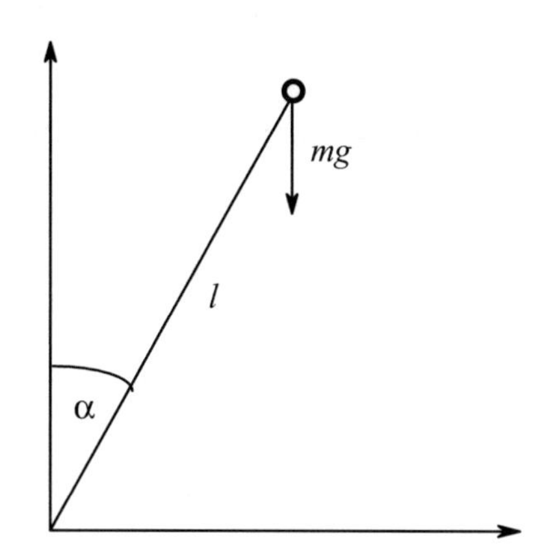

Fig. 3.6 Scheme of the inverted pendulum

$$P(s) = \frac{2(s+1)}{(s+1)(s+2)} = \frac{2}{s+2}$$

is already controllable and observable. ■

3.4.3 The Inverted Pendulum

Next the simplest case of the moving inverted pendulum shown in Chap. 2.6 is investigated, i.e., when the suspension of the pendulum is fixed. Via this example almost all of the methods of this Chapter from the linear modeling to the investigation of the controllability and observability issues can be demonstrated.

In order to determine the state-equation of the inverted pendulum, given in a simple schematic form in Fig. 3.6, introduce the following state variables: $x_2 = \mathrm{d}\alpha/\mathrm{d}t$ and $x_1 = \alpha$ (the angular velocity and the angular position). From the equality of the moments calculated for the center of the angular position it follows that

$$J\frac{\mathrm{d}^2\alpha}{\mathrm{d}t^2} = mgl\sin(\alpha) + mugl\cos(\alpha), \tag{3.94}$$

where it is assumed that the mass m is concentrated at the end of an ideal, weightless pendulum of length l, the inertia relating to the center of the rotation is denoted by J. The actuating signal is the horizontal acceleration of the value ug (measured in g), the output is the angular position α. The non-linear state-equation is obtained as

$$\begin{aligned} \frac{\mathrm{d}\boldsymbol{x}}{\mathrm{d}t} &= \dot{\boldsymbol{x}} = \boldsymbol{f}(\boldsymbol{x}, u) = \begin{bmatrix} \mathrm{d}\alpha/\mathrm{d}t \\ \frac{mgl}{J}\sin(\mathrm{d}\alpha/\mathrm{d}t) + \frac{mglu}{J}\cos(\alpha) \end{bmatrix} = \begin{bmatrix} x_2 \\ \sin(x_1) + u\cos(x_1) \end{bmatrix} \\ y &= \alpha = x_1 \end{aligned} \tag{3.95}$$

where choosing $\sqrt{J/mgl}$ as a time unit, the last, normalized form in Eq. (3.95) results. Thus the state-equation is a nonlinear, time-invariant, second order vector differential equation.

Let us linearize the equation in the case of zero actuating signal. The equilibrium point is

$$\dot{\boldsymbol{x}} = \boldsymbol{0} = f(u = 0) = \begin{bmatrix} x_2 \\ \sin(x_1) \end{bmatrix} = \begin{bmatrix} 0 \\ 0 \end{bmatrix}; \quad \begin{matrix} x_2 = \mathrm{d}\alpha/\mathrm{d}t = 0 \\ \sin(x_1) = \sin(\alpha) = 0 \end{matrix} \tag{3.96}$$

where $\alpha = 0$ and $\alpha = \pi$. At the first equilibrium point the pendulum is in the position upside, but in the second it is in down-side. Determining the derivatives with respect to $\boldsymbol{x}$ and u of the function $\boldsymbol{f}(\boldsymbol{x}, u)$ yields

$$\frac{\mathrm{d}\boldsymbol{f}(\boldsymbol{x},u)}{\mathrm{d}\,\boldsymbol{x}} = \begin{bmatrix} 0 & 1 \\ \cos(x_1) - u\sin(x_1) & 0 \end{bmatrix} \quad \text{and} \quad \frac{\mathrm{d}\boldsymbol{f}(\boldsymbol{x},u)}{\mathrm{d}\,u} = \begin{bmatrix} 0 \\ \cos(x_1) \end{bmatrix}. \tag{3.97}$$

Evaluating the derivates at the upper point ($u = 0$; $x_1 = 0$ and $x_2 = 0$) the parameter matrices

$$\boldsymbol{A} = \begin{bmatrix} 0 & 1 \\ 1 & 0 \end{bmatrix}; \quad \boldsymbol{b} = \begin{bmatrix} 0 \\ 1 \end{bmatrix} \quad \text{and} \quad \boldsymbol{c}^{\mathrm{T}} = [1 \quad 0] \tag{3.98}$$

are obtained. Computing the transfer function (by using A.1.10) yields

$$\begin{aligned} G(s) &= \boldsymbol{c}^{\mathrm{T}}(s\boldsymbol{I} - \boldsymbol{A})^{-1}\boldsymbol{b} = \frac{1}{\det\begin{bmatrix} s & -1 \\ -1 & s \end{bmatrix}}[1 \quad 0]\begin{bmatrix} s & 1 \\ 1 & s \end{bmatrix}\begin{bmatrix} 0 \\ 1 \end{bmatrix} \\ &= \frac{1}{s^2 - 1}[s \quad 1]\begin{bmatrix} 0 \\ 1 \end{bmatrix} = \frac{1}{s^2 - 1} = \frac{1}{(s+1)(s-1)} \end{aligned} \tag{3.99}$$

The root $s = 1$ shows that at this operating point the system is unstable. It can be easily checked that the controllability matrix

$$\boldsymbol{M}_{\mathrm{c}} = \begin{bmatrix} 0 & 1 \\ 1 & 0 \end{bmatrix} \tag{3.100}$$

is regular, thus the system is controllable. In this case the observability matrix, which is

$$\boldsymbol{M}_{\mathrm{o}} = \begin{bmatrix} 1 & 0 \\ 0 & 1 \end{bmatrix}. \tag{3.101}$$

is also regular, thus the system is observable. Coming from the simplicity of the above task, the DT control of the inverted pendulum is a typical and spectacular laboratory example all over the world for controlling an unstable process. (The complexity of the task increases drastically by placing more pendulums on top of each other.)

Evaluating the derivates in the lower point ($u = 0$; $x_1 = \pi$ and $x_2 = 0$), the parameter matrices

$$\boldsymbol{A} = \begin{bmatrix} 0 & 1 \\ -1 & 0 \end{bmatrix}; \quad \boldsymbol{b} = \begin{bmatrix} 0 \\ -1 \end{bmatrix} \quad \text{and} \quad \boldsymbol{c}^{\mathrm{T}} = [1 \quad 0] \tag{3.102}$$

are obtained. Calculate the transfer function [by using (A.1.10)] yields

$$
\begin{aligned}
G(s) = \boldsymbol{c}^{\mathrm{T}}(s\boldsymbol{I}-\boldsymbol{A})^{-1}\boldsymbol{b} &= \frac{1}{\det\begin{bmatrix} s & -1 \\ 1 & s \end{bmatrix}}[1 \quad 0]\begin{bmatrix} s & 1 \\ -1 & s \end{bmatrix}\begin{bmatrix} 0 \\ -1 \end{bmatrix} \\
&= \frac{1}{s^2+1}[s \quad 1]\begin{bmatrix} 0 \\ -1 \end{bmatrix} = \frac{-1}{s^2+1} = \frac{-1}{(s+j)(s-j)}
\end{aligned} \tag{3.103}
$$

The roots on the imaginary axis indicate that at this operating point the process is an oscillating system without any damping. Do not forget that no kind of damping (e.g., air or ordinary friction) is taken into consideration in the model. Simple computations similar to the above show that even at this operating point the system is controllable and observable.

Chapter 4
Negative Feedback

The aim of a control system is to ensure reference signal tracking as well as disturbance rejection. The control system must not be not very sensitive to measurement noise or to plant/model mismatch.

The designed control system has to ensure various quality specifications. Also, it has to be technically realizable and eligible in terms of economic and other (e.g., environmental protection or safety) viewpoints.

4.1 Control in Open- and Closed-Loop

If, when deciding whether intervention in a process is necessary, the information is taken not from the output of the process but from another source, or a priori knowledge about the process or its environment is used, then the realized structure is called open-loop control (Fig. 4.1). Here P denotes the transfer function of the process (plant), C is the transfer function of the controller (regulator), r denotes the reference signal, y is the output signal, while y_{ni} and y_{no} denote the input and the output disturbances, respectively.

The reference signal tracking would be ideal if the control device realized the inverse of the transfer function of the plant. With open-loop control, the reference signal tracking can be realizable, but open-loop control is not able to reject the effect of the disturbances.

The effect of the measurable output disturbance could be eliminated by feed forward of the disturbance according to Fig. 4.2.

Generally the perfect inverse of the transfer function of the plant is not realizable (If e.g., the process contains dead time, its inverse would mean the prediction of a future output value. The signal transfer is also non-realizable if the degree of the numerator of the inverse transfer function is higher than that of its denominator.). Approximating the inverse of the process can be realized by feeding back an amplifier of high gain K through the transfer function of the plant (Fig. 4.3).

L. Keviczky et al., *Control Engineering*, Advanced Textbooks in Control and Signal Processing, https://doi.org/10.1007/978-981-10-8297-9_4

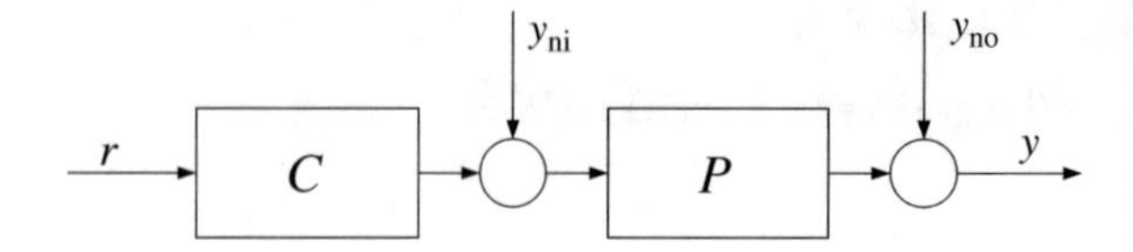

Fig. 4.1 Open-loop control

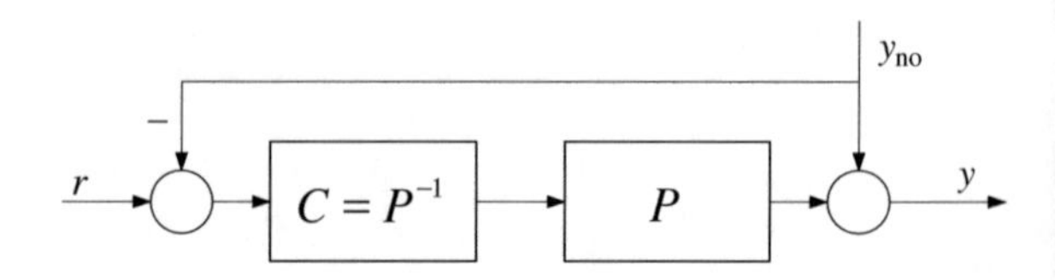

Fig. 4.2 Open-loop control with feed forward

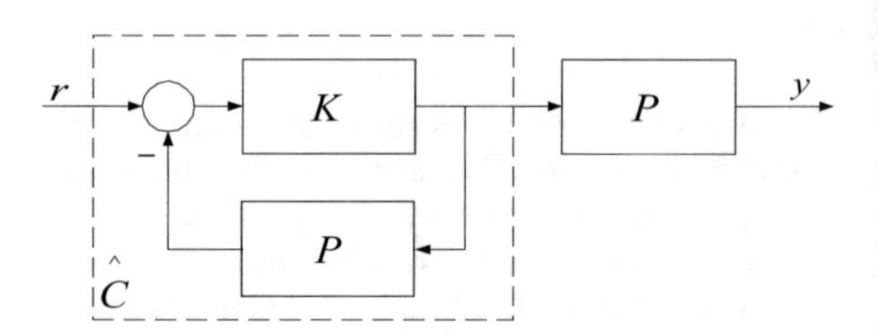

Fig. 4.3 Open-loop control with an element approximating the inverse of the process

Using the above structure the transfer function of the controller is:

$$\hat{C} = \frac{K}{1 + KP(s)} = \frac{1}{\frac{1}{K} + P(s)} \approx \frac{1}{P(s)}. \tag{4.1}$$

The control is realized through negative feedback if the input signal (the manipulated variable) of the process is affected by the difference between the measured output signal and its desired prescribed value. The measured output value is generally noisy because of the noise component y_z released by the measurement equipment. Based on the error signal e the controller C generates the manipulated variable u, which modifies the output signal of the process P. The output signal of the process changes according to the dynamics of the process until it reaches its desired value. Control via negative feedback is called closed-loop control. The block diagram of a closed-loop control system is given in Fig. 4.4. Often the reference signal is filtered by a filter element given by the transfer function F (denoted by the dotted line in the figure).

Comparing Figs. 4.3 and 4.4 shows that the two systems are the same if the disturbances and the measurement noise are not considered, the filter is supposed to be unity ($F = 1$) and in the closed-loop system the controller is proportional, chosen as $C = K$. But as in the closed-loop system the output signal is fed back, not the control signal, in addition to reference signal tracking, the closed-loop system is also able to reject the effects of the disturbances and the measurement noise. Whatever effect is causing a deviation of the output signal from its desired value,

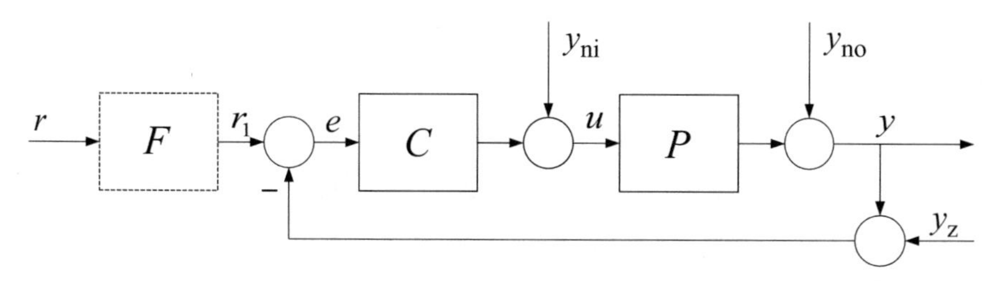

Fig. 4.4 Closed-loop control system

the error signal will be different from zero, creating a control signal to eliminate the deviation.

In a closed-loop control structure the best reference signal tracking is achieved by adjusting the controller C to ensure that the relationship between the control signal u and the reference signal r according to $U(s)/R(s) = C/(1+CP)$ would provide the inverse of the process model. If the exact inverse is non-realizable, then its best realizable approximation can be employed. (It has to be mentioned that generally the inverse of the process could be well approximated only within a given frequency range.)

A comparison of open-loop and closed-loop control was given in Chap. 1. In the sequel, the main properties of closed-loop control will be discussed.

4.2 The Basic Properties of the Closed Control Loop

The main properties of closed-loop control systems will be illustrated through some simple examples.

Stability. A basic requirement for a closed-loop control system is that for a finite change in the input signal it should respond with a finite change in the output signal, i.e. a steady state should be reached. In a control system realized by negative feedback oscillations with steady or increasing amplitudes may occur. The reason for this is that the execution of the decision to change the process output is delayed by the process dynamics. High gains in the control system may increase the unfavorable inertial change of the signals to such an extent that the control system will not be able to reach a steady state. Stability will be discussed in detail in Chap. 5.

Reference signal tracking. With a closed-loop control realized by negative feedback, the output signal should follow the reference signal as accurately as possible. In the control system in Fig. 4.5 the plant is described by a first-order lag and the controller is a proportional element. If a step reference signal is to be tracked, there will be a steady state error in the system, as only a steady constant input signal is able to produce a constant signal value at the output of the first-order lag. The value of the steady state error will be $e_{steady} = 1/(1+A_P A)$. This error will

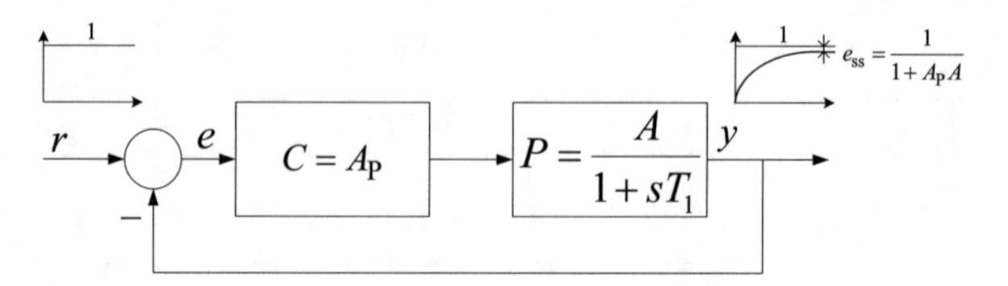

Fig. 4.5 Closed-loop control system with proportional controller

be small if the loop-gain $K = A_P A$ is high. Static accuracy in the steady state could be ensured by applying an integrating controller instead of a proportional controller. The property of an integrator is that its output can be constant only if its input (the error signal) finally has reached the value zero.

Stabilization of an unstable process. An unstable process can be stabilized by negative feedback.

Example 4.1 Let us consider the system given in Fig. 4.6. For unit step input, the output of the closed-loop system for $t \geq 0$ is $y(t) = \mathcal{L}^{-1}\{K/s(s-2)\} = K(e^{2t} - 1)/2$, which tends to infinity if $t \to \infty$. With a proportional negative feedback β, the resulting transfer function is

$$T(s) = \frac{Y(s)}{R(s)} = \frac{\frac{K}{s-2}}{1 + \frac{K\beta}{s-2}} = \frac{K}{s + K\beta - 2}.$$

The feedback system is stable, i.e. its transients decay for any β satisfying $K\beta > 2$. ■

Decreasing the effect of the disturbance in the output signal. In the open-loop control in Fig. 4.7 the disturbance appears entirely in the output. In the feedback system the effect of the disturbance in the output signal is decreased by $1/(1 + A\beta)$, i.e., the higher the value of the loop gain $A\beta$, the better the feedback reduces the effect of the disturbance.

Feedback can improve the transient response. Let us consider the first-order lag element in Fig. 4.8. With a constant feedback β the resulting transfer function is

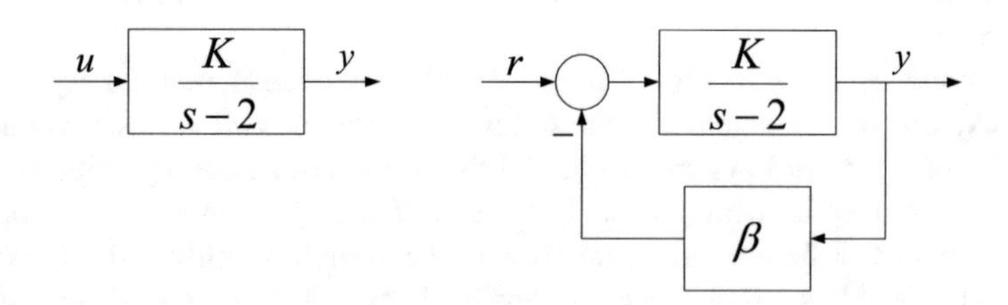

Fig. 4.6 An unstable system can be stabilized by negative feedback

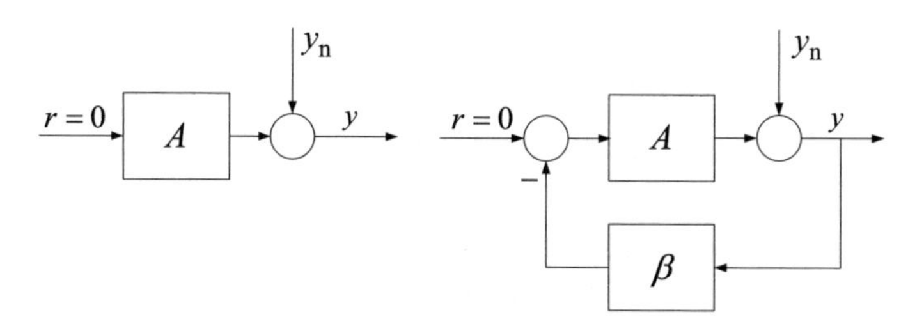

Fig. 4.7 Negative feedback decreases the effect of a disturbance

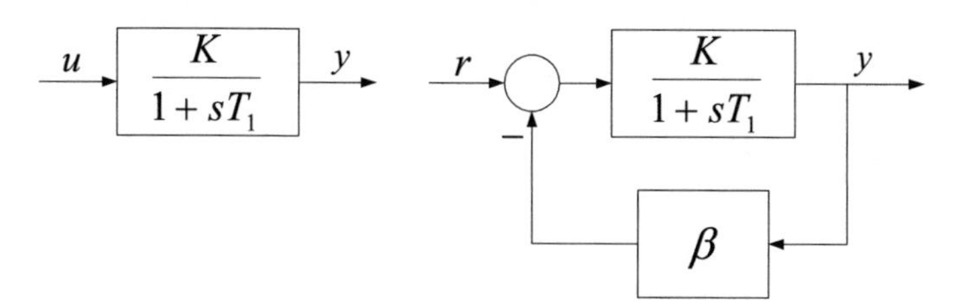

Fig. 4.8 Feedback modifies the transient response

$$T(s) = \frac{Y(s)}{R(s)} = \frac{K}{1+\beta K}\frac{1}{1+s\frac{T_1}{1+\beta K}} = K'\frac{1}{1+sT_1'}.$$

The time constant has been decreased, so the system is faster. At the same time the gain also has been decreased, which generally has to be compensated using a filter of the appropriate gain (as shown in Fig. 4.4).

Feedback decreases the sensitivity of the process to parameter changes.

In Fig. 4.9 the gain of the proportional element without feedback is 10. Suppose the gain of the feedback system is the same: $A_1/(1+A_1\beta) = 10$. Choose the value of A_1 to be 1000. With this value, $\beta = 0.099$ is obtained. If the value of the input signal is 10, then in both systems the value of the output signal is 100. Reduce both values A and A_1 by 2%. Then $A = 9.8$ and $A_1 = 980$. In the original system then the output value decreases to 98, while in the feedback system it remains quite close to

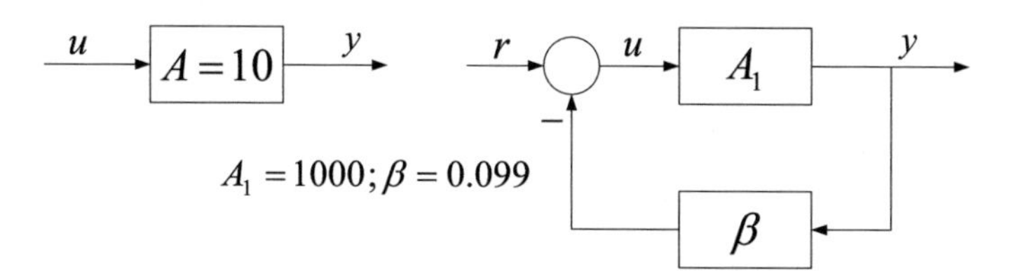

Fig. 4.9 With feedback the system becomes less sensitive to parameter changes

100 (99.98). In the feedback system the error appearing because of the parameter change is $(1 + A_1\beta)$ times less than in the original system.

In the range of high gains the feedback system creates the approximate inverse of the feedback element.

Let us consider the circuit in Fig. 4.10. The element given by the transfer function $H_1(s)$ is fed back with negative feedback through an element given by $H_2(s)$. The resulting transfer function is $H(s) = H_1(s)/[1 + H_1(s)H_2(s)]$. In the frequency range where $|H_1(j\omega)H_2(j\omega)|$ is much higher than 1, the resulting transfer function approximates the inverse of the transfer function H_2. In that portion of the frequency domain, where the absolute value of the loop frequency function is much less than 1, the resulting frequency function approximates the frequency function H_1 of the forward path.

Feedback has a linearizing effect.

Let us consider the static non-linear characteristics in Fig. 4.11a. The characteristics can be divided into three linearized ranges, where the linear transfer gain of the individual ranges is determined by the slope A of the straight line fitted to the curve at the given operating point. Suppose the value of the proportional feedback gain is β. In the feedback system the slope of the individual linearized ranges of the static characteristics is $A/(1 + A\beta)$. The bigger is $A\,\beta$, the better the transfer gain approximates the value of $1/\beta$, becoming independent of the slopes A of the individual ranges of the static characteristics. Figure 4.11b shows the gains of the linearized individual parts with feedback gain $\beta = 10$. It can be seen that the slopes in the different ranges are almost the same, the characteristic is approximately linear in the whole domain. For $\beta = 100$ the linearization is still better (with slopes 0.00998, 0.00995 and 0.0099).

It has to be emphasized that while the non-linear characteristics have been linearized to a great extent, considering the input u_1 the ranges of the linearized sections have been changed compared to the original sections. For example, if $\beta = 10$:

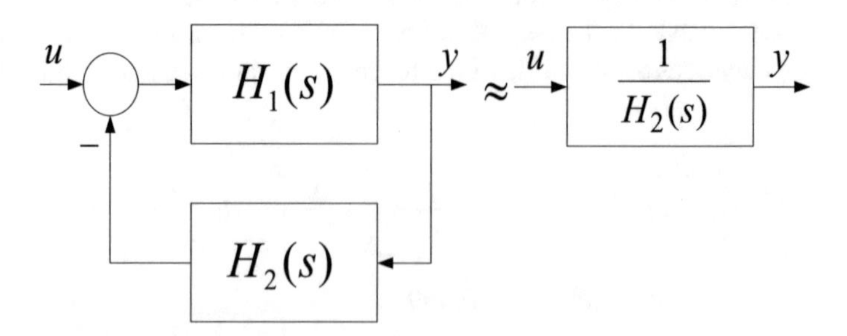

Fig. 4.10 In the range of high gains the feedback creates the approximate inverse of the feedback element

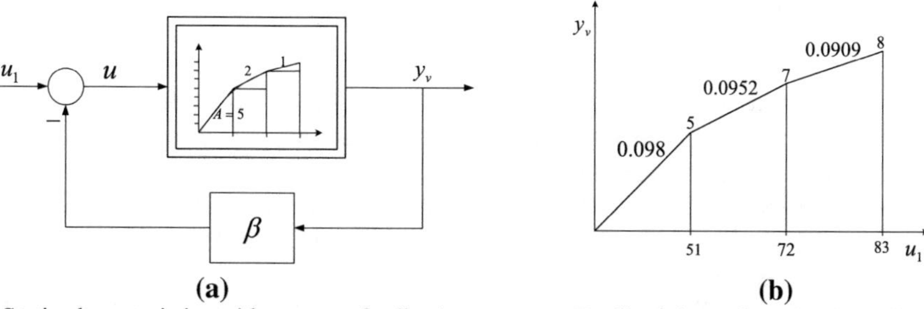

Fig. 4.11 Linearization by feedback.

If $0 \le u \le 1$, then $y_v = 5u$ and $u = u_1 - 50u$, hence $u_1 = 51u$ and $y_v = \frac{5}{51}u_1$.

If $1 < u \le 2$, then $y_v = 3 + 2u$ and $u = u_1 - 10(3 + 2u)$, hence $u = \frac{1}{21}u_1 - \frac{30}{21}$ and $y_v = \frac{3}{21} + \frac{2}{21}u_1$.

If $2 < u \le 3$, then $y_v = 5 + u$ and $u = u_1 - 10(5 + u)$, hence $u = \frac{1}{11}u_1 - \frac{50}{11}$ and $y_v = \frac{5}{11} + \frac{1}{11}u_1$.

Feeding back an integrator by a static non-linear element results in the inverse of the non-linear characteristics.

Let us consider the circuit given in Fig. 4.12. Negative feedback is applied to an integrator through a static quadratic non-linear element. As the output of the integrator can be constant only if its input, i.e. the error signal, becomes zero, $r = y^2$, and $y = \sqrt{r}$, i.e. the circuit realizes the inverse of the non-linearity in the feedback path.

Fig. 4.12 Feeding back an integrator by a non-linear static element realizes the inverse characteristics

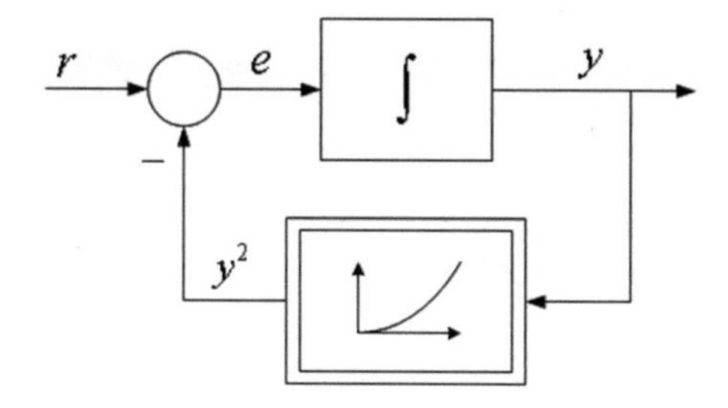

4.3 The Feedback Operational Amplifier

With the invention of the telephone and the development of telecommunication, high gain amplifiers were used to compensate the damping of signals over long transmission lines. Invented by BLACK, the amplifier with negative feedback (1927) ensured a stable solution to decrease the sensitivity of vacuum tube amplifiers to the change of their characteristics, and at the same time it linearized, to a great extent, the nonlinear characteristics of the amplifier.

Operational amplifiers built of integrated circuits are also used in control circuits for amplification and compensation.

Let us analyze the signal transfer properties of the feedback operational amplifier. Its circuit is shown in Fig. 4.13. In the input and feedback path, resistors, capacitors or an interconnection of resistors and capacitors can be employed. For the sake of simplicity let us consider resistors both in the forward and the feedback path. The gain G of the amplifier is of a very high value (in the range of $10^4 - 10^8$).

Let us determine the transfer function and the corresponding block diagram of the operational amplifier.

The output voltage can be expressed as

$$U_2 = -GU. \tag{4.2}$$

If the input current I can be neglected (e.g. the input resistance of the amplifier is high) then the following KIRCHHOFF voltage law equation can be written for the input point of the amplifier:

$$\frac{U_1 - U}{R_1} = \frac{U - U_2}{R_2}. \tag{4.3}$$

Let us express the variable U using this equation.

$$U = \frac{R_2}{R_1 + R_2}\left(U_1 + \frac{R_1}{R_2}U_2\right). \tag{4.4}$$

Substituting this expression into (4.2), the following equation is obtained after some manipulations:

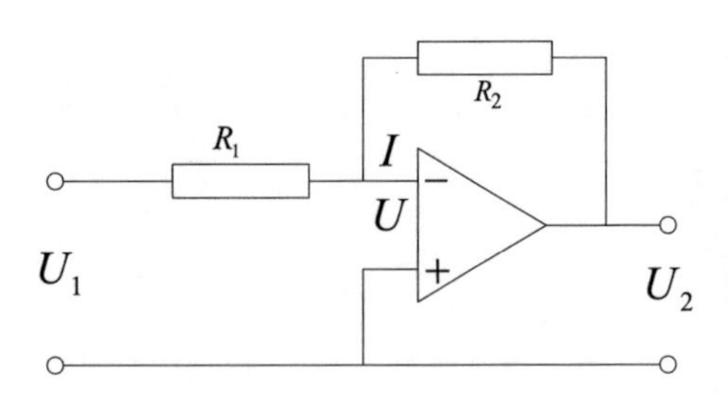

Fig. 4.13 Feedback operational amplifier

$$\frac{U_2}{U_1} = -\frac{R_2}{R_1}\frac{1}{1 + \frac{1}{G}\left(1 + \frac{R_2}{R_1}\right)}. \tag{4.5}$$

It can be seen that if $G \to \infty$, the resulting transfer gain is determined by the ratio of the two resistances. For high values of G, the transfer gain keeps its value quite close to its nominal value even in the case of possible changes in G. (Note that if impedances Z_1 and Z_2 are used in the input and the feedback path instead of resistors, the transfer function of the operational amplifier will be approximately $-Z_2/Z_1$, and depending on the representation of the impedances, different mathematical operations can be realized.)

Based on the above relationships, a block diagram of the feedback operational amplifier can be found. Figure 4.14 shows three equivalent schemes.

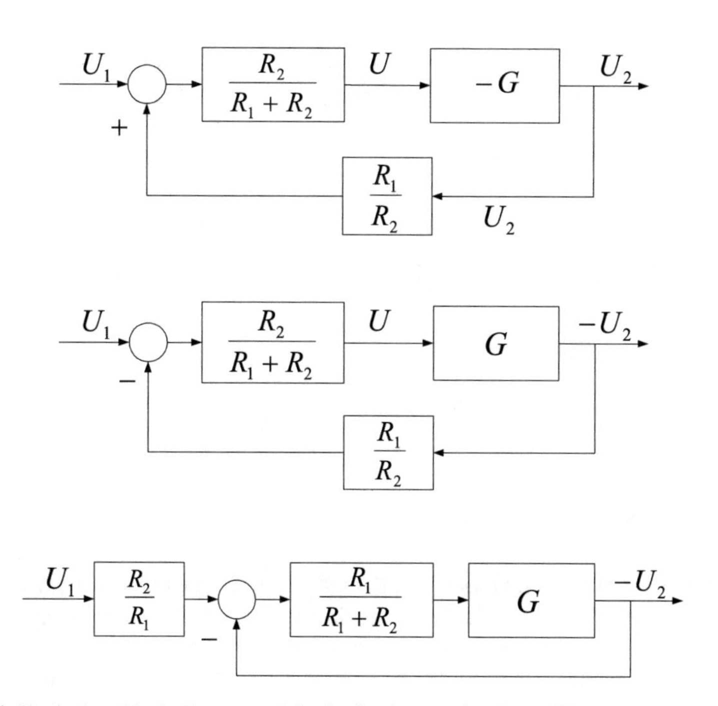

Fig. 4.14 Equivalent block diagrams of the feedback operational amplifier

4.4 The Transfer Characteristics of the Closed Control Loop

The behavior of a closed-loop control system can be investigated by the overall transfer functions exhibiting the relationships between the output and the input signals.

As the systems are assumed to be linear, the superposition theorem can be applied. The effect of the various external signals can simply be summed to obtain the output signal.

Let us determine the overall transfer functions between the controlled signal y, the error signal e, and the control signal u as the output signals, and the reference signal r, the output disturbance y_{no}, the input disturbance y_{ni}, and the measurement noise y_{z} as input signals.

According to Fig. 4.4, the relationships between these input and output signals are

$$Y(s) = \frac{F(s)C(s)P(s)}{1+C(s)P(s)}R(s) + \frac{1}{1+C(s)P(s)}Y_{\mathrm{no}}(s) + \frac{P(s)}{1+C(s)P(s)}Y_{\mathrm{ni}}(s) - \frac{C(s)P(s)}{1+C(s)P(s)}Y_{\mathrm{z}}(s), \tag{4.6}$$

$$E(s) = \frac{F(s)}{1+C(s)P(s)}R(s) - \frac{1}{1+C(s)P(s)}Y_{\mathrm{no}}(s) - \frac{P(s)}{1+C(s)P(s)}Y_{\mathrm{ni}}(s) - \frac{1}{1+C(s)P(s)}Y_{\mathrm{z}}(s), \tag{4.7}$$

$$U(s) = \frac{F(s)C(s)}{1+C(s)P(s)}R(s) - \frac{C(s)}{1+C(s)P(s)}Y_{\mathrm{no}}(s) - \frac{C(s)P(s)}{1+C(s)P(s)}Y_{\mathrm{ni}}(s) - \frac{C(s)}{1+C(s)P(s)}Y_{\mathrm{z}}(s), \tag{4.8}$$

On the basis of these relationships, the output signals can be determined with the knowledge of the input signals. From the time evolution of the output signals it can be verified whether the control system satisfies the quality specifications or not.

It has to be emphasized that the frequency ranges of the different input signals are generally different. The reference signal and the disturbances generally contain low frequency components, whereas the measurement noise generally is a zero mean signal containing high frequency components. If the absolute value of the frequency function obtained from an overall transfer function by substituting $s = j\omega$—considering a given input signal—is approximately unity over a significant frequency range, then the system tracks the signal, but if the transfer function approximates zero, the system attenuates the considered input signal.

It can be seen that all the overall transfer functions have the same denominator, namely $1 + C(s)P(s)$, which is the characteristic polynomial of the closed-loop control system. The roots of the characteristic polynomial determine the stability and the dynamic properties of the transients of the control system. For stable performance it is required that the transients of the output signal should decrease, i.e. the roots of the characteristic equation should be at the left hand side of the complex plane. Chapter 5 deals in detail with methods of stability investigation.

From Eq. (4.7) it can be seen, that if the filter $F(s)$ is a proportional element with gain unity, then the error of reference signal tracking and the error of output disturbance rejection are the same, i.e., the control system follows the reference signal with the same dynamics and the same static error as it rejects the effect of the output disturbance in the output signal. With the appropriate choice of filter $F(s)$ it can be ensured that the properties of reference signal tracking and of disturbance rejection would be different.

If $F(s) = 1$ the control system is called a *One-Degree of Freedom* (*ODOF*) system, while if $F(s)$ is given by a non-unity transfer function, it is called *Two-Degree of Freedom* (*TDF*) system. In the case of an *ODOF*, 4 overall transfer functions determine the overall signal transfer properties between the output signals (the controlled signal y and the manipulated variable u) and the input signals (the reference signal, the disturbances, and the measurement noise), but in the *TDOF* case, 6 overall transfer functions are needed for this determination.

As the disturbance y_{ni} can always be transformed to an equivalent output disturbance, and the signs do not have to be considered, it is sufficient to investigate the following 6 overall transfer functions.

$$\begin{aligned} \frac{Y}{R} &= \frac{FCP}{1+CP}; \quad \frac{Y}{Y_z} = \frac{-CP}{1+CP}; \quad \frac{Y}{Y_{\text{ni}}} = \frac{P}{1+CP} \\ \frac{U}{R} &= \frac{FC}{1+CP}; \quad \frac{U}{Y_z} = \frac{-C}{1+CP}; \quad \frac{E}{Y_{\text{no}}} = \frac{1}{1+CP} \end{aligned} \tag{4.9}$$

The first column characterizes reference signal tracking, the second column characterizes the properties of the disturbance rejection and the third column characterizes the rejection of measurement noise. If $F(s) = 1$, the second and third columns give the 4 characterizing transfer functions. Arranging these functions into matrix form, a transfer function matrix of the closed-loop control system is obtained. To ensure the stability of the closed-loop control system, all the overall transfer functions have to be stable. Also, all the overall transfer functions have to ensure the prescribed dynamic behavior between the given input and output signals.

One of the usual controller design procedures is the cancellation of the unfavorable poles of the plant P with the zeros of the controller C. But it can be seen that the dynamics of the plant P remains in the expression of the overall transfer function between the output signal and the input disturbance. It is not allowed to cancel the unstable poles of the plant, as though they become invisible in the relationship between the output signal and the reference signal, they do appear in the transfer relationship between the output signal and the input disturbance. (It has to be mentioned that even regarding the relationship between the output and the reference signal the unstable pole can not be cancelled quite accurately, as its value generally is obtained by measurements or modeling which certainly involves errors, or its value may change over time; therefore the pole cancellation is never perfectly accurate and instability persists in the system.)

It is reasonable to design a controller in two steps. First the controller C is to be designed to ensure the appropriate rejection of the disturbances and the measurement noise, then the filter F is to be designed for appropriate reference signal tracking.

For good reference signal tracking if $F(s) = 1$, the so called *complementary sensitivity function* $T = CP/(1 + CP)$ has to approximate 1 on those frequencies which characterize the input signal. This means that at these frequencies, the condition $CP \gg 1$ has to be fulfilled. Consider the overall transfer function $S(s) = 1/[1 + C(s)P(s)]$ giving the relationship between the error signal and the reference signal. $S(s)$ is also called the *sensitivity function*. Time domain analysis shows that the error signal contains signal components originating from the poles of the closed-loop and also from the poles of the reference signal. Once the transients decay in the error signal, the quasistationary components originating from the poles of the reference signal are maintained. If $C(s)$ contains the poles of the reference signal, in the error signal the poles of the controller cancel the poles of the reference signal. In this case tracking the reference signal $R(s)$ the steady-state error becomes zero. Thus $C(s) = K_c R(s)$, where $K_c \gg 1$. Considering the disturbance rejection, if the condition $CP \gg 1$ is fulfilled, then the LAPLACE transform of the error signal as a response for the input and the output disturbances is approximately: $E(s) \approx -[1/C(s)P(s)]Y_{no}(s) - [1/C(s)]Y_{ni}(s)$. For good rejection of the input disturbance it is suggested to choose $C(s) = K_c Y_{ni}(s)$ for the controller dynamics, where $K_c \gg 1$. Appropriate output disturbance rejection can be reached by choosing the controller dynamics according to $C(s) = K_c Y_{no}(s)$, again with $K_c \gg 1$ (supposing that the amplitudes of the frequency function of the plant are not too high in the characteristic frequency range of the disturbance). To ensure good reference signal tracking and good disturbance rejection the controller has to contain the dynamics of both the reference and the disturbance signals. The following example demonstrates the effects of the designed controller to the behaviour of the control system.

Example 4.2 The transfer function of the plant is $P(s) = 1/(1 + 0.5s)^3$. Suppose the transfer function of the controller is $C(s) = 0.5(1 + 0.5s)/s$. Let us accelerate the reference signal tracking of the system with an appropriate prefilter F. Its gain is 1, and let it compensate the complex conjugate poles of the closed-loop control system, replacing them with two identical (real and faster) poles. Apply an $F(s) = (s^2 + 1.161s + 0.7044)/\left(0.7044(1 + 0.4s)^2\right)$ transfer function as the prefilter. Figure 4.15 shows the unit step responses of the different output signals in the closed-loop control circuit. It can be seen that the dynamic behavior of the control system is different for the reference signal and for the input disturbance. It can be

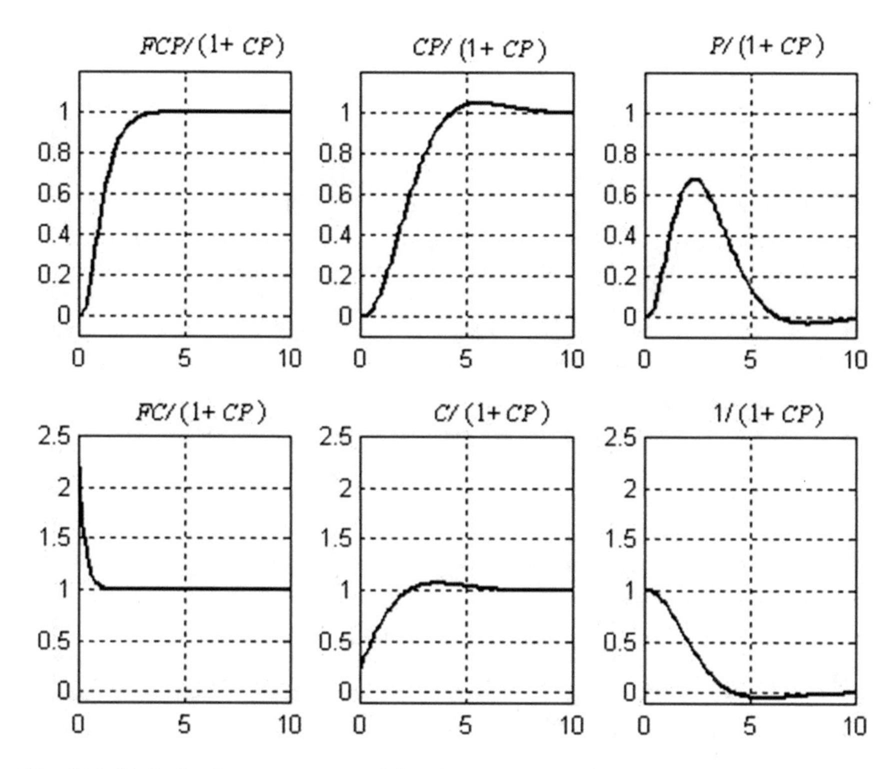

Fig. 4.15 Typical unit step responses of the closed-loop control system

observed that applying the filter accelerates the settling of the controlled output signal. The price paid for this is an overexcitation in the control signal. Figure 4.16 shows the frequency functions of the closed-loop control system. From the course of the frequency functions some evaluation of the time responses can also be derived. The frequency range where the disturbance rejection is efficient is also observable. For example, the third curve in the figure shows that the output will have its highest amplitude around frequency $\omega = 1$ for a sinusoidal input disturbance. From the sixth curve it can be concluded that the system attenuates the effect of the sinusoidal output disturbances up to the frequency $\omega = 1$, but beyond this frequency it tracks the disturbances.

From the second curve of Fig. 4.15 or from the equivalent left upper curve of Fig. 4.17 it can be seen that the control system tracks the unit step reference signal without steady state error. The controller contains an integrating element, whose pole is at the origin in the complex plane. Thus the controller contains the pole of the unit step signal (whose LAPLACE transform is $1/s$). Let us investigate the time evolution of the output signal with the given controller with prefilter $F = 1$ provided an exponential reference signal by $r(t) = \exp(-0.1t)$. The LAPLACE transform of the reference signal is $R(s) = 1/(s+0.1)$. The reference signal and the output signal are shown on the right upper curve of Fig. 4.17. It can be seen that the

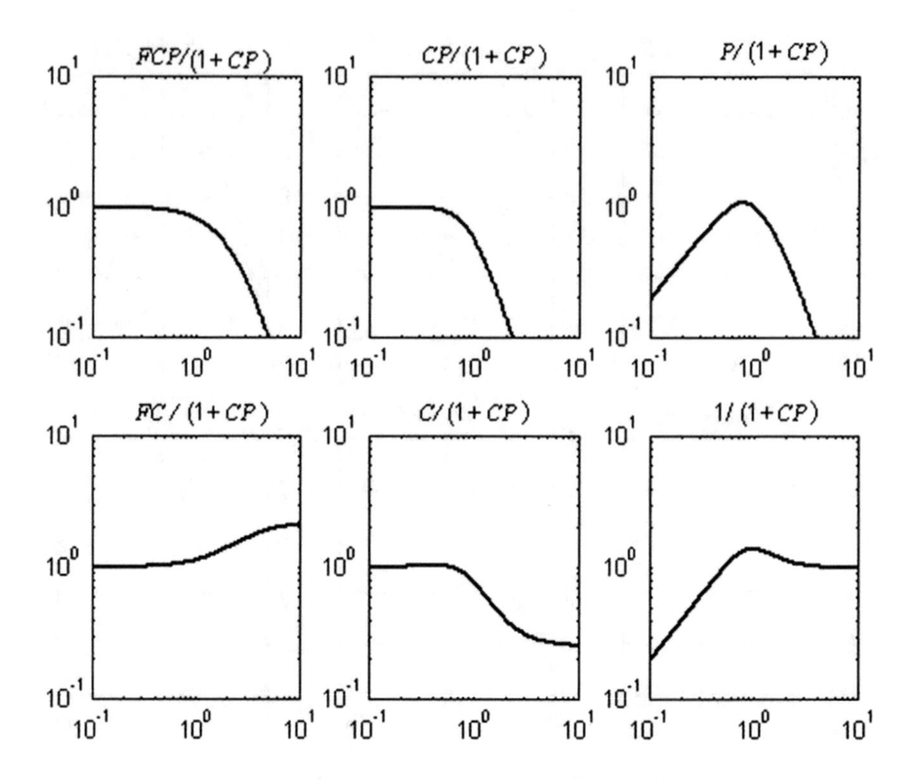

Fig. 4.16 Amplitude-frequency functions of the closed-loop control system

tracking is not accurate: after the transients decay, the output does not fit exactly the input signal. Let us change now the controller according to $C_1(s) = 4(1+0.5s)/(1+10s) = 0.4(1+0.5s)/(s+0.1)$. Now the pole of the controller is the same as the pole of the input signal. The right lower curve shows that after the transient period the output signal exactly tracks the input signal. But now the controller transfer function does not contain the pole of the unit step reference signal, therefore in the unit step response there will be a static deviation (left lower figure). ■

4.5 The Static Transfer Characteristics

If the closed-loop control system is stable, its steady state (or static in other words) properties can be determined on the basis of Eqs. (4.6)–(4.8) using the final value theorem of the LAPLACE transformation.

The signal transfer properties of closed-loop control circuits in steady state, i.e., the accuracy of reference signal tracking and of disturbance rejection in steady state depends on the so called type number and the loop gain of the system. The static

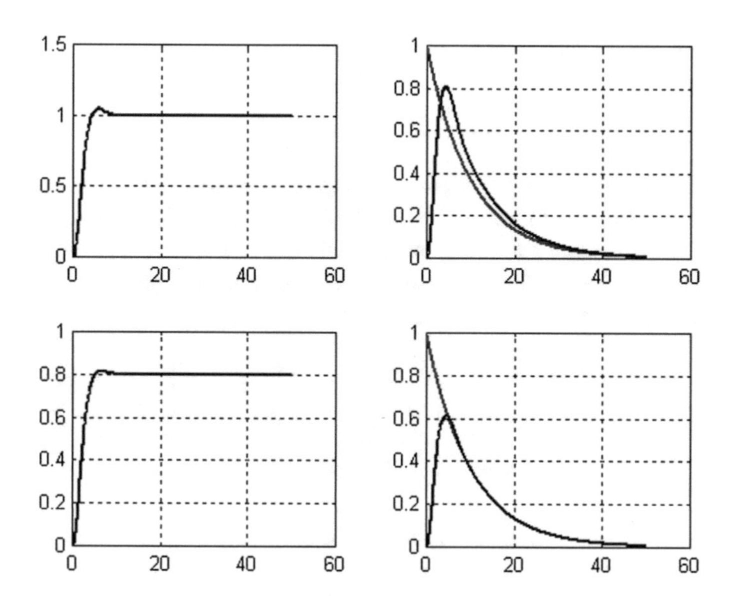

Fig. 4.17 Reference signal tracking is realized without steady-state error, if the controller contains the pole of the reference signal

accuracy also depends on the time evolution of the reference or the disturbance signal.

Let us suppose that $L(s) = C(s)P(s)$, the transfer function of the open-loop (the so-called loop transfer function) is given in its time constant form:

$$L(s) = C(s)P(s) = \frac{K}{s^i} \frac{\prod_{j=1}^{c}(1+s\tau_j)\prod_{j=1}^{d}\left(1+2\zeta_j\tau_{oj}s+s^2\tau_{oj}^2\right)}{\prod_{j=1}^{e}(1+sT_j)\prod_{j=1}^{f}\left(1+2\xi_j T_{oj}s+s^2T_{oj}^2\right)} e^{-sT_d} = \frac{K}{s^i}L_t(s). \tag{4.10}$$

Here the variable i is the type number, indicating the number of the integrators in the loop (in practice its value can be 0, 1 or 2), K denotes the loop gain. $L_t(s)$ represents the transfer function determining the transient response of the control circuit. Its important property is that it does not influence the steady state behavior, i.e., $L_t(s=0)=1$.

The overall transfer function between the error signal and the reference signal in the case where $F(s)=1$ is

$$E(s) = \frac{1}{1+L(s)}R(s) = \frac{s^i}{s^i+KL_t(s)}R(s). \tag{4.11}$$

The steady state value of the error signal is

$$\lim_{t\to\infty} e(t) = \lim_{s\to 0} sE(s). \tag{4.12}$$

Let us analyze the reference signal tracking properties of the closed-loop control system for unit step, unit ramp and parabolic input signals. The LAPLACE transforms of these reference signals are $R(s) = 1/s^j$, where $j = 1$ for the unit step, $j = 2$ for the unit ramp, and $j = 3$ for the parabolic reference input signal.

In case of a 0-type system, the steady state error is:

$$\begin{aligned}
&\text{for a unit step reference signal,} && \lim_{t\to\infty} e(t) = \lim_{s\to 0} s\frac{1}{s}\frac{1}{1+KL_1(s)} = \frac{1}{1+K};\\
&\text{for a unit ramp reference signal,} && \lim_{t\to\infty} e(t) = \lim_{s\to 0} s\frac{1}{s^2}\frac{1}{1+KL_1(s)} = \infty;\\
&\text{for a parabolic reference signal,} && \lim_{t\to\infty} e(t) = \lim_{s\to 0} s\frac{1}{s^3}\frac{1}{1+KL_1(s)} = \infty.
\end{aligned} \tag{4.13}$$

For a 1-type system, the steady state error is:

$$\begin{aligned}
&\text{for a unit step reference signal,} && \lim_{t\to\infty} e(t) = \lim_{s\to 0} s\frac{1}{s}\frac{s}{s+KL_1(s)} = 0;\\
&\text{for a unit ramp reference signal,} && \lim_{t\to\infty} e(t) = \lim_{s\to 0} s\frac{1}{s^2}\frac{s}{s+KL_1(s)} = \frac{1}{K};\\
&\text{for a parabolic reference signal,} && \lim_{t\to\infty} e(t) = \lim_{s\to 0} s\frac{1}{s^3}\frac{s}{s+KL_1(s)} = \infty.
\end{aligned} \tag{4.14}$$

For a 2-type system, the steady state error is:

$$\begin{aligned}
&\text{for a unit step reference signal,} && \lim_{t\to\infty} e(t) = \lim_{s\to 0} s\frac{1}{s}\frac{s^2}{s^2+KL_1(s)} = 0\\
&\text{for a unit ramp reference signal,} && \lim_{t\to\infty} e(t) = \lim_{s\to 0} s\frac{1}{s^2}\frac{s^2}{s^2+KL_1(s)} = 0\\
&\text{for a parabolic reference signal,} && \lim_{t\to\infty} e(t) = \lim_{s\to 0} s\frac{1}{s^3}\frac{s^2}{s^2+KL_1(s)} = \frac{1}{K}
\end{aligned} \tag{4.15}$$

In the following table, the values of the steady state errors are summarized.

Type number	$i = 0$	$i = 1$	$i = 2$
unit step reference signal, $j = 1$	$\frac{1}{1+K}$	0	0
unit ramp reference signal, $j = 2$	∞	$\frac{1}{K}$	0
parabolic reference signal, $j = 3$	∞	∞	$\frac{1}{K}$

A 0-type system tracks the step reference signal with steady state (static) error, whose value is less if the loop gain of the control circuit is higher (Fig. 4.18). But a high loop gain may cause an unstable behavior of the control system. A 0-type system is not able to track the ramp or the parabolic reference signals.

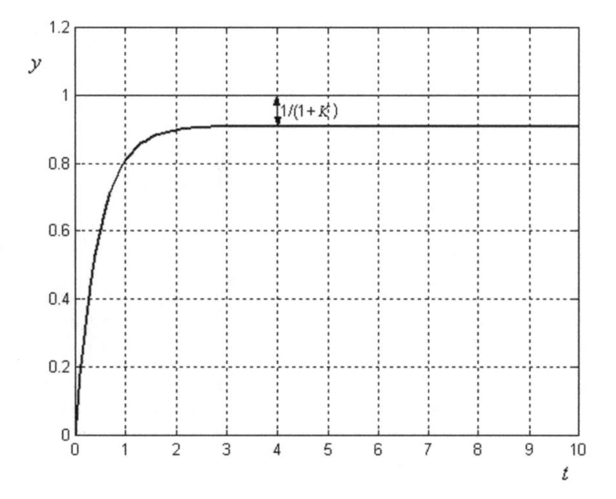

Fig. 4.18 A 0-type system tracks the unit step reference signal with steady state error

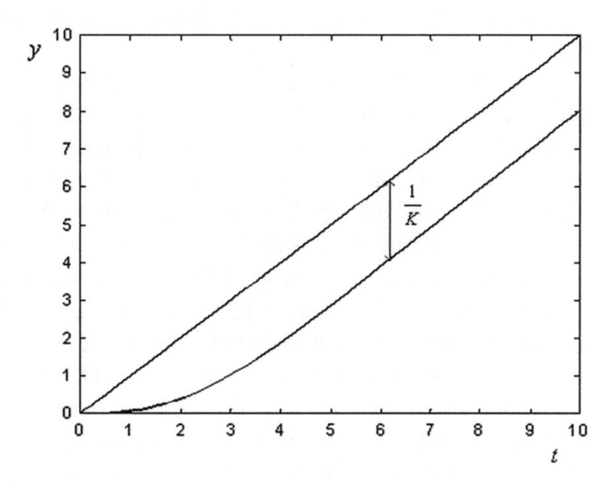

Fig. 4.19 A 1-type system tracks the ramp reference signal with steady state error

A 1-type control system containing one integrator tracks the step reference signal without steady state error. It can follow the ramp reference signal with a steady state error (Fig. 4.19). But it can not track the parabolic reference signal.

A 2-type system containing two integrators tracks the step and the ramp signals without steady state error (Fig. 4.20), and is able to follow the parabolic reference signal with a static error.

It can be seen, that coinciding with the previous statement related to the conditions of accurate reference signal tracking, the closed-loop control system is capable of tracking a reference signal whose Laplace transform contains poles at the origin of the complex plane without steady state error only if the loop transfer function contains as many poles at zero (integrators) as there are poles at zero of the Laplace transform of the reference signal. If the plant does not contain the required

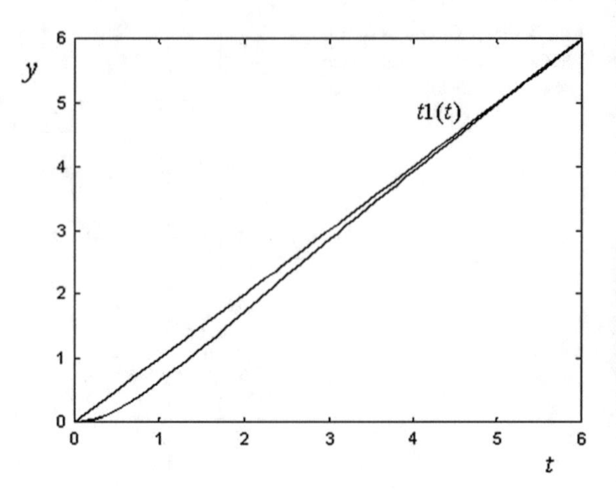

Fig. 4.20 A 2-type system tracks the ramp reference signal without static error

number of integrators ensuring the desired static accuracy, the integrators have to be put in the controller.

The effect of improving the static accuracy by inserting integrators in the control loop can be demonstrated by the following considerations. If the control system is of type 0, i.e. it is proportional, a constant signal value at its output can be maintained only by a constant input signal. Therefore it is necessary that also the error signal take a constant value. The property of the integrator is that its output reaches a constant value when its input finally becomes zero. If there is an integrator in the forward path of the closed-loop control circuit, then for a unit step reference signal the output signal will increase until the error signal—the input signal of the integrator—reaches zero. If the reference signal is a unit ramp, then at the output of the integrator a signal change with constant slope can be reached by a constant input signal, which means a constant error signal, i.e. a constant static deviation.

It can be seen that increasing the number of the integrators in the loop improves the static properties of the closed-loop control system. More specifically, increasing the loop gain reduces the static tracking error. But the number of integrators can not be increased to more than two, as this would lead to stability problems which could not be handled easily. Increasing the gain may also cause stability problems.

Static accuracy and stability are contradictory requirements. With controller design a satisfactory compromise has to be created to satisfy both requirements.

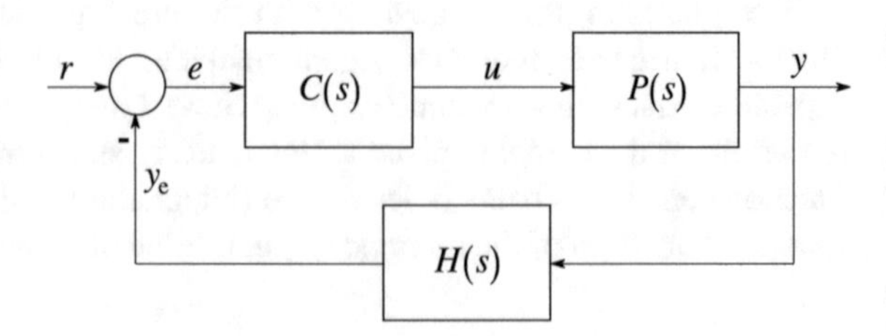

Fig. 4.21 Block diagram of a closed-loop control circuit

The evolution of the steady-state signal values in a closed-loop control circuit could also be demonstrated by a four-quarter-plane figure. On the four axes the error signal (e), the measured signal (y_e), the controlled signal (y) and the control signal (u) are indicated, respectively. Generally positive signal values are supposed. The block diagram is shown in Fig. 4.21. Generally the static characteristics of the plant and of the sensor are non-linear (but usually they are linearized in a vicinity of a given operating point). Stable behavior is supposed.

For a 0-type system, the four-quarter-plane curves are shown in Fig. 4.22. The right upper quarter represents the element creating the difference signal, where the location of the straight lines depends on the signal r. The static characteristic of the controller (left upper quarter) is generally linear, possibly saturating. The static characteristics of the plant (generally non-linear) is in the left lower quarter, here the effects of a disturbance and of parameter changes on the characteristics can be demonstrated. The right lower quarter shows the characteristic of the sensor, i.e. the measurement equipment, which sometimes is also non-linear. In the case of a 0-type system there is a steady state error ($e(\infty) \neq 0$).

In Fig. 4.22, which shows the four-quarter-plane static relations, it can be seen how the static states change if, e.g., the reference signal change. It can be investigated how the steady states change if $y(u)$, i.e., the static characteristic of the plant, changes as a consequence of parameter changes. (A similar four-quarter-plane representation was first introduced by SZILÁGYI.)

Figure 4.23 shows the static curves for a 1-type closed-loop control system. As now the static error for a step reference signal is zero, only the two lower quadrants of the plane are of interest. Now it would be sufficient to draw only these two, but for the sake of comparability the same coordinate system is given as before.

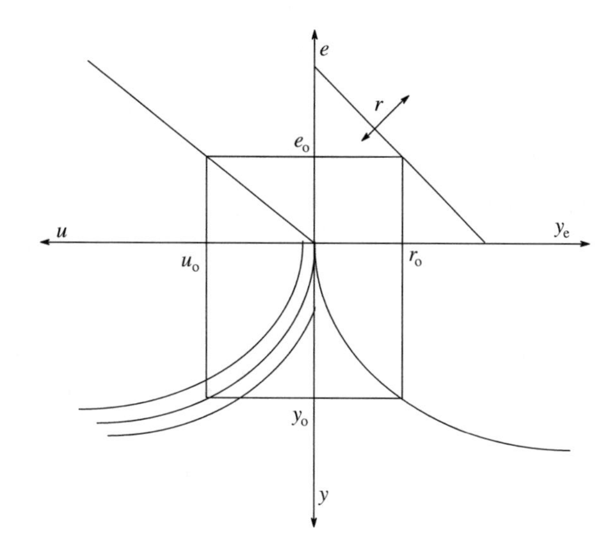

Fig. 4.22 Static characteristic curves of a 0-type closed-loop control system

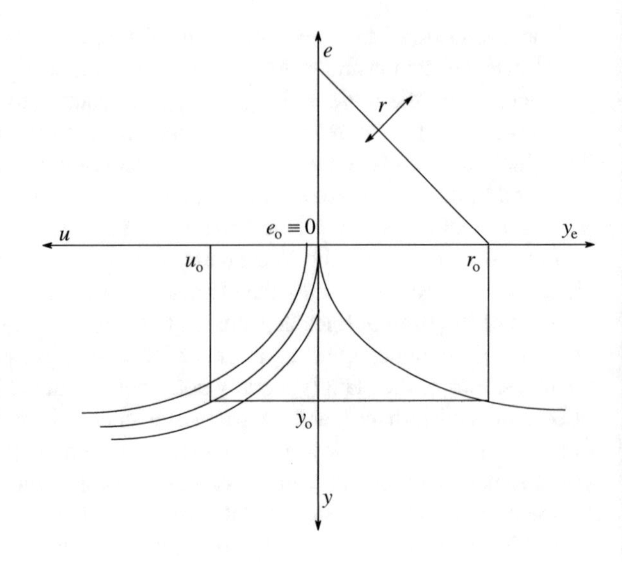

Fig. 4.23 Static characteristics curves of a 1-type closed-loop control system

4.6 Relationships Between Open- and Closed-Loop Frequency Characteristics

In a closed-loop control system $L = CP$ is called the loop transfer function (Fig. 4.24). The overall transfer function of a closed-loop realized by negative feedback (Fig. 4.25) calculated between the output signal and the reference signal is $T = CP/(1 + CP) = L/(1 + L)$, which is also called the complementary sensitivity function. Observe that $T = 1 - S = 1 - 1/(1 + L)$, where $S = 1/(1 + CP) = 1/(1 + L)$ is the sensitivity function. Regarding the frequency course of this function, approximate considerations can be given.

In the frequency range where

$$|L(j\,\omega)| \gg 1, \quad |T(j\,\omega)| \approx 1; \tag{4.16}$$

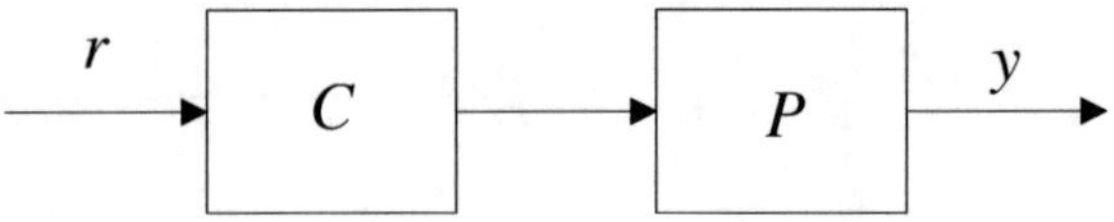

Fig. 4.24 Open-loop control

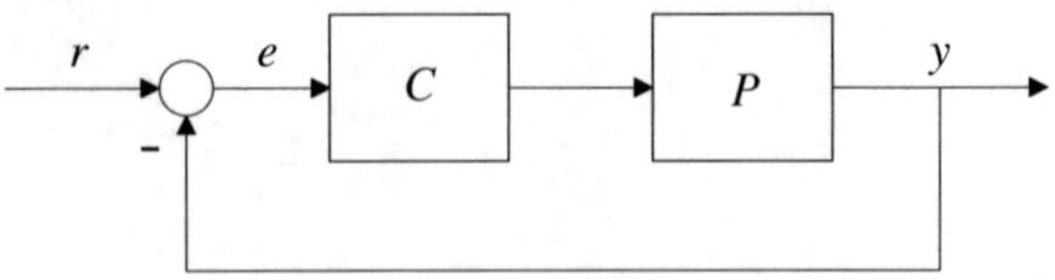

Fig. 4.25 Closed-loop control

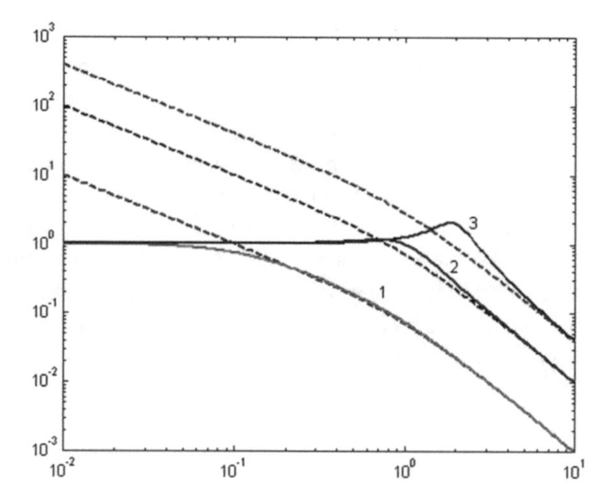

Fig. 4.26 Typical course of the amplitude-frequency functions of the open- and the closed-loop

while in the frequency range where

$$|L(j\omega)| \ll 1, \quad |T(j\omega)| \approx |L(j\omega)|. \tag{4.17}$$

The approximations are not valid in the vicinity of the cut-off frequency. Figure 4.26 shows typical amplitude-frequency curves of the open- and the closed-loop. The open-loop is a first order lag element serially connected to an integrating element. Curves 1, 2, 3 give the Bode amplitude-frequency diagrams of the open- and the closed-loop for three different loop gains. The highest gain is in the case of curve 3. It can be seen that the closed-loop diagrams approximate the value 1 up to the cut-off frequency of the open-loop, and then follow the course of the open-loop diagrams. For higher loop gains, the closed-loop curve shows an amplification in the vicinity of the cut-off frequency, which indicates the appearance of complex conjugate poles in the closed-loop transfer function and transients with decreasing oscillations in the unit step response. Figure 4.27 gives the unit step responses of the closed-loop system with the three different loop gains.

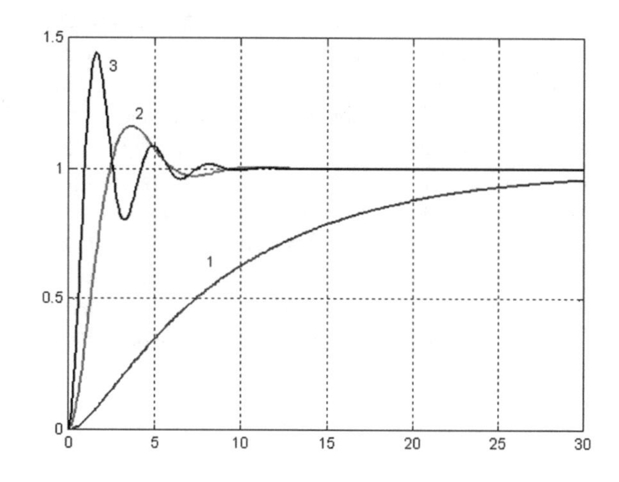

Fig. 4.27 Unit step responses of the closed-loop system

If there is no amplification in the amplitude diagram of the closed-loop, the closed-loop can be approximated by a first order lag element with unit gain and time constant reciprocal to the cut-off frequency ω_c: $T(s) \approx 1/(1+s/\omega_c)$. The next time constant which changes the slope of the approximating amplitude curve to −40 dB/decade can be neglected in this case. The unit step response approximates the steady state exponentially, and approximately within 3 time constants reaches its steady state within 5% accuracy. Increasing the loop gain the slope of the curve around the cut-off frequency will be −40 dB/decade, and the time of decaying of the oscillations can be approximated by 10 times the time constant $(1/\omega_c)$ of the second order oscillating element. Thus the settling time can be given by the following approximate relationship:

$$\frac{3}{\omega_c} < t_s < \frac{10}{\omega_c}. \tag{4.18}$$

To avoid oscillations a long section of slope −20 dB/decade has to be created around the cut-off frequency (before and after it) in the Bode amplitude-frequency diagram of the open-loop. To accelerate the system the cut-off frequency has to be set to higher values.

4.6.1 The $M-\alpha$ and $E-\beta$ Curves

For a deeper analysis of the relationship between the frequency functions of the open and the closed-loop systems, let us analyze the following considerations. The complementary sensitivity function of the closed-loop system is given by

$$T(s) = \frac{C(s)P(s)}{1+C(s)P(s)} = \frac{L(s)}{1+L(s)}. \tag{4.19}$$

In controller design, the relationship between the transfer function $T(s)$ of the closed-loop and the transfer function $L(s) = C(s)P(s)$ of the open-loop is taken into account. This relationship seems to be simple, but actually it means a conformal non-linear mapping from the $L(s)$ complex plane to the $T(s)$ complex plane. The complexity of this non-linear relationship is the reason why the controller can not always be designed unambiguously using simple methods.

For each point of the complex plane the mapping point (complex vector) according to relationship (4.19) can be determined. The absolute value of this vector is depicted on the vertical axis in Fig. 4.28. (The phase angle can also be visualized similarly.) Let us plot on the complex plane the Nyquist diagram of the open-loop (shown as a thick line in the figure). If the points of this Nyquist curve are projected to the three dimensional curve, the absolute values of the frequency function of the closed-loop are obtained. The Bode amplitude-frequency diagram of the closed-loop is visualized drawing these values versus the frequency.

The frequency function of the closed-loop can be given by its amplitude and phase angle:

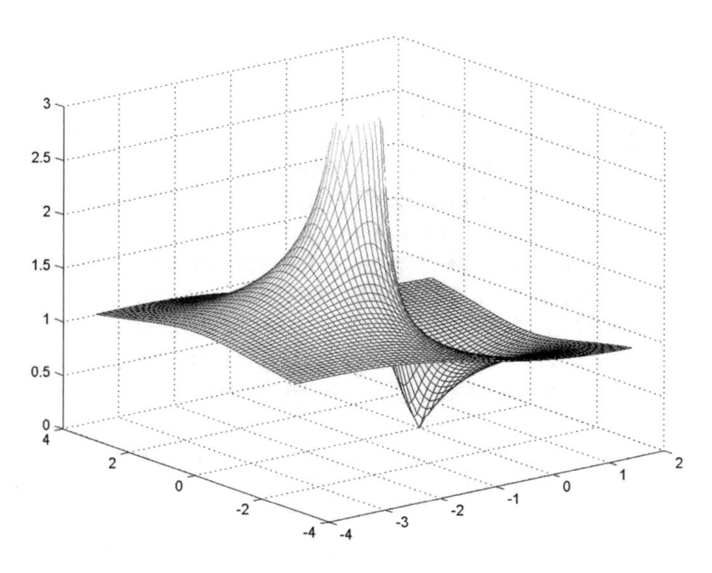

Fig. 4.28 The relationship between the amplitude diagrams of the open- and the closed-loop

$$T(j\omega) = \frac{L(j\omega)}{1 + L(j\omega)} = M(\omega)e^{j\alpha(\omega)}. \tag{4.20}$$

From Fig. 4.28 it can be seen that for high amplifications in the open-loop (in the horizontal plane, points which are far from the origin) the amplification of the closed-loop approximates the constant value 1. This relationship is seen also from the approximation

$$|T(j\omega)| = \left|\frac{L(j\omega)}{L(j\omega) + 1}\right|_{|L| \gg 1} \approx 1. \tag{4.21}$$

As in control systems the amplification of the open-loop in the low frequency domain is generally high, the amplification of the closed-loop here is approximately 1. Similarly it can also be seen that for points with low amplification in the open-loop (in the horizontal plane points close to the origin) the corresponding closed-loop points are of low amplification values, too.

$$|T(j\omega)| = \left|\frac{L(j\omega)}{L(j\omega) + 1}\right|_{|L| \ll 1} \approx |L(j\omega)|.$$

As the amplification of physical systems decreases at high frequencies, this relationship shows that at high frequencies the amplifications of the open- and the closed-loop are approximately the same, i.e. negative feedback at these frequencies does not change the open-loop.

It can also be seen that the curve has a singularity at the point $(-1+0j)$ of the complex plane, therefore for controller design the investigation of the neighborhood of this point will have a great importance. The closer we are to $(-1+0j)$, the higher the amplification of the closed-loop will be. When designing a control system, among the given quality specifications the prescribed value of the allowed overshoot is an important requirement. The overshoot in the step response of the closed-loop system is a time domain property, which is determined by the amplification of the amplitude in the frequency domain. Therefore it is important to investigate the location of the points in the complex plane where the closed-loop amplitudes $|T| = M$ are identical. The points of the frequency function of the closed-loop system where the amplitudes are identical are located on circles in the complex plane. This can be seen easily, solving the equation

$$M = \left|\frac{L(j\omega)}{1+L(j\omega)}\right| = \left|\frac{u+jv}{1+u+jv}\right| = \sqrt{\frac{u^2+v^2}{1+2u+u^2+v^2}}. \tag{4.22}$$

The equation of the curves belonging to a constant amplitude M are obtained by rearranging the above equation as

$$\left(u - \frac{M^2}{1-M^2}\right)^2 + v^2 = \left(\frac{M}{1-M^2}\right)^2. \tag{4.23}$$

This is the equation of a circle with radius r and center point (u_o, v_o), where

$$r = \left|\frac{M}{1-M^2}\right| \quad , \quad u_o = \frac{M^2}{1-M^2} \quad \text{and} \quad v_o = 0. \tag{4.24}$$

The circles belonging to different constant M amplitude values of the closed-loop are shown in Fig. 4.29.

The $M = 1$ constant curve is a vertical line at $u = -0.5$. For $M > 1$ the curves are to the left, and for $M < 1$ they are to the right of this line. If M tends to infinity, the curves shrink to point $(-1+0j)$, and if M tends to zero, the circle will be of infinitesimal radius around the origin.

Similarly to the circles belonging to constant M values, curves belonging to constant α values can also be given (4.20), which are also circles. These circles (both for constant M and constant α values) are called ARCHIMEDES circles. The two curve systems together are called $M - \alpha$ curves.

If the NYQUIST diagram of the open-loop is plotted in the complex plane where the constant M curves are also drawn, the amplitude-frequency diagram of the closed-loop can be obtained by reading the appropriate M values corresponding to the individual points of the NYQUIST diagram. The highest amplitude of the closed-loop is determined by how close the NYQUIST diagram of the open-loop approaches $(-1+0j)$. The highest value of M will be determined by the circle-tangential to the NYQUIST diagram.

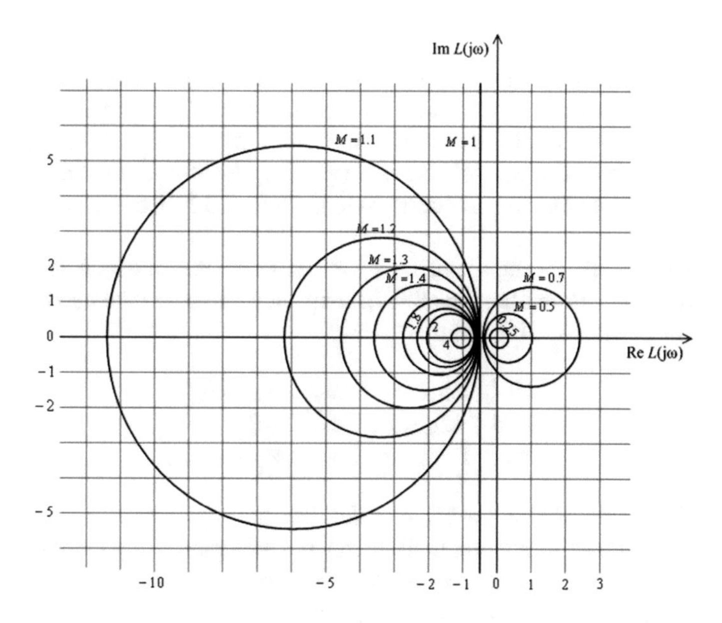

Fig. 4.29 Constant M curves

Some characteristic features of the M curves are shown on the complex plane in Fig. 4.30. The frequency ω_b, where the frequency function $L(j\omega)$ intercepts the circle of $M = 1/\sqrt{2}$, gives the so called bandwidth of the closed-loop system. In the figure the cut-off frequency ω_c and the frequency ω_a where $M = \sqrt{2}$ are also

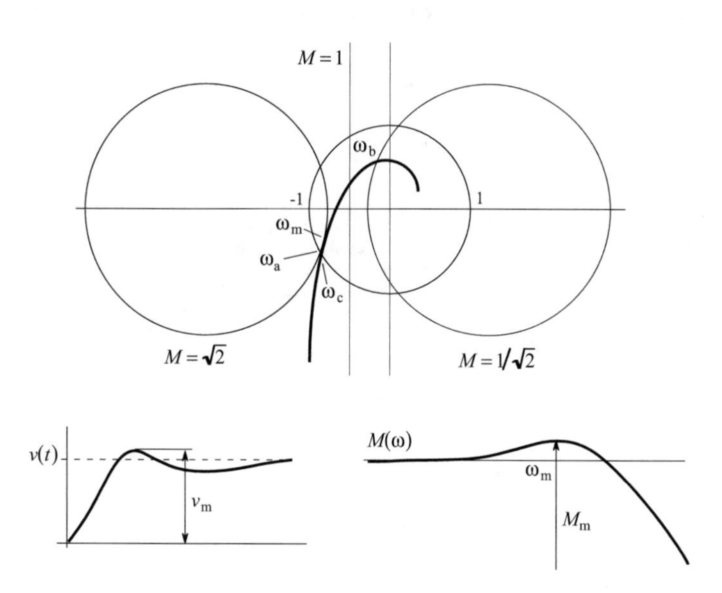

Fig. 4.30 Some characteristic features of the M curves in the complex plane: shape of the unit step response and the amplitude-frequency characteristics

indicated. At the frequency ω_m, the closed-loop system has its maximal amplitude value. The maximal value is the highest value of M belonging to the circle-tangential to the NYQUIST curve of the open-loop. The amplitude-frequency diagram of the closed-loop has amplification only if the NYQUIST diagram of the open-loop intersects the vertical line corresponding to $M = 1$, thus there is a frequency range where the NYQUIST curve is to the left of this line. At the intersection frequency $|T(j\omega)| = 1$.

Approximate relationships can be given between the maximal amplification $M_m = M_{max}$ of the closed-loop amplitude-frequency curve and the maximum value v_m of the step response (Fig. 4.30).

$$\begin{array}{ll} M_m \geq 1.5 & v_m \leq M_m - 0.1 \\ 1.25 \leq M_m \leq 1.5 & v_m \approx M_m \\ M_m \leq 1.25 & v_m < M_m \end{array} \tag{4.25}$$

To avoid oscillations and a big overshoot in the time response, high amplification is not allowed in the amplitude-frequency diagram of the closed-loop. The ideal and the real frequency curves of the closed-loop are shown in Fig. 4.31. Here ω_{cc} is the cut-off frequency of the closed-loop.

Similarly to the $M - \alpha$ curves of the $T(j\omega)$ frequency function, the so called $E - \beta$ curves can be constructed based on the overall error transfer function $S(j\omega)$ (sensitivity function).

$$S(j\omega) = \frac{1}{1 + L(j\omega)} = E(\omega)e^{j\beta(\omega)}. \tag{4.26}$$

Drawing the curves belonging to constant values of E is very simple, as

$$E = |S(j\omega)| = \frac{1}{|1 + L(j\omega)|} \tag{4.27}$$

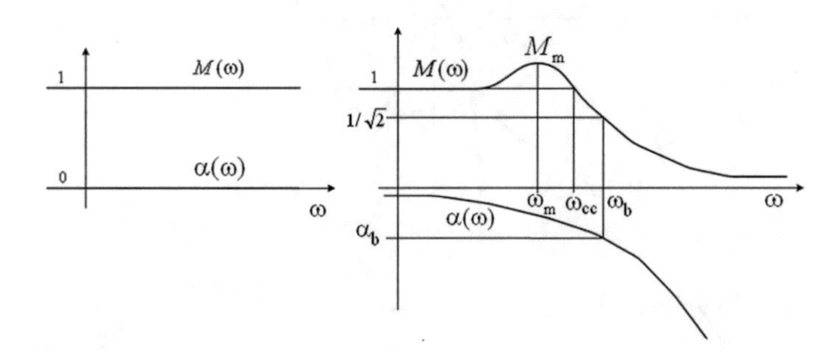

Fig. 4.31 Ideal and real frequency function of the closed-loop system

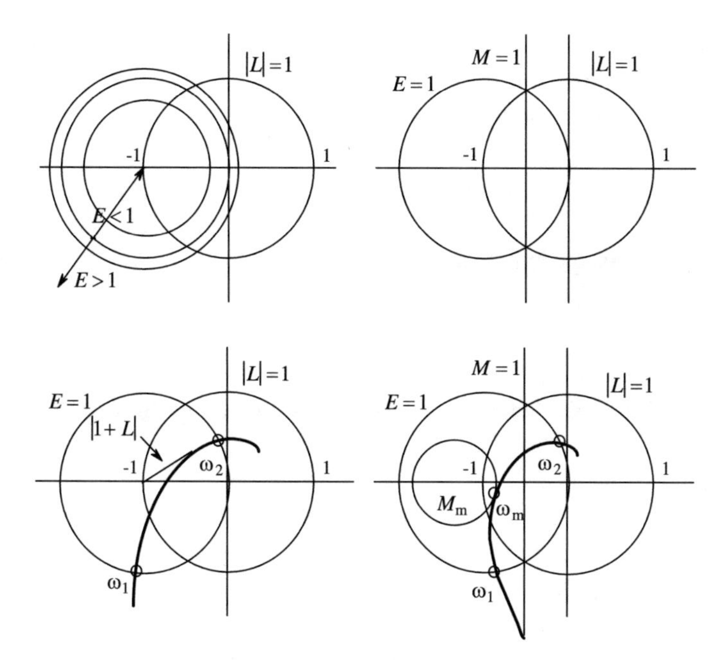

Fig. 4.32 Curves corresponding to $M = 1$, $E = 1$ and $|L| = 1$

and the $|1 + L(j\omega)|$ in the denominator is equal to the distance from the point $(-1 + j0)$. These curves are concentric circles around $(-1 + 0j)$ with radius $1/E$. The curve belonging to $E = 1$ has special significance.

In Fig. 4.32 the curves belonging to $M = 1$, $E = 1$, $|L| = 1$ also the distance $|1 + L|$ are indicated. It is shown how to determine the maximal value M_m with the open-loop Nyquist diagram. The interception points ω_1 and ω_2 of the $L(j\omega)$ characteristics and the $E = 1$ circle indicate the range where $|S(j\omega)| < 1$.

4.7 The Sensitivity of a Closed Control Loop to Parameter Uncertainties

The parameters of a process are never known quite accurately. Also, the process may change over time. The environment of the process may change and as a consequence the parameters of the process may also change within a given range. Negative feedback decreases the sensitivity of the system to parameter changes. In controller design it is advisable to take the possible parameter changes into consideration. The behavior of the control system has to be acceptable not only for the nominal parameter values, but throughout the whole possible range of the parameter changes.

Let us analyze the behavior of the system if the transfer function of the process changes from its nominal value $P_o(s)$ to $P(s) = P_o(s) + \Delta P(s)$. The overall transfer function of the open-loop is $L = CP$, whose small change is

$$\Delta L = \frac{\partial L}{\partial P}\Delta P = C\Delta P. \tag{4.28}$$

The relative change is expressed by

$$\frac{\Delta L}{L} = \frac{C\Delta P}{CP} = \frac{\Delta P}{P}. \tag{4.29}$$

The overall transfer function of the closed-loop realized by negative feedback (Fig. 4.25) is

$$T = \frac{CP}{1 + CP}, \tag{4.30}$$

whose small change is

$$\Delta T = \frac{\partial T}{\partial P}\Delta P = \frac{C}{(1 + CP)^2}\Delta P. \tag{4.31}$$

The relative value of the change is

$$\frac{\Delta T}{T} = \frac{1}{1 + CP}\frac{\Delta P}{P} = S\frac{\Delta P}{P}, \tag{4.32}$$

where S is the sensitivity function of the closed-loop:

$$S = \frac{\Delta T/T}{\Delta P/P} = \frac{1}{1 + CP}. \tag{4.33}$$

The sensitivity function shows how much a relative change of the process $(\Delta P/P)$ influences the relative change of the resulting transfer function $(\Delta T/T)$. In the frequency range where $|L(j\omega)| \to \infty$, the sensitivity function takes small values, thus even big parameter changes in the process have a small effect on the resulting closed-loop transfer function, and also on the output signal of the closed-loop.

For an infinitesimally small change ($\Delta P \to 0$):

$$\frac{\partial T}{T} = S\frac{\partial P}{P}, \tag{4.34}$$

whence

$$S = \frac{\partial T/T}{\partial P/P} = \frac{\partial \ln T}{\partial \ln P}. \tag{4.35}$$

The resulting transfer function T of the closed-loop is also called the complementary sensitivity function, as the following relationship holds:

$$S + T = 1. \tag{4.36}$$

Typical amplitude-frequency curves of the loop transfer function L, the sensitivity function S, and the complementary sensitivity function T are shown in Fig. 4.33.

Let us consider now the sensitivity of the control system with respect to parameter changes in the feedback element (Fig. 4.34). This sensitivity function can be defined by the following relationship:

$$S_{\mathrm{H}} = \frac{\Delta T/T}{\Delta H/H}. \tag{4.37}$$

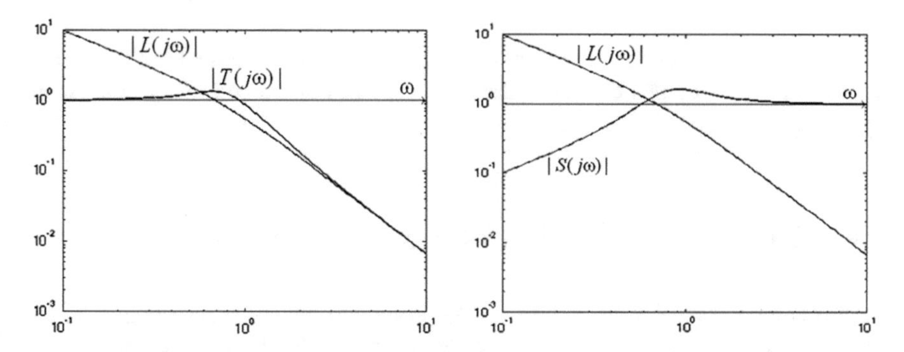

Fig. 4.33 Typical amplitude-frequency curves of the loop, the sensitivity function and the complementary sensitivity function

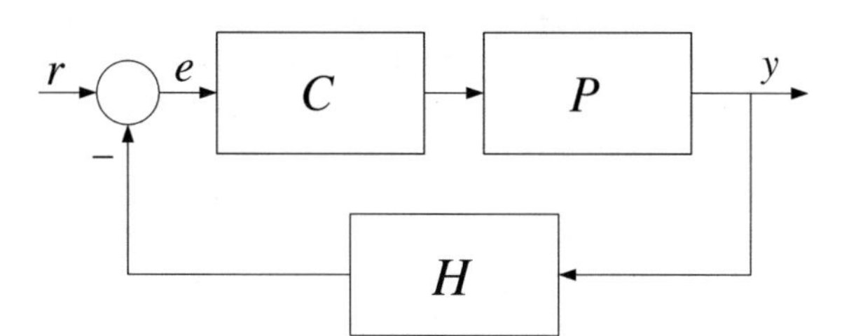

Fig. 4.34 Feedback control circuit, the sensor has dynamic characteristics

Now

$$T = \frac{CP}{1+CPH}, \quad \text{thus} \quad \Delta T = \frac{\partial T}{\partial H}\Delta H = -\frac{(CP)^2}{(1+CPH)^2}\Delta H, \tag{4.38}$$

and

$$\frac{\Delta T}{T} = S_{\mathrm{H}}\frac{\Delta H}{H} = -\frac{CPH}{1+CPH}\frac{\Delta H}{H} = -\frac{L}{1+L}\frac{\Delta H}{H}. \tag{4.39}$$

As $S_{\mathrm{H}} = -L/(1+L) = -T$ has to take approximately the value of 1 in a wide frequency range to ensure good reference signal tracking, the parameter changes in the feedback element may significantly influence the output signal. Therefore it is required to measure the output signal very accurately, or to realize unity feedback.

Formulas (4.6)–(4.8) giving the relationships between the input and the output signals can also be given by the sensitivity functions.

$$Y(s) = F(s)T(s)R(s) + S(s)Y_{\mathrm{no}}(s) + P(s)S(s)Y_{\mathrm{ni}}(s) - T(s)Y_{\mathrm{z}}(s) \tag{4.40}$$

$$E(s) = F(s)S(s)R(s) - S(s)Y_{\mathrm{no}}(s) - P(s)S(s)Y_{\mathrm{ni}}(s) - S(s)Y_{\mathrm{z}}(s) \tag{4.41}$$

$$U(s) = F(s)C(s)S(s)R(s) - C(s)S(s)Y_{\mathrm{no}}(s) - T(s)Y_{\mathrm{ni}}(s) - C(s)S(s)Y_{\mathrm{z}}(s) \tag{4.42}$$

So with the sensitivity functions, not only the effects of parameter changes can be investigated, but also the signal transfer properties of the control system can be analyzed.

4.8 Requirements for Closed-Loop Control Design

A closed-loop control system has to meet prescribed quality specifications. These specifications depend on the control aims, on the technology of the considered process and also on the process itself.

In a rolling-mill, e.g., the uniform thickness of the steel sheet has to be ensured with high accuracy. The aim of the utilization of the steel sheet will also influence the desired accuracy. In a heat treatment process the temperature has to be set according to a given program. In the treated material undesirable alterations should not happen. This requirement influences the prescribed accuracy of the reference signal tracking. The accuracy of directing an airplane into a path and then tracking the path is important to reach the destination station while ensuring the avoidance of other airplanes. The prescription of the required settling time is also important. This requirement has to consider the dynamics of the process. In case of a very slow process, a big acceleration can not be expected, as this would require too high, practically unrealizable manipulating input signals. The prescriptions should be tailored to the opportunities. The prescriptions consider both the static and the dynamic properties of the closed-loop control system.

The requirements set for a closed-loop control system are:

- stability
- appropriate static accuracy for reference signal tracking and disturbance rejection
- attenuation of the effect of measurement noise
- insensitivity to parameter changes
- prescribed dynamic (transient) behavior
- consideration of the restrictions stemming from the practicality of the realization.

A linear closed-loop control system is stable, its steady state is achieved, if the roots of the characteristic equation are on the left side of the complex plane (see Sects. 4.2 and 4.5).

Static accuracy of the control system for typical input signals (step, ramp, parabolic input) is determined by the number of the integrators in the open-loop (Sect. 4.5).

Disturbance rejection, attenuation of measurement noise, and the effects of parameter changes can be investigated by the sensitivity functions (Sects. 4.6 and 4.7).

The prescribed dynamic behavior is generally given by the characteristic parameters of the unit step response $v(t)$ of the closed-loop system (Fig. 4.35).

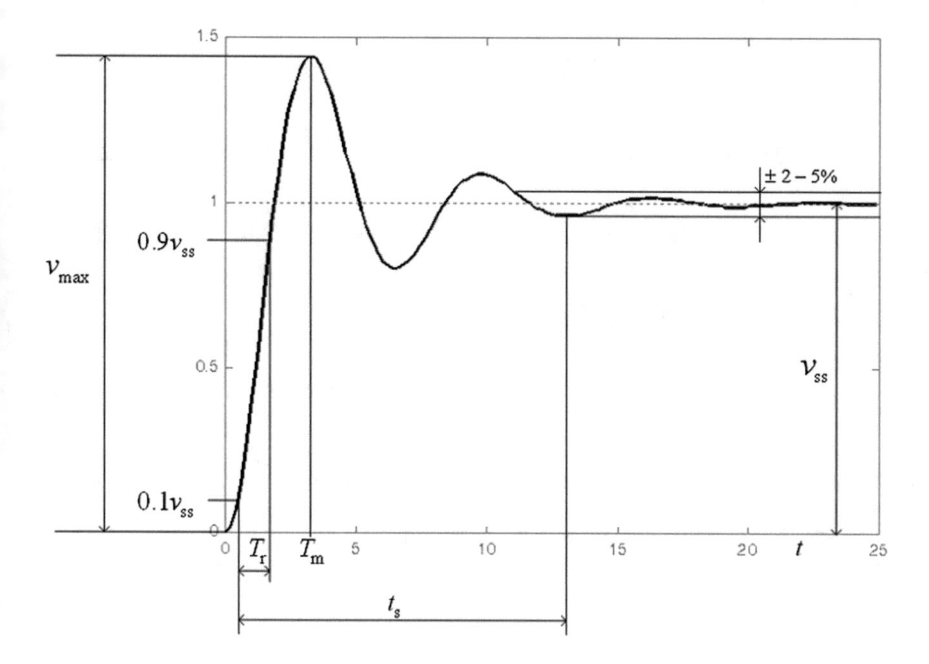

Fig. 4.35 Dynamic specifications of a control system

The static error for a unit step input is

$$1 - v_{ss}. \tag{4.43}$$

The overshoot of the unit step response is

$$\sigma = \frac{v_{max} - v_{ss}}{v_{ss}} 100\%. \tag{4.44}$$

The settling time t_s is the time when the unit step response of the closed-loop system reaches its steady state within $\pm(2-5)\%$ accuracy.

During the rise time T_r the step response starting from 10% reaches 90% of its steady state value. The time of reaching the maximum value is denoted by T_m.

In control techniques the main point of controller design is to find an acceptable compromise between a large overshoot and a long settling time. This compromise can be formulated on the one hand by prescribing the distance of the loop frequency function from the point $(-1+0j)$, which characterizes the stability limit (see Sect. 5.6). On the other hand, a quality index can be formulated, which can be the minimum (optimum) value of an integral criterion. This optimum value indicates a balance between the two extreme transients. In this case the quality of the control performance is evaluated on the basis of an integral of the error signal $e(t) = v(\infty) - v(t)$. The controller parameters are chosen to reach the minimum of this error integral.

The formulas for the different criteria involving integrals are as follows:

$$I_1 = \int_0^\infty e(t)\mathrm{d}t \quad \text{linear control error area} \tag{4.45}$$

(it can be applied only to aperiodic systems, it can be evaluated analytically)

$$I_2 = \int_0^\infty e^2(t)\mathrm{d}t \quad \text{quadratic control error area} \tag{4.46}$$

(it can be calculated analytically)

$$I_3 = \int_0^\infty |e(t)|\mathrm{d}t = IAE \quad \text{Integral of Absolute value Error} \tag{4.47}$$

$$I_4 = \int_0^\infty t|e(t)|\mathrm{d}t = ITAE \quad \text{Integral of Time multiplied by Absolute value Error} \tag{4.48}$$

I_3 and I_4 can be evaluated only by simulation.

Linear control error area.

With simple considerations the following relationship can be obtained:

$$I_1 = \lim_{t\to\infty} \int_0^t e(\tau)\mathrm{d}\tau = \lim_{s\to 0} s\frac{E(s)}{s} = E(0), \tag{4.49}$$

where $E(s) = \mathcal{L}\{e(t)\}$ is the LAPLACE transform of the error signal. For an aperiodic control system given by the transfer function $T(s)$ where $T(0) = A$ (see Fig. 4.36)

$$T(s) = A\frac{\prod_{k=1}^{m}(1+s\tau_k)}{\prod_{j=1}^{n}(1+sT_j)} = AT'(s), \tag{4.50}$$

let us calculate the linear control error area. According to (4.49),

$$\begin{aligned} I_1 &= \int_0^\infty (A - v(t))\,\mathrm{d}t = \left[(A - T(s))\frac{1}{s}\right]_{s=0} = A\left[\frac{1-T'(s)}{s}\right]_{s=0} \\ &= A\left[\frac{\prod_{j=1}^{n}(1+sT_j) - \prod_{k=1}^{m}(1+s\tau_k)}{s\prod_{j=1}^{n}(1+sT_j)}\right]_{s=0} = A\left(\sum_{j=1}^{n} T_j - \sum_{k=1}^{m}\tau_k\right) \end{aligned} \tag{4.51}$$

The time constants in the denominator of the transfer function increase the linear control error area, but the time constants in its numerator decrease it. Thus by introducing zeros, the system can be accelerated.

For aperiodic transients an equivalent dead time T_e can be defined, which is the dead time of the step function of amplitude A measured from time point $t = 0$, whose linear control error area is equal to the control error area of the step response of the considered transfer function.

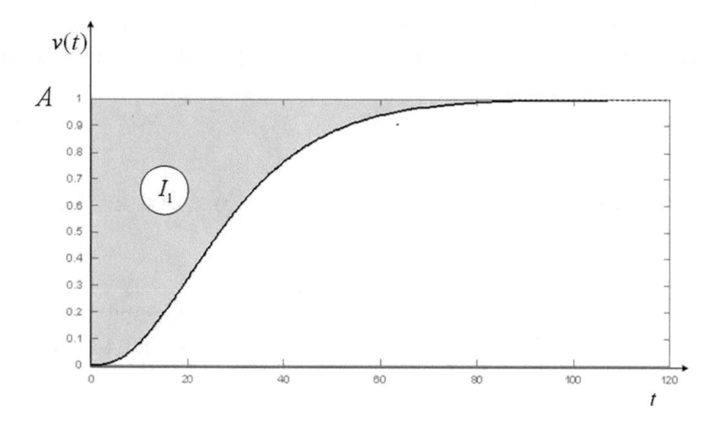

Fig. 4.36 Linear control area of an aperiodic process

$$T_{\mathrm{e}} = \frac{I_1}{A} = \sum_{j=1}^{n} T_j - \sum_{k=1}^{m} \tau_k. \tag{4.52}$$

Quadratic control error area

The quadratic control error area can be evaluated also in the frequency domain using the PARSEVAL theorem.

$$I_2 = \int_0^{\infty} e^2(t)\mathrm{d}t = \frac{1}{2\pi j} \int_{-\infty}^{\infty} E(-s)E(s)\mathrm{d}s = \frac{1}{\pi} \int_0^{\infty} |E(j\omega)|^2 \mathrm{d}\omega. \tag{4.53}$$

The quadratic control error area can be calculated analytically. For lower degree cases for strictly proper LAPLACE transforms of the error signal ($m < n$) of the form

$$E(s) = \frac{\sum_{i=0}^{m} c_i s^i}{\sum_{i=0}^{n} d_i s^i} \tag{4.54}$$

calculation formulae have been derived for evaluation of the I_2 integral for a given degree and for given c_i and d_i parameters. A general formula in algorithmic form can also be given, which provides a special, not too complex recursive algorithm. It should be mentioned that minimizing the quadratic error area as a function of a controller parameter generally results in a flat minimum. Unfortunately the optimal transient generally gives a quite high overshoot (20–25%), so this optimal controller can not be used in high quality control systems.

Absolute value criteria

It is difficult to evaluate a criterion using the absolute value of the error. Instead of analytical calculation the minimum can rather be determined by simulation or with searching optimization methods. The minimum of the cost function is generally sharp. The *Integral of Time multiplied Absolute value Error* (*ITAE*) criterion punishes the error values at the beginning of the time scale less than those occurring at later time points. The optimum (minimum) of this criterion provides beautiful transients with ~5% overshoot.

4.9 Improving the Disturbance Elimination Properties of the Closed-Loop

An adequately designed closed-loop control system ensures good reference signal tracking and also the rejection of the effects of input and output disturbances. If along the path from the disturbance to the output there are signal components with large time constants, then the disturbance rejection will be slow. Of course, in controller design considerations related to disturbance rejection have to be also taken into account.

Disturbance rejection can be improved if not only the effects caused by the disturbance in the output signal are utilized for disturbance rejection, but possibly some internal measurable signals are also used in which the effect of the disturbance appears already earlier than in the output signal. Utilizing the available information in the control circuit, better, deliberate decisions can be made, and thus the quality of the control system can be improved.

4.9.1 Disturbance Elimination Scheme (Feedforward)

If the disturbance is measurable, the quality of the control system, especially its disturbance rejection properties, can be significantly improved by letting it drive a feedforward. Based on the measured value of the disturbance it is possible to execute actions to reject it before its effect would appear in the controlled variable. The block diagram of feedforward control is shown in Fig. 4.37. With appropriate design of the feedforward controller $C_{\mathrm{n}}(s)$ the effect of the disturbance can be significantly decreased or even totally compensated. The disturbance acts on the output through two paths. The resulting transfer function between the output and the disturbance is

$$\frac{Y(s)}{Y_{\mathrm{n}}(s)} = \frac{P_{\mathrm{n}}(s) + C_{\mathrm{n}}(s)P(s)}{1 + C(s)P(s)}. \tag{4.55}$$

The effect of the disturbance will not appear in the output signal if the numerator of the above expression is zero, i.e. if

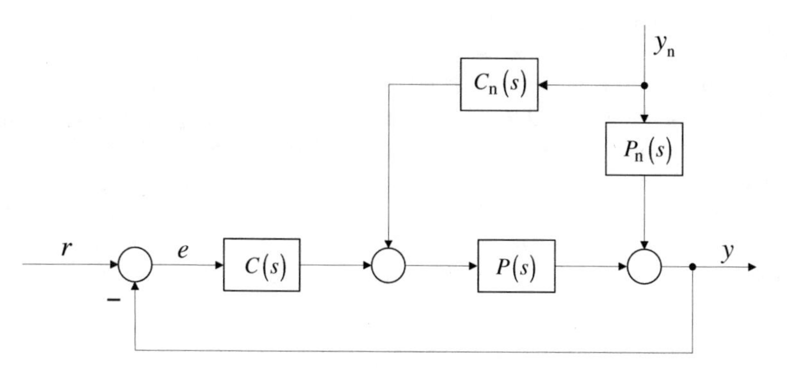

Fig. 4.37 Block diagram of feedforward control

$$P_{\mathrm{n}}(s) + C_{\mathrm{n}}(s)P(s) = 0. \tag{4.56}$$

Hence the transfer function of the feedforward compensator is:

$$C_{\mathrm{n}}(s) = -\frac{P_{\mathrm{n}}(s)}{P(s)}. \tag{4.57}$$

If this transfer function is realizable (i.e. if the degree of its numerator is not higher than the degree of its denominator and furthermore $P(s)$ does not contain dead time), the effect of the feedforward is perfect: the effect of the disturbance does not appear at all in the output signal. If $C_{\mathrm{n}}(s)$ is non-realizable, its transfer function has to be approximated by the best realizable controller.

Feedforward supplements the closed-loop control circuit with an open-loop path. The efficiency of the feedforward compensation depends on how accurately the effect of the disturbance on the output signal is known, and how much it is possible to compensate it with the available manipulations.

As an example let us consider the control scheme of a belt dryer furnace shown in Fig. 4.38. In the electrically heated furnace the material to be dried goes through the conveyor G driven by the motor M. The controlled signal is the moisture content of the material leaving the furnace. At a given conveyor speed the material abides in the furnace for a given time. The manipulated variable is the heating power, which can be changed by a voltage u across the resistance R. The humidity of the material leaving the furnace is measured. It is compared to the reference signal. In case of deviation, the heating power is modified through a *PI* controller. (A *PI* controller consists of a proportional (*P*) and an integrating (*I*) element connected in parallel, see Chap. 8). The main disturbance source is the change of the humidity of the incoming material. Time is needed to eliminate the effect of the disturbance. The control system comes into operation only after the effect of the disturbance has been detected at the output. Thus for a certain time the humidity of the outcoming material will differ from its desired value. If the humidity of the incoming material is measurable, then based on this measured value the heating power could be immediately set to a value which on the basis of a priori knowledge expectably would be needed to ensure the prescribed humidity value through the P part of the controller. (The integrating part of the controller can not be included in the feedforward path, as its output can not reach a finite steady state because of the constant input signal.) The feedforward part of the controller is denoted by the dashed line in Fig. 4.38. Then the closed-loop control circuit has to eliminate only the error component resulting from the inaccuracy of the a priori knowledge.

4.9.2 Cascade Control Schemes

Several times the processes can be separated into serially connected parts, and besides the output signal the intermediate signals can also be measured. Figure 4.39 shows the block diagram of a process which consists of two serially connected

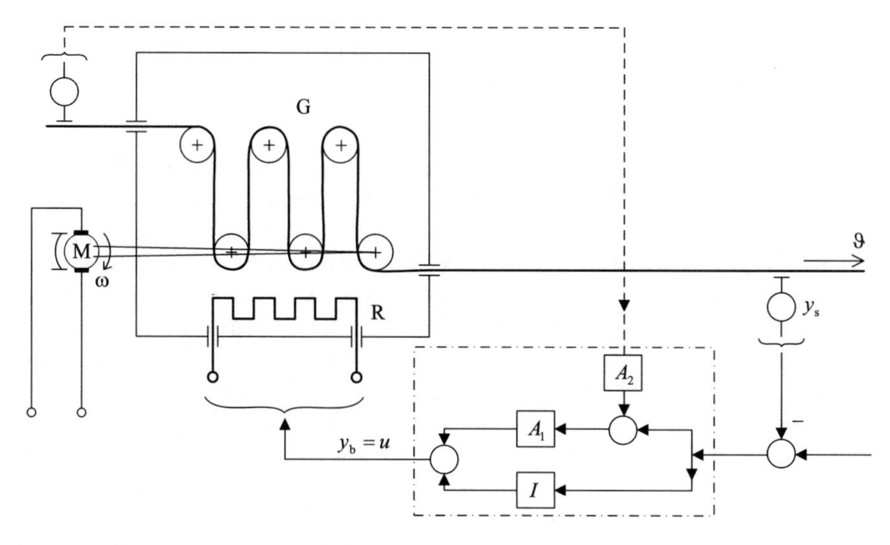

Fig. 4.38 Control scheme of a belt dryer furnace with feedforward

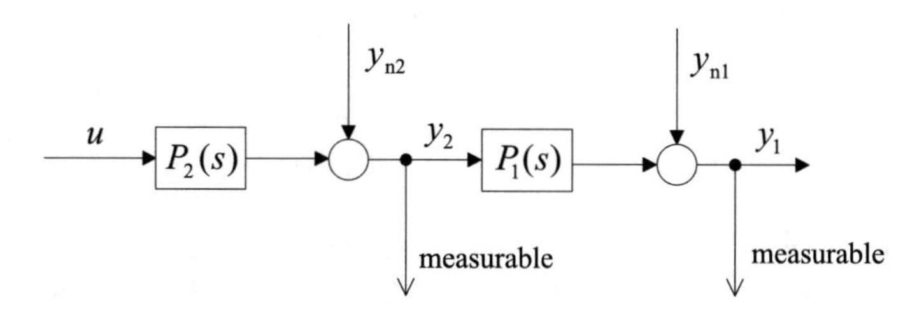

Fig. 4.39 A process which can be separated into two serially connected parts

parts. The disturbances may act on the output or between the two parts of the process. It is supposed that the disturbances themselves are not measurable.

The block diagram of the conventional feedback control is shown in Fig. 4.40. The closed-loop control system is able to track the reference signal and also to reject the effect of the disturbances. To activate the disturbance rejection it is necessary that the effect of the disturbance should appear at the output. Then an

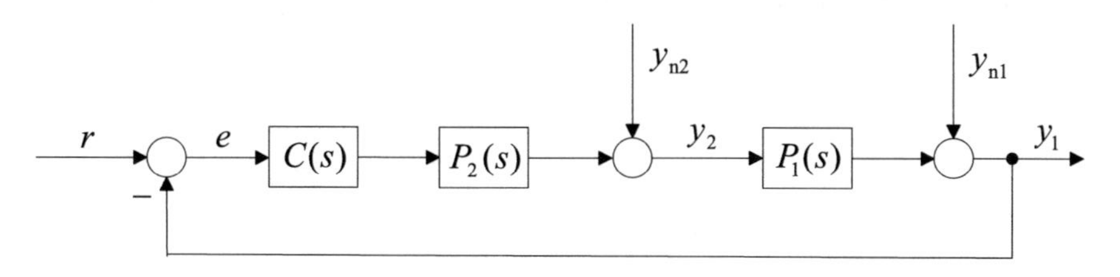

Fig. 4.40 Block diagram of feedback control

error signal appears in the closed-loop which activates the control circuit to eliminate the effect of the disturbance. If the $P_1(s)$ part of the process contains the larger time constants, the rejection of the disturbance y_{n2} acting between the two parts of the process will be slow.

It is worthwhile to create an inner loop using the measurable y_2 signal, which is capable of supporting a fast rejection of the inner disturbance. As the effect of the inner disturbance appears sooner in signal y_2 than in the output y_1, the inner loop can rather quickly decrease the effect of the inner disturbance. The outer loop ensures good reference signal tracking, the rejection of the output disturbance and further attenuation of the effect of the inner disturbance which has been already decreased by the inner loop. The block diagram of the control circuit with two loops —called a cascade control—is shown in Fig. 4.41. The advantage of cascade control compared to a single-loop feedback control is manifested if part $P_1(s)$ of the plant contains the large time constants and/or dead time, while part $P_2(s)$ contains the smaller time constants. The controller of the inner loop $C_2(s)$, is designed for fast performance of the internal loop, thus the inner loop will quickly reject the internal disturbance. With the controller $C_1(s)$ of the outer loop, good reference signal tracking and rejection of the external disturbance is to be ensured. The inner controller could be of structure *P* or *PD*. In the inner loop the feedback provides acceleration, thus because of the smaller time constants the compensation of the outer loop will be easier. The controller in the outer loop which ensures the quality specifications could be of structure *PI* or *PID*. (A *PID* controller consists of parallel connected proportional (*P*), integrating (*I*) and differentiating (*D*) elements, see Chap. 8.)

In some applications it is expedient to put a saturation after the external controller. As the output of the external controller provides the reference signal of the internal loop, by restricting its value, the internal signal y_2 can also be kept within prescribed limits.

Of course if the process can be separated into more than two components, where the internal signals are measurable, a cascade control can be realized with several nested control circuits.

Cascade control is applied generally in the speed or position control of electrical drives, where the output variable is the speed or the position, and the internal variable is the current. In this case the aim of cascade control is mainly the restriction of the armature current. Namely, the current may reach very high values when starting, breaking or loading the motor, while the speed is developed more slowly because of the mechanical inertia of the system. Thus it is not enough to feed back only the speed, the current also

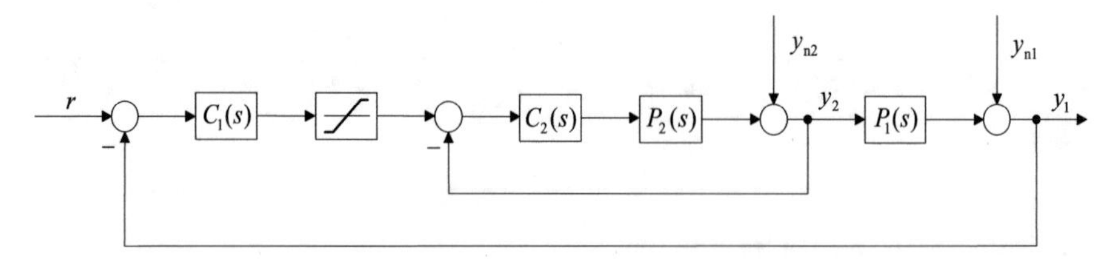

Fig. 4.41 Block diagram of cascade control

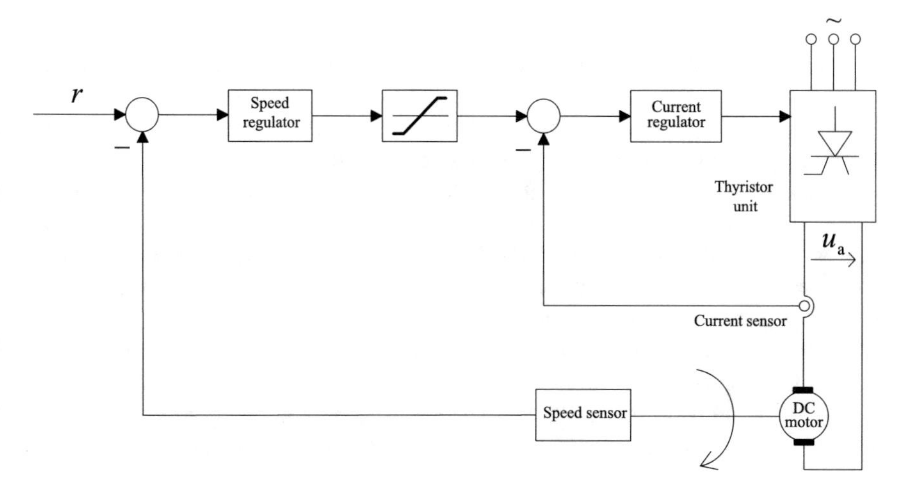

Fig. 4.42 Cascade control of a DC motor

has to be observed, and its value has to be kept within the allowed range. The cascade control of a DC motor is shown schematically in Fig. 4.42.

Figure 4.43 shows a cascade control solution for room temperature control. The controlled variable is the ϑ temperature of room T, which is set to the required value by the air blown across the steam heated heat exchanger H. The manipulated variable is the steam blowing through the heat exchanger, which is set by valve B. The main disturbance is the pressure of the steam, as the amount of the steam, i.e. the heating power entering the heat exchanger H depends on the pressure in a given valve position. For the cascade stage the internal controlled variable could be the temperature ϑ_k of the steam coming out of the heat exchanger, as the effect of the change of the heating power is observed sooner in ϑ_k than in the room temperature ϑ.

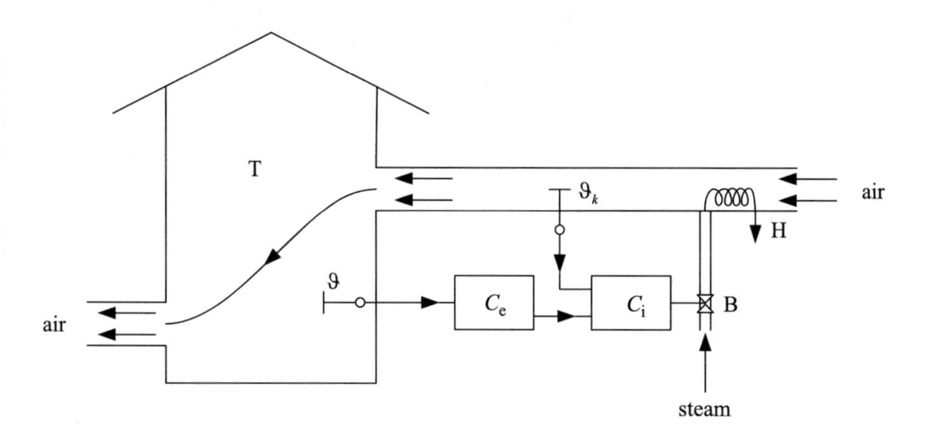

Fig. 4.43 Cascade solution for room temperature control

In a single loop closed-loop control circuit only the output signal is fed back. In cascade control, besides the output signal, one or more measurable internal signals are also fed back, thus improving the quality of the control. The control system will be faster, and could reject the internal disturbances more effectively. These internal variables generally are the state variables of the system.

In a system, the internal variables, the so called state variables, determine the dynamic behavior of the system. Their instantaneous values depend on the previous moves of the input signal. With the knowledge of the actual values of the state variables and the input signal, the states of the system and the output signal at the next time point can be determined.

When building a closed-loop control system not only the measurement of a single output signal or of some additional inner signals considered in cascade loops is important, but it is essential to measure and feed back all the state variables (in a system described by a differential equation of order n, their number is n). This control concept is called state feedback, which can be considered as a generalization of the cascade control concept. Chapter 10 deals with state feedback control in detail.

4.10 Compensation by Feedback Blocks

If some internal signals of a process are measurable, applying feedback on them the performance of the control system can favorably be influenced. The block diagram of feedback compensation is shown in Fig. 4.44. The equivalent series compensation can be determined in a straightforward way. The advantage of feedback compensation is that besides modifying the performance of the closed-loop system it linearizes the relationship between the output and the input signals of the internal loop and considerably decreases the effect of parameter changes. The internal loop is also effective in rejecting internal disturbances. Compensation by a feedback block may also show an advantageous behavior when the control signal is saturated.

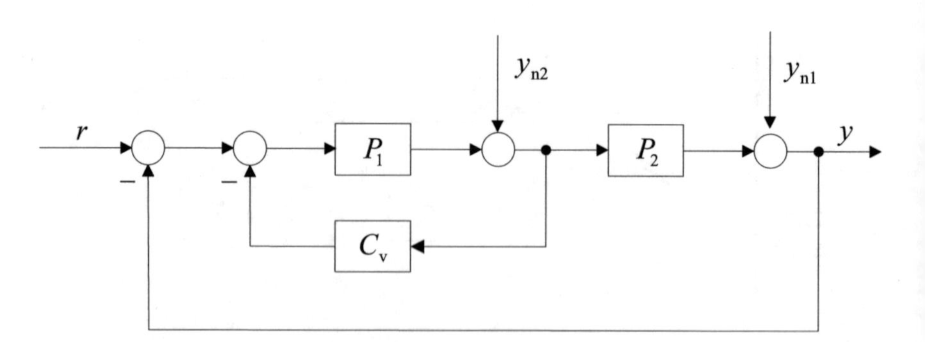

Fig. 4.44 Block diagram of compensation by a feedback block

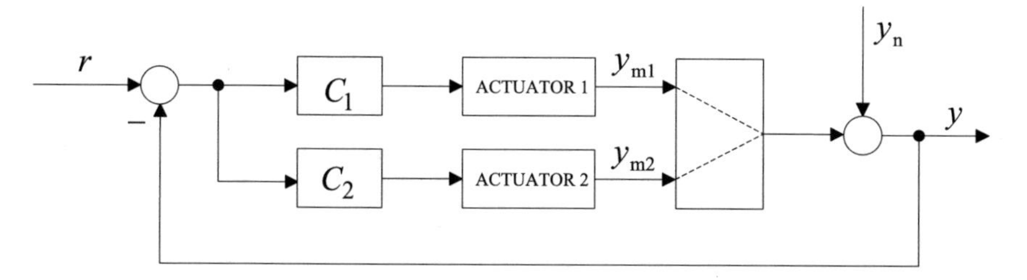

Fig. 4.45 Block diagram of a control system with two loops applying an auxiliary manipulated variable

With appropriate internal feedback, the inverse of the process can be generated, providing a favorable control solution.

Generally it is sufficient to have an internal loop working only in the transient state, while in the steady state only the external loop is efficient. This can be accomplished by feedback through a differentiating element combined with a first order lag: $C_v(s) = A_v s\tau/(1 + sT_1)$.

4.11 Control with Auxiliary Manipulated Variables

Typically there are several possibilities for manipulation at the process input. Depending on the properties of the process one of the manipulated variables is fundamental, while the others can be used as auxiliary possibilities applied, in general, only temporarily.

The block diagram of a control system with an auxiliary loop is shown in Fig. 4.45. As an application example, let us consider the control of the belt dryer furnace shown earlier in Fig. 4.38. If the humidity of the material coming into the furnace changes abruptly, without feedforward its effect on the output is recognized only when the furnace is already full of the material of changed quality. In this case, if only the heating of the furnace is modified, a relatively large amount of the material comes out of the furnace with humidity that differs from the required value, as the thermal inertia of the furnace is big and the temperature can only be changed slowly. The manipulation becomes more effective if the speed of the conveyor is also changed temporarily by changing the speed of the motor M. Thus the residence time of the material in the furnace is shortened. The auxiliary manipulation can only be temporary, which can be achieved by e.g. using a proportional controller in the auxiliary loop, that influences the armature voltage of the motor. In the main control loop, a *PI* controller is applied, thus in steady state the error signal is zero, and then the auxiliary circuit becomes inactive.

Chapter 5
Stability of Linear Control Systems

In practical applications, the stability of the control system is an important requirement. A control task can not be realized with an unstable control loop. The stability of a control system has to be distinguished from the stability of the process itself. There are cases when an unstable process has to be stabilized and controlled with a closed-loop control system. There are processes which would not even operate without control, the closed-loop control stabilizes the process. The best known examples of such systems are the control of an airplane, or in everyday life, riding a bicycle.

Closed-loop control circuits may present surprising phenomena. These phenomena are due to the process dynamics, inertia and time-delays. Therefore the processes can not follow immediately the commands acting on their input. Generally, the time of the response of a process is not within the time scale of human reactions (sometimes it is much slower, and sometimes much faster). In some cases the short time response does not agree with what we would experience waiting for a bit longer time (e.g., non-minimum phase processes). Therefore experimental investigation of the stability is not acceptable in operating and controlling real processes, which are generally very expensive. Precise mathematical methods are needed to analyze the stability of control systems.

5.1 The Concept of Stability

If a system has the property that it will get back into the equilibrium state again after moving away from its equilibrium state, then it is stable.

If the system is non-linear, its stability depends on the input signal and also on the operating point. In this case, stability is a characteristic of a state of the system, and not of system as a whole. In case of a linear system, stability is characteristic for

L. Keviczky et al., *Control Engineering*, Advanced Textbooks in Control and Signal Processing, https://doi.org/10.1007/978-981-10-8297-9_5

the system. Stability depends on the system's structure and parameters, but does not depend on the input signal. As far as stability is concerned, a number of various formulations exist.

Stability of an un-excited system

A system is stable if after removing it from its equilibrium state and allowing it to move freely, it returns to its original state. If the system moves away from its original state, its behavior is unstable. The system is on the boundary of stability and unstability if after removal from the equilibrium state it does not return to it, but remains in its close vicinity, which depends on the extent of the removal (e.g. it makes un-damped oscillations with bounded amplitude around the initial state). In non-linear systems, the system is also considered stable if in the boundary case removing it from the steady state it returns to an arbitrarily prescribed small vicinity of the steady state. The system is *asymptotically stable* if after removing it from its equilibrium state it returns to its original state. A stable linear system is asymptotically stable. In the case of asymptotic stability the weighting function $w(t)$ of a linear system is decreasing in the following sense:

$$\lim_{t\to\infty} w(t) = 0 \tag{5.1}$$

and furthermore $w(t)$ is absolutely integrable, i.e.

$$\int_0^\infty |w(t)|\mathrm{d}t < \infty \tag{5.2}$$

Stability of an excited system

A system is stable if it responds to any bounded input signal with a bounded output signal, from any initial condition. Stability of the excited system is called *Bounded-Input–Bounded-Output* (*BIBO*) stability.

For linear systems, stability is a system property. Stability does not depend on the magnitude of the excitation. Additionally, for linear systems, if the un-excited system is stable, then the excited system is also stable. Stability can be checked unambiguously from the system response to a simple input signal.

Internal stability

A closed-loop control system fulfills the requirement of *internal stability* if its output signal and all of its inner signals respond in a stable way to any outer excitation signal. Let us investigate the control system shown in Fig. 5.1. Besides tracking the reference signal r the rejection of the effect of disturbance y_{ni} and y_{no} acting at the input and the output of the plant P, respectively and the effect of the measurement noise y_{z} on the output are also investigated. The system is stable if for

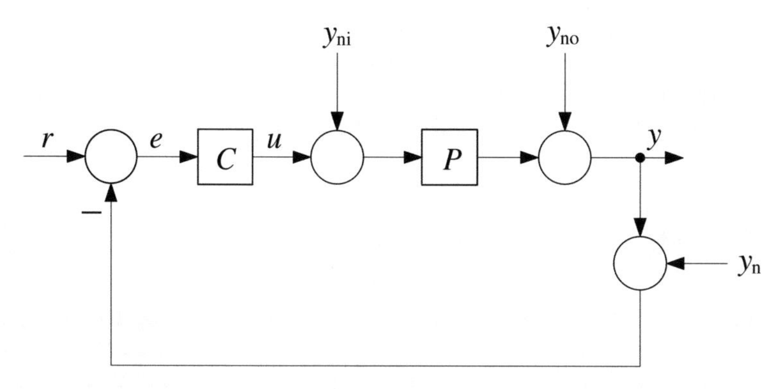

Fig. 5.1 Block diagram of a closed-loop control system

all the considered bounded input signals the controlled output signal y, the manipulated control variable u and the error signal e are bounded. It can be shown that in the structure of Fig. 5.1 it is always sufficient to choose two arbitrary external and two arbitrary internal signals. Internal stability requires the investigation of the stability of the following four overall transfer functions: $CP/(1+CP)$, $1/(1+CP)$, $P/(1+CP)$, $C/(1+CP)$. This can be characterized by the *transfer matrix* of the closed-loop control circuit

$$\boldsymbol{T}_{\mathrm{t}} = \begin{bmatrix} \frac{CP}{1+CP} & \frac{P}{1+CP} \\ \frac{C}{1+CP} & \frac{1}{1+CP} \end{bmatrix} \tag{5.3}$$

A closed-loop control system has the property of internal stability if $\boldsymbol{T}_{\mathrm{t}}$ is stable, i.e. all its elements are stable. Internal stability is equivalent to the stability of the excited system if the open-loop system has no non-observable or non-controllable right side poles (i.e. the zeros of the controller do not cancel the pole in the right half-planes of the plant). (It has to be emphasized that it is not allowed to cancel an unstable pole of the plant with a right side zero of the controller, as the unstable pole would become invisible only in the relationship between the output signal and the reference signal, but would remain in the relationship between the output signal and the disturbance acting at the input of the plant.)

LYAPUNOV stability

According to the LAGRANGE energy theorem a system is in balance if its potential energy is minimal. LYAPUNOV prescribes the determination of a scalar function of energy property (the so called LYAPUNOV function) belonging to the differential equation or state equation of a general nonlinear system with constant coefficients. If in the considered range of the state variables this function is positive and its derivative is negative, the system is asymptotically stable. The methods of LYAPUNOV provide sufficient conditions for the determination of the stability properties of nonlinear systems. Choosing a LYAPUNOV function is not always a simple task. LYAPUNOV suggests first the investigation of the stability of the linearized system at individual operating points. Of course, the method of LYAPUNOV can also be applied to investigate the stability of a linear system. But for linear systems, it is expedient to use simpler direct methods.

5.2 Stability of the Closed-Loop System

Negative feedback, which is the basic structure of a closed-loop control system, also involves the risk of instability. To demonstrate this let us consider the control loop shown in Fig. 5.2. Assuming a step-like abrupt change of the reference signal, the output signal starts to grow from zero. Then the error signal decreases starting from an initial value of 1. If the gain of the controller is high, first a large input signal appears at the plant input, which results in a sharp rise in the output signal. The dynamics of this change is determined by the dynamics of the process P and the controller C, i.e., by the gains and the time constants of the corresponding transfer functions. When the output signal reaches its required value, viz., the one prescribed by the reference signal, the error signal reaches zero. But because of the inertia of the system, the signals will not be settled immediately at their required values, but maintaining their trend they will continue changing further, according to their actual slope. If the output signal exceeds its prescribed value, the error signal becomes negative, and after a while the output signal will start decreasing. With large time constants of the process and high gains of the controller, the overshoot may be significant. Steady or increasing oscillations may appear in the control system. The problem of stability emerges because the system uses the information supplied by the error signal in a delayed manner, and if the gains are high, during the delay time the output signal "runs away" so much that the control system will not be able to bring it back to its required value. The parameters of the controller always have to be chosen in such a way that the control system is stable.

The instability of a feedback control system is caused by large time constants and high gains. This phenomenon is illustrated by the behavior of the control system shown in Fig. 5.3. The process is represented by a pure dead-time with unit gain, given by its step response. The dead-time element follows its input signal u after a time specified by T_{d}. The controller is a pure proportional element with gain A, thus the loop gain is $K = A$. Let us investigate the signals in the control circuit for $K = 0.5$, 1 and 2. The evolution of the signals can be easily followed.

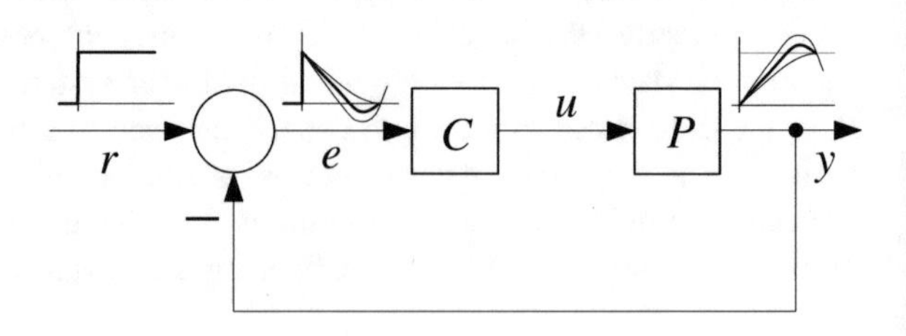

Fig. 5.2 Dynamics of a control system

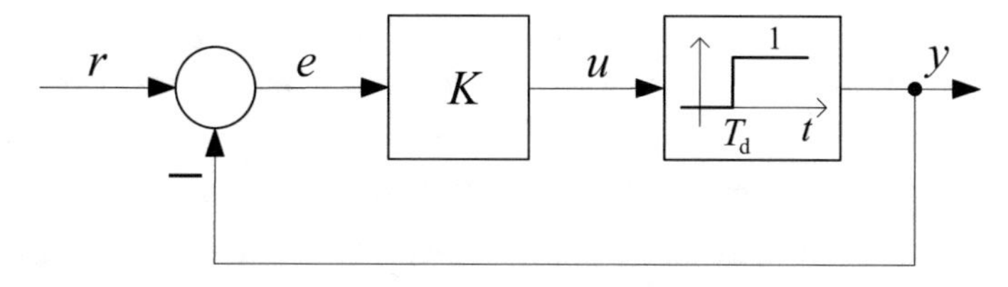

Fig. 5.3 Control system with dead-time

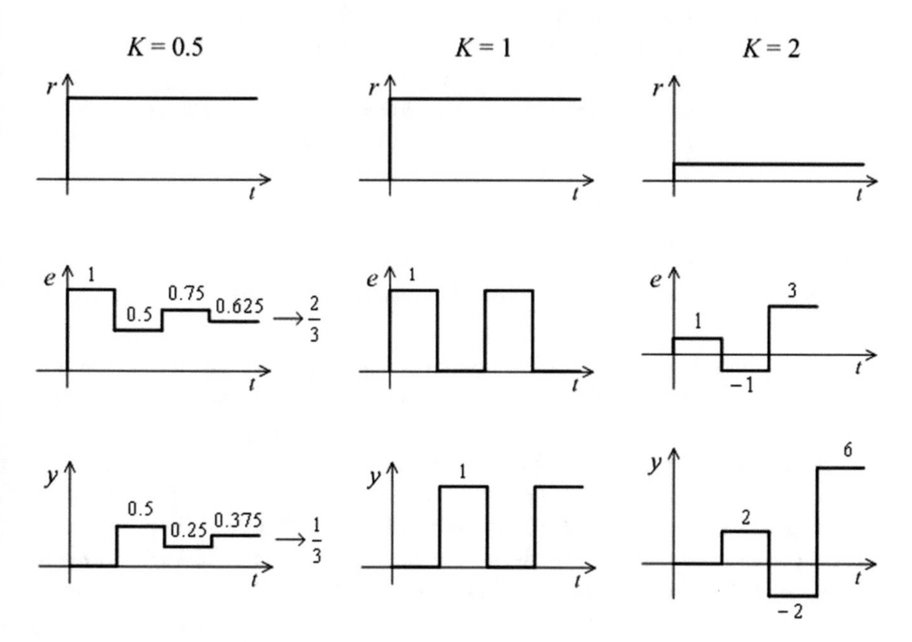

Fig. 5.4 Signals in a control circuit with dead-time

Figure 5.4 shows the reference signal, the error signal, and the controlled output signal. With $K = 0.5$ the control system is stable (but it is inaccurate: the output signal settles at 1/3 instead of the required value 1). In the case of $K = 1$, steady oscillations appear: the system is on the borderline between stability and instability. With $K = 2$, the system is unstable.

The values of the individual signals can also be given analytically in the considered time ranges according to Table 5.1.

Table 5.1 Signal values in a closed-loop control system with dead-time

Time range t	Error signal e	Output signal y
$0 - T_d$	1	0
$T_d - 2T_d$	$1 - K$	K
$2T_d - 3T_d$	$1 - K(1 - K)$	$K(1 - K)$
$3T_d - 4T_d$	$1 - K[1 - K(1 - K)]$	$K[1 - K(1 - K)]$

It is seen that with the progress of time the error signal e can be given by a geometrical series with quotient $-K$. If $K<1$, the series converges to $\lim_{t\to\infty} e(t) = 1/(1+K)$, and the limit value of the output signal is $\lim_{t\to\infty} y(t) = K/(1+K)$. Thus the stability limit is $K = 1$. The higher the value of K, the smaller the steady error in the control circuit, but the requirement of stability sets a limit for increasing K. Stability and static accuracy are often contradictory requirements. In the design of a control system, an appropriate compromise has to be realized to ensure both stability and the required static accuracy.

Stability is an important property for a linear system. In the case of instability the control system "runs away" even if it is excited only temporarily by some noise, e.g. an impulse acts at its input. Figure 5.5 shows the signals in the case of $K = 2$ when the reference signal is a short time impulse of amplitude unity.

5.3 Mathematical Formulation of the Stability of Continuous Time Linear Control Systems

If an un-excited closed-loop control system is asymptotically stable then the time function describing its transients contains components that are decreasing functions of time. The transient time function is a combination of exponential components whose exponents are the roots of the characteristic equation of the system.

In a controllable and observable control system (when the zeros of the controller do not cancel the poles of the plant) the roots of the characteristic equation are identical to the poles of the overall transfer function of the closed-loop. Formally, the characteristic equation of the differential equation describing the system is equivalent to the denominator of the overall transfer function of the closed-loop system.

That is, the overall transfer function of the closed-loop between the output signal y and the reference signal r is

$$T(s) = \frac{Y(s)}{R(s)} = \frac{C(s)P(s)}{1+C(s)P(s)} = \frac{C(s)P(s)}{1+L(s)}. \tag{5.4}$$

The differential equation of the system is the inverse LAPLACE transform of

$$[1+L(s)]Y(s) = C(s)P(s)R(s) \tag{5.5}$$

and the characteristic equation is, formally,

$$1+L(s) = 0. \tag{5.6}$$

Thus the roots of the characteristic equation are the same as the poles of the overall transfer function of the closed-loop system.

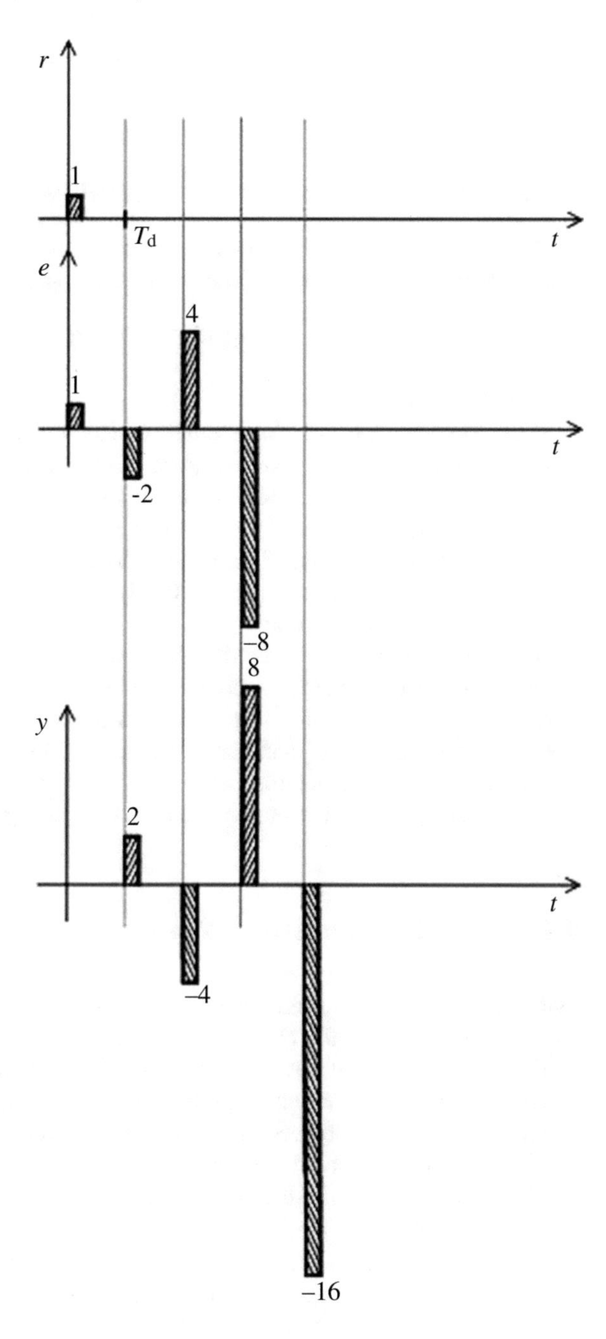

Fig. 5.5 The signals at the output of an unstable system "run away" even if a short time signal acts at its input

If the loop transfer function is a rational fraction, i.e. $L(s) = \mathcal{N}(s)/\mathcal{D}(s)$ where $\mathcal{N}(s)$ and $\mathcal{D}(s)$ are polynomials, then the characteristic equation can also be given in the following form:

$$\mathcal{A}(s) = \mathcal{D}(s) + \mathcal{N}(s) = 0 \tag{5.7}$$

or

$$a_n s^n + a_{n-1} s^{n-1} + \cdots + a_1 s + a_o = a_n (s - p_1)(s - p_2)\ldots(s - p_n) = 0. \tag{5.8}$$

If the system is described by its state equation with state matrix $\boldsymbol{A}$, then the characteristic equation can be given by the relationship

$$\det(s\boldsymbol{I} - \boldsymbol{A}) = 0. \tag{5.9}$$

(see also Chap. 3).

If the coefficients of the characteristic equation are real numbers, then the roots of the equation are real numbers or pairs of complex conjugate numbers.

The condition for asymptotic stability is that the real part of the poles p_i of the closed-loop have negative real parts, as this condition ensures that the transients are decreasing function of time. This condition can also be formulated as follows: a closed-loop control system is asymptotically stable if all of its poles lie in the left half-plane of the complex plane.

If any of the poles lies in the right half-plane, the system is unstable. If besides the poles in the left half-plane there are poles in the origin, then there is an integrating effect in the system, for step input its output signal goes to infinity. If there are pairs of complex conjugate simple poles on the imaginary axis, then steady oscillations do appear in the transients. In the case of multiple poles, the amplitudes of the oscillations are increasing. In practice, only asymptotic stability is acceptable.

5.4 Analytical Stability Criteria

Stability can be decided from the location of the roots of the characteristic equation which are the poles of the closed-loop system.

If there is no dead-time, the characteristic equation is an algebraic equation, whose roots can be given analytically provided the degree is less than 5 (Galois theorem). For higher degrees, numerical root searching methods can be applied which determine the roots with a given accuracy.

If the system contains dead-time, then the characteristic equation is a transcendental equation $\mathcal{D}(s) + \mathcal{N}(s)e^{-sT_d} = 0$, whose solution is not simple, and in the case of instability it is difficult to decide how to stabilize the system. In this case, the characteristic equation can be approximated by a rational functional approximation of the dead-time, or the investigation has to be done in the frequency domain (see Sect. 5.6).

Several procedures have been elaborated to determine the stability without solving the characteristic equation. These procedures are referred to as stability criteria.

If there is no dead-time, then based on the relationships between the roots and the coefficients of the algebraic equation it can be checked with analytical stability criteria whether all the roots lie in the left half of the complex plane, i.e., whether the system is stable or not.

A necessary condition for stability is that all the coefficients of the characteristic equation must be of the same sign and none of the coefficients can be zero. This can be seen easily based on Eq. (5.8). That is, if all the poles have negative real parts, then by multiplying the root factors, all the coefficients will be positive. If there are also pairs of complex conjugate roots with negative real parts, then multiplying the root factors the obtained coefficients are also positive. Suppose $p_{1,2} = -\alpha \pm j\beta$, where $\alpha > 0$ and $\beta > 0$. Let us multiply together the two corresponding factors $[s-(-\alpha+j\beta)][s-(-\alpha-j\beta)] = s^2 + 2\alpha s + \alpha^2 + \beta^2$. The coefficients are evidently positive. In the first- and second-degree cases the sameness of the signs of the coefficients is not only a necessary, but also a sufficient condition for stability. In the sequel, two analytical methods will be given, without proof for checking stability.

5.4.1 *Stability Analysis Using the* Routh *Scheme*

Let us build the following scheme from the coefficients of the characteristic polynomial given in (5.8):

$$\begin{array}{ccccc} a_n & a_{n-2} & a_{n-4} & a_{n-6} & \cdots \\ a_{n-1} & a_{n-3} & a_{n-5} & a_{n-7} & \cdots \\ b_{n-2} & b_{n-4} & b_{n-6} & b_{n-8} & \cdots \\ c_{n-3} & c_{n-5} & c_{n-7} & c_{n-9} & \cdots \\ \vdots & & & & \end{array} \tag{5.10}$$

where

$$\begin{aligned} b_{n-2} &= \frac{a_{n-1}a_{n-2} - a_n a_{n-3}}{a_{n-1}}, \; b_{n-4} = \frac{a_{n-1}a_{n-4} - a_n a_{n-5}}{a_{n-1}}, \; b_{n-6} = \frac{a_{n-1}a_{n-6} - a_n a_{n-7}}{a_{n-1}}, \ldots \\ c_{n-3} &= \frac{b_{n-2}a_{n-3} - a_{n-1}b_{n-4}}{b_{n-2}}, \; c_{n-5} = \frac{b_{n-2}a_{n-5} - a_{n-1}b_{n-6}}{b_{n-2}}, \ldots \end{aligned} \tag{5.11}$$

The length of the rows is decreasing. If the degree of the characteristic polynomial is n, the scheme consists of $n+1$ rows. The arrangement given by (5.10) and, (5.11) is called the Routh scheme.

A system is stable if all the coefficients of its characteristic equation are positive and all the elements of the first column of its Routh scheme are positive. If not all the elements in the first column are positive, the system is unstable, and the number of the changes in the signs gives the number of poles of the closed-loop system that lie in the right half-plane. A zero in the first column indicates that the characteristic equation has a root on the imaginary axis. In this case, the scheme can be continued by taking an arbitrarily small ε value instead of zero.

Example 5.1 Suppose the loop transfer function of a control circuit is $L(s) = K/s(1+s)(1+5s)$. In a closed-loop circuit, a unit negative feedback is applied. Let us determine the value of the critical gain K that brings the control system to the stability limit. The characteristic equation is

$$1+L(s) = 1+\frac{K}{s(1+s)(1+5s)} = 0,$$

or

$$5s^3+6s^2+s+K = 0.$$

As all the coefficients have to be positive, the necessary condition for stability is $K>0$.

$$\text{The ROUTH scheme is :}\quad \begin{matrix} 5 & 1 \\ 6 & K \\ \frac{6-5K}{6} & 0 \\ K & \end{matrix}.$$

To ensure stability, all the elements of the first column have to be positive. Thus the condition for stability is

$$0<K<1.2.$$

■

5.4.2 *Stability Analysis Using the* Hurwitz *Determinant*

Let us build the following Hurwitz determinant of dimension $n \times n$ from the coefficients of the characteristic polynomial (5.8)

$$\begin{vmatrix} a_{n-1} & a_{n-3} & a_{n-5} & a_{n-7} & \cdots \\ a_n & a_{n-2} & a_{n-4} & a_{n-6} & \cdots \\ 0 & a_{n-1} & a_{n-3} & a_{n-5} & \cdots \\ 0 & a_n & a_{n-2} & a_{n-4} & \cdots \\ 0 & 0 & a_{n-1} & a_{n-3} & \cdots \\ \vdots & & & & \end{vmatrix} \tag{5.12}$$

Elements with negative indices are taken to be zeros. The system is stable if all the coefficients of the characteristic equation are positive and all the subdeterminants along the main diagonal are also positive: $\Delta_i > 0$. The subdeterminants are:

$$\Delta_1 = |a_{n-1}|,\ \Delta_2 = \begin{vmatrix} a_{n-1} & a_{n-3} \\ a_n & a_{n-2} \end{vmatrix},\ \Delta_3 = \begin{vmatrix} a_{n-1} & a_{n-3} & a_{n-5} \\ a_n & a_{n-2} & a_{n-4} \\ 0 & a_{n-1} & a_{n-3} \end{vmatrix}, \ldots, \Delta_n. \tag{5.13}$$

Example 5.2 Let us investigate the stability of the system analyzed in Example 5.1 on the basis of the HURWITZ determinant. The characteristic equation is

$$5s^3 + 6s^2 + s + K = 0.$$

As all the coefficients have to be positive, $K > 0$.
The HURWITZ determinant is

$$\begin{vmatrix} 6 & K & 0 \\ 5 & 1 & 0 \\ 0 & 6 & K \end{vmatrix}. \tag{5.14}$$

The subdeterminants along the main diagonal are

$$\Delta_1 = 6 > 0;\ \Delta_2 = 6 - 5K > 0 \text{ and } \Delta_3 = K\Delta_2 > 0.$$

Thus the condition of stability is $0 < K < 1.2$. ■

5.5 Stability Analysis Using the Root Locus Method

The root locus gives the location of the roots of the characteristic equation of the closed-loop system in the complex plane as a parameter (generally the loop gain) changes between zero and infinity.

If the roots are in the left half-plane, the system is stable. At the critical gain the root locus crosses the imaginary axis. At gains where the root locus has moved to the right half-plane, the system becomes unstable.

From the root locus, not only the stability of the closed-loop system can be checked, but from the location of the roots, also the dynamic properties can be determined approximately.

For drawing the root locus, the characteristic equation has to be solved for different parameter values. Today's computer techniques and *CAD* programs provide considerable help in drawing the root locus branches. But often there is the need for a rapid qualitative analysis to assist the designer in design considerations. Therefore, several rules have been elaborated to support the quick sketching of the root locus. (It is also called the EVANS method, after the name of the developer of the method.)

5.5.1 Basic Relationships of the Root Locus Method

The characteristic equation of the closed-loop control circuit $1 + L(s) = 0$ can be written also in the following form:

$$L(s) = -1 = k\frac{\prod_{j=1}^{Z}(s - z_j)}{\prod_{i=1}^{P}(s - p_i)}, \tag{5.15}$$

where Z denotes the number of zeros, P is the number of poles and k, is the loop gain factor of the pole-zero form.

For all the points of the root locus the *absolute value condition*

$$|L(s)| = 1 \tag{5.16}$$

and the phase condition

$$\varphi = \pm N180°; \quad N = 1, 3, 5, \ldots \tag{5.17}$$

have to be fulfilled.

This means that for the construction of the root locus, those points in the complex plane are to be looked for that fulfill both the phase condition and the absolute value condition.

Let us denote the absolute value of the vector connecting the zero z_j with an arbitrary point s of the complex plane by C_j, and its phase angle with the positive real axis by γ_j. The absolute value of the vector connecting the pole p_i with the same point s is denoted by D_i, while its phase angle is denoted by δ_i (Fig. 5.6). That is

$$s - z_j = C_j e^{j\gamma_j} \tag{5.18}$$

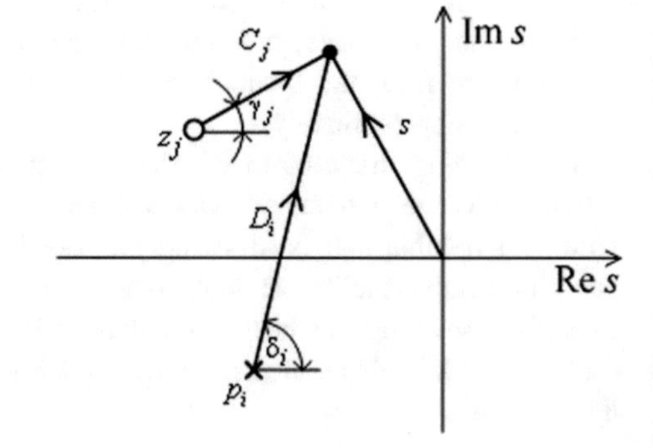

Fig. 5.6 Notation for the vectors connecting the points of the root locus with the poles and zeros of the open-loop

and

$$s - p_i = D_i e^{j\delta_i}. \tag{5.19}$$

The phase condition can be given in the following form:

$$\sum_{j=1}^{Z} \gamma_j - \sum_{i=1}^{P} \delta_i = \pm N 180^\circ; \quad N = 1, 3, 5, \ldots \tag{5.20}$$

or

$$\sum_{i=1}^{P} \delta_i - \sum_{j=1}^{Z} \gamma_j = \mp N 180^\circ; \quad N = 1, 3, 5, \ldots \tag{5.21}$$

For the absolute value condition, the following relationship holds:

$$\frac{\prod_{i=1}^{P} D_i}{\prod_{j=1}^{Z} C_j} = k. \tag{5.22}$$

A point in the complex plane is a point of the root locus if for that point both the phase condition and the absolute value condition are fulfilled.

The phase condition can also be formulated as follows: a point s on the complex plane is the point of the root locus if from the sum of the angles of the vectors connecting the zeros of the open-loop with that point s one subtracts the sum of the angles of the vectors connecting the poles of the open-loop with s and gets an odd multiple of $\pm 180^\circ$.

The absolute value condition states that a point s is the point of the root locus if dividing the product of the absolute values of the vectors connecting the poles with point s by the product of the absolute values of the vectors connecting the zeros with point s yields the loop gain factor.

Generally, the points of the root locus are determined from the phase condition, and the value of the loop gain factor corresponding to the considered point is obtained from the absolute value condition. Then from the loop gain factor, the loop gain K belonging to the time constant form of the transfer function of the open-loop is calculated by

$$K = L(s)|_{s=0} = k \frac{\prod_{j=1}^{Z} (-z_j)}{\prod_{i=1}^{P} (-p_i)}. \tag{5.23}$$

5.5.2 Rules for Drawing Root Locus

There are some simple rules which facilitate drawing the root locus:

1. The root locus is symmetrical with respect to the real axis.
2. The number of its branches is equal to the number of poles of the open-loop transfer function.
3. The root locus starts from the poles of the open-loop when $K=0$ and runs to the zeros or to infinity when $K \to \infty$. If the number of poles is P and the number of zeros is Z, then Z branches of the root locus run to the zeros and $P-Z$ branches run to infinity. If $P=Z$, the whole root locus is located in a finite range of the complex plane.
4. Sections of the root locus will be on the real axis if to the right of the considered point the sum of the poles and zeros is odd. (It is sufficient to count the real poles and zeros, as the complex poles or complex zeros appear in pairs.)
5. The direction of the asymptotes of the root locus is given by the angles

$$\alpha = \mp \frac{N180^\circ}{P-Z}; \quad N = 1, 3, 5, \ldots \tag{5.24}$$

6. The asymptotes of the root locus cross the real axis at the point calculated by the following relationship:

$$x_o = \frac{\sum_{i=1}^{P} p_i - \sum_{j=1}^{Z} z_j}{P-Z} = \frac{\sum_{i=1}^{P} \mathrm{Re}p_i - \sum_{j=1}^{Z} \mathrm{Re}z_j}{P-Z}. \tag{5.25}$$

7. The location of leaving or entering the real axis can be determined by the equation

$$\sum_{i=1}^{P} \frac{1}{x-p_i} - \sum_{j=1}^{Z} \frac{1}{x-z_j} = 0 \tag{5.26}$$

8. The critical gain factor can be determined from the characteristic equation by the Routh scheme or the Hurwitz determinant. The crossing points with the imaginary axis can be calculated from the characteristic equation assuming that in this case two of its roots are pure imaginary complex conjugate roots.

Explanation of the drawing rules

1. As the coefficients of the characteristic equation are real numbers, its roots are real or complex conjugate pairs. Therefore the root locus is symmetrical to the real axis.
2. The degree of the characteristic equation is equal to the number of the poles of the open-loop. Namely, if the transfer function of the open-loop is a rational fraction, $L(s) = \mathcal{N}(s)/\mathcal{D}(s)$, the characteristic equation is $1+L(s) = 1+\mathcal{N}(s)/\mathcal{D}(s) = 0$, or $\mathcal{D}(s)+\mathcal{N}(s) = 0$. As the degree of $\mathcal{D}(s)$ is

greater than or equal to the degree of $\mathcal{N}(s)$, the degree of the characteristic equation will be equal to the degree of $\mathcal{D}(s)$, and the number of its roots will be equal to the number of poles of the open-loop. Thus with a change of the loop gain the root locus will have as many branches as the number of poles of the open-loop.

3. From relationship (5.15)

$$-k = \frac{\prod_{i=1}^{P}(s-p_i)}{\prod_{j=1}^{Z}(s-z_j)}; \quad P \geq Z. \tag{5.27}$$

$k = 0$ holds if $s = p_i$. Thus the root locus starts from the poles of the open-loop when $k = 0$. $k = \infty$ holds if $s = z_j$ or $s \to \infty$. Thus if $k \to \infty$ the roots of the characteristic equation run into the zeros of the open-loop, or if $P > Z$ then the number $P - Z$ of the roots goes to infinity.

4. If a point s of the root locus is on the real axis, the vectors connecting it with the complex conjugate poles (or zeros) make an angle of 0° or 360° considering the pairs, therefore they can be disregarded. The real poles or zeros if they are to the left of the considered point s, make an angle of 0°, while if they are to the right of the point, they make an angle of 180°. To fulfill the phase condition (5.17), the sum of the number of the poles and the zeros to the right of s has to be odd.
5. The asymptotes approach the very distant points of the root locus, from where the p_i poles and the z_j zeros of the open-loop are all seen under the same angle α:

$$P\alpha - Z\alpha = \mp N 180° \tag{5.28}$$

hence the angles of the asymptotes are

$$\alpha = \mp \frac{N180°}{P-Z}; \quad N = 1, 3, 5, \ldots \tag{5.29}$$

6. Taking into consideration the poles with weight +1 and the zeros with weight −1, the crossing point of the asymptotes with the real axis is just at the center of gravity, as looking at the system from a longer distance it can be replaced by its center of gravity. The rule can be derived analytically as well.
7. The phase condition is fulfilled also for a point x where the root locus steps out or arrives at the real axis. According to Fig. 5.7 leaving the real axis with a small

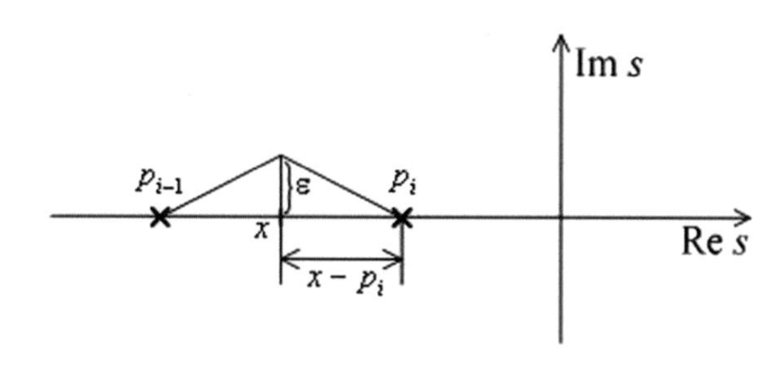

Fig. 5.7 Determination of the place where the root locus leaves the real axis

ε distance perpendicularly and replacing the small angles with their tangents the following relationship can be written:

$$\sum_{i=1}^{P} \frac{\varepsilon}{x - p_i} - \sum_{j=1}^{Z} \frac{\varepsilon}{x - z_j} = 0 \tag{5.30}$$

Now (5.26) follows from (5.30).

8. On the borderline between stability and unstability, the characteristic equation has roots on the imaginary axis.

5.5.3 *Examples of the Root Locus Method*

Example 5.3 Let us consider the system given in Examples 5.1 and 5.2. The loop transfer function is $L(s) = K/s(1+s)(1+5s)$. A negative feedback of unity is applied. The loop transfer function in zero-pole form is

$$L(s) = \frac{k}{s(s+1)(s+0.2)},$$

where $k = 0.2K$ is the loop gain. Determine the root locus. The varying parameter is the loop gain k.

On the basis of the construction rules it can be seen that the root locus has three branches. The branches start from $s_1 = 0$, $s_2 = -0.2$, and $s_3 = -1$, the poles of the loop transfer function, and go to infinity. On the real axis the root locus has a section between the points 0 and -0.2, and in the range between -1 and $-\infty$. Between the points 0 and -0.2 the root locus steps off of the real axis. The angle of the asymptotes going to infinity is $\alpha = \pm N180°/(3-0)$: at $N = 1$ the angle is $\pm 60°$, and at $N = 3$ it is $180°$. The asymptotes cross the real axis at $-1.2/3 = -0.4$. The point where the root locus steps out of the real axis is calculated by solving equation $\frac{1}{x} + \frac{1}{x+1} + \frac{1}{x+0.2} = 0$. The solutions are: $x_1 = -0.7055$ and $x_2 = -0.0945$. Only x_2 can be a solution, since the root locus may not have a point at x_1. Figure 5.8 shows the root locus. The critical loop gain k_{cr} can be determined from the characteristic equation by either the Routh or the Hurwitz criterion. The root locus crosses the imaginary axis at this gain. In Examples 5.1 and 5.2 its value was calculated by both methods. The stability range of the system is $0<K<1.2$ or $0<k<0.24$, respectively. The characteristic equation at the critical value $k_{cr} = k = 0.24$ is:

$$s(s+1)(s+0.2) + 0.24 = s^3 + 1.2s^2 + 0.2s + 0.24 = 0$$

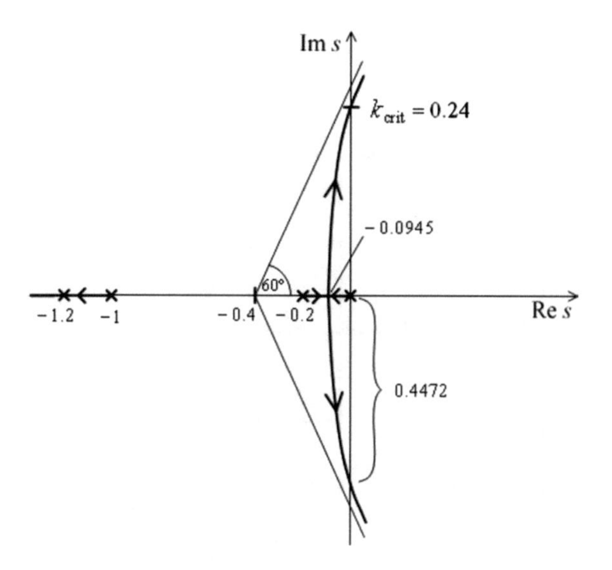

Fig. 5.8 Root locus of an integrating two lag element with negative unity feedback

Two of the roots are on the imaginary axis. Thus

$$\begin{aligned} s^3 + 1.2s^2 + 0.2s + 0.24 &= (s+\gamma)(s+j\eta)(s-j\eta) = (s+\gamma)\left(s^2+\eta^2\right) \\ &= s^3 + \gamma s^2 + \eta^2 s + \gamma\eta^2. \end{aligned}$$

Comparing the coefficients, we obtain

$$\gamma = 1.2 \quad \text{and} \quad \eta = \sqrt{0.2} = 0.4472.$$

The oscillation frequency is determined by the η interception with the imaginary axis. ■

Further examples for root loci

The root loci of some systems (without proper scaling) are shown in Table 5.2. Comparing the figures, it can be seen that a new pole pushes away the branches of the root locus, while a new zero attracts them.

Figure 5.9 shows that in the case of three poles, the introduction of a zero, modifies the shape of the root locus. By appropriate location of the zero the closed-loop system can be stabilized over the whole range of the gain factor.

Figure 5.10 gives the root locus of an unstable open-loop system. The transfer function of the open-loop is

$$L(s) = \frac{k(s+1)}{s(s-1)(s+6)}.$$

This open-loop system has an unstable pole. Inserting an additional zero can ensure that the closed-loop system becomes stable with for appropriate choice of the gain $(k > k_{\text{cr}} = 7.5)$.

Table 5.2 Root loci of typical systems

P	Z			
	0		1	
	Transfer function	Root locus	Transfer functions	Root locus
1	$\frac{K}{s}$	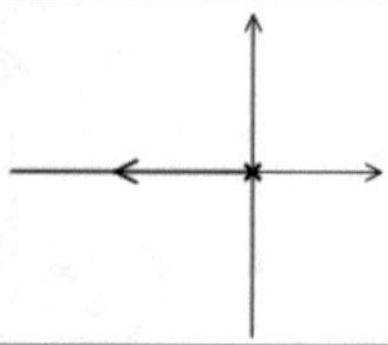	$K\frac{1+sT_1}{1+sT_2}$ $T_1 > T_2$	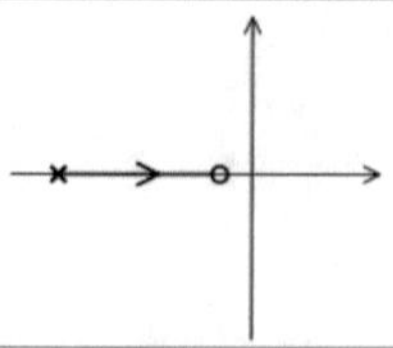
	$\frac{K}{1+sT}$	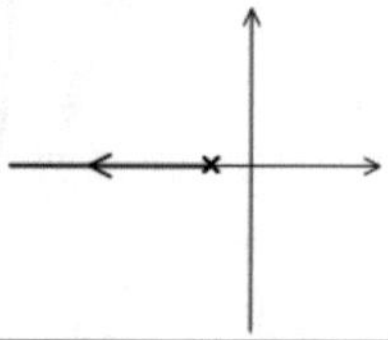	$K\frac{1+sT_1}{1+sT_2}$ $T_2 > T_1$	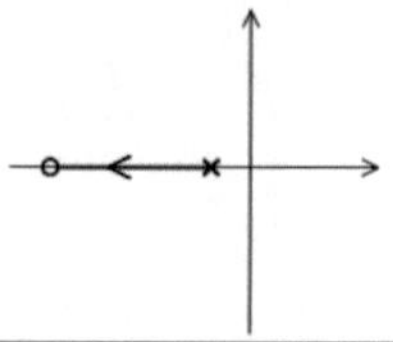
2	$\frac{K}{(1+sT_1)(1+sT_2)}$	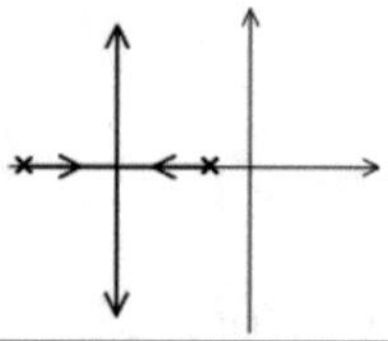	$\frac{K(1+sT_2)}{(1+sT_1)(1+sT_3)}$	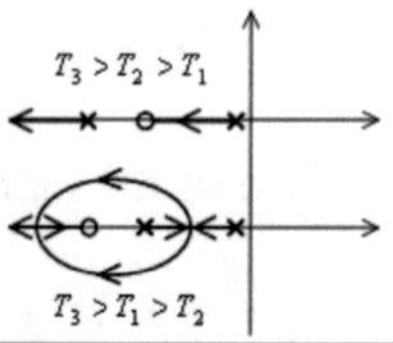 $T_3 > T_2 > T_1$ $T_3 > T_1 > T_2$
	$\frac{K}{1+s2\xi T+s^2T^2}$	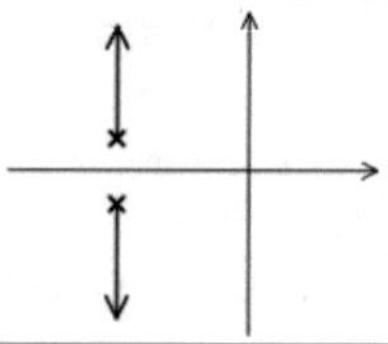	$\frac{K(1+sT_1)}{1+s2\xi T+s^2T^2}$	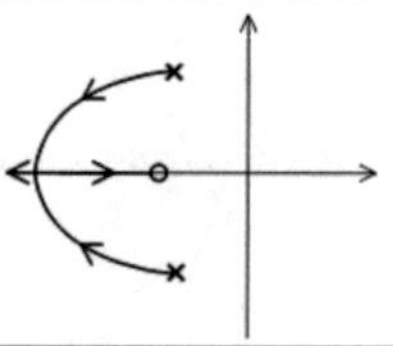
3	$\frac{K}{(1+sT_1)(1+sT_2)(1+sT_3)}$	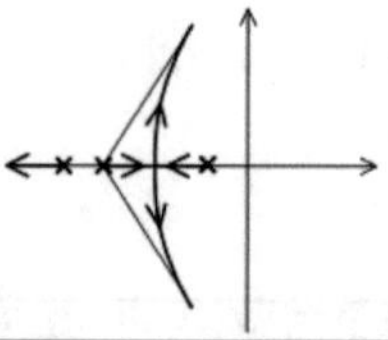	$\frac{K(1+sT_4)}{(1+sT_1)(1+sT_2)(1+sT_3)}$	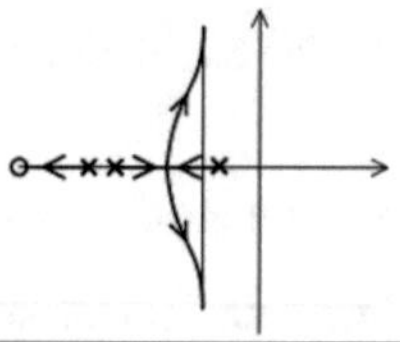
	$\frac{K}{(1+s2\xi T_1+s^2T_1^2)(1+sT_2)}$	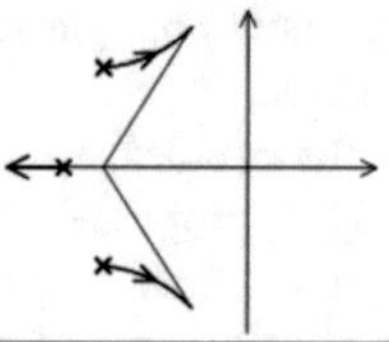	$\frac{K(1+sT_2)}{(1+s2\xi T_1+s^2T_1^2)(1+sT_3)}$	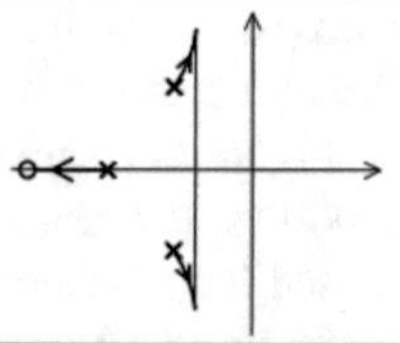

The shape of the root locus shows an analogy to the electrostatic field. If positive and negative charges are located in a plane, the asymptotes of the electrostatic field take the shape of the root locus, if the positive charges are replaced by the poles and the negative charges by the zeros. (Generally the analogy with the potential field of sources and sinks can be considered.)

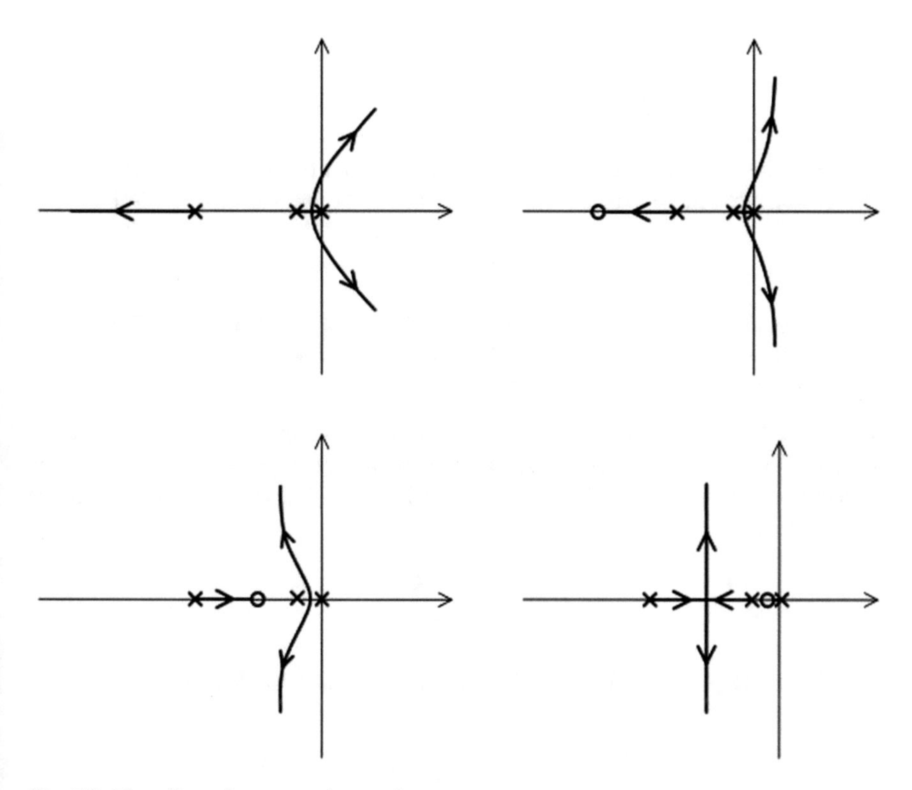

Fig. 5.9 The effect of a zero to the root locus

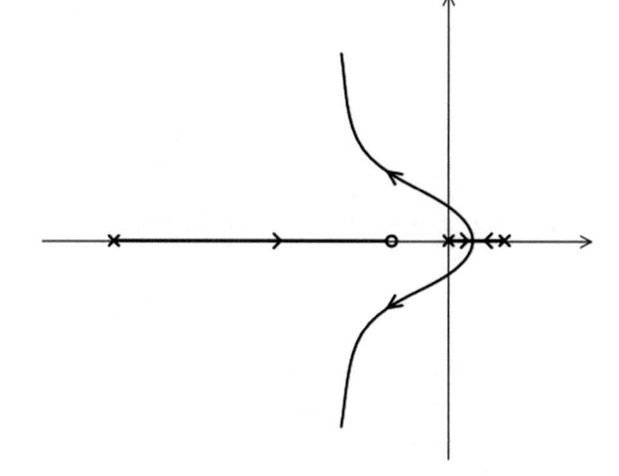

Fig. 5.10 Stabilization of an unstable open-loop with negative feedback by inserting a zero

5.5.4 Root Locus in the Case of Varying a Parameter Different from the Gain

If the root locus is to be determined as a function of a parameter different from the gain factor, then the characteristic equation has to be transformed to the form

$$\alpha H(s) = -1$$

where α is the varying parameter and $H(s)$ is the transfer function obtained as a result of the transformation. Drawing the root locus, α takes the role of the gain and $H(s)$ is a constructed loop transfer function.

Example 5.4 The procedure will be presented when the open-loop is a proportional element with two time lags where, instead of the gain factor, a pole (the time constant) of the system varies from zero to infinity. The transfer function of the open-loop is

$$L(s) = \frac{10}{(s+\alpha)(s+2)}.$$

The varying parameter is now alpha (the pole is $-\alpha$). The characteristic equation is

$$(s+\alpha)(s+2) + 10 = 0$$

or

$$s(s+2) + \alpha(s+2) + 10 = 0.$$

Rearranging yields

$$1 + \alpha \frac{s+2}{s^2+2s+10} = 0.$$

The root locus is determined for the transfer function

$$H(s) = \alpha \frac{s+2}{s^2+2s+10}$$

(see Fig. 5.11). It can be seen that for small values of α $(0<\alpha<8.3246)$ there are decaying oscillations in the closed-loop system. If α increases further, the transients is aperiodic. ■

Today's modern computer techniques make possible—beside the effect of the change of the loop gain—to observe the effect of an additional parameter as well. In this case the usual root locus is calculated for the discrete values of the other parameter (e.g., α), and an array (in layers) of curves is drawn in three dimensions

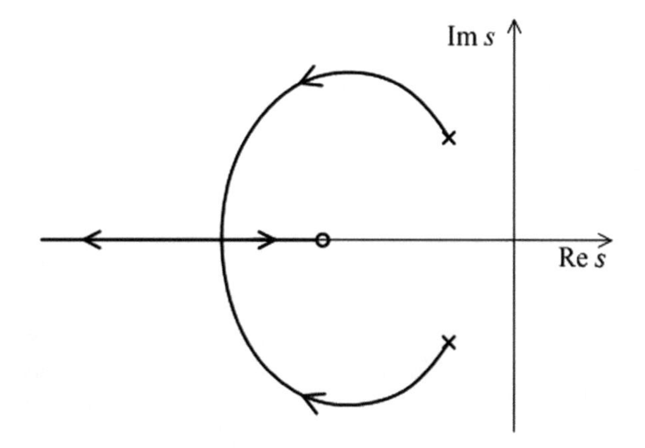

Fig. 5.11 Root locus of a proportional system with two time lags when one of its poles is varied

(3D). The fundamental two dimensions are represented by the complex plane itself, above it the further root-loci are plotted "in layers". Thus the third axis is for the variable α. For 3D graphical representation, a variety of powerful software tools are known, which makes it possible to depict very useful surfaces.

5.6 The Nyquist Stability Criteria

With the analytical Routh-Hurwitz stability criteria, the stability of a closed-loop control system can be determined based on the coefficients of the characteristic equation, but in the case of instability it is difficult to tell how to change the parameters of the system to ensure the appropriate dynamical performance.

The root locus gives an expressive picture of the change of the location of the roots of the closed-loop characteristic equation in the complex plane versus a parameter, thus a comprehensive view can be obtained of the stability and dynamical properties of the system.

With the Nyquist stability criterion, the stability of the *closed-loop* control system can be determined based on the frequency diagram of the *open-loop*. The method is expressive, and in the case of instability it can be easily determined how to modify expediently the structure and the parameters of the system. By appropriately forming the frequency function—i.e., introducing new zeros and poles—the prescribed properties of the closed-loop system, in addition to its stability, as well as its required static and dynamical properties can be ensured.

5.6.1 *Illustration of the Evolution of Undamped Oscillations in the Frequency Domain*

The characteristic equation of a closed-loop control system is $1 + L(s) = 0$, where $L(s)$ is the open-loop transfer function. Substituting $s = j\omega$ it can be checked

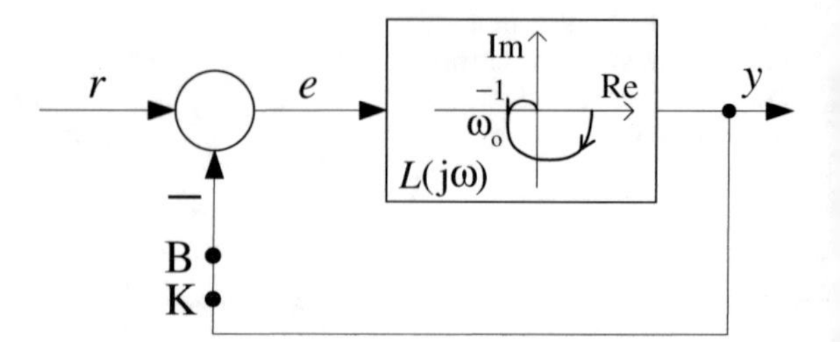

Fig. 5.12 NYQUIST diagram of an open-loop control system, where the closed-loop is working at the stability limit

whether the equation has a solution on the imaginary axis. If there exists a frequency ω_o fulfilling the condition $1 + L(j\omega_o) = 0$, that is $L(j\omega_o) = -1$, then in the closed-loop system, an un-damped oscillation arises with this frequency, thus the system gets to the borderline of stability. In this case the NYQUIST diagram of the open-loop goes through the point $-1 + 0j$ of the complex plane.

The evolution of un-damped oscillations can be illustrated as follows. Let us consider the control loop in Fig. 5.12. The NYQUIST diagram of the open-loop goes through the $-1 + 0j$ point at frequency ω_o. Imagine that the system is opened at points B-K. Let the reference signal r be a sinusoidal signal with frequency ω_o. The system transfers this signal with the same amplitude but with opposite sign. If now the points B-K are connected again, because of the negative feedback the error signal e coincides with the sinusoidal input signal. This un-damped sinusoidal signal will be maintained in the system even if the reference signal is removed. Oscillations with this frequency do appear in the system even in the case when the reference signal is not the considered sinusoidal signal, but a different deterministic signal, e.g., a unit step. That is, since in the frequency spectrum of the reference signal all the frequencies do appear, the reference signal can be built from these sinusoidal components. A component of frequency ω_o is maintained in the system.

5.6.2 The Simple NYQUIST Stability Criterion

Let us suppose that the transfer function of the open-loop has no poles on the right half of the complex plane, thus the open-loop is stable.

Let us draw the frequency function in the complex plane for the domain $-\infty < \omega < \infty$ (the complete NYQUIST diagram). Go through the NYQUIST diagram in the direction of increasing frequencies.

If the NYQUIST diagram does not encircle the point $-1 + 0j$, the closed-loop control system is stable.
If the NYQUIST diagram crosses the point $-1 + 0j$, the system is at the stability limit.
If the NYQUIST diagram encircles the point $-1 + 0j$, the system is unstable.

In a simpler formulation, it is sufficient to draw the NYQUIST diagram only for positive ω. If we go through the diagram from $\omega = 0$ to ∞, and the point $-1 + 0j$ is to the left of the curve, the closed-loop control system is stable. If the curve crosses

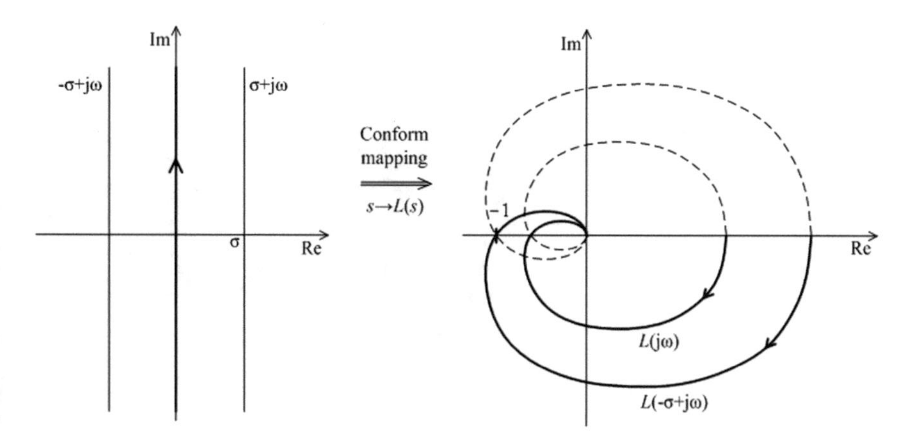

Fig. 5.13 The simple NYQUIST stability criterion can be proved by conformal mapping

the point $-1+0j$, the system is at the stability limit. If the point $-1+0j$ is to the right of the curve, the system is unstable.

The simple NYQUIST stability criterion can be proved based on conformal mapping. The NYQUIST diagram of $L(j\omega)$ is the conformal mapping of the imaginary axis by the function $L(s)$ as ω changes between $-\infty$ and $+\infty$ (Fig. 5.13). Let us consider the straight lines $-\sigma+j\omega$ and $\sigma+j\omega$, which are parallel to the imaginary axis. Here, σ is a given positive number.

Conformal mapping preserves the angles and ratios. Therefore a conformal mapping of the straight line $-\sigma+j\omega$ according to $L(-\sigma+j\omega)$ lies to the left of the curve $L(j\omega)$, while conformal mapping of the straight line $\sigma+j\omega$ according to $L(\sigma+j\omega)$ lies to its right. So if the curve $L(j\omega)$ crosses the real axis to the right of the point $-1+0j$, and thus does not encircle it, then the equation $L(s_i)=-1$ can be fulfilled only for roots with negative real part, i.e. the transients are decreasing. In this case the closed-loop control system is stable. Similarly, if the curve $L(j\omega)$ crosses the real axis to the left of the point $-1+0j$, and thus encircles it, then the equation $L(s_i)=-1$ can be fulfilled only for roots with positive real part, therefore the amplitude of the transients is increasing and the system is unstable.

Example 5.5 Let us consider the closed-loop control circuit in Fig. 5.14. Let us determine the critical loop gain based on the NYQUIST stability criterion.

Figure 5.15 shows the NYQUIST diagram of the open-loop for the case of the stability limit. The NYQUIST diagram goes through the point $-1+0j$ of the complex

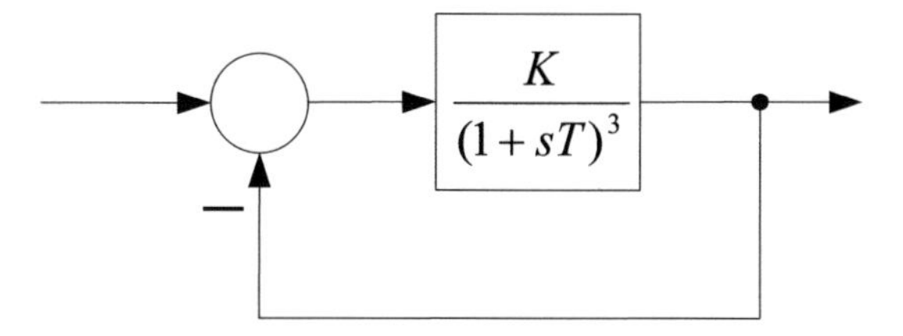

Fig. 5.14 Stability analysis of a proportional system with three time lags

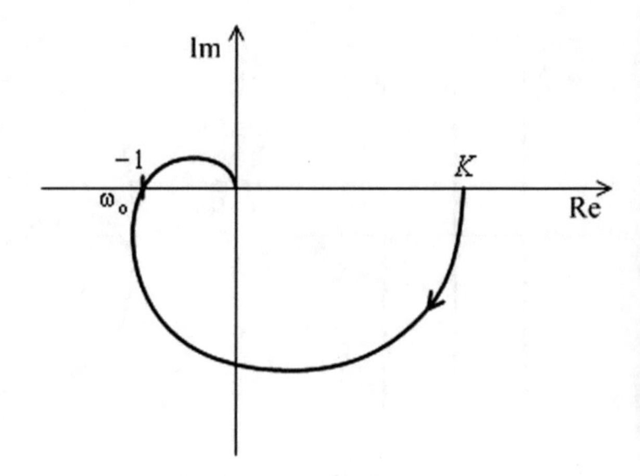

Fig. 5.15 Nyquist diagram of a proportional system with three time lags at the stability limit

plane at frequency ω_o. At this frequency the phase angle of the frequency function is $-180°$ and its absolute value is 1. So we have

$$\varphi(\omega_o) = -3\text{arctg}(\omega_o T) = -180°,$$

whence $\omega_o T = \sqrt{3}$. Thus $K_{\text{krit}} = \left(\sqrt{1 + \omega_o^2 T^2}\right)^3 = 8$, which does not depend on the value of the time constant T. ■

Example 5.6 The Nyquist stability criterion can also be applied to systems with dead-time. Let us consider a control system containing dead-time (see Fig. 5.3). The Nyquist diagram of the open-loop is a circle with radius K which keeps on circling itself infinitely many times as the frequency increases (Fig. 5.16). At the

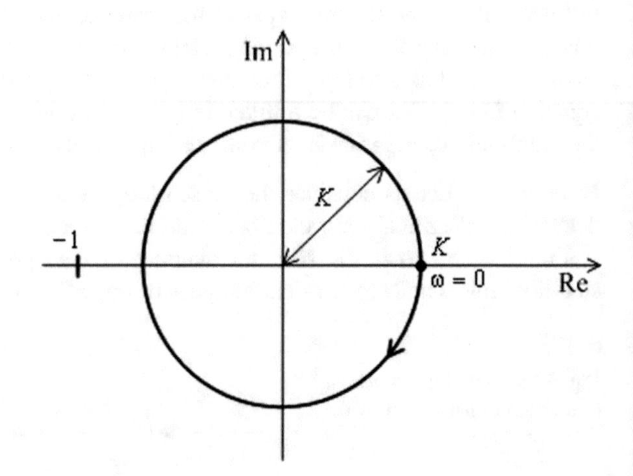

Fig. 5.16 Nyquist diagram of a pure dead-time system

stability limit, it crosses the point $-1+0j$, thus $K_{\text{crit}}=1$, in agreement with Fig. 5.4., and the convergence condition given in Table 5.1. ■

5.6.3 The Generalized Nyquist Stability Criterion

The generalized Nyquist stability criterion gives a condition for stability even for the case when the open-loop has poles in the right half-plane, i.e., the open-loop is unstable. The question is whether the closed-loop can be stabilized with negative feedback.

The generalized Nyquist stability criterion can be formulated as follows: If the open-loop is unstable and the number of its poles lying in the right half-plane is P, then the closed-loop control system is asymptotically stable if the complete Nyquist diagram $(-\infty<\omega<\infty)$ of the open-loop encircles the point $-1+0j$ counter-clockwise (considered as the positive direction) as the number of the poles of the open-loop is in the right half-plane (i.e., P times).

The complete Nyquist diagram is given now more precisely than in the formulation in the previous subsection. In the s plane the straight line $s=j\omega$ $(-\infty<\omega<\infty)$ is closed with a half-circle on the right side with infinite radius, as in Fig. 5.17. The conformal mapping of this closed curve by the function $L(s)$ gives the complete Nyquist diagram of the open-loop. (If the degree of the denominator of the rational fraction $L(s)$ is higher than the degree of its numerator, then the half-circle of infinite radius is mapped into the zero point.) If $L(s)$ has a pole on the

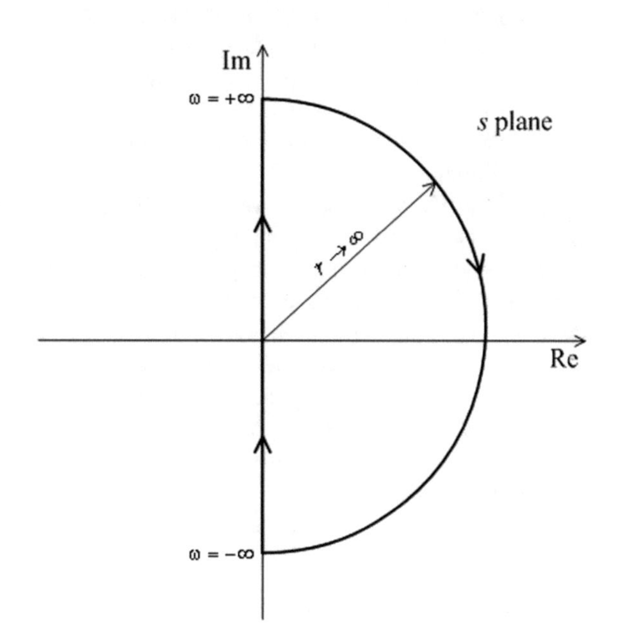

Fig. 5.17 Creating the complete Nyquist diagram

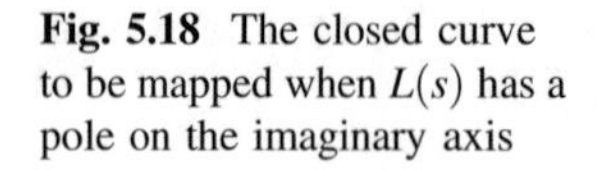

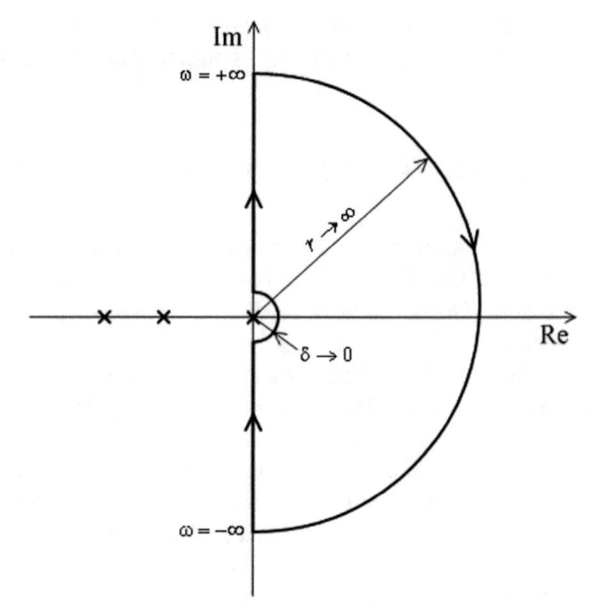

Fig. 5.18 The closed curve to be mapped when $L(s)$ has a pole on the imaginary axis

imaginary axis, then the closed curve is modified to get around the given point from the right or from the left with a half-circle of infinitesimal radius δ. If the curve gets around the pole on the imaginary axis from the right as pictured in Fig. 5.18., then the pole can be considered as a pole in the left half-plane. If the roundabout is executed from the left, then the pole is considered as being on the right side.

The generalized NYQUIST stability criterion can be demonstrated through considerations related to complex functions. Let $f(s)$ be the following function of the complex variable s: $f(s) = (s - s_o)^m$, where s_o is a given point. Let us investigate how the vector $f(s)$ changes if the final point of the vector $s - s_o$ goes through a closed curve on the s-plane clockwise, where on this curve the function $f(s)$ is regular (differentiable).

If s_o is inside the closed curve (Fig. 5.19a), then the vector $s - s_o$ starting from an initial point and passing through the curve clockwise gets into its original position, and its phase angle changes by -2π. In the meantime the mapping by the function $f(s)$ rotates from the starting point by an angle of $-m2\pi$ on the curve determined by $f(s)$. This curve encircles the origin a total of m turns clockwise (m is positive) or counter-clockwise (m negative) (Fig. 5.19b). But if the point s_o is outside the closed curve (Fig. 5.19c), passing through the closed curve the angle of vector $s - s_o$ first is increasing in one direction, then it is decreasing with the same value in the other direction, and finally the curve described by $f(s)$ does not encircle the origin (Fig. 5.19d).

Let us apply the above considerations to the characteristic function of a closed-loop control system. Let the transfer function of the open-loop be a rational

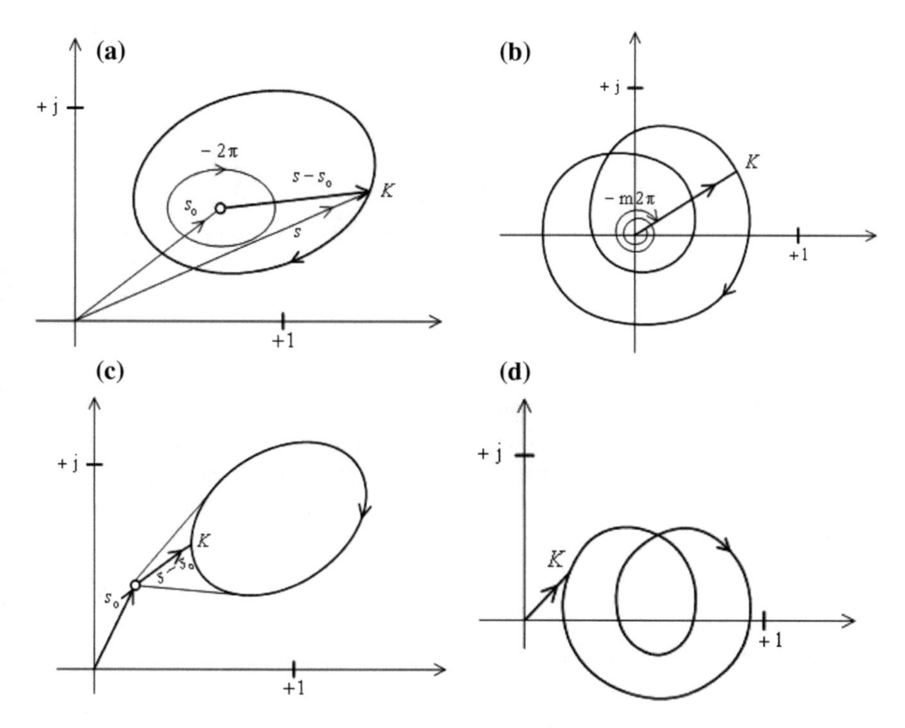

Fig. 5.19 Considerations in the complex plane

fraction whose numerator and denominator are the polynomials $\mathcal{N}(s)$ and $\mathcal{D}(s)$, so that, $L(s) = \mathcal{N}(s)/\mathcal{D}(s)$. The characteristic function is then

$$1 + L(s) = 1 + \frac{\mathcal{N}(s)}{\mathcal{D}(s)} = \frac{\mathcal{D}(s) + \mathcal{N}(s)}{\mathcal{D}(s)} = k\frac{(s - z_1)(s - z_2)\ldots(s - z_n)}{(s - p_1)(s - p_2)\ldots(s - p_n)}. \quad (5.31)$$

The roots of the numerator are denoted by z_i, which are the zeros of $1 + L(s)$. The roots of the denominator are denoted by p_i, which are the poles of $1 + L(s)$. Here, k is a constant. The poles of $1 + L(s)$ coincide with the poles of the transfer function of the open-loop. (Multiple poles appear in the expression when there are multiple, i.e. repeated factors.)

Let us consider the closed curve on the complex plane shown in Fig. 5.17. Go through the curve on the imaginary axis from $-\infty$ to $+\infty$, then close the curve with a half-circle on the right half plane whose radius tends to infinity. Map this curve according to the characteristic function given by (5.31). For all the factors in Eq. (5.31), the above considerations related to complex functions are valid. (For the zeros $s_o = z_1, z_2, \ldots, z_n$ and $m = 1$, while for the poles $s_o = p_1, p_2, \ldots, p_n$ and $m = -1$.) The phase angle of $1 + L(s)$ is the sum of the phase angles of the individual factors taken with the appropriate signs. If the function $1 + L(s)$ has Z zeros and P poles in the right half-plane, inside of the curve in Fig. 5.17, then the

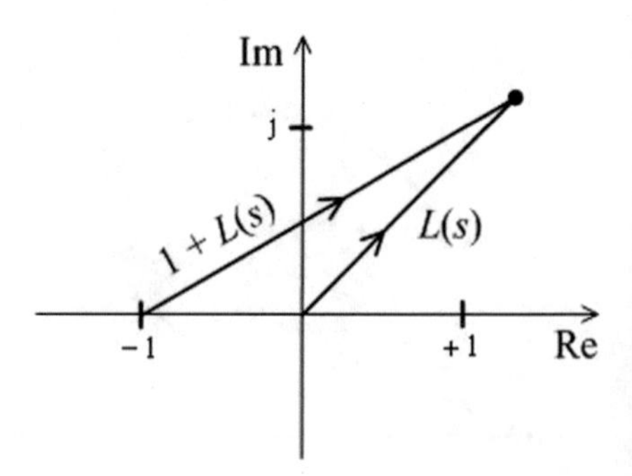

Fig. 5.20 Relationship of vectors $L(s)$ and $1 + L(s)$

number of times the conformal mapping by the function $1 + L(s)$ of the considered closed curve encircles the origin clockwise is the difference of the number of zeros and the number of poles inside this curve. The difference between the phase angles of the initial and the final states is $-2\pi(Z - P)$, and the number R of windings around the origin is

$$R = P - Z \tag{5.32}$$

where a counterclockwise encirclement is defined as positive (see the detailed derivation in A.5.1. of Appendix A.5).

Simply consider the function $1 + L(s)$ as if looking at the curve produced by $L(s)$ from $-1 + 0j$ (Fig. 5.20). The mapping of $L(s)$ along the closed curve in Fig. 5.17 (the so-called complete Nyquist diagram) encircles the point $-1 + 0j$ point $R = P - Z$ times.

Now, P is the number of the poles of the characteristic function in the right half-plane. But these poles, according to (5.31), coincide with the right side unstable poles of the open-loop. Also, Z is the number of the zeros of the characteristic equation in the right half-plane. In the case of stable behavior, the characteristic equation has no zeros in the right half-plane. Thus the condition for stability is

$$Z = 0 \quad \text{i.e.} \quad R = P. \tag{5.33}$$

The simple Nyquist stability criterion can be derived from the generalized Nyquist stability criterion. If the open-loop system has no poles in the right half-plane, i.e. if, $P = 0$, the closed-loop is stable if $R = 0$, so the Nyquist diagram does not encircle the point $-1 + 0j$. In most practical cases the open-loop is stable, and it is in the closed-loop system that the feedback may cause unstable behavior. But sometimes unstable processes have to be dealt with, that is they are to be

Fig. 5.21 A juggler can balance the rod underpinned at its bottom edge

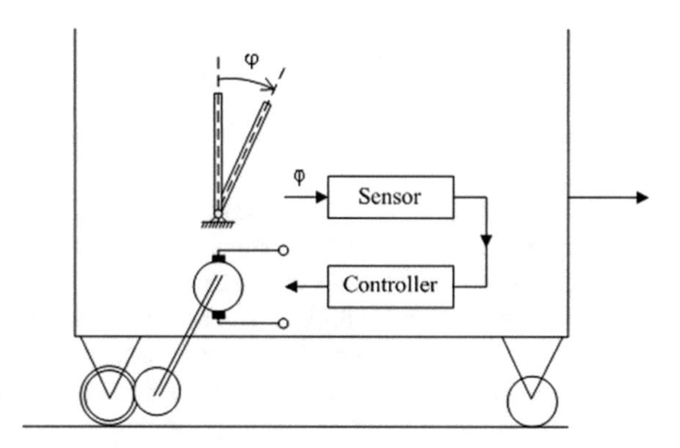

Fig. 5.22 Stabilizing the motion of an inverted pendulum

stabilized by control systems with negative feedback. For example the inverted pendulum is an unstable process. A juggler in the circus is able to balance the leaning rod using the appropriate motions, which are faster than the dynamics of the rod (Fig. 5.21). Thus his body realizes a controller in a closed-loop control system. The automatic solution for stabilizing the motion of the inverted pendulum is shown in Fig. 5.22.

5.6.4 Examples of the Application of the Nyquist Stability Criteria

Example 5.7 Consider the open-loop transfer function

$$L(s) = \frac{5}{1-s} = -\frac{5}{s-1}.$$

Let us analyze the stability of the closed-loop control system.

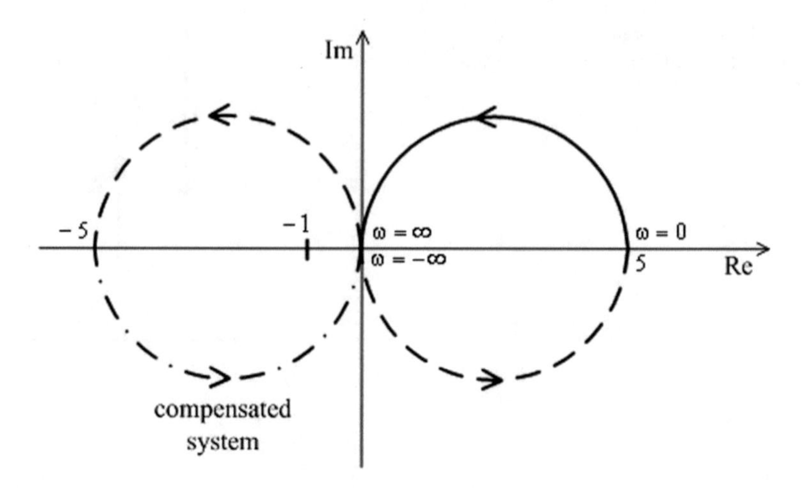

Fig. 5.23 Stability analysis of an unstable system with negative feedback

The system has a pole in the right half-plane, thus $P = 1$. The NYQUIST diagram is shown in Fig. 5.23. As the NYQUIST diagram does not encircle the point $-1 + 0j$, $R = 0$, thus the closed-loop is unstable.

The system can be stabilized if a so called compensation element of a constant gain by $A = -1$ is connected into the forward path. This element changes the sign of the points of the NYQUIST diagram reflecting it about the origin (dashed-dotted curve). Thus the number of windings around $-1 + j0$ will be $R = P = 1$. ■

Example 5.8 Let us consider for example the case when the open-loop is an $L(s) = K_I/s$ integrator, whose pole is at the origin. The closed curve is created by getting around the pole from the right. By mapping this curve according to $L(s)$ the complete NYQUIST diagram shown in Fig. 5.24a is obtained. The case involving getting around the pole from the left is demonstrated in Fig. 5.24b. In the s-plane, the points denoted by 1, 2 and 3 on the small circle surrounding the pole are mapped into the points $1'$, $2'$ and $3'$ in the $L(s)$-plane. In case (a) $P = 0$ and $R = 0$, in case (b) $P = 1$ and $R = 1$, thus in both cases the stable behavior of the system can be established. ■

Example 5.9 Let the transfer function of an open-loop be a proportional element with three time lags, $L(s) = K/[(1 + sT_1)(1 + sT_2)(1 + sT_3)]$. The poles $p_1 = -1/T_1, p_2 = -1/T_2, p_3 = -1/T_3$ are all in the left half-plane, thus $P = 0$. Let us apply the generalized NYQUIST stability criterion. The complete NYQUIST diagram obtained by mapping of the curve given in Fig. 5.17 is shown in Fig. 5.25. If the NYQUIST diagram goes through $-1 + j0$, the system is at the stability limit. If the NYQUIST diagram does not include the point $-1 + j0$ (K_1 loop gain), $R = P = 0$, thus the control system is stable. If the NYQUIST diagram includes the point $-1 + 0j$ (K_2 loop gain), $R \neq P$, thus the control system is unstable. To determine the number of windings R, let us put the spike of an imaginary compass on the point $-1 + 0j$, and with the other end of the compass pass through the NYQUIST diagram from $\omega = -\infty$ to $+\infty$. The number of windings is $R = -2$ (clockwise). The

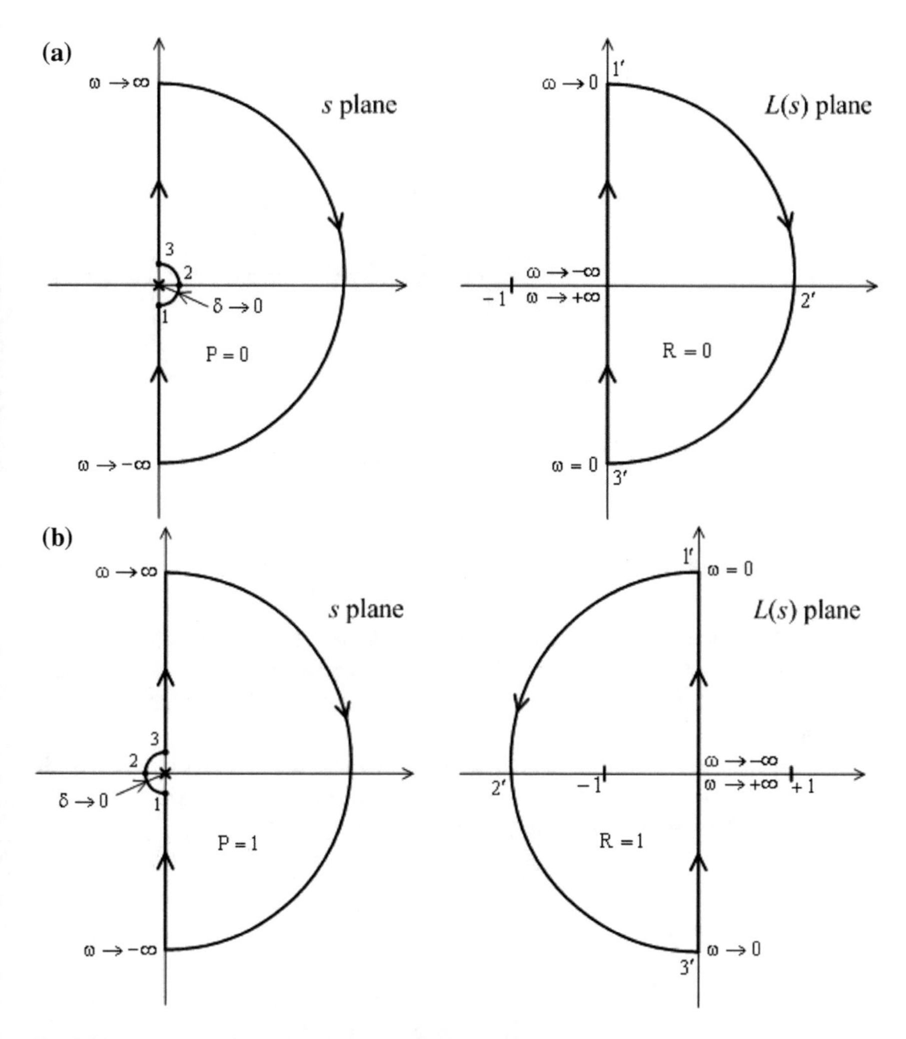

Fig. 5.24 Stability analysis of a control circuit (an integrator is fed back by a unity constant gain)

characteristic equation has two roots in the right half-plane, so, $Z = 2$, and as $R = -2 = P - Z = 0 - Z$, in this case the system is unstable. ■

In the case of a stable open-loop, it is sufficient to use the simple Nyquist stability criterion. In the stable case $-1 + 0j$ lies to the left of the Nyquist diagram drawn for positive frequencies, whereas in the unstable case it is to the right of that curve. The simplified stability investigation can be applied also to the cases when the open-loop contains integrators, and thus there are poles at the origin.

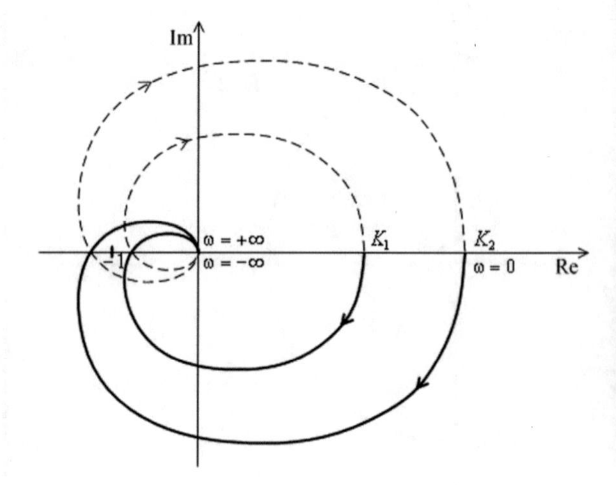

Fig. 5.25 Stability analysis of a proportional system with three time lags

5.6.5 *Practical Stability Measures*

In case of a stable open-loop, the closed-loop is stable if the NYQUIST diagram of the open-loop does not encircle the point $-1+0j$. It can be said that the system has a certain amount of stability reserve, if the NYQUIST diagram is kept sufficiently far from the point $-1+0j$.

Some measures can be defined indicating how far is the NYQUIST diagram of the open-loop from the point $-1+0j$. Such measures include the *phase margin*, the *gain margin*, the *modulus margin* and the *delay margin*.

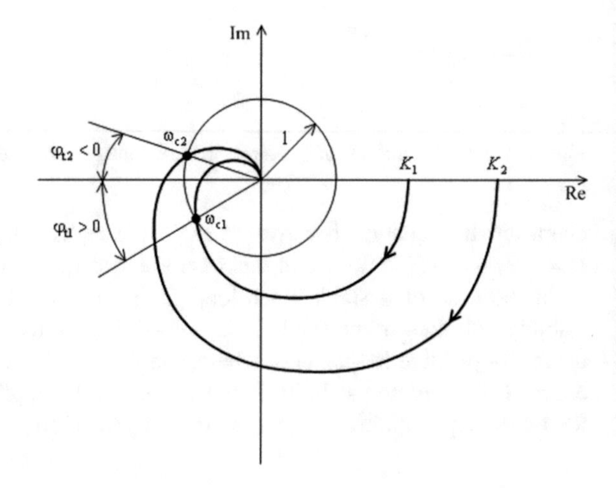

Fig. 5.26 Interpretation of the phase margin

Phase Margin

Let us draw the NYQUIST diagram of the open-loop for positive frequencies. Let us determine the intersection point of the NYQUIST diagram with the circle of unit radius. The frequency belonging to this point is called the *cut-off frequency* and is denoted by ω_c. Let us connect the origin and the intersection point with a straight line. The angle formed by this straight line with the negative real axis is called the *phase margin* (Fig. 5.26):

$$\varphi_t = \varphi(\omega_c) + 180^\circ = \arg L(j\omega_c) + 180^\circ. \quad (5.34)$$

If the phase margin is positive, the system is stable. If the phase margin is zero, the system is at the stability limit. If the phase margin is negative, the system is unstable.

Thus for the stability of the control system the following statements can be made:

$$\begin{array}{ll} \varphi_t > 0 & \text{Stable system} \\ \varphi_t = 0 & \text{Boundary of stability} \\ \varphi_t < 0 & \text{Unstable system} \end{array} \quad (5.35)$$

The stability of the system can be evaluated based on the phase margin as a single measure only if the NYQUIST diagram of the open-loop crosses the unit circle only once.

Gain Margin

Let us determine the intersection point of the NYQUIST diagram with the negative real axis and also the distance $\kappa = |1 + L(j\omega_{180})|$ of this point from the point $-1 + 0j$ (Fig. 5.27). the distance κ is called the *gain margin*. It is apparent that for $\kappa > 0$ the stability domain of the simple NYQUIST criterion is obtained. The stability

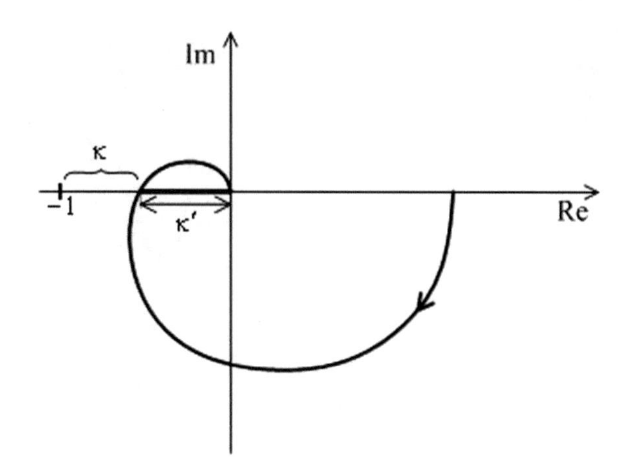

Fig. 5.27 Interpretation of the gain margins

of the system can be evaluated based on the gain margin as a single measure only if the NYQUIST diagram of the open-loop crosses the negative real axis only once.

The *modified gain margin* κ', is defined by the intercept $\kappa' = L(j\omega_{180}) = 1 - \kappa$ seen in Fig. 5.27. If $\kappa' < 1$, the system is stable. If $\kappa' = 1$, the system is at the stability limit. If $\kappa' > 1$, the system is unstable. Thus for the stability of the control system the following statements can be made:

$$\begin{array}{ll} \kappa' < 1 & \text{Stable system} \\ \kappa' = 1 & \text{Boundary of stability} \\ \kappa' > 1 & \text{Unstable system} \end{array} \tag{5.36}$$

The meaning of κ is more expressive than that of κ', however the reciprocal of κ' specifies the factor by which multiplying the actual loop gain the system reaches the stability limit. Therefore it is straightforward to also use the measure $g_t = 1/\kappa' = 1/|L(j\omega_{180})|$ as the relative gain margin. Multiplying the loop gain by g_t the value of the critical gain is obtained. With simple considerations, the inequalities $g_t \geq M_m/(M_m - 1)$ and $\varphi_t \geq 2\arcsin(1/M_m)$ can be derived. (See (4.25) for the interpretation of M_m.)

Figure 5.28 shows the NYQUIST diagram of a system where neither the phase margin nor the gain margin can be interpreted. (Such a NYQUIST diagram is formed if oscillating elements and zeros are included in the transfer function of the system.) In this case the whole NYQUIST diagram has to be considered. Based on the simple NYQUIST stability criterion the stability can be evaluated: as going through the curve the point $-1 + 0j$ is to the right side of the curve, the system is unstable.

Besides stability, the relevant transient performance is also required. To ensure an overshoot less than 10% in the step response of a closed-loop system, the desired phase margin is about 60°, and the desired relative gain margin g_t is about 2 ($\kappa \approx \kappa' \approx 0.5$). These values can be considered characteristic if there are no resonant frequencies in the loop frequency function.

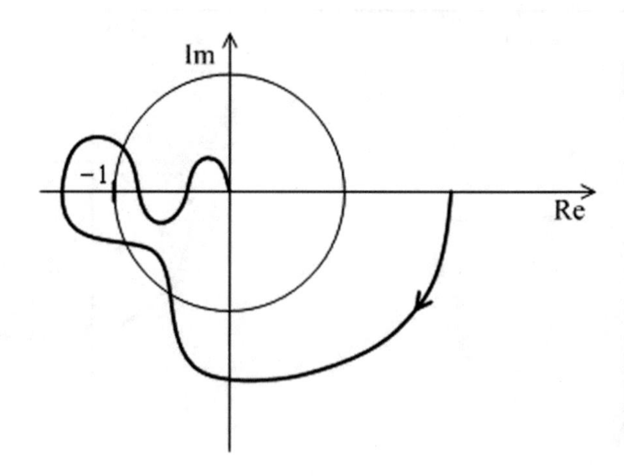

Fig. 5.28 When the phase margin and the gain margin can not be interpreted

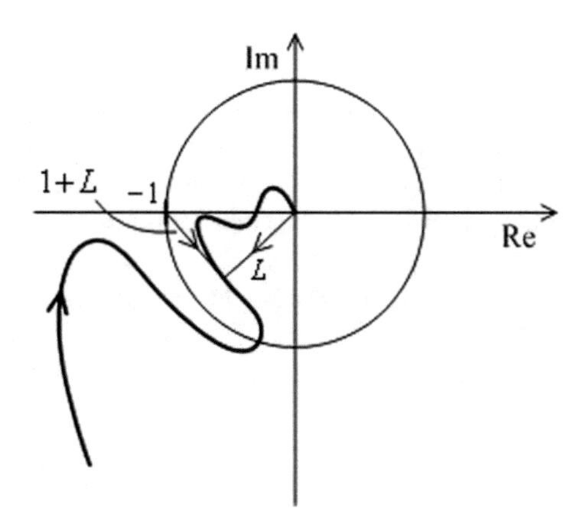

Fig. 5.29 When the dynamic behavior of the system is not satisfactory even if the phase margin and the gain margin are of appropriate values

The appropriate phase margin and gain margin do not give reliable information about the stability margin of the system in every case. Let us consider, e.g. the Nyquist diagram in Fig. 5.29. In spite of the fact that both the phase margin and the gain margin are appropriate, there may occur high amplifications in the frequency function $L/(1+L)$ of the closed-loop in the vicinity of the cut-off frequency, as with increasing frequency the amplitude of L is only slightly changed, while the amplitude of $1+L$ decreases significantly. High amplification in the amplitude-frequency curve of the closed-loop in the vicinity of the cut-off frequency indicates oscillations in the unit step response. Furthermore, if the parameters of the plant change a little, the closed-loop system may even become unstable. The phase margin and the gain margin characterize the stability properties of the system only if the Nyquist diagram does not go too close to the unit circle before and after the cut-off frequency.

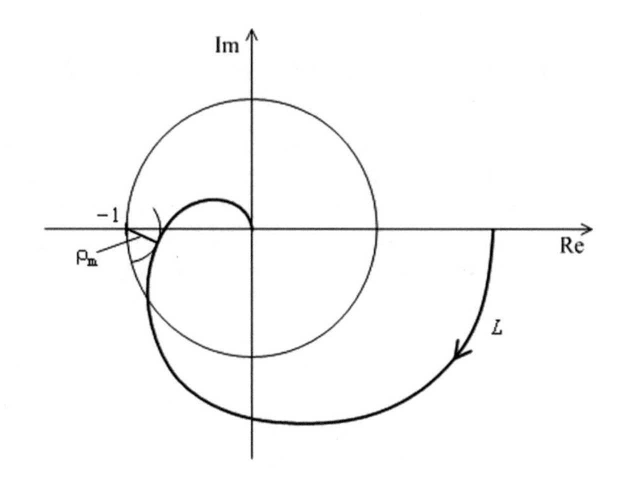

Fig. 5.30 Interpretation of the modulus margin

Modulus Margin

The ρ_m is the minimum of the distance of point $-1+0j$ from the NYQUIST diagram, i.e., it is the radius of the smallest circle tangential to the diagram and centered at $-1+0j$ (Fig. 5.30). The modulus margin shows how far the most sensitive point of the system is from the stability limit. As a reasonable prescription, let the modulus margin be $\rho_m > 0.5$.

The modulus margin is also called NYQUIST stability margin. An important formula is that ρ_m can be expressed as the reciprocal of the maximum of the absolute value of the sensitivity function (see Chap. 6):

$$\rho_m = \frac{1}{\max\limits_{\omega}|S(j\omega)|} = \min_{\omega}\left|S^{-1}(j\omega)\right| = \min_{\omega}|1+L(j\omega)| \tag{5.37}$$

The three margins $\varphi_t, \kappa, \rho_m$ are analogous concepts, as each of them tries to guarantee somehow the distance from the point $-1+0j$.

Delay Margin

The *delay margin* gives the smallest value of the dead-time T_{min} by which—inserting it serially as an extra dead-time into the loop—the closed-loop control system would reach the stability limit. The delay margin can be calculated from the phase margin measured in radians by the following formula:

$$T_{min} = \frac{\varphi_t}{\omega_c}, \tag{5.38}$$

where ω_c denotes the cut-off frequency.

With these stability margins, not only can the stability be evaluated, but it can also be established, "how far" the system is from the stability limit.

5.6.6 Structural and Conditional Stability

Let us suppose that the open-loop is stable, thus the stability of the closed-loop can be evaluated according to the simplified NYQUIST stability criterion.

Most systems generally are stable for small loop gains: they reach the stability limit at a given critical gain, then increasing the gain further they show unstable behavior. (Such a control system is obtained by negative feedback of a proportional plant with three time lags, see Examples 5.5 and 5.9.)

But there are also systems which—because of their structure—remain stable at any value of the loop gain. Such systems are called structurally stable systems. For example a first order or a second order lag element or a pure integrator or an integrator serially connected to a first order lag with negative constant feedback have this property, as their NYQUIST diagram does not encircle the point $-1+0j$ even if the loop gain is arbitrarily increased. By increasing the loop gain the system

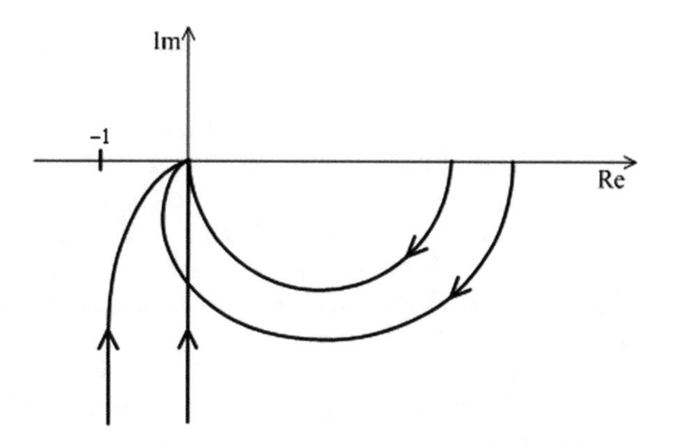

Fig. 5.31 NYQUIST diagrams of structurally stable systems

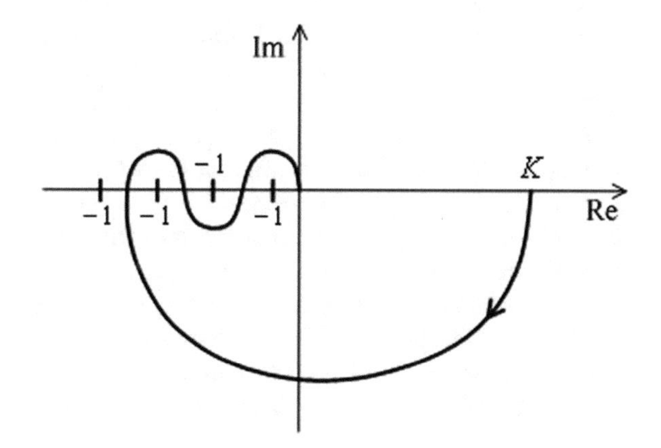

Fig. 5.32 NYQUIST diagram of a conditionally stable system

will not become unstable, but its stability margins do decrease. The NYQUIST diagrams of such systems are shown in Fig. 5.31.

There are systems which are stable in given regions of the loop gain, while in other regions they show unstable behavior. In the case of such systems the loop gain has to be set carefully. These systems are called conditionally stable systems. Figure 5.32 shows an example of the NYQUIST diagram of a conditionally stable system. Besides being influenced by the poles, the course of the NYQUIST diagram is influenced by the zeros as well. If the gain is small, the point $-1+0j$ is to the left of the NYQUIST diagram, so for small gains the control system is stable. By increasing the gain, the point $-1+0j$ will be to the right of the diagram, so the control system becomes unstable. By increasing the gain further, the point $-1+0j$ will get to the left of the diagram, so the control system will become stable again. Increasing the gain even more the diagram will encircle again the point $-1+0j$, causing again unstable performance.

5.6.7 Stability Criteria Based on the BODE Diagrams

The phase margin and the gain margin can also be read from the BODE diagram. The absolute value of the frequency function at the cut-off frequency ω_c is 1. The BODE amplitude-frequency diagram crosses the horizontal 0 dB axis at this frequency. The deviation of the phase angle from $-180°$ at this frequency gives the phase margin. The absolute value at the frequency where the phase angle is $\varphi = -180°$ gives the value of the parameter κ in dB-s, and from this the gain margin can be determined (Fig. 5.33).

If the open-loop is of minimum phase (i.e. its transfer function does not contain zeros or poles in the right half-plane), and furthermore the control system does not contain dead-time, the stability can be determined very simply from the approximate BODE amplitude-frequency curve of the open-loop.

Then from the BODE amplitude diagram the course of the phase angle follows unambiguously, as the phase angle belonging to the poles is negative and the phase angle belonging to the zeros is positive, changing according to arctangent curves.

A minimum phase system which does not contain dead-time is stable if the asymptotic BODE amplitude diagram of the open-loop crosses the frequency axis at a straight line section of slope −20 dB/decade. The system is surely unstable if the slope of the crossing is equal to or greater than −60 dB/decade. If the slope of the intersection is −40 dB/decade, then the system can be stable or unstable depending on the phase margin, which in this case is surely very small (Fig. 5.34).

The above statement can be shown based on the following deliberations. Let us consider the asymptotic BODE diagram in Fig. 5.35. The cut-off frequency lies on a straight line of slope −20dB/decade. The phase angle in the vicinity of the cut-off

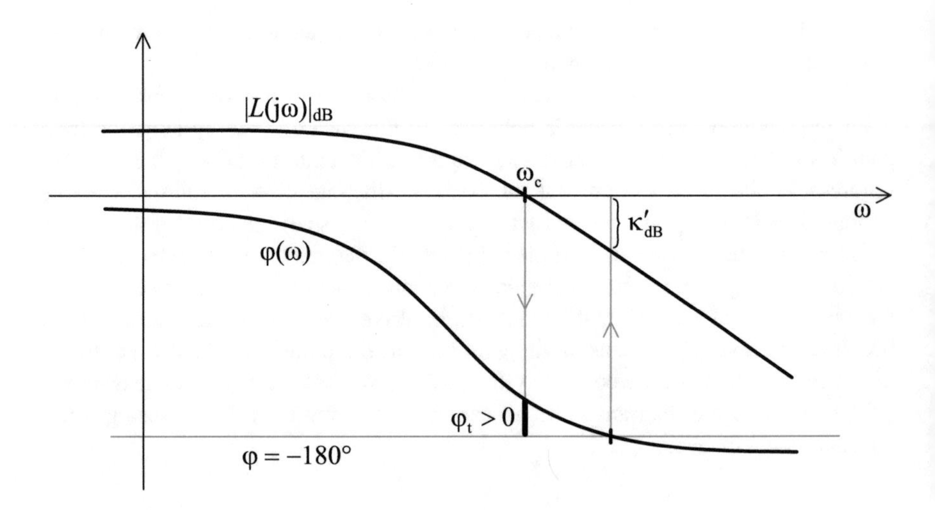

Fig. 5.33 Reading the phase margin and the gain margin from the BODE diagram

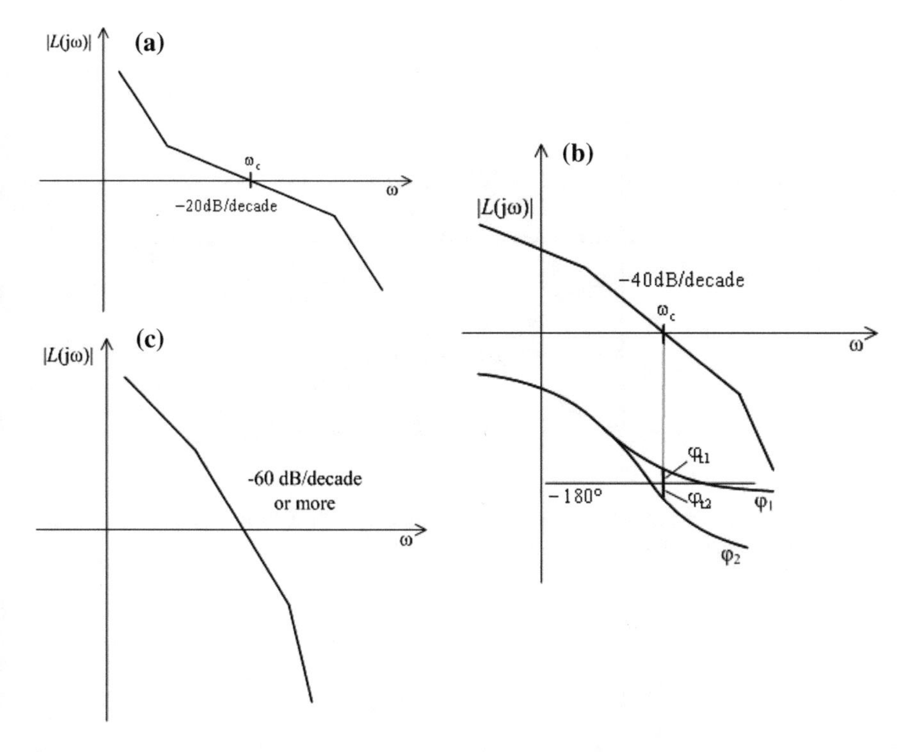

Fig. 5.34 Stability of a minimum phase system without dead-time can be determined from the approximate Bode amplitude-frequency diagram

frequency ω_c approximates $-90°$. The phase angle resulting from a breakpoint which is to the right of the cut-off frequency (especially if it is far away from ω_c, at least located at $5\omega_c$ or at a still higher frequency) will only slightly affect the phase angle at ω_c. Before the straight line of slope -20dB/decade the approximate Bode diagram might have a horizontal section or a section with slope −20 or −40 dB/decade. The courses of the phase angle resulting from these parts of the Bode diagram are indicated in the figure by dashed, dotted and dashed-dotted lines, respectively. The effect of these sections on the phase angle at frequency ω_c is small (especially if the breakpoint before ω_c is far away, located at less than $\omega_c/5$). Thus the system surely has a positive phase margin, which is expected to be satisfactory not only for ensuring stability, but also for guaranteeing the appropriate transient behavior.

If the cut-off frequency is located at a straight line of slope −40 dB/decade, the phase angle at frequency ω_c may approach or even exceed $-180°$ with the phase angle resulting from the previous breakpoints. Thus the system will get close to the stability limit (Fig. 5.36). In this case, evaluating the stability requires calculating the value of the phase margin.

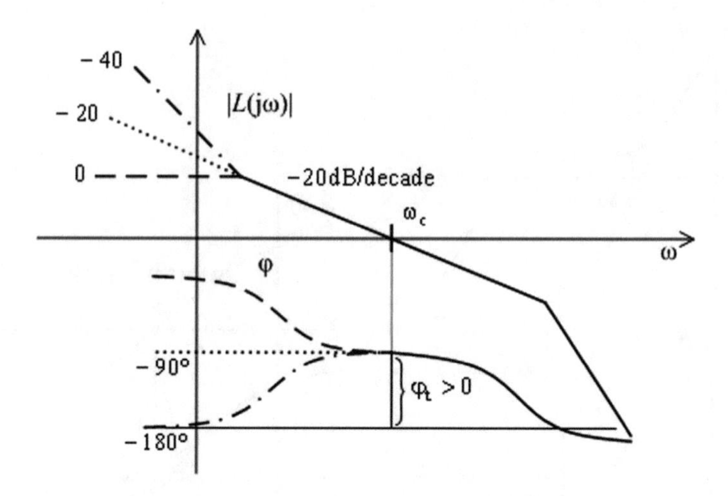

Fig. 5.35 The system surely has positive phase margin if the cut-off frequency is located on a straight line of slope −20 dB/decade

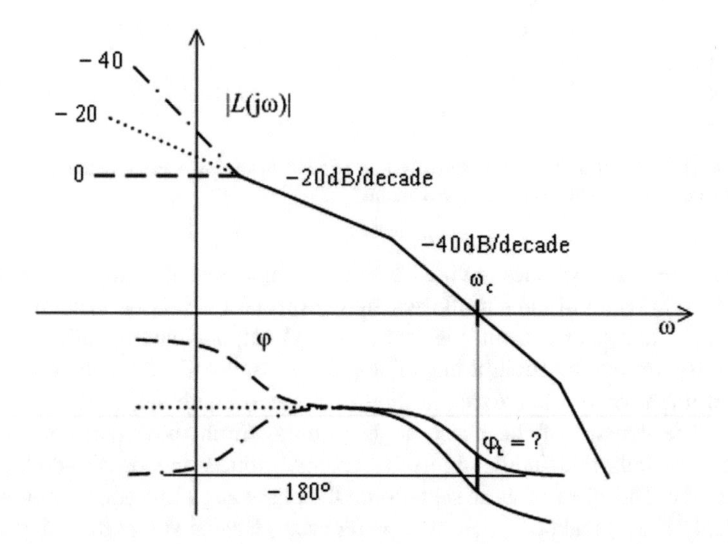

Fig. 5.36 The system is close to the stability limit if the cut-off frequency is located on a straight line of slope −40 dB/decade

If the cut-off frequency is located on a straight line of slope −60 dB/decade or more, then the phase margin surely will become negative.

Thus, to ensure stability, the cut-off frequency has to be located on a straight line of slope −20 dB/decade. (This section has to be long enough to ensure a satisfactory phase margin of about 60°.)

If the open-loop is of non-minimum phase type, or contains also dead-time, then the stability can not be evaluated considering only the BODE amplitude diagram. In this case the BODE amplitude and phase diagram have to be jointly taken into account.

5.7 Robust Stability

Generally, the parameters of the plant are determined from measurement data. The parameters may change around their nominal values in a given range. The closed-loop control system has to be stable throughout the given uncertainty ranges of the parameters.

Suppose that the open-loop is stable. The controller designed for the nominal plant ensures the stability of the nominal closed-loop control system. Let us analyze whether the system remains stable with the parameter uncertainties of the open-loop. Stability is maintained if the NYQUIST diagram of the modified open-loop does not encircle the point $-1+0j$.

The uncertainty of the plant is expressed by the absolute model error

$$\Delta P = P - \hat{P} \tag{5.39}$$

and the relative model error

$$\ell = \frac{\Delta P}{\hat{P}} = \frac{P - \hat{P}}{\hat{P}}, \tag{5.40}$$

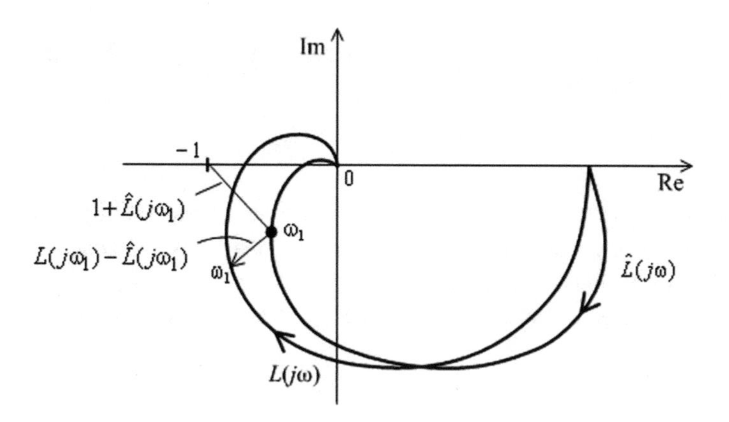

Fig. 5.37 Change of the NYQUIST diagram of an uncertain system

where $\hat{P}$ is the available nominal model used for the design, and P is the real plant. If there is an uncertainty of ΔP (or parameter change) in the transfer function of the plant, then applying the same controller this uncertainty appears in the absolute error $\Delta L = C\Delta P$ of the loop transfer function, while its relative error

$$\ell_{\mathrm{L}} = \frac{\Delta L}{\hat{L}} = \frac{L - \hat{L}}{\hat{L}} = \frac{CP - C\hat{P}}{C\hat{P}} = \frac{P - \hat{P}}{\hat{P}} = \ell \tag{5.41}$$

is equal to the relative model error. Here $\hat{L}$ denotes the nominal, while L denotes the real loop transfer function.

Robust stability means, that the closed-loop control system should not reach an unstable behavior even in the worst case of the parameter changes. The bound for ΔL can be formulated based on Fig. 5.37 taking into account simple geometrical considerations: the Nyquist diagram will not encircle the point $-1 + 0j$, if the following relationship is satisfied for all the frequencies:

$$|\Delta L(j\omega)| = |\ell(j\omega)|\left|\hat{L}(j\omega)\right| < \left|1 + \hat{L}(j\omega)\right| \quad \forall\omega. \tag{5.42}$$

With further straightforward manipulations on (5.42) the necessary and sufficient condition for robust stability can be obtained as

$$|\ell(j\omega)| < \left|\frac{1 + \hat{L}(j\omega)}{\hat{L}(j\omega)}\right| = \frac{1}{\left|\hat{T}(j\omega)\right|} \quad \forall\omega, \tag{5.43}$$

where $\hat{T} = \hat{L}/\left(1 + \hat{L}\right)$ is the nominal supplementary sensitivity function. Condition (5.43) can also be expressed as

$$\left|\hat{T}(j\omega)\right| < \frac{1}{|\ell|} \quad \forall\omega. \tag{5.44}$$

It is a common practice to express the above inequalities for robust stability also in the following form:

$$\left|\hat{T}(j\omega)\right||\ell| < 1 \quad \forall\omega. \tag{5.45}$$

This form is also called the dialectic relationship of robust stability. In the design process, the first factor $\left|\hat{T}(j\omega)\right|$ is calculated for the supposed (known) nominal parameters of the plant, thus it depends on the designer. The second factor $|\ell|$ does not (or only partly) depend on the designer, as it contains the uncertainties in the knowledge of the plant or its unexpected parameter changes. In those frequency ranges where the uncertainty is large, unfortunately only a small transfer gain can be designed for the closed-loop. Where $|\hat{T}(j\omega)|$ is high, very accurate information has to be available to guarantee a small error. The higher the absolute value of the complementary sensitivity function, the smaller the permissible parameter uncertainty.

Condition (5.43), which considers the whole frequency range, is fairly strict, therefore generally it is replaced by a more practical condition if the maximum value of $|\hat{T}(j\omega)|$ is known. Suppose

$$\hat{T}_{\mathrm{m}} = \max_{\omega}|\hat{T}(j\omega)| \tag{5.46}$$

With this value, (5.43) can be simplified to the following satisfactory condition:

$$|\ell(j\omega)| < \frac{1}{\hat{T}_{\mathrm{m}}} \quad \forall\omega \tag{5.47}$$

(Let us refer to Chap. 4, where $M(\omega)$ is defined as $M(\omega) = |T(j\omega)|$.)

If the open-loop is unstable, and the feedback stabilizes the nominal system, then the closed-loop system remains stable with the parameter uncertainties if the number of the poles of the open-loop in the right half-plane does not change, and the number of windings of the Nyquist diagram around the point $-1+0j$ does not change either.

Chapter 6
Regulator Design in the Frequency Domain

A closed-loop control system has to meet several prescribed quality specifications. These specifications are:

- stability
- prescribed static accuracy for reference signal tracking and disturbance rejection
- attenuation of the effect of measurement noise
- insensitivity to parameter changes
- prescribed dynamical (transient) behavior
- consideration of constraints resulting from practical realization.

Generally the performance of a control system does not meet all these requirements. The required operation can be ensured by an appropriate design of the control system. The most frequent control scheme is the serial control loop shown in Fig. 6.1. For the plant to be controlled a serially connected controller is to be designed which ensures the stability of the closed-loop control system and fulfills the prescribed quality specifications.

In the control loop $P(s)$ is the transfer function of the plant, while $C(s)$ is the transfer function of the controller (regulator). The controlled output signal y is fed back with a negative sign and compared to the reference signal. The disturbance is taken into account by a signal $y_{\mathrm{n}}(t)$ acting on the output of the plant. This disturbance model is generally satisfactory to handle practical situations.

In the sequel, some considerations are given for controller design based on relationships in the frequency domain.

L. Keviczky et al., *Control Engineering*, Advanced Textbooks in Control and Signal Processing, https://doi.org/10.1007/978-981-10-8297-9_6

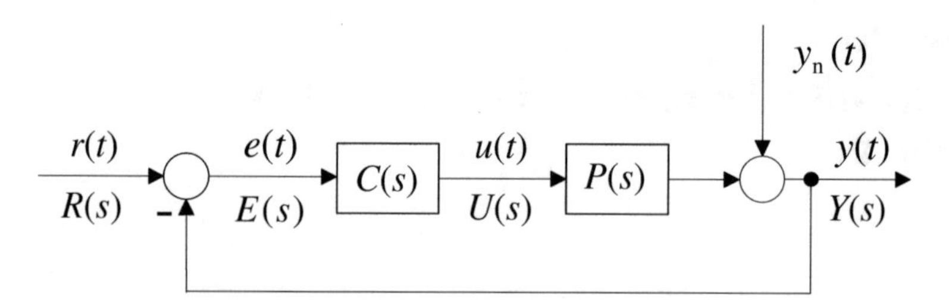

Fig. 6.1 Closed loop control system with serially connected controller

6.1 On the Relationships Between Properties in the Time- and Frequency-Domain

The quality specifications can be demonstrated in the frequency domain, looking as well at the course of the frequency functions of the open-loop and of the closed-loop.

The stability and the susceptibility of the system to oscillations is characterized by the amplification of the Bode amplitude-frequency curve of the closed-loop. Amplification may occur in the vicinity of the cut-off frequency in the open-loop diagram. To determine the stability, the oscillatory behavior, the transient response the course of the frequency function has to be investigated in the middle frequency range (Sect. 4.6). As shown previously the settling time can be estimated from the cut-off frequency. Stability investigation on the basis of the Nyquist and the Bode diagrams has been dealt with in detail in Chap. 5.

Static properties of the control system can be determined from the course of the frequency function in the low frequency range. As seen, the error of the reference signal tracking and of the disturbance rejection in the case of step, ramp and quadratic input signals depends on the type number (the number of integrators in the open-loop) of the control system and on the value of the loop gain. In the case of a type 0 control, the approximate Bode amplitude diagram of the open-loop starts in the low frequency range with a straight line parallel to the 0 dB axis with the value K_o of the loop gain. For control systems of type 1, the diagram starts with a straight line of slope −20 dB/decade. Extending this line it crosses the 0 dB axis at the frequency K_1 corresponding to the loop gain. In the case of control system of type 2, the Bode diagram starts with a straight line of slope −40 dB/decade, whose extension crosses the horizontal axis at $\sqrt{K_2}$. To ensure the static requirements, the low frequency range of the open-loop Bode diagram has to be shaped appropriately (Fig. 6.2).

Based on the relationship of the open-loop and the closed-loop frequency functions (Sect. 4.6) several statements can be given. These statements have to be considered as criteria to design control systems able to ensure good reference signal tracking, disturbance and noise rejection, and to compensate the effect of parameter uncertainties:

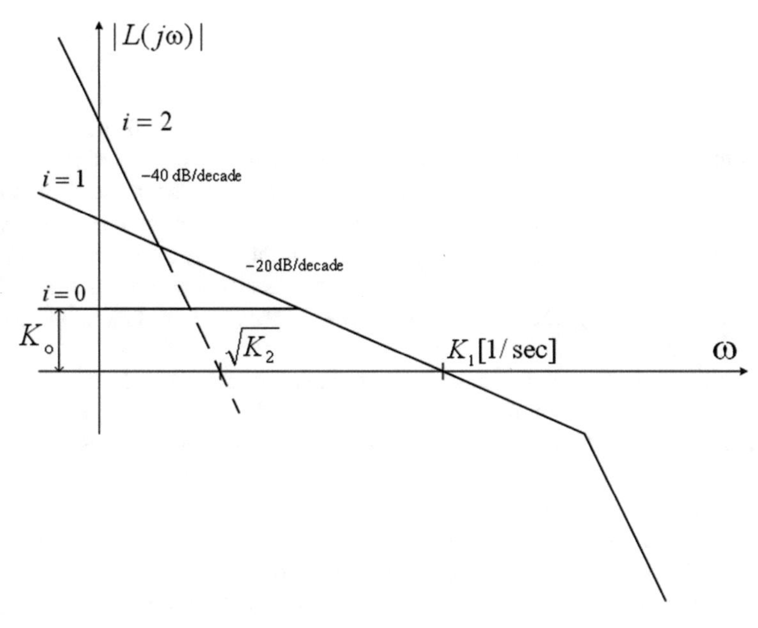

Fig. 6.2 Static accuracy of the control system is characterized by the low frequency range of the open-loop amplitude-frequency curve

- for good reference signal tracking, $|L(j\omega)|$ should be large.
- for effective rejection of the input and output disturbances, $|L(j\omega)|$ should be large.
- for good rejection of measurement noise, $|L(j\omega)|$ should be small.
- to compensate the effect of the parameter uncertainties of the plant, $|L(j\omega)|$ should be large.
- to avoid too high actuating control signals, the acceleration should be moderate and, ω_c should not be too high.

Some requirements are contradictory, and can not be ensured simultaneously for a given frequency range. Therefore different prescriptions have to be given for different frequency ranges. Section 6.2. formulates the quality requirements in the frequency domain in more detail.

6.2 Quality Requirements in the Frequency Domain

As already seen in Chaps. 4 and 5, the quality specifications can also be demonstrated in the frequency domain on the course of the closed-loop and the open-loop frequency functions. The behavior of the closed-loop control system is characterized by the overall frequency function of the closed-loop system. As previously seen, the maximal amplification M_m of the amplitude-frequency function of the closed-loop characterizes the overshoot of the step response in the time domain. An experimental observation is that if $M_m < 1.25$, then there is no significant overshoot in the step response. Many times a closed-loop system can be well approximated by a dominant pair of poles, thus it can be replaced by a second order oscillating element, whose damping factor determines the maximal value of the overshoot. If there is no amplification in the amplitude-frequency curve of the closed-loop, then there is no overshoot in the step response. If the damping factor of the approximating oscillating element is higher than 0.5, the transients will be aperiodic. If the value of the damping factor is around 0.6–0.7, then the overshoot of the step response is about 5–10%.

On the basis of the relationships between the transfer functions of the open and the closed-loop, instead of analyzing the frequency function of the closed-loop we may analyze the amplitude-frequency curve of the open-loop, as well. The overshoot is related to the phase margin. If $\varphi_t \approx 60°$, the damping factor is $\xi \approx 0.7$; ξ also increases by increasing φ_t. With $\varphi_t > 90°$, the transient behavior becomes aperiodic. If the phase margin decreases, the damping factor also decreases. For $\varphi_t \approx 30°$, the damping factor range of $\xi \approx 0.2 - 0.3$, which results in significant, but still decaying oscillations. These simple considerations are satisfactory for orientation.

The control system has to be designed to meet the quality specifications. This can be accomplished by designing an appropriate controller. The Bode diagram of the open-loop has to be shaped to ensure the prescriptions (this procedure is called *loop-shaping*, [see Fig. 6.3]).

The prescriptions for reference signal tracking and for disturbance rejection, and for attenuating the effects of measurement noise, are contradictory in relation to the shape of the Bode amplitude diagram $|L(j\omega)|$. But in practice, generally, the characteristic frequency range of the reference signal and that of the measurement noise are different: the reference signal contains components in a lower frequency range, while the measurement noise mainly contains high frequency components. Thus $|L(j\omega)|$ can be large in the low frequency domain, and small in the high frequency domain.

The course of the amplitude diagram in the low frequency domain determines the static properties. In case of type number 0 the Bode diagram starts horizontally, with type number 1 the initial slope is −20 dB/decade, while in case of type 2 it is −40 dB/decade. The prescribed static accuracy determines the required type number of the control system.

The middle frequency range determines the stability and the dynamical properties of the control system. As shown in Chap. 5, to ensure stability the cut-off frequency ω_c has to be located on a quite long straight line with a slope of −20 dB/decade. This condition is sufficient to ensure stability if the system is of minimum-phase and does not contain dead-time. Otherwise the value of the phase margin or the gain margin has to be calculated. A positive phase margin, or gain margin whose value is higher than 1, ensures stability. Besides stability, the appropriate dynamic behavior (an overshoot less than 10%) can be provided if the

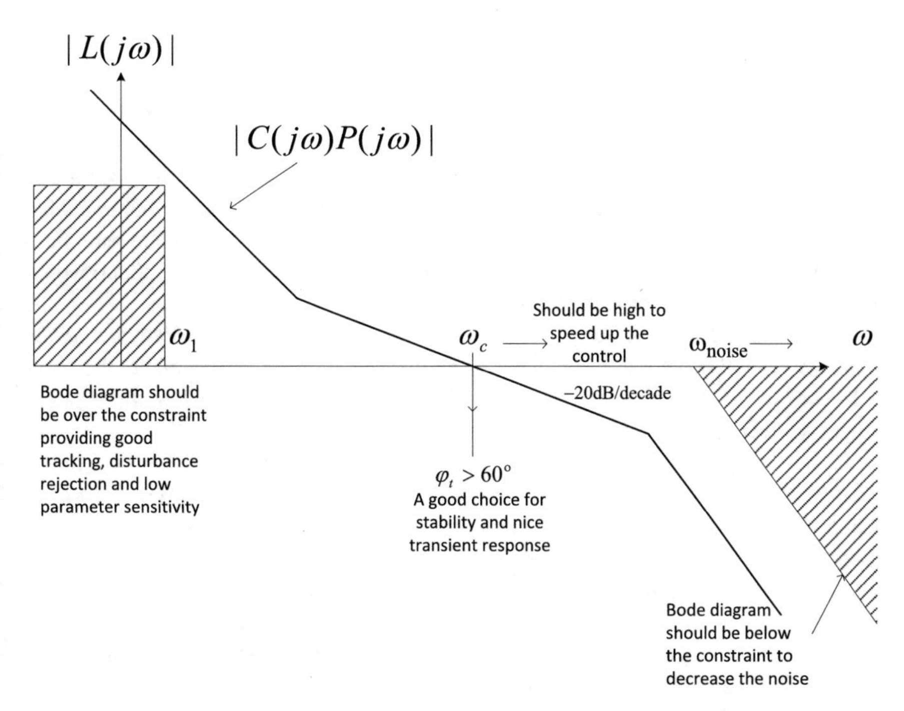

Fig. 6.3 Considerations for loop-shaping on the BODE amplitude-frequency diagram

phase margin is about 60°. Later it will be seen that for minimum-phase systems containing dead-time T_d the phase margin will be about 60° if the cut-off frequency is chosen to be $\omega_c \approx 1/2T_d = 0.5/T_d$. The settling time is related to the cut-off frequency. The higher ω_c is, the shorter the settling time is. As seen in Chap. 4, the settling time can be estimated as $3/\omega_c < t_s < 10/\omega_c$.

In a conventional control structure (Fig. 6.1) in open-loop systems containing dead-time, the cut-off frequency can not be increased beyond one half of the reciprocal of the dead-time. This sets a limit to the acceleration of the system. To achieve a higher ω_c, an advanced control scheme should be considered (e.g. a SMITH predictor, see Chap. 12).

By shaping the low frequency and the high frequency sections of the BODE amplitude diagram according to Fig. 6.3 the suppression of the measurement noise and the insensitivity of the control system to the parameter uncertainties can be ensured, as well.

Often the different requirements are contradictory. A satisfactory compromise has to be reached to ensure stability, a good dynamic response, fast performance and the required constraints on the value of the control signal. The control system can be accelerated only by applying a large control action. This high control signal is to be ensured by the actuator which generally has limits due to its physical realization.

In practice the controller is able to provide signals only within a given range. It is important that during the operation the control signal $u(t)$ remains within this range, that is, its maximum value should not exceed the physically reachable maximum value. If a command is given that would require a control signal value exceeding the maximum, the actuator will be saturated: its output will be at the maximum value until—with an increase of the output signal—the error signal reaches such a small value that the output signal of the controller would go out of the saturation domain. The designer of the control system has to consider this phenomenon already in the controller design phase.

Thus during the controller design process the shape of the open-loop frequency function around the point $-1+0j$ has to be formed. In the complex plane, the point $-1+0j$ can be described by two conditions, namely the gain is 1 and the phase angle is $-180°$, that is, $|L(j\omega)|=1$ and $\varphi=-180°$.

In the design procedure these two conditions can be handled as follows: in the case of the fulfillment of one of these conditions, how far is the second one from the value given above? If the gain is 1, then the phase deviation is given by the phase margin at the cut-off frequency, whereas if the phase angle is $-180°$, then the gain margin can be calculated from the value of the gain. If the controller is designed for a prescribed phase margin, then the phase angle of the frequency function has to be set at the frequency belonging to the unity gain (which is the cut-off frequency).

Thus the shape of the open-loop frequency function around the cut-off frequency ω_c is critical. In some cases it is not sufficient to determine the phase angle only at this frequency, but the shape of the frequency function has to be examined also in its vicinity.

Figure 6.4 shows two Nyquist diagrams with identical phase margins. The diagram of L_1 after the cut-off frequency goes quickly to the origin, whereas the diagram of L_2 approximates the point $-1+0j$. For the first system the amplitude of the closed-loop system $|T|=|L/(1+L)|=|L|/|1+L|$ quickly decreases after ω_c, whereas in the second case it may show significant amplification. In this latter case the absolute value of the sensitivity function $|S|=1/|1+L|$ also has amplification in this frequency range.

6.3 Methods to Shape the Open-Loop Frequency Characteristics

Employing the frequency function $H(j\omega)$ for control system analysis let us consider the Bode theorems. The first Bode theorem states that under some conditions (stable and minimum-phase system without dead-time) the amplitude-frequency function of $H(j\omega)$ unambiguously determines the phase-frequency function. That is at a given frequency ω_o

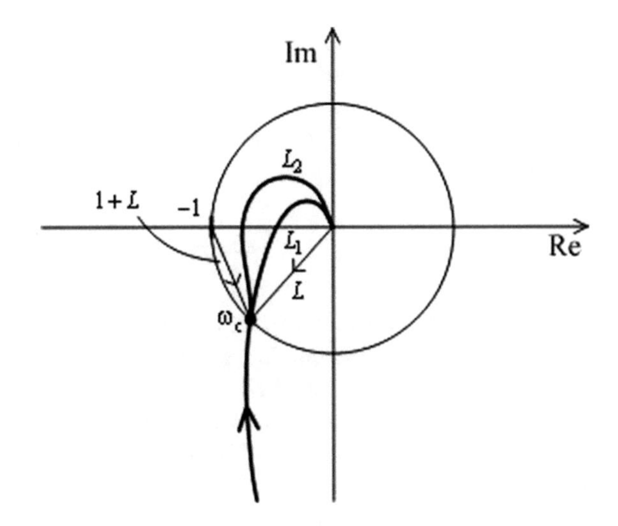

Fig. 6.4 The shape of the Nyquist diagram around ω_c and around $-1+0j$ affects significantly the overshoot

$$\arg H(j\omega_o) = \frac{2\omega_o}{\pi}\int_0^\infty \frac{\log|H(j\omega)| - \log|H(j\omega_o)|}{\omega^2 - \omega_o^2}\,\mathrm{d}\omega$$
$$= \frac{1}{\pi}\int_0^\infty \frac{\mathrm{d}\log|H(j\omega)|}{\mathrm{d}\log\omega}\log\left|\frac{\omega+\omega_o}{\omega-\omega_o}\right|\mathrm{d}\omega \approx \frac{\pi}{2}\frac{\mathrm{d}\log|H(j\omega)|}{\mathrm{d}\log\omega}. \tag{6.1}$$

According to this relationship, at a given $\omega = \omega_o$, the complete shape of the phase-frequency curve contributes to form the value $\varphi(\omega_o)$ through the expression of the definite integral above. At the same time, because of the inner weighting function

$$\log\left|\frac{\omega+\omega_o}{\omega-\omega_o}\right|, \tag{6.2}$$

the far points have only a barely noticeable effect on the function in the vicinity of ω_o. Equation (6.1) means that if in the vicinity of a point, the slope (in logarithmic scale) of the approximate amplitude curve is $+1$, then the corresponding phase angle is $+\pi/2$. (In decibels +20 dB/decade corresponds to a slope of $+1$).

A similar relationship can be given for how the amplitude-frequency function relates to the phase-frequency function, that is, the relationship is unique and mutual. The relationship is not unique for non-minimum-phase systems.

If the breakpoint frequencies of the frequency function are far enough from each other, then the phase-frequency curve belonging to the long straight lines of the approximated amplitude-frequency curve is horizontal with a value of $\varphi(\omega) \approx \pm j\pi/2$, whereas close to the breakpoints, it is not horizontal and the phase angle can be approximated by $\varphi(\omega) \approx \pm(2j+1)\pi/4$.

In the previous chapters it was shown that the stability and robustness properties, static and dynamic behavior of the closed-loop system all can be designed by the *loop-shaping* of the frequency function of the open-loop. BODE, partly on an empirical basis, came to the conclusion that the optimal (ideal) form of the loop frequency function $L(j\omega)$ is

$$L_{\mathrm{id}}(j\omega) = \frac{1}{(j\omega)^{\eta}}. \tag{6.3}$$

If η is not an integer, then in the complex plane the NYQUIST diagram is a straight line going through the origin. Integrators of order n are special cases of this general form when η is an integer. But non-integer forms cannot be realized by lumped parameter linear systems, therefore this expectation is only of theoretical significance and we can only try to approximate it. If at the cut-off frequency ω_c the phase margin φ_t were set according to a characteristic given by $L_{\mathrm{id}}(j\omega)$, then any uncertainty in the gain of $L_{\mathrm{id}}(j\omega)$ would not influence the design condition (i.e., the prescribed φ_t). The ideal case is best approximated if the slope of the phase characteristic $\mathrm{d}\varphi/\mathrm{d}\omega$ is minimal at the cut-off frequency ω_c. This can be ensured if the breakpoint frequencies in the neighborhood of ω_c are far away from it (see BODE's first theorem).

Discussing the interpretation of the modulus margin, it was seen that its value is the reciprocal of the maximum absolute value of the sensitivity function. According to the geometric representation its value is the closest distance of the NYQUIST diagram from the point $-1+0j$. For NYQUIST curves which have parts in the third and the fourth quadrant (these are the so-called positive real systems) the absolute value of the sensitivity function can not be higher than 1. But among the non positive real systems there can also be such systems where the sensitivity is less than 1. On the basis of the $M-\alpha$ and $E-\beta$ curves (see Sect. 4.6.) the range where both conditions, $|S(j\omega)| = E(\omega) \leq 1$ and $|T(j\omega)| = M(\omega) \leq 1$ are fulfilled at the same time can be given simply. Figure 6.5 shows the restricted area (crosshatched), where there will already be amplification in $|S(j\omega)|$ or in $|T(j\omega)|$. In the figure the sensitivity function $S(j\omega)$ belonging to $L(j\omega)$ has no amplification. The necessary condition to achieve this is that the pole excess of $L(s)$ be 1, that is, if $\omega \to \infty$ the condition $L(j\omega) \approx -j\omega$ is fulfilled. (In the figure as a curiosity it is also shown, that for systems of integrating type generally $\mathrm{Re}[L(j\omega)]_{\omega\to 0} \neq 0$, that is, the curve starts not from the imaginary axis, a fact not widely known). On the basis of the geometrical interpretation it can easily be understood that for the so-called *positive real* frequency functions ($\mathrm{Re}[H(j\omega)] \geq 0$) (which have less significance in control theory, but more importance in telecommunication) the condition of avoiding amplification in $S(j\omega)$ is automatically fulfilled.

The intersection points of $L(j\omega)$ with the unit circle $E = 1$ and the straight line $M = 1$ also contain significant information about the course of the frequency characteristics of the open and the closed-loop.

According to Fig. 6.6 the sensitivity function $S(j\omega)$ at low frequencies starts from zero. At ω_1 it reaches the value 1, then at ω_c the absolute value of $L(j\omega)$ will

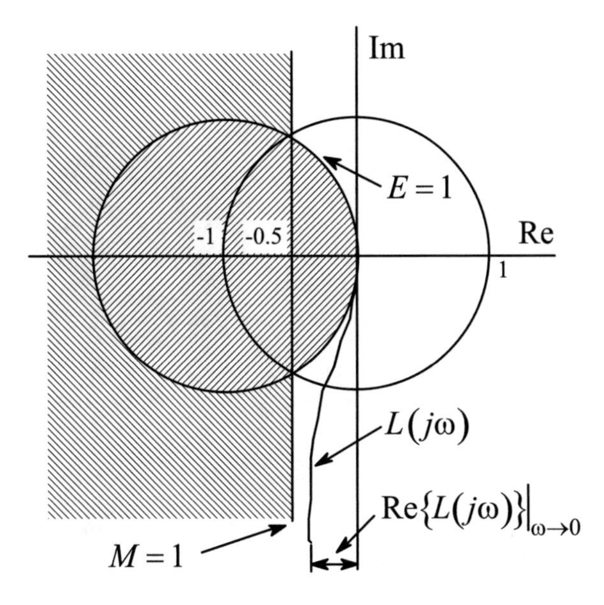

Fig. 6.5 Areas restricting amplification of the sensitivity and the supplementary sensitivity functions

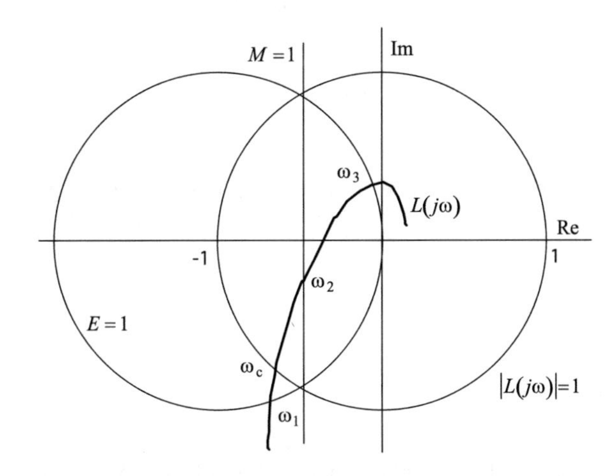

Fig. 6.6 Following the course of $S(j\omega)$ and $T(j\omega)$

be 1. Exceeding its maximum value at ω_3 again it will satisfy $|S(j\omega)| = 1$. If at high frequencies $L(j\omega)$ tends to zero, then $S(j\omega)$ tends to 1. From this analysis we can conclude that the frequency functions of dead-time systems (which may also contain lag elements, integrators, etc.) can never avoid the circle $E = 1$, therefore $|S(j\omega)|$ always has amplification. (The condition $E = 1$ is also called the KALMAN-HO condition).

Based on the construction rules of the $M - \alpha$ and $E - \beta$ curves it is easy to construct the circles (crosshatched areas) shown in Fig. 6.7 to ensure the conditions $|S(j\omega)| \leq 2$ and $|T(j\omega)| \leq 2$. The frequency belonging to $M = \sqrt{2}/2$ is the bandwidth of the closed-loop system.

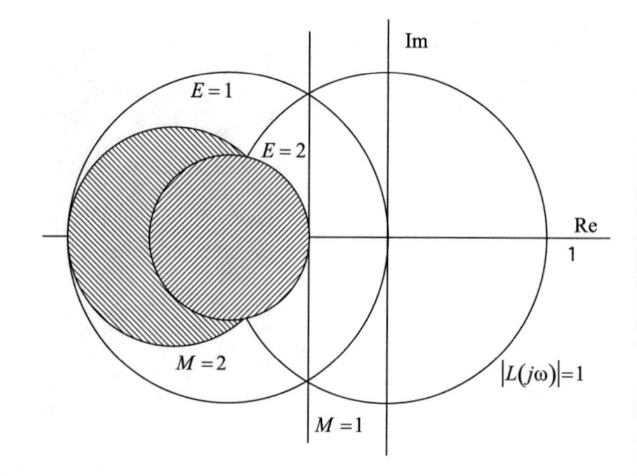

Fig. 6.7 Restricted areas ensuring the conditions $|S(j\omega)| \leq 2$ and $|T(j\omega)| \leq 2$

Unfortunately there are also theoretical limits to the arbitrary shaping of the sensitivity function. According to the second Bode theorem, if $L(j\omega)$ tends to zero more quickly than $1/j\omega$ for high values of ω (that is, the pole excess is at least 2), then the following integral equation is valid:

$$\int_0^\infty \log|S(j\omega)|\mathrm{d}\omega = \int_0^\infty \log\frac{1}{|1+L(j\omega)|}\mathrm{d}\omega = \pi\sum_i \mathrm{Re}\,p_i, \tag{6.4}$$

where the p_i denote the unstable poles. Consider the simplest case, when there are no poles in the right half-plane, so $L(s)$ is stable or is on the boundary of stability. Then

$$\int_0^\infty \log|S(j\omega)|\mathrm{d}\omega = 0. \tag{6.5}$$

This equation can be interpreted more easily, as it represents the so-called water-bed effect, which means that if $|S(j\omega)|$ is pressed at one point, it will bulge at another one. If for example for low frequencies we make an effort to decrease the values of $|S(j\omega)|$, then at high frequencies it will be high. This is because we calculate the integral of the logarithm of an absolute value, thus the area of $|S(j\omega)|$ below 1 will be equal to the area above 1 (in logarithmic scale the area below the zero dB scale will be equal to the area above it).

Note that drawing the Nyquist diagram and also considering the relationships discussed above can convince one, that significant properties, such as the stability and/or robustness of the control system, can be determined from the course of the frequency function. These properties can be recognized in a somewhat more involved way from the Bode diagram. Unfortunately the quantitative properties cannot be concluded exactly from either the Nyquist diagram nor the Bode diagram.

Example 6.1 Let us consider the control system with the open-loop transfer function

$$L(s) = \frac{1 + s\tau}{sT_I(1 + sT_1)}. \tag{6.6}$$

The frequency function is

$$L(j\omega) = \frac{1 + j\omega\tau}{j\omega T_I(1 + j\omega T_1)} = \frac{\tau - T_1}{T_I\left(1 + \omega^2 T_1^2\right)} - j\frac{1 + \omega^2 \tau T_1}{\omega T_I\left(1 + \omega^2 T_1^2\right)} = \mathrm{Re}(\omega) + j\,\mathrm{Im}(\omega). \tag{6.7}$$

The value of the real part at zero frequency is

$$\mathrm{Re}(\omega \to 0) = \frac{\tau - T_1}{T_I} = \frac{\tau}{T_I} - \frac{T_1}{T_I} = \begin{cases} > 0, & \text{if} \quad \tau > T_1 \\ < 0, & \text{if} \quad \tau < T_1 \end{cases} \tag{6.8}$$

and

$$\mathrm{Re}(\omega \to 0) = \frac{\tau - T_1}{T_I} < 1, \quad \text{if} \quad \tau < T_1 + T_I \tag{6.9}$$

which is a necessary condition to avoid amplification in the supplementary sensitivity function $T(j\omega)$. The pole excess of the loop transfer function is 1, which is the necessary condition to avoid amplification in the sensitivity function. Let us examine the sufficient condition as well. The sensitivity function is obtained by a simple calculation according to

$$S = \frac{1}{1 + L} = \frac{sT_I(1 + sT_1)}{sT_I(1 + sT_1) + (1 + s\tau)}. \tag{6.10}$$

The frequency function of the sensitivity function is

$$S(j\omega) = \frac{j\omega T_I(1 + j\omega T_1)}{j\omega T_I(1 + j\omega T_1) + (1 + j\omega\tau)} = \frac{-\omega^2 T_I T_1 + j\omega T_I}{(1 - \omega^2 T_I T_1) + j\omega(T_I + \tau)}. \tag{6.11}$$

The function $S(j\omega)$ has no amplification if

$$|S(j\omega)|^2 = E^2 = \frac{\omega^2 T_I^2\left(1 + \omega^2 T_1^2\right)}{\left(1 - \omega^2 T_I T_1\right)^2 + \omega^2 (T_I + \tau)^2} \le 1 \tag{6.12}$$

Solving the inequality above, the following sufficient condition is obtained:

$$\frac{\tau}{T_{\mathrm{I}}} \geq \sqrt{1+\frac{2T_1}{T_{\mathrm{I}}}} - 1. \tag{6.13}$$

The complementary sensitivity function is

$$T = \frac{L}{1+L} = \frac{1+s\tau}{sT_{\mathrm{I}}(1+sT_1)+(1+s\tau)} \tag{6.14}$$

and its frequency function is

$$T(j\omega) = \frac{1+j\omega\tau}{j\omega T_{\mathrm{I}}(1+j\omega T_1)+(1+j\omega\tau)} = \frac{1+j\omega\tau}{(1-\omega^2 T_{\mathrm{I}} T_1)+j\omega(T_{\mathrm{I}}+\tau)}. \tag{6.15}$$

Now, $T(j\omega)$ has no amplification if

$$|T(j\omega)|^2 = M^2 = \frac{1+\omega^2\tau^2}{(1-\omega^2 T_{\mathrm{I}} T_1)^2+\omega^2(T_{\mathrm{I}}+\tau)^2} \leq 1. \tag{6.16}$$

From the solution of the inequality $\tau \geq T_1 - T_{\mathrm{I}}/2$ is obtained as a sufficient condition. So there is no amplification if

$$T_1 - T_{\mathrm{I}}/2 \leq \tau \leq T_1 + T_{\mathrm{I}} \tag{6.17}$$

With the root locus method it can be seen that the closed-loop system is structurally stable. If $\tau > T_1$, then there are only real poles. If $\tau < T_1$, then real poles are obtained only for the ranges $K_{\mathrm{I}} < K_1$ and $K_{\mathrm{I}} > K_2$, and in the range $K_1 < K_{\mathrm{I}} < K_2$ the poles are complex conjugate pairs, where

$$K_{1,2} = \left(\frac{2T_1}{\tau} - 1\right) \pm \sqrt{\left(\frac{2T_1}{\tau} - 1\right)^2 - 1}. \quad \blacksquare \tag{6.18}$$

Chapter 7
Control of Stable Processes

In the initial period, heuristic, trial-and-error methods based on rule of thumb were widely used for controller design. At the same time a great effort was made to elaborate a general mathematical methodology for a theoretical approach to design methods. The transfer function of an mth-order controller has $(2m+1)$ unknown parameters. While tuning the higher order controllers it was noticed that a certain design goal can be reached by many parameter sets, moreover, as a consequence, in many cases the parameters were not independent, so the parameterization of the controller was redundant. The main question is how to parameterize a general, stable controller to solve the basic design tasks of a closed-loop control system with a minimum number of non-redundant parameters. The most important solution is provided by the so-called YOULA-parameterization. The YOULA-parameter, as a matter of fact, is a stable (following from its definition), regular transfer function.

$$Q(s) = \frac{C(s)}{1+C(s)P(s)} \quad \text{or, for simplicity,} \quad Q = \frac{C}{1+CP}, \tag{7.1}$$

where $C(s)$ is a stabilizing controller and $P(s)$ is the transfer function of the stable process. The inner stability of a system is defined by the fact, that introducing a bounded input signal at any point of the system provides a bounded output signal at any other point of the loop (see Sect. 5.2). For the investigation of inner stability, the so-called transfer matrix of the closed-loop has to be constructed

$$\mathbf{T}_{\mathrm{t}}(P,C) = \begin{bmatrix} \dfrac{CP}{1+CP} & \dfrac{P}{1+CP} \\ \dfrac{C}{1+CP} & \dfrac{1}{1+CP} \end{bmatrix} = \frac{1}{1+CP}\begin{bmatrix} CP & P \\ C & 1 \end{bmatrix} \tag{7.2}$$

The transfer matrix represents the connection between two independent outer and two inner signals. The closed-loop is inner stable, if and only if, all elements of $\mathbf{T}_{\mathrm{t}}(P,C)$ are stable.

L. Keviczky et al., *Control Engineering*, Advanced Textbooks in Control and Signal Processing, https://doi.org/10.1007/978-981-10-8297-9_7

7.1 The YOULA-Parameterization

The transfer matrix can also be expressed by the YOULA parameter $Q(s)$ instead of the controller $C(s)$:

$$\mathbf{T}_t(P,Q) = \begin{bmatrix} QP & P(1-QP) \\ Q & 1-QP \end{bmatrix}. \tag{7.3}$$

Here it can be easily seen that the inner stability is ensured by any stable $Q(s)$ for a stable process.

It follows from the definition of the YOULA parameter that the structure of the realizable and stabilizable controller is fixed in the control loop parameterized in this way, i.e.,

$$C(s) = \frac{Q(s)}{1-Q(s)P(s)} \quad \text{or, for simplicity} \quad C = \frac{Q}{1-QP}. \tag{7.4}$$

The YOULA-parameterized (*YP*) control loop is shown in Fig. 7.1, where r is the reference signal, e is the error signal, u is the output of the controller (the actuating signal), y_n is the disturbance signal affecting the output, and y is the output signal of the process, i.e., the controlled variable.

The overall transfer function of the closed-loop (the complementary sensitivity function) is

$$T(s) = \frac{C(s)P(s)}{1+C(s)P(s)} = Q(s)P(s) \quad \text{or, more simply,} \quad T = \frac{CP}{1+CP} = QP, \tag{7.5}$$

which is linear in $Q(s)$. (This linearity, as will be seen later, will facilitate to a great extent the design of the required dynamics of the one degree of freedom (*ODOF*) closed-loop.) The sensitivity function has the form

$$S(s) = \frac{1}{1+C(s)P(s)} = 1-Q(s)P(s) \quad \text{or, for simplicity,} \qquad S = \frac{1}{1+CP} = 1-QP. \tag{7.6}$$

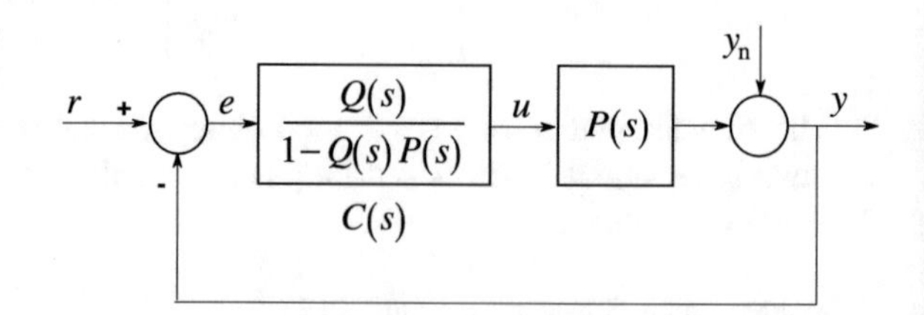

Fig. 7.1 The *YP* control loop

The relationship between the most important signals of the closed-loop can be obtained by simple calculations:

$$\begin{aligned} u &= Qr - Qy_{\mathrm{n}} \\ e &= (1 - QP)r - (1 - QP)y_{\mathrm{n}} = Sr - Sy_{\mathrm{n}} \\ y &= QPr + (1 - QP)y_{\mathrm{n}} = Tr + Sy_{\mathrm{n}} \end{aligned} \tag{7.7}$$

The effect of r and y_{n} on u and e is completely symmetrical (not considering the sign). Thus in this system the input of the process depends only on the outer signals and on $Q(s)$.

It is interesting to see that the *YP* controller of (7.4) can be realized by the simple control loop of positive feedback shown in Fig. 7.2. Using this scheme the control loop of Fig. 7.1 can be transferred to the equivalent block scheme of Fig. 7.3 by identical conversions. This latter scheme is called an *internal model control* (*IMC*). The basic principle of this control is that only the deviation (ε) (i.e. the error signal) of the process output and the model output is fed back. This error signal is zero in the ideal case when the inner model is completely equivalent to the process. This case is presented in Fig. 7.3. In reality, however, the transfer function $\hat{P}(s)$ of the inner model is only a good approximation of the true process $P(s)$, since the original system is not known. For simplicity, only the ideal case is investigated here.

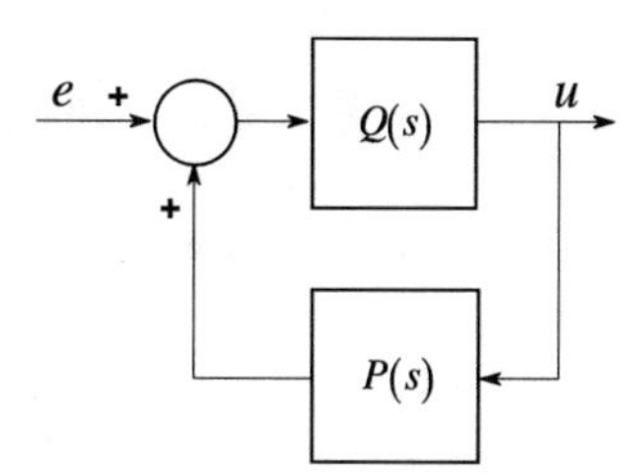

Fig. 7.2 The realization of the *YP* controller

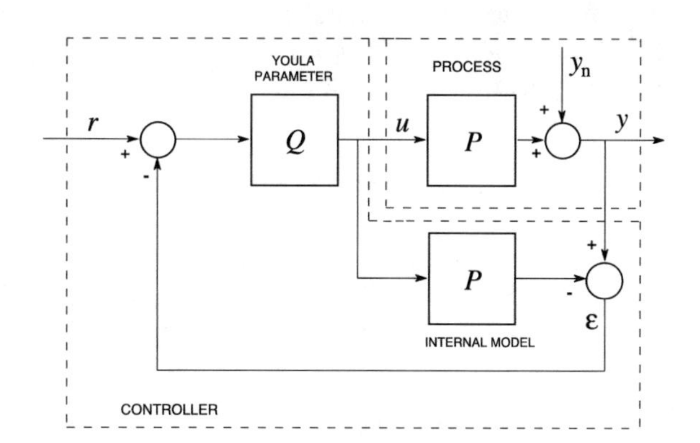

Fig. 7.3 The equivalent *IMC* loop

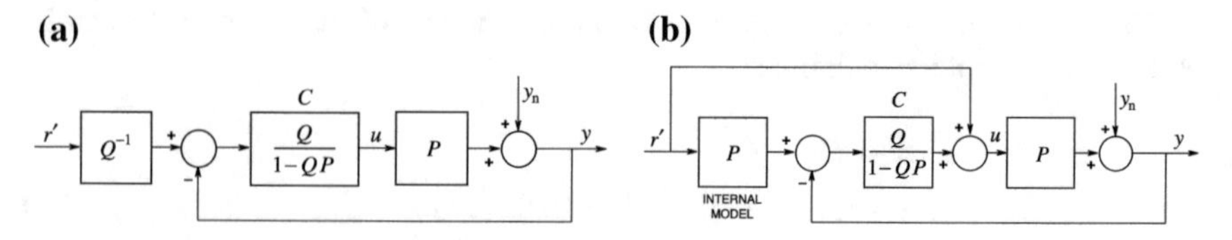

Fig. 7.4 Block schemes for "opening" the closed-loop

Based on the last equation of (7.7) it can be seen that the *IMC* has the transfer function QPr for the tracking of the reference signal. If the inverse of Q is connected in series to the control loop of Fig. 7.1 according to Fig. 7.4a, then the tracking performance becomes independent of Q, i.e., it becomes that given by Pr', which means that practically the closed-loop is "opened". It can be easily checked that this block scheme is equivalent to that of Fig. 7.4b. It has to be noted that here the reference signal has a direct effect on the input of the process: it does not go through the controller and the whole closed-loop. The effect of the controller (concerning the reference signal) is in operation only when the inner model is not equal to the real process.

Following the above train of thought, the extension of the YOULA-parameterization can also be introduced for two-degree of freedom (*TDOF*) control loops. To do this let us simply introduce the parameter Q_r for designing the tracking behavior and connect it in series to the control loop of Fig. 7.4. Then we get the block scheme of Fig. 7.5. The resulting transfer characteristics of this system are

$$\begin{aligned} u &= Q_r y_r - Q y_n \\ e &= (1 - Q_r P) y_r - (1 - QP) y_n = (1 - T_r) y_r - S y_n \\ y &= Q_r P y_r + (1 - QP) y_n = T_r y_r + (1 - T) y_n = T_r y_r + S y_n \end{aligned} \tag{7.8}$$

where the tracking performance can be designed by choosing the parameter Q_r in $T_r = Q_r P$, while the performance of the disturbance rejection can be designed by choosing Q in $T = QP$. Thus these two performances can be handled separately. The reference signal of the whole system is noted by y_r. The same preconditions are valid for Q_r and Q. The transfer function of the *TDOF* closed-loop referred to the reference signal T_r is analogous to the complementary sensitivity function $T$of the *ODOF* system for the tracking.

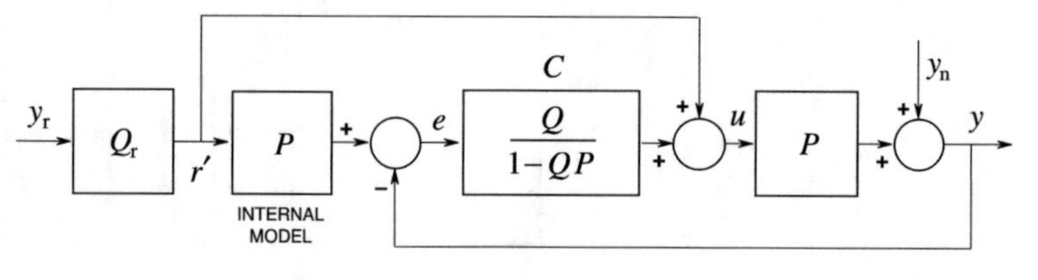

Fig. 7.5 Two degree of freedom version of the *YP-controller*

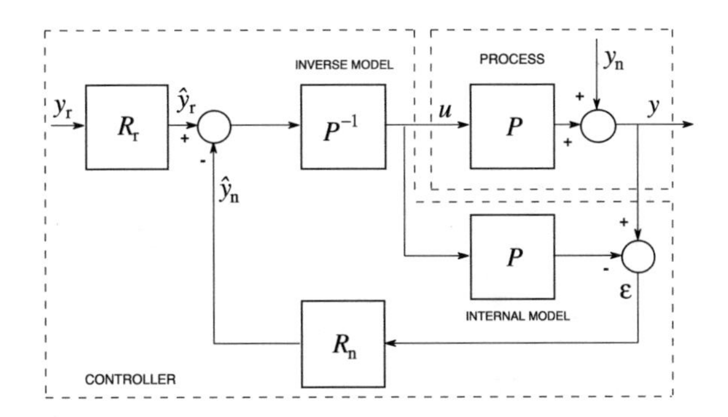

Fig. 7.6 The extension of the control loop based on the ideal *IMC*

The *IMC* of Fig. 7.3 can be further developed according to Fig. 7.6. Here the predicted value $\hat{y}_n$ of the output noise y_n is constructed from the difference ε between the output of the process and the model by the predictor R_n. Similarly the predictor R_r provides the prediction $\hat{y}_r$ of the reference signal y_r. The disturbance compensation of the loop works by giving the predicted value $-\hat{y}_n$ through the inverse of the process to the input of the process, thus in the case of an exact estimation, the disturbance is eliminated. The tracking works in a similar way. Here the operation of R_r can be referred to as a reference model (the desired system dynamics), therefore the introduced predictor is also referred to as a reference model. It is generally required that these predictors have to be strictly proper with unit static gain, i.e., $R_n(\omega = 0) = 1$ and $R_r(\omega = 0) = 1$.

The best operation of a *TDOF* control loop can be attained by the special conditions $R_r = R_n = 1$ or $1 - R_n = 0$, but,—as will be shown later,—it can not be realized in most practical systems.

The block scheme of Fig. 7.6 can be redrawn to the equivalent form of Fig. 7.7. Note that the transfer function—in the ideal case, i.e., when the inverse of the process is realizable and stable—is

$$C_{id} = \frac{(R_n P^{-1})}{1 - (R_n P^{-1})P} = \frac{Q}{1 - QP} = \frac{R_n}{1 - R_n} P^{-1}, \tag{7.9}$$

which is the *YP*-controller with the Youla parameter

$$Q = R_n P^{-1}. \tag{7.10}$$

For the tracking, however, the parameter is

$$Q_r = R_r P^{-1}. \tag{7.11}$$

It can be seen that the controller C_{id} is realizable if the pole excess of R_n is greater than or equal to that of the process, i.e., a pole excess of j can be easily ensured by the reference model $R_n = 1/(1 + sT_n)^j$.

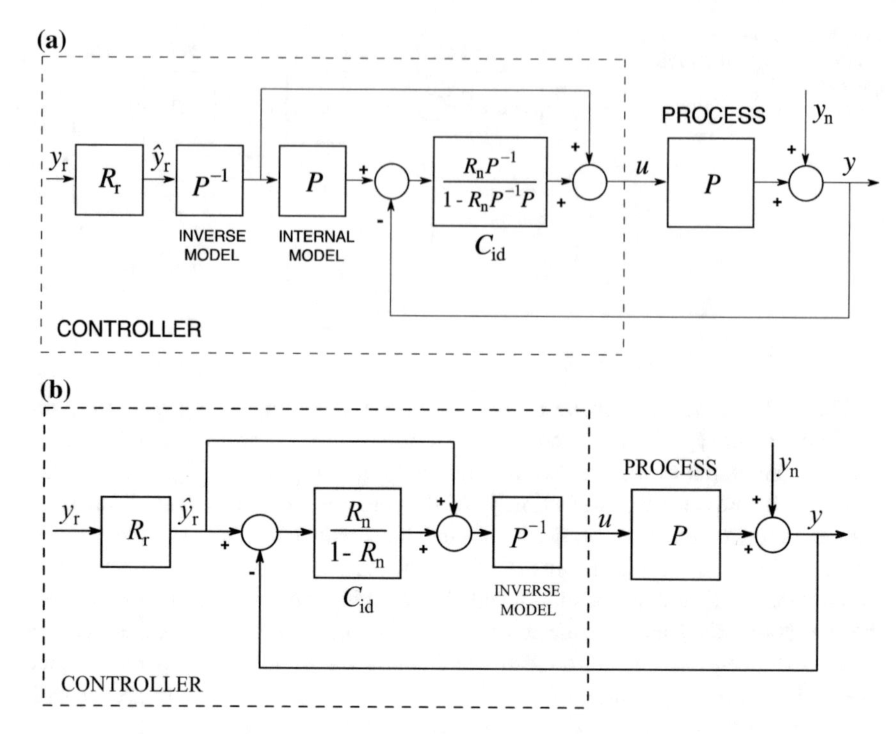

Fig. 7.7 Equivalent ideal control loops using the extended *IMC* principle

The most important signals of the closed-loop for the ideal case are

$$\begin{aligned} u_{\mathrm{id}} &= R_{\mathrm{r}} P^{-1} y_{\mathrm{r}} - R_{\mathrm{n}} P^{-1} y_{\mathrm{n}} \\ e_{\mathrm{id}} &= (1 - R_{\mathrm{r}}) y_{\mathrm{r}} - (1 - R_{\mathrm{n}}) y_{\mathrm{n}} = \left(1 - T_{\mathrm{r}}^{\mathrm{id}}\right) y_{\mathrm{r}} - S_{\mathrm{id}} y_{\mathrm{n}} \\ y_{\mathrm{id}} &= R_{\mathrm{r}} y_{\mathrm{r}} + (1 - R_{\mathrm{n}}) y_{\mathrm{n}} = T_{\mathrm{r}}^{\mathrm{id}} y_{\mathrm{r}} + (1 - T_{\mathrm{id}}) y_{\mathrm{n}} = T_{\mathrm{r}}^{\mathrm{id}} y_{\mathrm{r}} + S_{\mathrm{id}} y_{\mathrm{n}} \end{aligned} \tag{7.12}$$

thus in the ideal case, $T_{\mathrm{r}}^{\mathrm{id}} = R_{\mathrm{r}}$ and $T_{\mathrm{id}} = R_{\mathrm{n}}$, which are our design goals.

Note that the previously followed thoughts presented only the main idea of the YOULA-parameterization and its equivalency with the control based on *IMC*. There is, however, a very critical point of the realizability of the resulting schemes, namely the realizability of the inverse of the process P. Unfortunately, this, in general—disregarding some rare exceptions—is not true for *continuous-time* (CT) systems. For practical applications, versions of the above approach have to be found where all elements of the *TDOF* system are realizable.

To introduce a generally applicable controller, let us assume the transfer function of the process has the following factored form

$$P(s) = P_+(s)\bar{P}(s)_- = P_+(s)P_-(s)e^{-sT_d}, \quad \text{or, for short,} \qquad (7.13)$$
$$P = P_+\bar{P}_- = P_+P_-e^{-sT_d},$$

where P_+ is stable, and its inverse is also stable and realizable (*ISR*). The inverse of $\bar{P}_-$ is unstable (*Inverse Unstable*: *IU*) and not realizable (*IUNR*). P_- is inverse unstable (*IU*). Here, in general, the inverse of the dead-time part e^{-sT_d} is not realizable, because it would be an ideal predictor. The generalized *IMC* principle can also be applied to the general process structure that is shown in Fig. 7.8.

The block scheme of Fig. 7.8 can be redrawn into the equivalent form of Fig. 7.9, where the realizable *YP*-controller of optimal structure obtained for the general case has the form

$$C_{\text{opt}} = \frac{Q_{\text{opt}}}{1 - Q_{\text{opt}}P} = \frac{R_n G_n P_+^{-1}}{1 - R_n G_n P_- e^{-sT_d}} = \frac{R_n K_n}{1 - R_n K_n P} = R_n G_n C'_{\text{opt}}, \qquad (7.14)$$

where the optimal YOULA parameter is

$$Q_{\text{opt}} = R_n G_n P_+^{-1} = R_n K_n \quad \text{where} \quad K_n = G_n P_+^{-1} \qquad (7.15)$$

and

$$Q_r = R_r G_r P_+^{-1} = R_r K_r \quad \text{where} \quad K_r = G_r P_+^{-1}. \qquad (7.16)$$

The obtained general control loop—due to the *YP* – gives structurally the best controller for stable processes. Further optimality of the controller can be set by the embedded transfer functions G_r and G_n. To understand this, let us consider again the most important signals of the *TDOF* closed-loop in the optimal case:

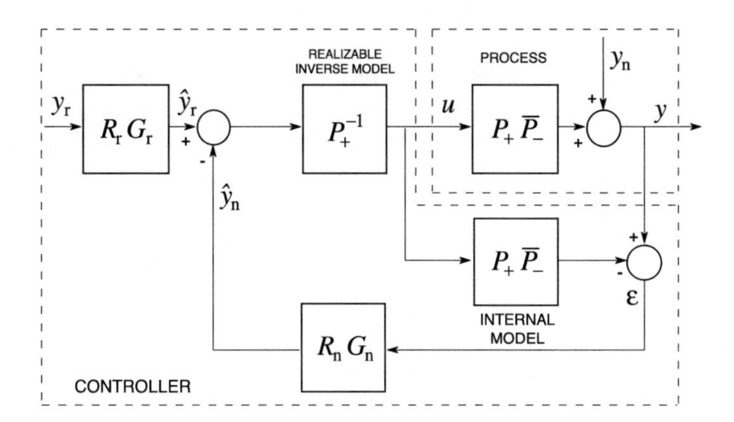

Fig. 7.8 The optimal control loop based on the generalized *IMC* principle

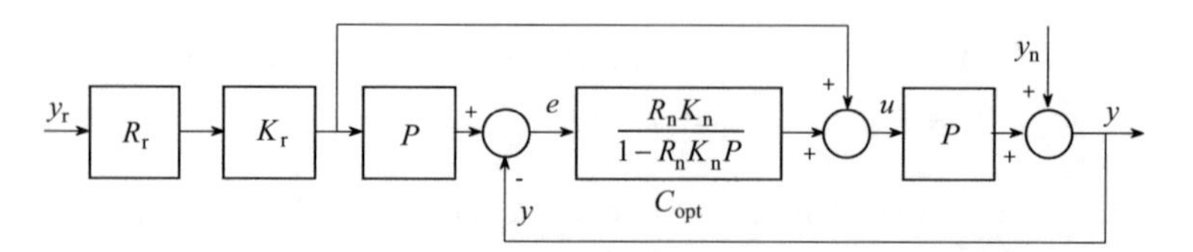

Fig. 7.9 The equivalent optimal control loop corresponding to the generalized *IMC* principle

$$\begin{aligned} u_{opt} &= R_r G_r P_+^{-1} y_r - R_n G_n P_+^{-1} y_n \\ e_{opt} &= \left(1 - R_r G_r P_- e^{-sT_d}\right) y_r - \left(1 - R_n G_n P_- e^{-sT_d}\right) y_n = \left(1 - T_r^{opt}\right) y_r - S_n^{opt} y_n \\ y_{opt} &= R_r G_r P_- e^{-sT_d} y_r + \left(1 - R_n G_n P_- e^{-sT_d}\right) y_n = T_r^{opt} y_r + \left(1 - T_n^{opt}\right) y_n = T_r^{opt} y_r + S_n^{opt} y_n \end{aligned} \tag{7.17}$$

where the equalities $T_r^{opt} = R_r G_r P_- e^{-sT_d}$ and $T_n^{opt} = R_n G_n P_- e^{-sT_d}$ occur.

Compare the ideal output y_{id} of (7.12) with the optimal output y_{opt} of (7.17). It can be easily seen that the ideal and the designed transfer functions determined by the reference models R_r and R_n can not be reached, only approximated. The element $P_- e^{-sT_d}$ appearing in the approximate transfer functions can not be eliminated, therefore it is called an invariant factor. Thus the dead-time e^{-sT_d} and the *inverse unstable* (*IU*) term P_- of the process can not be eliminated by any controller. In case of CT processes, this latter term contains the unstable zeros of the non-minimum phase processes and those poles of the stable poles which could not get into the invertible P_+ of (7.13). In practice, only the necessary number of the slowest poles (whose number corresponds to the number of the stable zeros in P) of P are usually included in P_+, the rest should be added to P_-. The effect of the invariant P_- can only be attenuated by the transfer functions G_r and G_n.

The formulation of the deviation of the outputs of the ideal and best reachable (optimal) control loops is

$$\Delta y = y_{id} - y_{opt} = R_r \left(1 - G_r P_- e^{-sT_d}\right) y_r + R_n \left(1 - G_n P_- e^{-sT_d}\right) y_n, \tag{7.18}$$

where the error comes from the transfer function $R_x \left(1 - G_x P_- e^{-sT_d}\right)\big|_{x=r,n}$ both for the tracking and noise rejection. The minimization of this error in terms of different criteria (in theoretical investigations, the so-called $\mathcal{H}_2$ and $\mathcal{H}_\infty$ optimality) can be accomplished by the optimal choice of $G_x|_{x=r,n}$. (The discussion of optimality is not the subject of this book.)

Further equivalent forms of the best reachable optimal control loop are shown in Fig. 7.10. From these the simplest one that is realizable has to be chosen. Figure 7.10b gives advice for the realization of a system having dead-time. The control loop shown in Fig. 7.9 is called the most general (generic) form of a *TDOF* systems. (The YOULA-parameterization has been extended to *TDOF* systems by KEVICZKY and BÁNYÁSZ by introducing two further parameters, R_r and R_n instead of

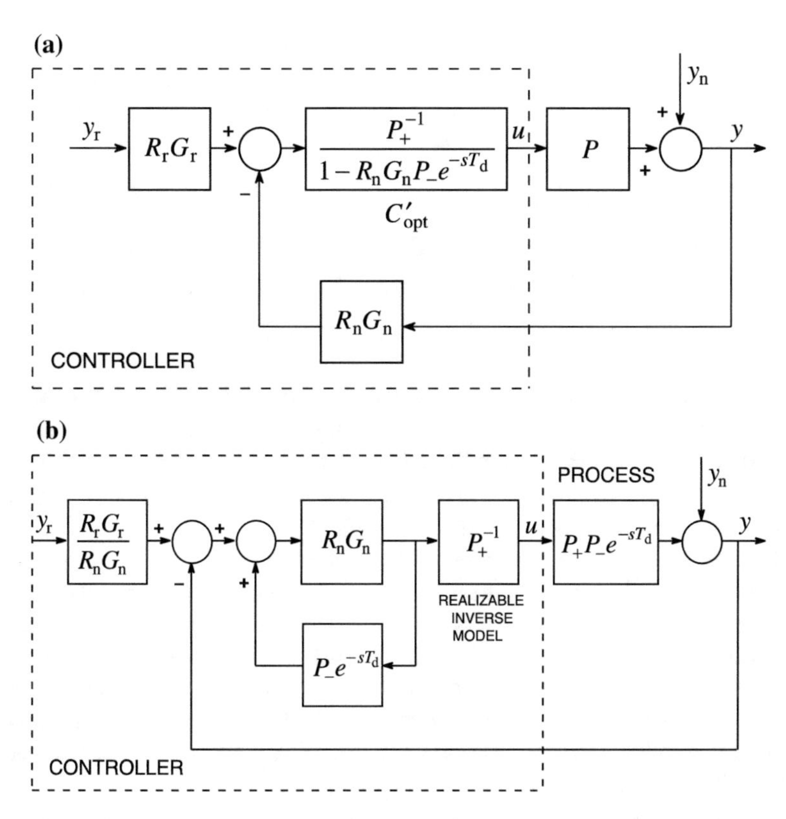

Fig. 7.10 Equivalent forms of the best reachable (optimal) control loops

Q (called the K-B-parameterization). The derivation of the generic schemes and their optimization possibilities are also connected to their names).

The questions of realizability can be dealt with in long discussions. If the design of the optimal controller includes also the optimization of G_r and G_n, then the procedure itself must also ensure the realizability of the transfer functions $G_r P_-$ and $G_n P_-$, and the realizability of the other factors, (like $R_r G_r P_+^{-1}$, $R_n G_n P_+^{-1}$ and

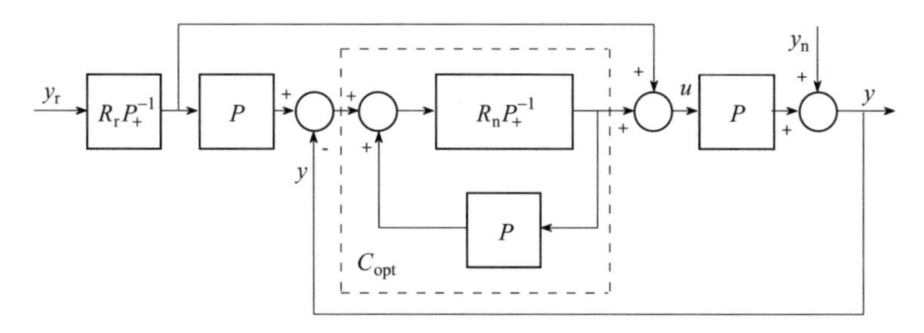

Fig. 7.11 A realizable Youla-parameterized control loop with the choice $G_r = G_n = 1$

$R_nG_nP_-$), has to also be considered. As was mentioned earlier, the theory concerning the optimality of G_r and G_n is not discussed here. In this case the choice $G_r = G_n = 1$ does not change the invariant P_-, i.e., it appears, as a consequence, unchanged in the signals of the system

$$u = R_r P_+^{-1} y_r - R_n P_+^{-1} y_n$$
$$e = \left(1 - R_r P_- e^{-sT_d}\right) y_r - \left(1 - R_n P_- e^{-sT_d}\right) y_n = (1 - T_r) y_r - S_n y_n$$
$$y = R_r P_- e^{-sT_d} y_r + \left(1 - R_n P_- e^{-sT_d}\right) y_n = T_r y_r + (1 - T_n) y_n = T_r y_r + S_n y_n \tag{7.19}$$

furthermore the realizability of the transfer functions $R_r P_+^{-1}$, $R_n P_+^{-1}$ and $R_n P_-$ is required. It is evident that in this case the realizability can be simply handled by the appropriate choice of the order of the reference models R_r and R_n, and of the pole excess, e.g., by prescribing $R_r = 1/(1 + sT_r)^j$ (and the same for R_n). A realizable, but not optimal control loop can be seen in Fig. 7.11.

Although the controller is theoretically realizable, it can not be expected in practice that for CT systems an ideal dead-time element modeling the time-delay of the process can be realized in the inner positive feedback loop of the controller and in the serial compensator. Therefore in the case of time-delay CT systems, the above discussed optimal control scheme has only theoretical importance. In some cases, the time-delay term can be approximated by higher order PADE-series. In computer controlled cases (for sampled DT controls), however, the method can be fully applied (see Chap. 12).

Example 7.1 Let the controlled system be a first order time-delay lag

$$P = \frac{1}{1 + 10s} e^{-5s} \quad \text{i.e.} \quad P_+ = \frac{1}{1 + 10s}; \quad \bar{P}_- = e^{-5s} \quad \text{and} \quad P_- = 1, \tag{7.20}$$

which should be sped up by the control. Let the tracking and disturbance cancellation reference models be

$$R_r = \frac{1}{1 + 4s} \quad \text{and} \quad R_n = \frac{1}{1 + 2s}. \tag{7.21}$$

Since $P_- = 1$, there is nothing to be optimized, i.e., $G_r = 1$ and $G_n = 1$ can be chosen. The optimal controller is

$$C_{opt} = \frac{R_n G_n P_+^{-1}}{1 - R_n G_n P_- e^{-sT_d}} = \frac{1}{1 - R_n e^{-sT_d}} R_n P_+^{-1} = \frac{1}{1 - \frac{1}{1+2s} e^{-5s}} \frac{1 + 10s}{1 + 2s}$$
$$= \frac{1 + 10s}{1 + 2s - e^{-5s}} \tag{7.22}$$

and the serial compensator has the form

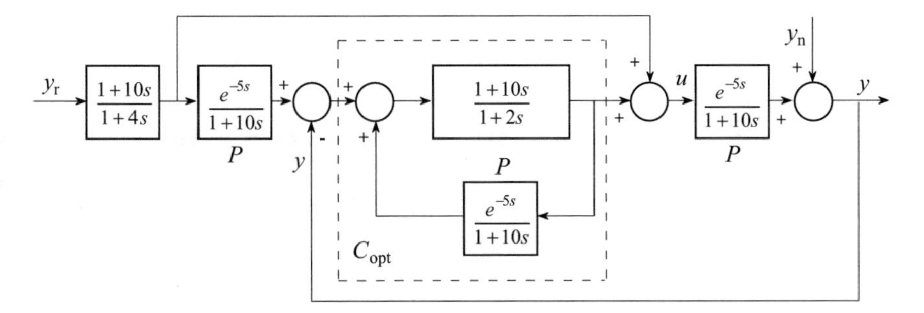

Fig. 7.12 The optimal control loop of Example 7.1

$$R_r K_r = R_r G_r P_+^{-1} = R_r P_+^{-1} = (1+10s)/(1+4s), \tag{7.23}$$

thus the optimal *TDOF* control loop has the structure shown in Fig. 7.12. Observe that $C_{opt}(s=0) = \infty$, i.e., the controller has integrating behavior, which results from the condition $R_n(s=0) = 1$.

It can be easily checked, that the output of the closed system is

$$y_{opt} = R_r e^{-sT_d} y_r + \left(1 - R_n e^{-sT_d}\right) y_n = \frac{1}{1+4s} e^{-5s} y_r + \left(1 - \frac{1}{1+2s} e^{-5s}\right) y_n, \tag{7.24}$$

which completely corresponds to the designed *TDOF* control loop. ∎

Example 7.2 Let the controlled process be a second order lag

$$P = \frac{(1+5s)(1+6s)}{(1+10s)(1+8s)} = P_+ \quad \text{i.e.,} \quad P_- = 1. \tag{7.25}$$

Suppose that the tracking and disturbance cancellation models are again of the form (7.21). Since $P_- = 1$, there is nothing to be optimally compensated, i.e., $G_r = 1$ and $G_n = 1$ can be chosen. Now the optimal controller is

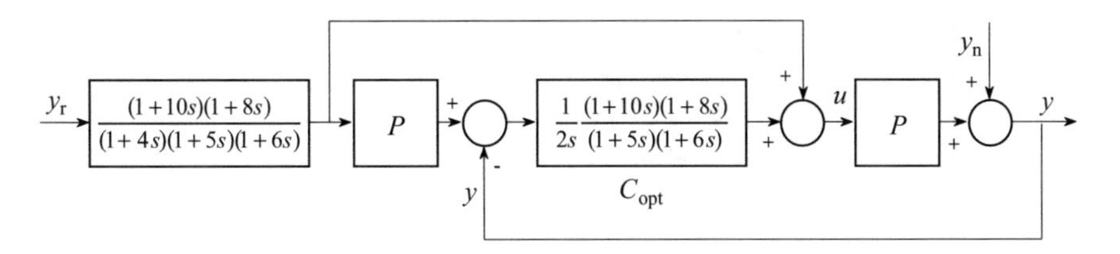

Fig. 7.13 The optimal control loop of Example 7.2

$$C_{\text{opt}} = \frac{R_{\text{n}} G_{\text{n}} P_{+}^{-1}}{1 - R_{\text{n}} G_{\text{n}} P_{-} e^{-sT_{\text{d}}}} = \frac{R_{\text{n}}}{1 - R_{\text{n}}} P^{-1} = \frac{1}{2s} \frac{(1+10s)(1+8s)}{(1+5s)(1+6s)} = C_{\text{id}} \tag{7.26}$$

thus it corresponds to the ideal controller. The serial compensator, however, has the form

$$R_{\text{r}} K_{\text{r}} = R_{\text{r}} G_{\text{r}} P_{+}^{-1} = R_{\text{r}} P^{-1} = \frac{(1+10s)(1+8s)}{(1+4s)(1+5s)(1+6s)}. \tag{7.27}$$

It is evident, that the controller is of integrating type, as is also shown in Fig. 7.13.

Note that in ideal case the term $R_{\text{n}}/(1 - R_{\text{n}})$ in the controller corresponds to an integrator, whose integrating time is equal to the time constant of the first order reference model R_{n}. ∎

7.2 The Smith Controller

The handling of the time-delay of the processes has required the special attention of the designers of the control loops from the beginning. First Otto Smith suggested a technique by means of which it was thought for a long time that the controller can be designed without the consideration of the dead-time. To understand his method let us consider a simple dead-time process of (7.13)

$$P(s) = P_{+}(s)\bar{P}_{-}(s) = P_{+}(s)e^{-sT_{\text{d}}} \quad \text{or, more simply,} \quad P = P_{+}\bar{P}_{-} = P_{+}e^{-sT_{\text{d}}}, \tag{7.28}$$

where P_{+} is stable. Figure 7.14a shows the original idea of Smith. Since this figure is equivalent to Fig. 7.14b, his main goal can be clearly seen, namely to separate the original dead-time loop into a closed-loop which does not contain the time-delay

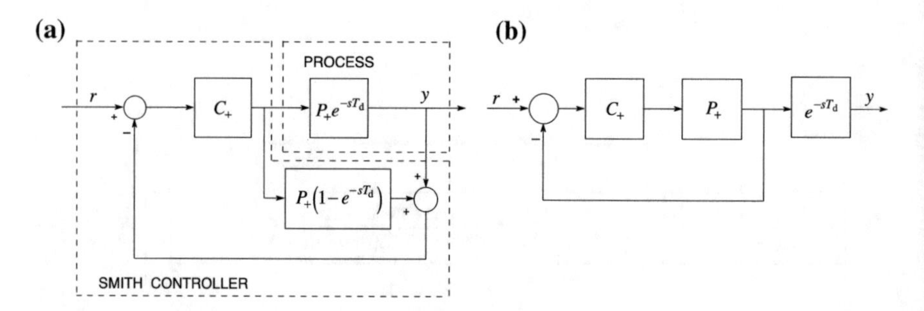

Fig. 7.14 The block scheme of the Smith controller

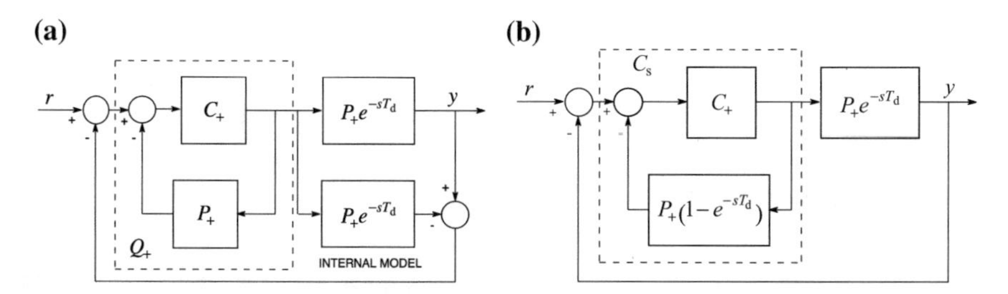

Fig. 7.15 Equivalent SMITH controller block schemes

and a serially connected dead-time. So the controller C_+ regulating the process P_+ can be designed by a conventional method.

By simple block manipulations, Fig. 7.14a can be redrawn to the equivalent forms of Fig. 7.15a, b.

The *IMC* structure of the Fig. 7.15a clearly shows that the SMITH controller is a *YP*-controller with a special YOULA parameter

$$Q_+ = \frac{C_+}{1+C_+P_+} = \frac{C_+P_+}{1+C_+P_+}P_+^{-1} = \frac{L_+}{1+L_+}P_+^{-1} = R_+P_+^{-1} \tag{7.29}$$

if the controller C_+ stabilizes the delay free part of the process P_+. Here $L_+ = C_+P_+$ is the loop transfer function of the closed system of Fig. 7.14b, furthermore the complementary sensitivity function

$$T_+ = R_+ = \frac{L_+}{1+L_+} \tag{7.30}$$

will be the reference model R_+. Since in the *IMC* structure the inner model predicts the output of the process, the name SMITH-predictor derives from this phenomenon. At the time of its introduction the *IMC* principle and the YOULA-parameterization were not yet known.

Figure 7.15b shows the equivalent complete closed control loop, where the serial (YOULA-parameterized) controller is

$$C_s = \frac{Q_+}{1-Q_+P_+e^{-sT_d}} = \frac{C_+}{1+C_+P_+(1-e^{-sT_d})} = C_+K_S, \tag{7.31}$$

which, at the same time, also shows the inner closed-loop referring to the realization. Here K_S means the serial transfer function by which the SMITH controller modifies the effect of the original controller C_+.

Thus

$$K_{\mathrm{S}} = \frac{1}{1 + C_+ P_+ \left(1 - e^{-sT_{\mathrm{d}}}\right)} = \frac{1}{1 + L_+ \left(1 - e^{-sT_{\mathrm{d}}}\right)}. \tag{7.32}$$

At the stability limit $L_+ = -1$, we get

$$K_{\mathrm{S}} = \frac{1}{1 + (-1)\left(1 - e^{-sT_{\mathrm{d}}}\right)} = \frac{1}{1 - 1 + e^{-sT_{\mathrm{d}}}} = e^{sT_{\mathrm{d}}}\Big|_{\omega_{\mathrm{c}}} = e^{j\omega_{\mathrm{c}} T_{\mathrm{d}}}, \tag{7.33}$$

which causes the Smith controller to add a significant positive phase advance to the original closed-loop, which is why it can be applied very successfully for stabilization in many cases. At the same time it is very sensitive to a change of the parameters.

Unfortunately it should be repeated that in the practice for CT systems one cannot expect to realize an ideal dead-time element only its higher order lag approximation can be implemented for the application of a Smith controller (see what was stated about Example 7.1 of the previous chapter).

To complete the evaluation of the Smith controller it has to be also mentioned that it can be used only for the design of a one-degree of freedom (*ODOF*) system, i.e., for tracking. The controller designs the tracking of the reference signal only in an indirect way, as the expression $T_+ = R_+$ of (7.30) shows. From the elaboration of the concept of Youla-parameterization it has been known that simple design method is also available for *TDOF* systems both for tracking and disturbance rejections via the design of the reference models.

7.3 The Truxal-Guillemin Controller

Prior to the Youla-parameterization, Truxal and Guillemin recommended a simple algebraic method for the control design of *ODOF* systems. According to the method the required design goal has to be formulated for the transfer function of the closed-loop

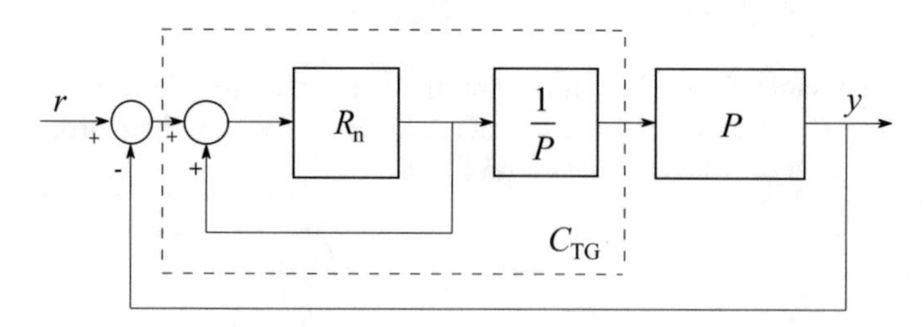

Fig. 7.16 The realization of the Truxal-Guillemin controller

$$R_n = T = \frac{CP}{1+CP} \tag{7.34}$$

from which a simple algebraic equation results for C:

$$CP = R_n + CPR_n. \tag{7.35}$$

Solving for the controller we get

$$C = \frac{R_n}{1-R_n}\frac{1}{P} = C_{TG}. \tag{7.36}$$

Note that this form is the same as the simple case of the YOULA controller C_{id} in (7.9). The realization of the controller can be made according to the Fig. 7.16.

Thus R_n corresponds to one of the reference models of the YOULA method. For the *ODOF* case, however, $R_n = R_r$. Let the reference model be $R_n = \mathcal{B}_n/\mathcal{A}_n$, and the process be $P = \mathcal{B}/\mathcal{A}$. So the polynomial form of the controller is

$$C_{TG} = \frac{\mathcal{B}_n}{\mathcal{A}_n - \mathcal{B}_n}\frac{\mathcal{A}}{\mathcal{B}}. \tag{7.37}$$

The controller is realizable if the pole excess of R_n is greater than or equal to that of the process. If R_n has unity gain $(R_n(0) = 1)$, then the type of the controller is one. TRUXAL observed that the loop transfer function $L = \mathcal{F}/s^k\mathcal{D} = \mathcal{CP}$ of type k can be established by the reference model

$$\begin{aligned} R_n = T = \frac{\mathcal{N}}{\mathcal{N}+s^k\mathcal{D}} &= \frac{f_o + f_1 s + \ldots + f_{k-1}s^{k-1}}{f_o + f_1 s + \ldots + f_{k-1}s^{k-1} + s^k + \ldots + d_{n_R+k}s^{n_R+k}} \\ &= \frac{f_o + f_1 s + \ldots + f_{k-1}s^{k-1}}{f_o + f_1 s + \ldots + f_{k-1}s^{k-1} + s^k(1 + \ldots + d_{n_R}s^{n_R})}; n_R - k - 1 \geq n - m \end{aligned} \tag{7.38}$$

where the first k terms of the denominator are equal to the numerator.

7.4 The Effect of a Constrained Actuator Output

The control signals applied in the closed control systems, or the output of the actuator whose task is to increase that signal to the proper level, are always amplitude constrained.

$$|u(t)| \leq U_{max} \tag{7.39}$$

This means that a jump of any size in u, or a significant change in the starting value of the signal related to its final value, i.e., arbitrary overexcitation is

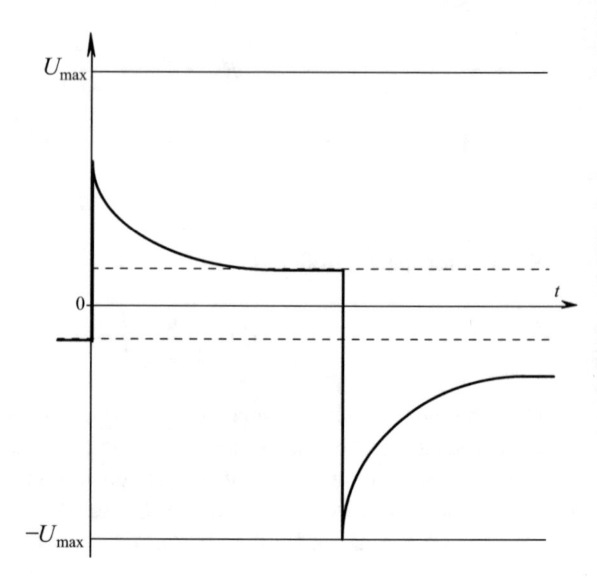

Fig. 7.17 Typical control output (actuator signal) in the case of overexcitation

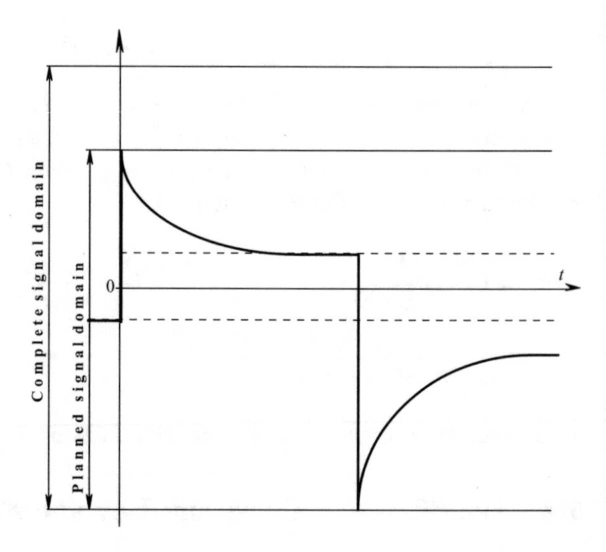

Fig. 7.18 Design of the signal domain for the control output

impossible. It was shown in Sect. 2.4 that a pole cancellation can be made by adding extra zeros which results in speeding up the system. This speeding up always requires overexcitation (energy surplus). The optimal control methods discussed in this chapter almost always applied a certain kind of pole cancellation, i.e., overexcitation (see Fig. 7.17). The above mentioned amplitude constraints in practice mean, that in spite of computing the optimal control parameters, the provided output cannot be realized because of the constraints. The reachable speed-up really depends on the applicable overexcitation.

The design of the signal domain for the control output needs special care and the knowledge of the equipment employed. In general, the working point is set to the center of the signal domain, and the possible changes compared to this point have to be designed to perform without saturation (Fig. 7.18).

In spite of the most careful design it may happen that the control output violates the signal domain. In this case the original design goal has to be reduced. The advantage of the *KB*-parameterization of the generic *TDOF* control loops is that in this case it is enough to redesign only the problematic (very demanding) reference models R_r or R_n by less demanding design conditions. This process can usually be made in small steps by iteration. The iteration steps may include both the model simulation and an experiment on the real system. (In case of lower order reference models it is possible to elaborate explicit design formulas for determining the time constant of the model (bandwidth) with the knowledge of the process model and the amplitude constraints U_{max}.)

In many cases not only the amplitude of the control output has constraints but its changing velocity is also limited in practice. Let us think of the control valves of big pipes, where the motor needs time to transfer the valve from one position to another. Handling these so-called velocity constraints by analytic methods is more difficult, so only simulation and practical experiment remain as a solution. The applied method is the same as earlier: the demand required by the design goal has to be reduced.

Summarizing, it can be stated that the fastest reachable control depends, primarily and to a great extent, on the limitations of the control output. This limitation, however, does not depend on the control design method, but on the type of the equipment used in the given technology. So any improvement can be made only by changing the equipment itself or by redesigning the whole technology.

The Concept and Computation of Dynamic Overexcitation

In control systems the actuator signal $u(t)$ has an important role because this is the input of the process and the physical constraints appear here. Due to the change of the reference signal $r(t)$ transient processes take place according to the dynamics of the system. During a transient, in general, the actuator signal might have a higher value than its static one. The extent or magnitude of this can be described by the dynamic overexcitation. The definition of dynamic overexcitation is

$$u_t = \frac{U_{max}}{u(t \to \infty)} \cong \frac{u(0)}{u_\infty}.$$

In general the maximum value is the initial value, therefore, for the sake of the simplicity of the computations, this is used as an approximation. The dynamic overexcitation can be interpreted this way only when u_∞ is not equal to zero. This happens when the process contains an integrator, since in this case the system can get to a steady state only if the input of the integrator becomes zero. Here the dynamic overexcitation is replaced by U_{max}.

7.5 The Concept of the Best Reachable Control

7.5.1 General Theory

In the previous section, the basic importance of the constraints relating to the output of the actuator for ensuring the reachable/attainable/obtainable best control has been emphasized. Based on the Sects. 7.1–7.3, however, further constraints not depending on us have to be mentioned. Of these, the dead-time of the process is the most important, which is invariant for any kind of regulation method, i.e., its effect can not be eliminated.

Other such factors are the unstable zeros of the process, which also can not be eliminated by any method. The effect of the invariant zeros on the transients can be compensated (decreased or attenuated) to a certain extent. (This compensation can be made by the optimal choice of the embedded filters G_{r} and G_{n}, which is not the topic of this book.)

Thus both the dead-times and the unstable zeros are considered independent features of the process that can not be influenced by the control design methods, only by the redesign of the whole process or technology.

So far, in the long discussion of the different control design methods, it was supposed that the transfer function P of the process is known. In reality the exact transfer function of the process is not known: only its model $\hat{P}$ is available. This distinction was used, until now, only in the discussion of the concept of robust stability in Sect. 5.7. For the closed-loop control design it should be noted that the complementary sensitivity function

$$\hat{T} = \frac{C\hat{P}}{1 + C\hat{P}} \tag{7.40}$$

resulting from the *ODOF* model based design is not equal to the real one

$$T = \frac{CP}{1 + CP} = \hat{T}\frac{1 + \ell}{1 + \hat{T}\ell}. \tag{7.41}$$

Here ℓ means the relative uncertainty of the process model of (5.40). The sensitivity function of the real closed-loop can be written in the following decomposed form:

$$\begin{aligned}S &= \underbrace{(1-R_\mathrm{n})}_{S_\mathrm{des}} + \overbrace{\underbrace{\left(R_\mathrm{n}-\hat{T}\right)}_{S_\mathrm{real}} - \underbrace{\left(T-\hat{T}\right)}_{S_\mathrm{mod}}}^{S_\mathrm{perf}} = S_\mathrm{des} + S_\mathrm{real} + S_\mathrm{mod} \\ &= \underbrace{(1-R_\mathrm{n})}_{S_\mathrm{des}} + \underbrace{(R_\mathrm{n}-T)}_{S_\mathrm{perf}} = S_\mathrm{des} + S_\mathrm{perf} = \underbrace{\left(1-\hat{T}\right)}_{S_\mathrm{contr}} + S_\mathrm{mod} \\ &= \underbrace{\left(1-\hat{T}\right)}_{S_\mathrm{des}+S_\mathrm{real}} + S_\mathrm{mod} = S_\mathrm{contr} + S_\mathrm{mod}\end{aligned} \tag{7.42}$$

Here $S_\mathrm{des} = (1-R_\mathrm{n})$ is the design loss, $S_\mathrm{real} = \left(R_\mathrm{n}-\hat{T}\right)$ is the realizability loss, and $S_\mathrm{mod} = -\left(T-\hat{T}\right) = \hat{T}-T$ represents the modeling loss part of the sensitivity function. In the other form $S_\mathrm{contr} = \left(1-\hat{T}\right)$ is the term referring to the control loss, and $S_\mathrm{perf} = (R_\mathrm{n}-T)$ is the performance loss. Each term can be simply interpreted and explained very easily. The meaning of the reference model R_n has been discussed in the previous chapters. It is obvious that there are trivial equalities $S = 1-T$ and $\hat{S} = 1-\hat{T}$, where

$$\hat{S} = \frac{1}{1+C\hat{P}} \quad \text{and} \quad S = \frac{1}{1+CP} = \hat{S}\frac{1}{1+\hat{T}\ell} = \hat{S} + S_\mathrm{mod}. \tag{7.43}$$

The term S_mod can be further simplified:

$$S_\mathrm{mod} = S - \hat{S} = \hat{T} - T = -\frac{\hat{T}\hat{S}\ell}{1+\hat{T}\ell} = -\hat{T}S\ell\big|_{\ell\to 0} \approx -\hat{T}\hat{S}\ell. \tag{7.44}$$

It can be seen easily that $\left|\hat{T}\hat{S}\right|$ has its maximum at the cut-off frequency ω_c, so the model must be the most accurate in the vicinity of this frequency.

For *TDOF* control loops the complete transfer function corresponding to the concept of the complementary sensitivity function is obtained by adding an extension as $T_\mathrm{r} = FT$, in general. For the model based control

$$T_\mathrm{r} = \hat{T}_\mathrm{r}\frac{1+\ell}{1+\hat{T}\ell} \tag{7.45}$$

again, as it was in (7.41).

Obviously the triviality $S_\mathrm{r} = 1 - T_\mathrm{r}$ and the triple decomposition introduced in (7.42)

$$S_\mathrm{r} = (1-R_\mathrm{r}) + \left(R_\mathrm{r}-\hat{T}_\mathrm{r}\right) - \left(T_\mathrm{r}-\hat{T}_\mathrm{r}\right) = S^\mathrm{r}_\mathrm{des} + S^\mathrm{r}_\mathrm{real} + S^\mathrm{r}_\mathrm{mod} \tag{7.46}$$

also exist.

The term S^{r}_{mod} can be further simplified

$$S^{r}_{mod} = \hat{T}_{r} - T_{r} = -\frac{\hat{T}_{r}\hat{S}\ell}{1+\hat{T}\ell} = -\hat{T}_{r}S\ell\big|_{\ell\to 0} \approx -\hat{T}_{r}\hat{S}\ell. \tag{7.47}$$

The ideal control loop has to follow the signals prescribed by R_{r} and R_{n} (more exactly $1 - R_{n}$), thus the ideal output of the closed-loop is

$$y_{id} = y^{o} = R_{r}y_{r} - (1 - R_{n})y_{n} = y^{o}_{r} + y^{o}_{n} \tag{7.48}$$

according to (7.12).

Theoretically, instead of (7.48) only

$$y = T_{r}y_{r} - Sy_{n} = T_{r}y_{r} - (1 - T)y_{n} \tag{7.49}$$

can be obtained, and even this has to be modified according to the model based design

$$\hat{y} = \hat{T}_{r}y_{r} - \hat{S}y_{n} = \hat{T}_{r}y_{r} - \left(1 - \hat{T}\right)y_{n}. \tag{7.50}$$

The deviation between the ideal and the theoretically reachable output is

$$\Delta y = y^{o} - y = (R_{r} - T_{r})y_{r} - (R_{w} - T)y_{n} = S^{r}_{perf}y_{r} - S^{n}_{perf}y_{n}, \tag{7.51}$$

where S^{r}_{perf} refers to the performance loss concerning the tracking and $S^{n}_{perf} = S_{perf}$ means the performance loss concerning the disturbance rejection. A similar expression can be obtained for the deviation between the ideal output (y^{o}) and $\hat{y}$ obtained by the model based design

$$\Delta\hat{y} = y^{o} - \hat{y} = \left(R_{r} - \hat{T}_{r}\right)y_{r} - \left(R_{n} - \hat{T}\right)y_{n} = S^{r}_{real}y_{r} - S^{n}_{real}y_{n}. \tag{7.52}$$

Note that in the above expressions the terms S_{real} and S^{r}_{real} can be made zero only in the case of inverse stable systems, while these terms can never be made zero for inverse unstable systems.

The above triple decomposition of the sensitivity functions gives a good insight into the limit-optimality of the closed-loop control systems, i.e., to the characterization of the best reachable control. To this end, distinct optimality criteria have to be created for each term, i.e.,

$$\begin{aligned} J_{tracking} &\le J^{r}_{des} + J^{r}_{real} + J^{r}_{mod} = \left\|S^{r}_{des}\right\| + \left\|S^{r}_{real}\right\| + \left\|S^{r}_{mod}\right\| \\ J_{control} &\le J^{n}_{des} + J^{n}_{real} + J^{n}_{mod} = \left\|S_{des}\right\| + \left\|S_{real}\right\| + \left\|S_{mod}\right\| \end{aligned} \tag{7.53}$$

both for the tracking and disturbance rejection behaviors. Here the notation $\|\ldots\|$ is used for expressing the optimality criterion. (Strictly speaking, in mathematical analysis this notation is used to refer to the chosen norm of the transfer function.)

Optimization of the Design Loss

The optimization of the first term primarily means the determination of the best (fastest) reachable reference models $R_{\mathrm{r}} = R_{\mathrm{r}}^{\mathrm{opt}}$ and $R_{\mathrm{n}} = R_{\mathrm{n}}^{\mathrm{opt}}$, i.e. the solution of the optimization task under the following limitations

$$
\begin{aligned}
R_{\mathrm{r}}^{\mathrm{opt}} &= \arg\left\{ \min_{R_{\mathrm{r}}} \left(J_{\mathrm{des}}^{\mathrm{r}} \right) \Big|_{u \in U} \right\} = \arg\left\{ \min_{R_{\mathrm{r}}} \| 1 - R_{\mathrm{r}} \| \Big|_{u \in U} \right\} \\
R_{\mathrm{n}}^{\mathrm{opt}} &= \arg\left\{ \min_{R_{\mathrm{n}}} \left(J_{\mathrm{des}}^{\mathrm{n}} \right) \Big|_{u \in U} \right\} = \arg\left\{ \min_{R_{\mathrm{n}}} \| 1 - R_{\mathrm{n}} \| \Big|_{u \in U} \right\}
\end{aligned}
\tag{7.54}
$$

where the chosen criteria $J_{\mathrm{des}}^{\mathrm{r}} = \|1 - R_{\mathrm{r}}\|$ and $J_{\mathrm{des}}^{\mathrm{n}} = \|1 - R_{\mathrm{n}}\|$ express that each reference model has to approach as closely as possible the ideal term. This task has to be solved under the limitation $u \in \mathbb{U}$ concerning the output of the controller. Here $\mathbb{U}$ usually means the allowed domain for u, e.g. the amplitude constraints $\mathbb{U} : |u| \leq 1$ [see Sect. 7.4].

Equation (7.54) constitutes a very difficult optimization task because the solution is always on the border of the constrained domain. An analytical solution, except for some low order simple cases, can not be found. The optimal reference models are usually determined by simulation *CAD* tools. Note that for the solution of the task (7.54), under the given constraints, faster reference models can not be applied. Quite the opposite: if no solution is obtained for a reference model providing the required goals under a given constraint, then there is no other possibility than to choose a less demanding (usually slower) reference model. Thus the best (fastest) reachable output of the closed-loop basically depends on the limitations of the controller output. In (7.54), of course, both the controller and the process, i.e., the whole real closed-loop, appear in a very complicated way, therefore its optimality depends on the process, on the model and also on the invariant factors.

Since the reference model is an important parameter of the general Youla design, the condition of robust stability shown in (5.45) can also be guaranteed with it. Based on (7.5) it can be seen easily that the condition (5.45) for *YP* control loops is

$$
\left| Q \hat{P} \ell \right| < 1 \quad \forall \omega . \tag{7.55}
$$

This condition can be further simplified to the condition

$$
|R_{\mathrm{n}}| < \frac{1}{|\ell|} \quad \text{or} \quad |\ell| < \frac{1}{|R_{\mathrm{n}}|} \quad \forall \omega . \tag{7.56}
$$

Based on this condition it can be stated that by choosing a first order reference model R_{n}, robust stability can be ensured even for the case of 100% relative model error.

Optimization of the Realizability Loss

The purpose of this task is to optimize the realizability losses $J^{\mathrm{r}}_{\mathrm{real}}$ and $J^{\mathrm{n}}_{\mathrm{real}}$ in terms of

$$
\begin{aligned}
G_{\mathrm{r}}^{\mathrm{opt}} &= \arg\left\{\min_{G_{\mathrm{r}}}\left(J^{\mathrm{r}}_{\mathrm{real}}\right)\right\} = \arg\left\{\min_{G_{\mathrm{r}}}\left\|R_{\mathrm{r}} - \hat{T}_{\mathrm{r}}\right\|\right\} \\
G_{\mathrm{n}}^{\mathrm{opt}} &= \arg\left\{\min_{G_{\mathrm{n}}}\left(J^{\mathrm{n}}_{\mathrm{real}}\right)\right\} = \arg\left\{\min_{G_{\mathrm{n}}}\left\|R_{\mathrm{n}} - \hat{T}\right\|\right\}
\end{aligned}
\tag{7.57}
$$

which can be ensured by the optimal choice of $G_{\mathrm{r}} = G_{\mathrm{r}}^{\mathrm{opt}}$ and $G_{\mathrm{n}} = G_{\mathrm{n}}^{\mathrm{opt}}$ (see Sect. 7.1). As was mentioned earlier the conditions $R_{\mathrm{r}} = \hat{T}_{\mathrm{r}}$ and $R_{\mathrm{n}} = \hat{T}_{\mathrm{n}}$ can be theoretically reached in the *ISR* case, which means the trivial solution $G_{\mathrm{r}} = G_{\mathrm{n}} = 1$. For the more general *IU* case, the optimal transfer functions have to be determined.

The Optimization of the Modeling Loss

The optimization of the modeling loss $J^{\mathrm{r}}_{\mathrm{mod}}$ means the determination of a special optimal excitation $y_{\mathrm{r}} = y_{\mathrm{r}}^{\mathrm{opt}}$ applied as a reference signal, and the optimal process model $\hat{P} = \hat{P}^{\mathrm{opt}}$ obtained as the solution of the so-called minimax problem below

$$
\hat{P}^{\mathrm{opt}} = \arg\left\{\min_{\hat{P}}\left[\max_{y_{\mathrm{r}}}\left(J^{\mathrm{r}}_{\mathrm{mod}}\right)\right]\right\} = \arg\left\{\min_{\hat{P}}\left[\max_{y_{\mathrm{r}}}\left\|S^{\mathrm{r}}_{\mathrm{mod}}\right\|\right]\right\}. \tag{7.58}
$$

This task has two steps: The optimal reference signal (depending on the criterion) usually provides the maximum variance in the output in the case of an amplitude constrained y_{r}. Measuring the output of the closed-loop the process model ensuring the minimum of the optimality criterion $\left\|S^{\mathrm{r}}_{\mathrm{mod}}\right\|$ of the modeling loss has to be determined by a proper modeling (identification) method. The task (7.58) is called *worst case* identification task.

It is not an easy task to optimize all the three terms simultaneously. In practice, an iteration technique is used where in a particular step the solution of only one optimality problem is solved.

(As a criterion $\|\ldots\|$ mentioned above, usually the $\mathcal{H}_2$ or $\mathcal{H}_\infty$ norms, frequently applied in the higher level control theory, are chosen. As it was mentioned earlier, these tasks are not discussed in this book, although a short description can be found in Chap. 16.)

7.5.2 Empirical rules

It has already been seen in the investigation of the best reachable control systems that the basic constraint derives from the saturation of the actuator signal and the process dynamics itself. One of the most important dynamical limits is the dead-time of the process, which can not be eliminated, since the system can not respond in a shorter time than the dead-time. The first-order PADE approximation of a dead-time lag has already been discussed, according to which

$$e^{-sT_d} \approx \frac{1 - sT_d/2}{1 + sT_d/2} = \frac{s - 2/T_d}{s + 2/T_d} = \frac{s - z_j}{s + z_j}, \tag{7.59}$$

i.e., by the right-hand zero z_j approximation it corresponds to a time-delay lag $T_d = 2/z_j$. This also means, at the same time, that the right-hand zeros of the process correspond, in any way, to certain limits. An unstable zero with a small value corresponds to a high dead-time.

It can be assumed that the unstable poles of the process can also result in constraints. It can be expected that in order to stabilize an unstable process, a sufficiently fast controller is required.

Summarizing, the constraints can derive from the dead-time and the unstable zeros (z_j) and poles (p_j) (located in the right half-plane) of the process dynamics. Based on fundamental theoretical considerations and practice, the constraints are the following:

- a right-half-plane unstable (RU) zero z_j has the following limit for the cross-frequency

$$\omega_c < 0.5 z_j \tag{7.60}$$

  A slow RU zero has especially a very bad effect.
- the dead-time also limits the cross-frequency, based on the practice

$$\omega_c < 0.5 \frac{1}{T_d} \tag{7.61}$$

- an *RU* pole requires a high cross-frequency

$$\omega_c > 2p_j \tag{7.62}$$

- systems, having both RU zeros and poles simultaneously can not be controlled, in general, only if the poles and zeros are far enough from each other.

$$p_{\mathrm{j}} > 6z_{\mathrm{j}} \tag{7.63}$$

- unstable dead-time systems can not be controlled, unless the separation (distance) condition

$$p_{\mathrm{j}} < 0.16\frac{1}{T_{\mathrm{d}}} \tag{7.64}$$

is fulfilled.

Chapter 8
Design of Conventional Regulators

The aim of controller design is to ensure that the closed-loop control system meets the quality specifications. A block diagram of a control system is shown in Fig. 6.1. This structure is called a series control structure, as the regulator (sometimes called controller) is serially connected to the plant. The regulator is designed considering the properties of the plant and the specifications.

After the choice of the control structure, the parameters of the regulator have to be set. In the frequency domain with the appropriate choice of the regulator the frequency function of the open-loop is shaped according to the quality specifications (Fig. 6.2).

In Chap. 7 we could see, that for stable processes, precise theoretical methods are available to determine the optimal structure and the optimal parameters of the regulator for different cases. But already long before the elaboration of these theoretical methods, a well established class of control equipment was widely used in the control of industrial processes. This type of regulators has its determining significance even today. The development of the technology of electronic devices—which made possible the realization of more and more complicated transfer functions—played a significant role in the development of these conventional or classical regulators. Passive R-L-C circuits and precise operational amplifiers provided the technological basis for the development of the simple so-called *PID* regulator family. (In the case of regulators using mechanical, pneumatic, etc., signals only some restricted forms were realized, mainly because of realization constraints.)

PID regulators react proportionally to the current error value, take into consideration the past error signal history with the integral of the error signal, and count the future error signal trend by the differential of the error. Figure 8.1 shows [1] that the *PID* regulator calculates the manipulated variable (the control signal) with the *P* effect which is proportional to the error, with the *I* effect, the integral of the error, and the *D* effect, the differential of the error.

Application of *PID* regulators is quite general: more than 90% of the realized industrial control systems work with this type of regulator. In industrial process control the most frequently applied controllers are the *PID* regulators, as the quality

L. Keviczky et al., *Control Engineering*, Advanced Textbooks in Control and Signal Processing, https://doi.org/10.1007/978-981-10-8297-9_8

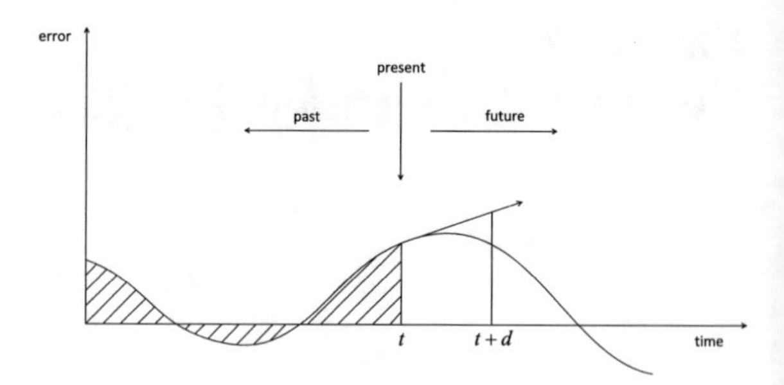

Fig. 8.1 The *PID* regulator calculates the manipulated variable from the current, the past and the future trend (the slope) of the error signal

specifications mostly can be fulfilled applying them, they have an easily realizable simple structure, and the effect of the parameter changes can easily be evaluated. The design of *PID* regulators can be discussed as a design method in the frequency domain or as a pole-zero cancellation technique.

8.1 The *PID* Regulator Family and Design Methods

The transfer function of the ideal *PID* regulator can be given in the following two forms:

$$C_{\mathrm{PID}} = A_{\mathrm{P}}\left(1 + \frac{1}{sT_{\mathrm{I}}} + sT_{\mathrm{D}}\right) = A_{\mathrm{P}} + k_{\mathrm{I}}\frac{1}{s} + k_{\mathrm{D}}s = \frac{A_{\mathrm{P}}}{T_{\mathrm{I}}}\frac{1 + sT_{\mathrm{I}} + s^2 T_{\mathrm{I}} T_{\mathrm{D}}}{s}. \quad (8.1)$$

Here the regulator parameters are A_{P}, the proportional transfer gain, T_{I} the integrating time constant, and T_{D} the differentiating time constant. The unit step response of the regulator is expressed as

$$v(t) = \mathcal{L}^{-1}\left[\frac{1}{s}C_{\mathrm{PID}}(s)\right] = A_{\mathrm{P}} + \frac{A_{\mathrm{P}}}{T_{\mathrm{I}}}t + A_{\mathrm{P}}T_{\mathrm{D}}\delta(t) \quad t \geq 0, \quad (8.2)$$

which is seen in Fig. 8.2.

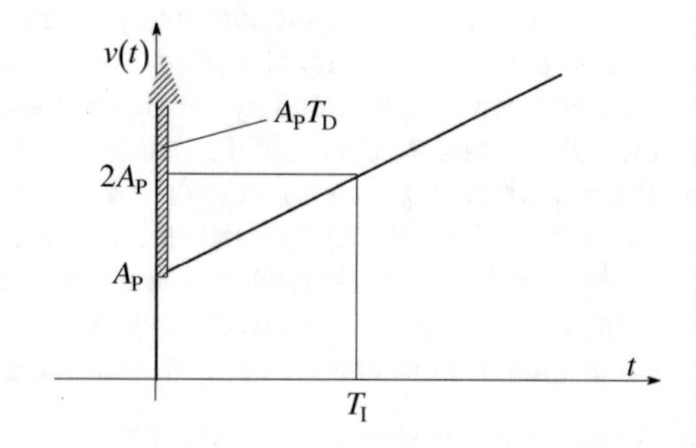

Fig. 8.2 Unit step response of the ideal *PID* regulator

The *PID* regulator consists of parallel connected proportional (*P*), integrating (*I*) and differentiating (*D*) effects. The analytical expression (time function) of the ideal regulator, that is the operation executed on the error signal, $e(t)$ is given by

$$u(t) = A_{\mathrm{P}} e(t) + k_{\mathrm{I}} \int_0^t e(\tau)\mathrm{d}\tau + k_{\mathrm{D}} \frac{\mathrm{d}e(t)}{\mathrm{d}t}. \tag{8.3}$$

This expression clearly shows the mentioned characteristics of the three regulator channels.

The regulator has a pole at the origin. It also has two zeros, which in the case of $T_{\mathrm{I}} \geq 4T_{\mathrm{D}}$, are located on the negative real axis. This condition requires a significant separation of the integrating and differentiating time constants, which requires a fourfold distance at least of the corresponding breakpoints in the amplitude-frequency diagram.

The location of the poles and zeros of the *PID* regulator in the complex plane is shown in Fig. 8.3. Its asymptotic Bode amplitude-frequency and phase-frequency diagram is seen in Fig. 8.4. The ideal *PID* regulator is non-realizable, it is used only for theoretical considerations and explanations. By ensuring the realizability of the *D* effect (combining it with a serially connected lag element with a small time constant) an approximate, but realizable *PID* regulator is obtained.

The transfer function of the approximate *PID* regulator can be given in the following form:

$$\widehat{C}_{\mathrm{PID}} = A_{\mathrm{P}} \left(1 + \frac{1}{sT_{\mathrm{I}}} + \frac{sT_{\mathrm{D}}}{1+sT}\right) = \frac{A_{\mathrm{P}}}{T_{\mathrm{I}}} \frac{1 + s(T_{\mathrm{I}} + T) + s^2 T_{\mathrm{I}}(T_{\mathrm{D}} + T)}{s(1+sT)}. \tag{8.4}$$

The unit step response is expressed as

$$\widehat{v}(t) = \mathcal{L}^{-1}\left[\frac{1}{s}\widehat{C}_{\mathrm{PID}}(s)\right] = A_{\mathrm{P}} + \frac{A_{\mathrm{P}}}{T_{\mathrm{I}}} t + \frac{A_{\mathrm{P}} T_{\mathrm{D}}}{T} e^{-t/T} \quad t \geq 0 \tag{8.5}$$

and is shown in Fig. 8.5. The location of the poles and zeros of the approximate *PID* regulator is shown in Fig. 8.6. The asymptotic Bode amplitude-frequency and phase-frequency diagram is given in Fig. 8.7.

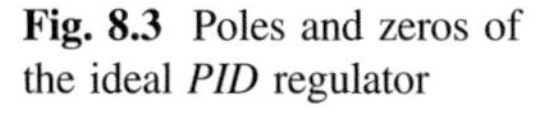

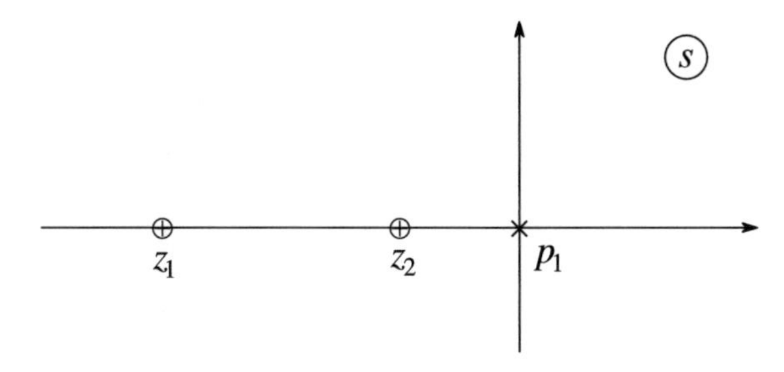

Fig. 8.3 Poles and zeros of the ideal *PID* regulator

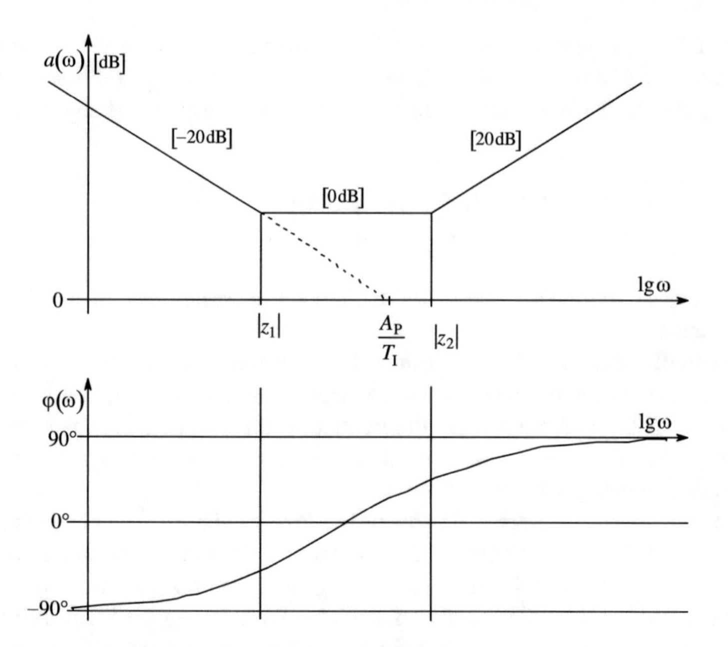

Fig. 8.4 Asymptotic Bode diagram of the ideal *PID* regulator

Fig. 8.5 Unit step response of the approximate *PID* regulator

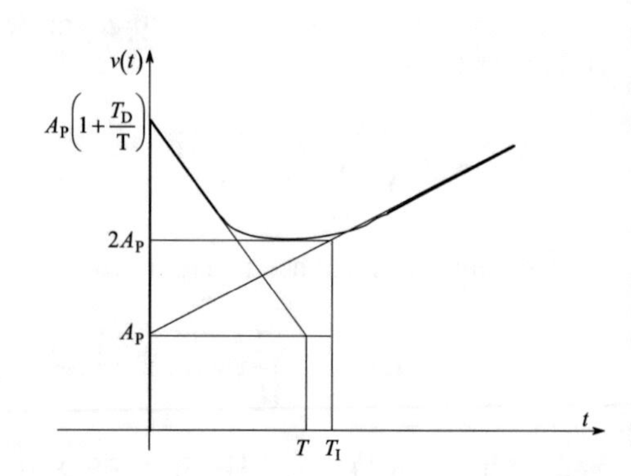

Fig. 8.6 Poles and zeros of the approximate *PID* regulator

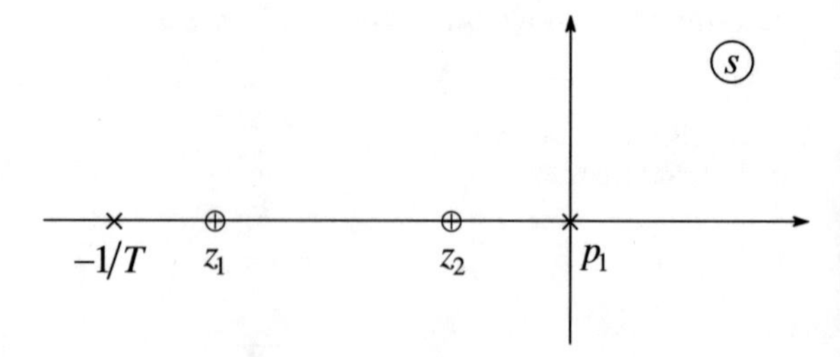

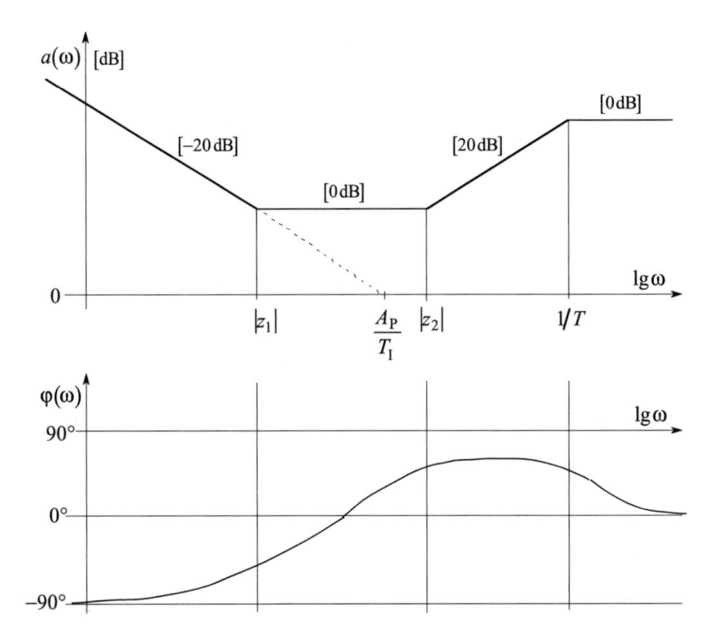

Fig. 8.7 Asymptotic Bode diagram of the approximate *PID* regulator

Now the regulator has two poles, one at the origin and the second at $-1/T$ in the complex plane, and its two zeros located on the negative real axis if $T_D \leq (T_I - T)^2/4T_I$. Again, this condition requires a significant separation of the integrating and the differentiating time constants, that is, a significant distance between the corresponding breakpoint frequencies.

The unit step response of the approximate *PID* regulator does not now start with the $\delta(t)$ (Dirac delta) function at $t = 0$, nevertheless the value of the initial jump expressed by $A_P(1 + T_D/T)$ may exceed the linearity range of the actuator and the equipment could be saturated. This situation is to be avoided, as on the one hand the normal operation mode of the actuator is ensured only in the linear range, and on the other hand the tuning of the regulator for stable performance, for the prescribed quality specifications, accuracy, etc., ensures the required behavior only in the case of linear models. The ratio T_D/T which determines the initial jump is called the overexcitation. Its value in real control systems should not exceed the limit of $T_D/T \leq 10$, and frequently should be even lower, with an allowable upper value of $4 \leq T_D/T \leq 6$. This realization limit determines first of all the fastest reachable transient (the cut-off frequency) for a control system of a given process.

Note that a modified version of *PID* regulators is often used, which significantly attenuates the effect of a possible abrupt change of the reference signal on the output signal in the closed-loop control system through a simple mechanism. In these regulators the manipulated variable—instead of the usual relationship (8.3)—is calculated by the expression $u(t) = A_P e_p(t) + k_I \int_0^t e(\tau)\mathrm{d}\tau + k_D \mathrm{d}e_d(t)/\mathrm{d}t$, where

$e_p(t) = br(t)$ and $e_d(t) = cr(t)$, setting parameters b and c whose values generally are between 0.2 and 0.8.

From the ideal *PID* regulator, the following different types of regulators can be obtained:

$$\begin{array}{ll} P & A_P \\ I & \dfrac{1}{sT_I} = \dfrac{K_I}{s} \\ PI & A_P\left(1 + \dfrac{1}{sT_I}\right) \\ PD & A_P(1 + sT_D) \end{array} \tag{8.6}$$

In the approximate *PID* regulator, the *PD* effect is considered by the transfer function

$$PD \qquad A_P\frac{1 + sT_D}{1 + sT}. \tag{8.7}$$

This element is also called a phase-lead or phase-lag element, as in the case of $T_D > T$ it realizes an approximate differentiating (*PD*), whereas if $T_D < T$, it provides an approximate *PI* effect.

In many practical cases the parameterization of the approximate *PID* regulator can be given in the following form:

$$\widehat{C}_{PID} = A_P\frac{(1 + sT_I)(1 + sT_D)}{sT_I(1 + sT)}. \tag{8.8}$$

The advantage of this form is that it locates the breakpoint frequencies exactly at $1/T_I$, $1/T_D$ and $1/T$, which belong to the given integrating, differentiating time constants and to the time constant of the lag element. The location of the poles and the zeros is seen in Fig. 8.8.

Now the poles of the regulator are at the origin ($p_1 = 0$) and at $p_2 = -1/T$ and the zeros are at $-1/T_I$ and $-1/T_D$.

The *PID* regulator can also be considered as a general second order regulator, which has sufficient degrees of freedom to provide an appropriate solution for many simple control applications.

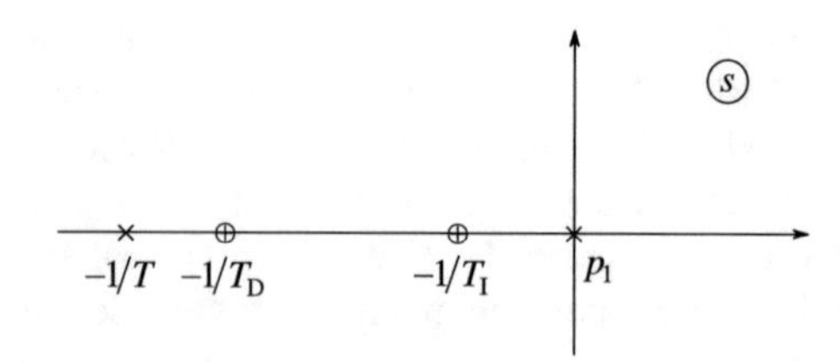

Fig. 8.8 Location of the poles and zeros of the approximate *PID* regulator according to (8.8)

8.1.1 Tuning of P Regulators

The P regulator $(C_P(s) = A_P)$ generally results in a 0-type control system (except if the process itself contains an integrating effect). The 0-type control system has a finite steady error with a value of $1/(1+K)$, and the gain of the closed-loop transfer function is $K/(1+K)$, where $K = A_P P(0)$ is the gain of the open-loop (and not of the regulator!). It is also called the loop-gain and $P(0)$ is the gain of the process. In this case, when realizing the regulator, a static compensation factor (a calibration factor) with a value of $(1+K)/K$ is applied to ensure accurate reference signal tracking. The proportional regulator means the multiplication of the transfer function of the process by a constant factor, which influences only the amplitude of the frequency function, and has no effect on the phase characteristic: $|C_P(j\omega)| = A_P$; $\varphi(\omega) = \arg\{C_P(j\omega)\} = 0$.

By changing the value of the constant A_P the loop gain and thus the cut-off frequency can be set. The Bode amplitude diagram can be shifted parallel, thus the cut-off frequency can be changed to ensure the appropriate phase margin for the control system.

The control system can be stabilized by a P regulator, the overshoot of the unit step response can be kept within the permissible limit by setting the appropriate phase margin. As the cut-off frequency can not be increased significantly, the control system will work slowly. If the process does not contain integrating elements, the control system is of 0-type, which can track only the unit step reference signal. The control system works with a constant steady state error.

In industrial regulators, instead of the A_P gain of the proportional regulator, the so-called proportional band (*PB*) is set, which is defined as $PB = [1/A_P]100\%$. The proportional band is that range of the input signal in which the output signal runs through its entire range.

In the P regulator only one parameter, the A_P gain, can be set. This means that the prescribed static error and the prescribed phase margin can not always be ensured at the same time. These two prescriptions can be fulfilled together only in fortunate cases. In the usual design procedure first the maximal loop gain $K_{max}(\varphi_{to})$ is determined belonging to the prescribed phase margin φ_{to}. The maximal reachable loop gain gives the minimal reachable static error

$$e_{min} = \min\{e(s)|_{s=0}\} = 1/[1 + K_{max}(\varphi_{to})]. \tag{8.9}$$

If the task is given in a reverse way, that is, if K_{min} has to be calculated from the maximal allowed error, then

$$K_{min} = (1/e_{max}) - 1; \quad e_{max} < 1. \tag{8.10}$$

If the condition $K_{min} < K_{max}$ is fulfilled, then the twofold criterion has a solution, otherwise it does not. P compensation shifts the Bode amplitude-frequency diagram of the process parallel to the horizontal axis with the value of the regulator gain $A_P = K/P(0)$. The value of the shift initially is determined by K_{max} or K_{min}.

8.1.2 Tuning of I Regulators

The aim of applying an I regulator is to ensure a 1-type control system which provides zero steady state error in the case of a step reference signal. The design of the *I* regulator given by the transfer function

$$C_{\mathrm{I}} = \frac{1}{sT_{\mathrm{I}}} = \frac{K_{\mathrm{I}}}{s} \tag{8.11}$$

is relatively simple, as it has only one parameter. Thus the maximal loop gain $K_{\max}(\varphi_{\mathrm{to}})$ which provides the prescribed phase margin φ_{to} can be determined easily with the usual design methods, then $K_{\mathrm{I}} = K_{\max}/P(0)$.

8.1.3 Tuning of PI Regulators

In the case of *P* regulators one could see that the allowable maximal loop gain $K_{\max}(\varphi_{\mathrm{to}})$, whose value depends on the performance criterion generally is not high enough to ensure a small enough steady state error. To eliminate this problem *PI* regulators can be used, which ensure at least 1-type number for the control system, thus the steady state error will be zero in the case of a step-like reference or disturbance signal. The transfer function of the regulator is

$$C_{\mathrm{PI}}(s) = A_{\mathrm{P}}\left(1 + \frac{1}{sT_{\mathrm{I}}}\right) = K_{\mathrm{I}}\frac{1 + sT_{\mathrm{I}}}{s}. \tag{8.12}$$

With the choice $T_{\mathrm{I}} = \max\{T_i\} = T_1$, the largest time constant of the process can be cancelled. Then the maximal loop gain $K_{\max}(\varphi_{\mathrm{to}})$ that provides the prescribed phase margin φ_{to} can be determined by the usual design methods, and then the integral gain of the regulator is calculated by $K_{\mathrm{I}} = A_{\mathrm{P}}/T_{\mathrm{I}} = K_{\max}/P(0)$. The amplitude-frequency diagram of the open-loop $L(j\omega)$ starts with a slope of −20 dB/decade at low frequencies, as the *PI* regulator reallocates the smallest breakpoint frequency $(1/T_1)$ belonging to the largest time constant of the process to the origin. Furthermore it shifts the amplitude-frequency diagram of the process parallel to the horizontal axis by K_{I}.

Based on the parallel form of the transfer function, the unit step response of the regulator can be drawn easily, whereas the product form provides the possibility of easily sketching the asymptotic Bode amplitude diagram. Figure 8.9 gives the unit step response and the Bode diagram of the *PI* regulator, the latter is plotted for $A_{\mathrm{P}} = 1$. If the gain is different, the Bode amplitude diagram is shifted parallel.

PI regulators ensure a high gain in the low frequency domain. The integrating effect increases the type number of the control system by 1. In case of a proportional process, the static error will be zero for a step-like reference signal. By appropriately placing the cut-off frequency, a stable control system can be obtained with the required phase margin. As the cut-off frequency can not be put into the high frequency domain, the control system will be slow.

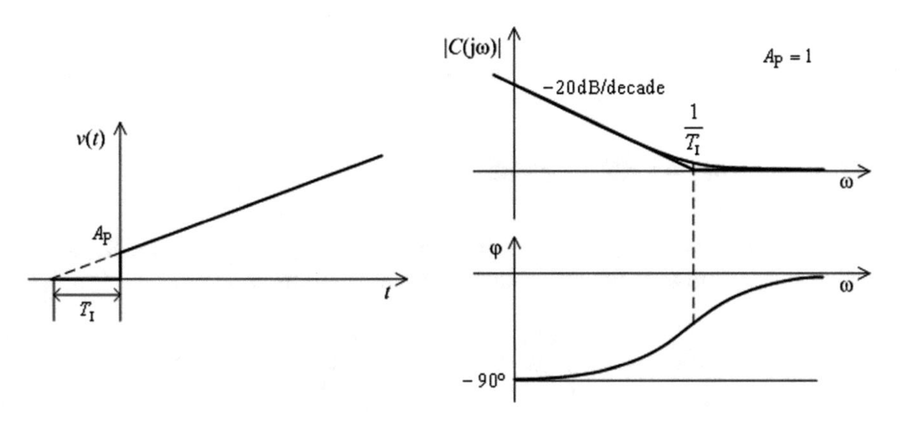

Fig. 8.9 The unit step response and the approximating Bode diagram of the *PI* regulator

An approximate realization of the *PI* regulator is the phase-lag element described by the transfer function

$$\tilde{C}_{\mathrm{PI}}(s) = A_{\mathrm{P}} \frac{1 + sT_{\mathrm{I}}}{1 + sT} = C_{\mathrm{PL}}(s), \tag{8.13}$$

where $T > T_{\mathrm{I}}$. Now by the choice $T_{\mathrm{I}} = \max\{T_i\} = T_1$ the smallest breakpoint frequency $(1/T_1)$ is shifted left to the point $1/T$. The amplitude-frequency diagram of the loop transfer function $L(j\omega)$ starts now with 0 slope, thus the control system remains of 0 type. By applying $\tilde{C}_{\mathrm{PI}}$ the straight line section of slope −20 dB/decade becomes longer, thus a higher allowable $K_{\max}(\varphi_{\mathrm{to}})$ gain can be ensured than with a simple *P* regulator. A single approximate *PI* regulator can reallocate a single breakpoint to a lower frequency.

Figure 8.10 gives the unit step response and the Bode diagram of the phase-lag element. It can be seen that this element approximates the characteristics of the *PI* element for small time instants in the time domain and in the high frequency range

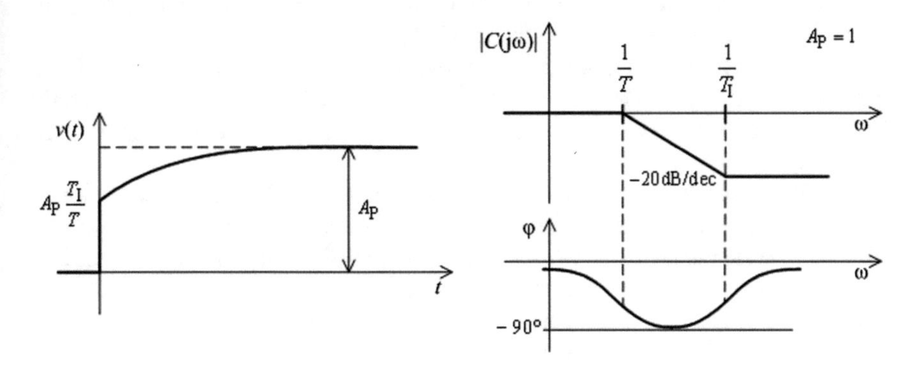

Fig. 8.10 Unit step response and the approximate Bode diagram of the phase-lag element

in the frequency domain. For sinusoidal inputs the output signal is phase-delayed related to the input signal. As this regulator does not contain an integrating effect, it does not ensure zero steady state error.

8.1.4 Tuning of PD Regulators

The ideal *PD* regulator given by the transfer function $C_{\mathrm{PD}}(s) = 1 + sT_{\mathrm{D}}$ can not be realized, therefore only the approximate realizable phase-lead form

$$\tilde{C}_{\mathrm{PD}}(s) = A_{\mathrm{P}}\frac{1+sT_{\mathrm{D}}}{1+sT} = A_{\mathrm{P}}\left(1+\frac{s\tau}{1+sT}\right); \quad T_{\mathrm{D}} = T+\tau \tag{8.14}$$

has practical applications, which formally is the same as $\tilde{C}_{\mathrm{PI}}$. The difference is that here $T_{\mathrm{D}} > T$, and the time constant of the differentiating channel is τ. With the choice $T_{\mathrm{D}} = T_2$, where T_2 is the second largest time constant of the process, the $\tilde{C}_{\mathrm{PD}}$ regulator lengthens the higher frequency part of the straight line of slope −20 dB/decade by shifting the breakpoint frequency $1/T_2$ to the right to point $1/T$. This improves the phase conditions, to reach a given phase margin the value of ω_{c} can be increased, which results in a faster settling process. There is a limit for the choice of the time constant T, as in the unit step response of the regulator at $t = 0$ there is a jump of value $A_{\mathrm{P}}T_{\mathrm{D}}/T$, which decreases asymptotically, its final value is A_{P}. So the value of the overexcitation is $\eta = T_{\mathrm{D}}/T$, which is the same as the pole placement ratio. Not every actuator can execute this jump. The big mechanical actuators can tolerate 2-3-fold, whereas the more advanced electronic devices can bear at most a 10-fold jump. Therefore during regulator design an average 3-5-fold jump is allowed, then in practice it is tested whether the regulator is saturated (that is reaches the limit of its signal range). In some cases the choice of $T_{\mathrm{D}} = T_3$ (where T_3 is the third largest time constant) is also possible, but the effect of this has smaller significance.

Figure 8.11 gives the unit step response and the Bode diagram of the *PD* regulator. The latter is drawn for $A_{\mathrm{P}} = 1$. If the gain is different, the Bode amplitude diagram is shifted parallel. This element is also called a phase-lead element, as its phase angle is positive, that is in case of a sinusoidal input signal the output signal is accelerated in phase relative to the input signal.

The accelerating effect of the control system can be understood the most easily in case of a *PD* regulator. The inertial behavior of the processes can be overtaken only by conveying extra accelerating energy. This is ensured by the overexcitation (see Sect. 2.4).

The *PD* regulator is used if the system has to be accelerated. This acceleration is reached if the straight line section of slope +20 dB/decade of the *PD* regulator is placed in the frequency range where the slope of the Bode diagram of the process is −40 dB/decade. Thus the straight line section of the loop transfer function with slope −20 dB/decade is lengthened, and the cut-off frequency can be placed at a higher frequency. The higher the value of the parameter η, the bigger the acceleration.

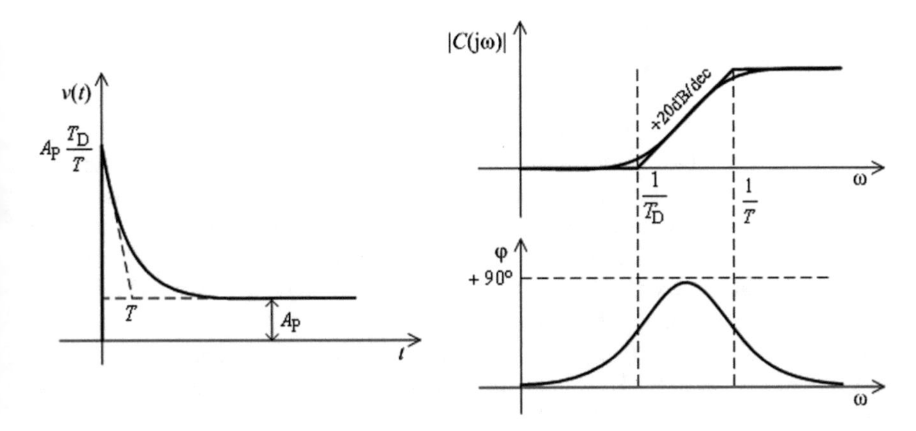

Fig. 8.11 Unit step response and the approximate Bode diagram of the *PD* regulator

Thus the *PD* regulator places one single breakpoint (generally the second one) of the Bode diagram at a higher frequency, but at the same time an overexcitation is produced which is equal to the ratio of the new and the old breakpoint. The amplitude-frequency diagram of the process is shifted parallel to the horizontal axis by A_P.

With the *PD* regulator the system can be stabilized. The system can be accelerated. Setting an appropriate phase margin the prescribed dynamic behavior can be reached. But as the regulator is of proportional type, with proportional processes the control system will have a static error for unit step reference signal.

The acceleration effect can be explained by the fact that at the beginning the error signal—exciting the *PD* regulator—produces a high signal at the regulator output, and the process temporarily starts to track this higher signal with its time constant. Thus the output signal starts with a big slope. In the control system then the error signal decreases, and the output signal settles at its steady value.

8.1.5 Tuning of PID Regulators

The simplest *PID* regulator parameterization is given by (8.8), that is

$$\tilde{C}_{PID} = A_P \frac{(1+sT_I)(1+sT_D)}{sT_I(1+sT)}. \tag{8.15}$$

This regulator is the combination of the previous two (*PI* and *PD*) regulators, resulting in their series connection. Thus the design procedure shown previously has to be repeated here. The integrating time is set by the choice $T_I = \max\{T_i\} = T_1$, whereas $T_D = T_2$ is chosen for the differentiating time. By this design the straight line section of slope −20 dB/decade on the Bode diagram of the loop transfer function is lengthened by the maximal possible extent provided by the structure of the regulator. In the control circuit after determining $K_{max}(\varphi_{to})$, the maximal integrating gain belonging to the prescribed phase margin φ_{to}, the integrating gain factor of the regulator is obtained according to $K_I = A_P/T_I = K_{max}/P(0)$. Regarding the choice of the time constant T of the approximate differentiating effect the considerations discussed in the design of *PD* regulators are also valid here, but the overexcitation in control circuits containing integrating effects is calculated in a different way. As the output of an integrator is changing until its input reaches the zero value, therefore the steady state of the control system is reached if the error signal has settled to zero. The initial jump of the PID regulator in the case of a unit step reference signal is $A_P T_D/T$. The steady state value of the process input is $1/P(0)$. The overexcitation now is calculated by $\eta = A_P P(0) T_D/T = K T_D/T$. Thus the overexcitation is obtained by the product of the loop gain and the pole placement ratio.

The form (8.15) of the regulator can be used straightforwardly for the analysis in the frequency domain. Figure 8.12 gives the unit step response and the Bode diagram of the *PID* regulator, the latter is drawn for $A_P = 1$. If the gain is different, the Bode amplitude diagram is shifted parallel.

PID regulators are used when the static accuracy of the control system has to be increased and the system also has to be accelerated. With the initial −20 dB/decade slope of the Bode diagram of the *PID* regulator, the low frequency range of the open-loop Bode diagram is modified, thus increasing the type number and the static accuracy. With the straight line section of slope +20 dB/decade of the Bode diagram of the regulator, the performance in the middle frequency range is modified. As by ensuring the appropriate phase margin, now the cut-off frequency can be placed to a higher value, thus a faster behavior of the control system can be reached.

The performance of the *PID* regulator can be approximated by the so-called phase-lag-lead element. Its transfer function is

$$C_{FKS}(s) = A_P \frac{1+sT_1}{1+sT_3}\frac{1+sT_2}{1+sT_4}, \quad \text{where} \quad T_3 > T_1 > T_2 > T_4.$$

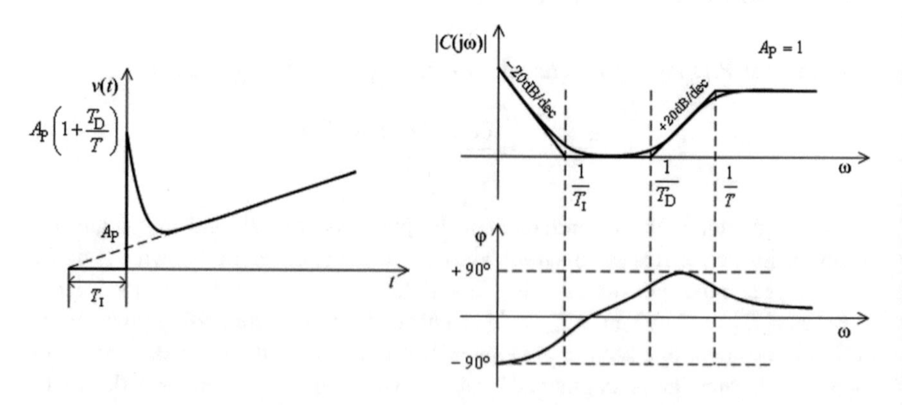

Fig. 8.12 Unit step response and the approximate Bode diagram of the *PID* regulator

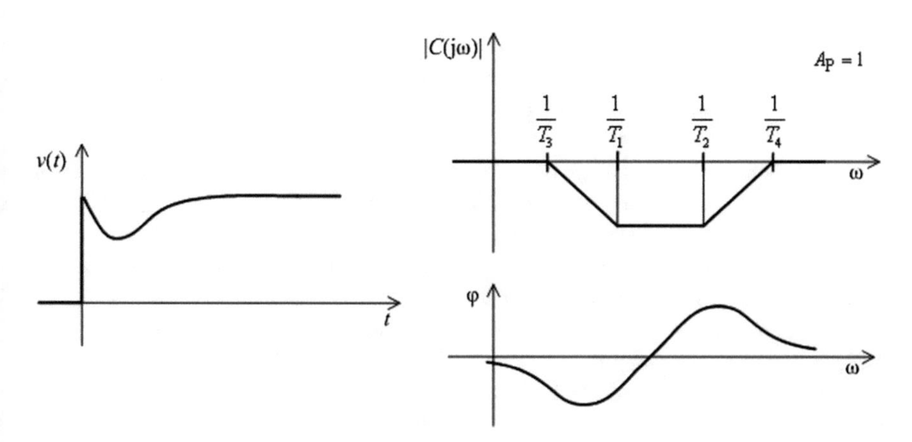

Fig. 8.13 Unit step response and the approximating Bode diagram of the phase-lag-lead element

The unit step response and the approximate Bode diagram of the phase-lag-lead element is given in Fig. 8.13. As this element does not contain an integrating effect, it will not ensure a zero steady state error in the control system.

The regulator is designed for the process (or for its model) to fulfill the quality specifications. First a decision has to be made about the regulator structure considering the given process and the prescriptions. Then the parameters of the regulator are chosen. For example in a *PID* regulator there are four free parameters, A_P, T_I, T_D and T. Regulator design means the choice of the parameters. The design can be executed in the time or in the frequency domain. Several procedures and methods have been elaborated to execute the design. The diversity of the design methods comes also from the fact that in the individual applications the design specifications may differ significantly. In many cases the prescriptions create contradictory requirements. In the general case, this corresponds to an optimization problem, when a proper compromise is formed to satisfy the contradictory requirements. In practice often some iteration procedure (intelligent guessing) is applied instead of executing the optimization (Table 8.1).

The presented *P*, *PI*, *PD* and *PID* regulators are also called compensators.

Let us summarize the practical rules for the design of the *PID*-like regulators using pole cancellation.

Table 8.1 Summarizing the regulator design

Regulator	T_I	T_D	A	K_I
P			$A_P = K/P(0)$	
I				$K_I = K_{max}/P(0)$
PI	T_1			$K_{PI} = K_{max}/P(0)$
PD		T_2		$K_{PD} = K_{max}/P(0)$
PID	T_1	T_2		$K_{PID} = K_{max}/P(0)$

The *P* regulator can be used if there are no high requirements for the static accuracy of the control system, and the control system can be slow. If the process contains an integrating effect, then the static accuracy will be appropriate also with the proportional regulator.

PI regulators are used if accurate tracking is required in steady state for a unit step reference signal. The integrating effect will ensure an accurate settling. With *PI* regulators, the control system will be slow.

PD regulators accelerate the control system. This effect is reached by the overexcitation provided by the regulator.

PID regulators are used if both the static accuracy and the speed of the control system have to be increased.

In the case of a proportional process, the parameters of the regulator applying pole cancellation technique are chosen as follows: the parameter T_{I}, the integrating time constant is chosen equal to the largest time constant of the process (this is the pole belonging to the smallest breakpoint frequency), and the parameter T_{D} is taken equal to the second largest time constant. Thus the zeros of the regulator cancel the poles of the process. The parameter T which appears in the denominator of the realizable differentiating element is given in by $T = T_{\mathrm{D}}/\eta$, where η is the pole placement ratio, which specifies the frequency shift of the compensated pole realized by the *PD* element. If it is chosen to be a higher value, the control system will be faster at the price of a higher maximum value of the control signal. As A_{P} does not influence the phase-frequency course of the open-loop, it is used to set the prescribed phase margin.

If the process is not proportional, the type of the regulator can be decided on the basis of the approximate Bode diagram to fulfill the quality specifications. Pole cancellation can be applied expediently also in this case.

8.1.6 Influence of the Dead-Time

The effect of the dead-time can be considered relatively simply in the case of series compensation, as

$$H_{\mathrm{H}}(s) = e^{-sT_{\mathrm{d}}} \Rightarrow H_{\mathrm{H}}(j\omega) = e^{-j\omega T_{\mathrm{d}}} = e^{-j\varphi_{\mathrm{d}}} \tag{8.16}$$

This means that $L(j\omega)$, the frequency characteristic of the loop transfer function, is modified by an element of unit amplitude and of phase angle $\varphi_{\mathrm{d}} = -\omega T_{\mathrm{d}}$, thus only the phase characteristic is changed. This can be taken into account by prescribing the required phase margin to be $\varphi_{\mathrm{to}}^{\mathrm{d}} = \varphi_{\mathrm{to}} + \omega T_{\mathrm{d}}$ instead of the original φ_{to}.

As the transfer function of the dead-time is a non-rational function, computer programs which are not prepared for handling such functions (e.g., MATLAB®) can not easily take its effect into account. In this case the transfer function of the dead-time can be approximated by a rational fraction.

Rational fractional approximations of the dead-time were discussed in Sect. 2.5.

8.1.7 Realization of PID Regulators

Analog regulators can be realized based on different physical conceptions (electronic, pneumatic, hydraulic, etc.). The electronic realization is built by a feedback operational amplifier (Fig. 8.14). The operational amplifier has a very high voltage amplification (10^5–10^7), its input resistance is also high. The relationship between its output and input is: $C(s) = -Z_v(s)/Z_b(s)$, where $Z_v(s)$ is the feedback impedance and $Z_b(s)$ is the input impedance. Figure 8.15 shows a realization of a pure integrator, and of a *PI* circuit.

Different versions of the circuits can be given. For example an aspect can be the realization of a circuit where changing the value of one element (generally a resistance) sets only one regulator parameter, and has no effect on the others.

Approximate phase-lag, phase-lead and phase-lag-lead regulators can be built from passive elements, the circuit does not contain operational amplifier. Figure 8.16 shows the circuits of these regulators with resistors and capacitors.

A compact regulator is produced by firms that manufacture automatic elements. In this regulator, the structure (*P*, *PI*, *PD* or *PID*) can be set by a switch, and the parameters generally can be tuned by setting potentiometers.

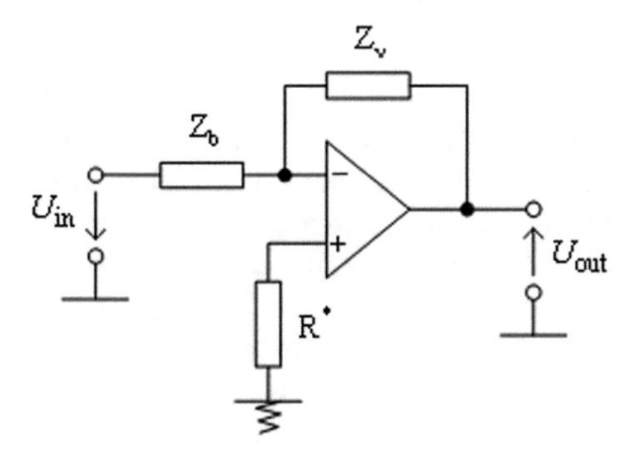

Fig. 8.14 Realization of a regulator with an operational amplifier

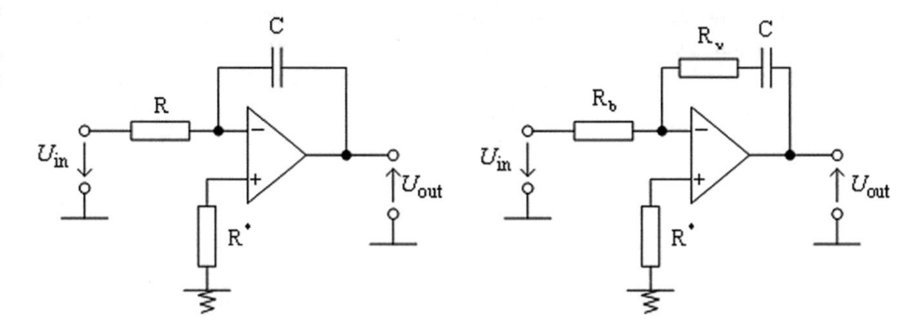

Fig. 8.15 Realization of an *I* and a *PI* regulator

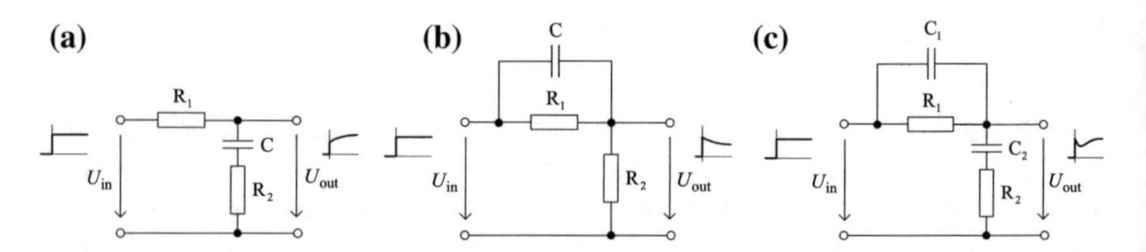

Fig. 8.16 Realization of phase-lag, phase-lead and phase-lag-lead regulator with resistors and capacitors

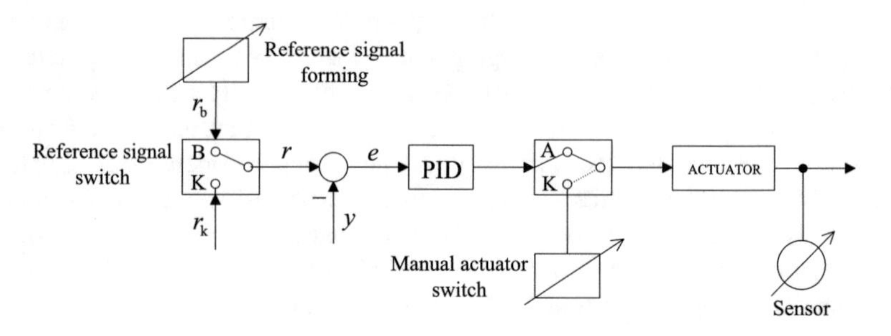

Fig. 8.17 Block-scheme of a *PID* regulator with manual-automatic switch-over

The regulators realize a bumpless transfer when switching from manual mode to automatic mode, and in the case of saturation, also handle integrator windup. Figure 8.17 gives a block diagram of a *PID* regulator, also showing the manual-automatic switch-over. The reference signal can be switched over between two signal generators. The system can be switched over between manual and automatic modes of operation. In the case of slow processes, when switching over from automatic operation mode to manual mode, the operator sets the value of the manipulated variable shown by the measuring instrument with the potentiometer to create the manual manipulated variable, and then executes the switch-over. Switching-over from manual mode to automatic mode is more critical, as during manual operation the signal on the integrating channel of the regulator is likely to "run away". Figure 8.18 shows a possible solution for manual-automatic switch-over in the case of a *PI* regulator, avoiding windup. During the manual mode, the capacitor C is charged to the voltage of the manual manipulated signal, and after the switch-over the integration starts from this initial value. In manual mode the resistor R ensures the feedback of the operational amplifier, thus it will not be saturated. Handling of saturation will be discussed in Sect. 8.4.

Nowadays, instead of analog techniques, it is more and more the *programmable logic controllers* (*PLC*-s) or process control computers that realize the regulator. The process is connected to the computer via an A/D converter. The *PID* control algorithm is realized by a computer program. The regulator output is connected to the process input through a D/A converter. The program has to handle saturation

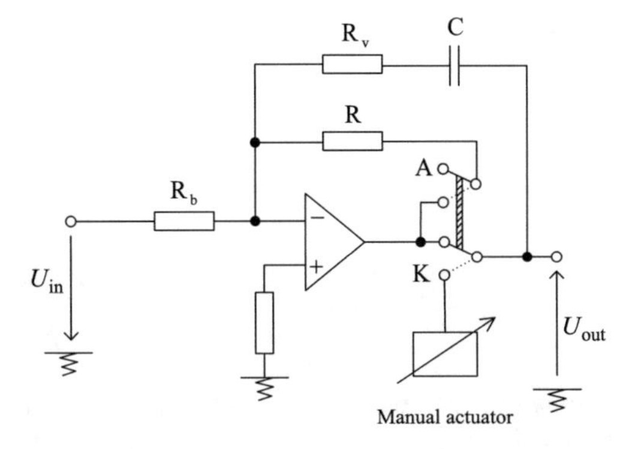

Fig. 8.18 *PI* regulator with manual-automatic switch-over

effects and also ensure bumpless transfer between the manual and the automatic operation modes. Digital *PID* regulators will be discussed in detail in Chap. 13.

8.2 Design of Residual Systems

For some typical cases, regulator design can also be executed analytically. In the design of the usual regulators one can see that in the generally used simplest method, the breakpoints corresponding to the integrating and differentiating time constants of the *PID* regulator are fitted to the breakpoints belonging to the two largest time constants of the process. Then the only free parameter of the regulator is the gain, which has to be tuned appropriately. The gain is chosen to ensure a prescribed phase margin, gain margin, or NYQUIST stability margin. If after cancellation of the two dominant time constants the order of the so-called residual or reduced system is low, then the design can be executed easily, several times providing an explicit analytical result. If the residual system is a higher order, more complicated system, then only numerical methods (MATLAB®) can be applied. In the sequel, the design of the loop gain will be executed for some typical residual systems.

8.2.1 Simple Residual System with Dead-Time and Integrator

As seen previously, a pure dead-time element with negative feedback is at the borderline case of stability with unit loop gain. The control system can be stabilized and also its static accuracy can be improved applying an integrator instead of a proportional regulator. The considered residual system is shown in Fig. 8.19; its loop transfer function is

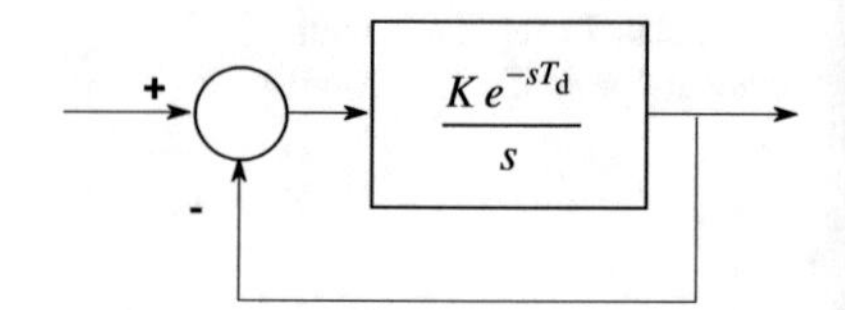

Fig. 8.19 Residual system containing an integrator and dead-time

$$L(s) = \frac{Ke^{-sT_d}}{s} = \frac{e^{-sT_d}}{sT_I}; \quad L(j\omega) = \frac{Ke^{-j\omega T_d}}{j\omega} = a(\omega)e^{j\varphi(\omega)}. \tag{8.17}$$

The amplitude-frequency curve of the open-loop is a straight line of slope −20 dB/decade, and the loop gain $K = 1/T_I$ gives the cut-off frequency of the open-loop. If φ_t is the prescribed phase margin, then both the phase condition

$$-\pi/2 - \omega T_d = -\pi + \varphi_t \tag{8.18}$$

and the absolute value condition

$$\frac{K}{\omega} = 1 \tag{8.19}$$

have to be fulfilled. Solving for the two equations the loop gain yields

$$K = \frac{\pi/2 - \varphi_t}{T_d} = \frac{\pi - 2\varphi_t}{2T_d} = K_{\varphi_t}. \tag{8.20}$$

If the gain of the regulator is A_c and the gain of the process is A_p, then the gain of the regulator is

$$A_c = \frac{K_{\varphi_t}}{A_p}. \tag{8.21}$$

The resulting formula (8.20) can be used also for the design of a pure integrating regulator if the process contains only a pure dead-time. Thus the integrating time constant of an integrating regulator used for compensation of a pure dead-time process is designed to be

$$T_I = \frac{1}{K_I} = \frac{1}{K} = \frac{1}{K_{\varphi_t}}. \tag{8.22}$$

Relating it to the dead-time the following relationship is obtained:

$$\frac{T_I}{T_d} = \frac{1}{\frac{\pi}{2} - \varphi_t} = \frac{2}{\pi - 2\varphi_t}. \tag{8.23}$$

Some typical values: if $\varphi_t = 30° = \pi/6$, then $T_I/T_d = 3/\pi \approx 1$; if $\varphi_t = 60° = \pi/3$, then $T_I/T_d = 6/\pi \approx 2$. At the borderline of stability $\varphi_t = 0$, and $T_I/T_d = 2/\pi \approx 0.637$.

Thus in the case of a minimum phase system which also contains dead-time it is not sufficient to place the cut-off frequency on the straight line section of the Bode amplitude diagram of slope −20 dB/decade: to ensure a phase margin of about $\varphi_t = 60°$ it has to be located around one half of the reciprocal of the dead-time.

Let κ be the prescribed gain margin (see Sect. 5.6). Now the phase condition is given by

$$-\frac{\pi}{2} - \omega T_d = -\pi, \tag{8.24}$$

whence the intersection frequency of the open-loop Nyquist diagram with the negative real axis is expressed as

$$\omega_a = \frac{\pi}{2T_d} \tag{8.25}$$

and the absolute value of the loop frequency function at this point is

$$a(\omega_a) = \left.\frac{K}{\omega}\right|_{\omega=\omega_a} = \frac{K}{\omega_a} = 1 - \kappa. \tag{8.26}$$

The loop gain is obtained from the solution of the last two equations:

$$K = \frac{\pi(1-\kappa)}{2T_d}. \tag{8.27}$$

A typical value: if $a_t = 0.5$, then $T_I/T_d = 4/\pi \approx 1.2$. The borderline of stability $(a_t = 0)$ again is obtained at $T_I/T_d = 2/\pi \approx 0.637$.

The tuning relationship of the integrating time constant of an I integrating regulator used for compensating a pure dead-time process is given by

$$\frac{T_I}{T_d} = \frac{2}{\pi(1-\kappa)}. \tag{8.28}$$

The Nyquist stability margin $\rho_m = \rho_{min}$ is defined as the smallest distance of the loop frequency function $L(j\omega)$ from the point $-1 + 0j$. Generally it is not easy to give this distance as an explicit algebraic expression, as it can be derived from the solution of an extremum seeking problem. Therefore generally its graphical representation is employed. In Fig. 8.20 ρ_{min} is plotted versus the ratio T_I/T_d.

As seen in Chap. 5, a control system built of a pure dead-time element with unity negative feedback is stable only if its loop gain is less than 1. But then the static error is very high. Therefore the above considerations are frequently used not only when the pole cancellation regulator design technique is applied, but also in the

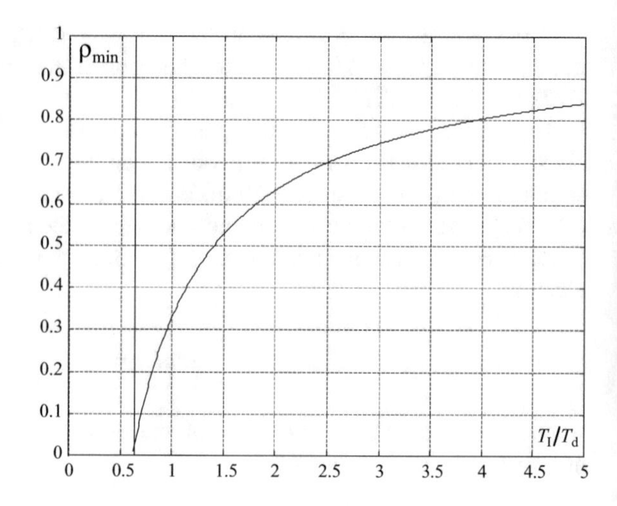

Fig. 8.20 NYQUIST stability margin ρ_{min} versus T_I/T_d

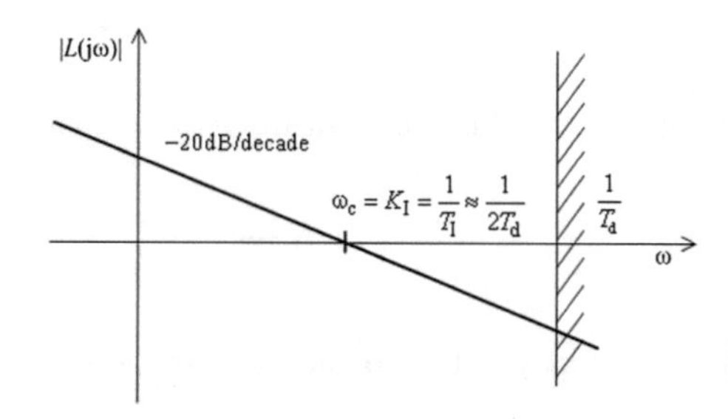

Fig. 8.21 BODE amplitude-frequency diagram of a compensated dead-time system

case of a pure *I* regulator. The BODE amplitude-frequency diagram of the open-loop for the case of $\varphi_t = 60° = \pi/3$ is shown in Fig. 8.21. From (8.23) it can be seen, that when compensating a dead-time system, the cut-off frequency has to be placed at about on the half of the reciprocal of the dead-time on the long straight line section of slope −20 dB/decade. At the borderline of stability $\varphi_t = 0$ and then $K_I = \pi/2T_d \approx 1.57/T_d$.

An aperiodic process can be quite well approximated by a first-order (or second-order) lag element (see Sect. 8.3). To meet higher quality requirements, instead of an *I* regulator, a *PI* regulator can be used. With the zero of the *PI* element, the pole of the process can be cancelled, and instead an integrating effect is introduced. With this *PI* compensation the open-loop is given as an integrating element with dead-time:

$$L(s) = A_P \frac{1 + sT_1}{s} \frac{K_P e^{-sT_d}}{1 + sT_1} = \frac{A_P K_P e^{-sT_d}}{s} = \frac{K e^{-sT_d}}{s}. \tag{8.29}$$

The parameter *K*, which also gives the cut-off frequency of the open-loop, can be designed to ensure 60° for the phase margin according to Sect. 6.2:

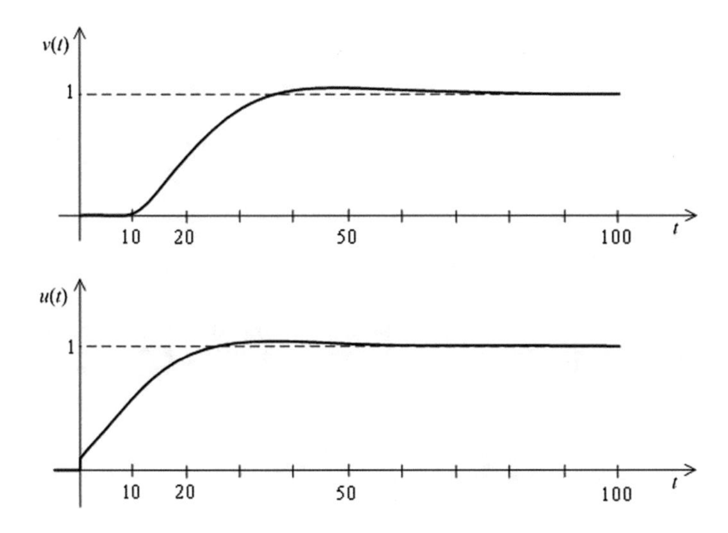

Fig. 8.22 Output and control signals of a dead-time system compensated by a *PI* regulator in the case of a unit step reference signal

$K = \omega_c \approx 1/2T_d$. The output signal and the control signal of the control system for a unit step reference signal is shown in Fig. 8.22 in the case of $T_d = 10$ and $T_I = 2$. The dead-time can not be eliminated from the system; the output signal starts to change only after the dead-time.

(Note that in the case of a dead-time process with first order lag it makes no sense to add a *PD* compensation as well. In case of a dead-time process with two time lags, adding a *PD* effect will accelerate the system only if the time constants of the lag elements are the dominant ones. If the dead-time is intermediate or the largest time constant, then because of the significant phase shift of the dead-time, the phase margin can not be increased significantly by the effect of the *PD*, thus it makes no sense to apply it. As the increase of the cut-off frequency is limited by the dead-time, the behavior of a control system containing dead-time will be slow.)

8.2.2 Simple Residual System with Integrator and Time Lag

The residual system is shown in Fig. 8.23. Its loop transfer function is

Fig. 8.23 Residual system consisting of an integrator and a first order lag

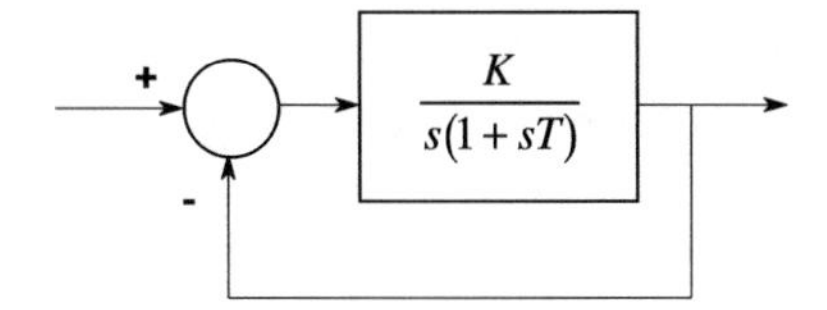

$$L(s) = \frac{K}{s(1+sT)} = \frac{1}{sT_{\mathrm{I}}(1+sT)}; \quad L(j\omega) = \frac{K}{j\omega(1+j\omega T)} = a(\omega)e^{j\varphi(\omega)}. \quad (8.30)$$

Using (8.30) the overall transfer function of the closed-loop system, that is the supplementary sensitivity function, is

$$T(s) = \frac{K}{K+s+Ts^2} = \frac{1}{1+\frac{1}{K}s+\frac{T}{K}s^2} = \frac{1}{1+2\xi\tau s+\tau^2 s^2}, \quad (8.31)$$

which is a second order oscillating element whose damping factor can be set accurately by the loop gain. Comparing the coefficients the following relationships are obtained:

$$K = \frac{1}{4\xi^2 T}; \quad \frac{T_{\mathrm{I}}}{T} = 4\xi^2. \quad (8.32)$$

A damping factor, $\xi = \sqrt{2}/2 \cong 0.707$, which provides a "nice" transient response is obtained by a loop gain of $K = 0.5/T$. In some applications, an aperiodic transient is required. The limiting case of aperiodic response is $\xi \geq 1$, which corresponds to $K \leq 0.25/T$.

Let φ_t be the prescribed phase margin. Then based on the loop transfer function the following relationship can be written for the phase condition:

$$-\frac{\pi}{2} - \mathrm{arctg}(\omega T) = -\pi + \varphi_t; \quad (8.33)$$

and for the absolute value condition,

$$\frac{K}{\omega\sqrt{1+\omega^2 T^2}} = 1. \quad (8.34)$$

From the solution of these two equations, the loop gain is obtained as

$$K = \frac{1}{T}\mathrm{tg}(90-\varphi_t)\sqrt{1+\mathrm{tg}^2(90-\varphi_t)} = \frac{1}{T}\frac{\sin(90-\varphi_t)}{\cos^2(90-\varphi_t)}, \quad (8.35)$$

where the trigonometric identity

$$\mathrm{tg}(x)\sqrt{1+\mathrm{tg}^2(x)} = \frac{\sin(x)}{1-\sin^2(x)} = \frac{\sin(x)}{\cos^2(x)} \quad (8.36)$$

was taken into account.

Some typical values: if $\varphi_t = 45° = \pi/4$, then $KT = \sqrt{2} \cong 1.414$; if $\varphi_t = 60° = \pi/3$, then $KT = 2/3 \cong 0.667$. Also, the φ_t corresponding to a given ξ can be

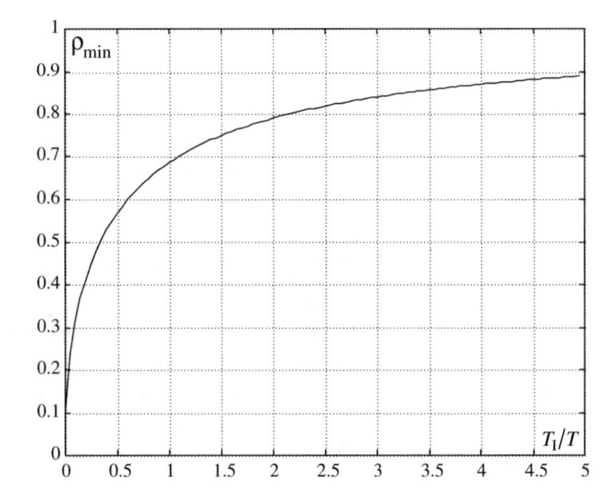

Fig. 8.24 ρ_{min}, the NYQUIST stability margin versus T_I/T

calculated. If $\xi = \sqrt{2}/2 \cong 0.707$ is required, then the corresponding phase margin is obtained as $\varphi_t = 65.53°$.

As the NYQUIST diagram of this residual system does not step across the third quadrant of the complex plane, therefore now the gain margin can not be designed ($\kappa_t \equiv 1.0$). This system is structurally stable.

Here ρ_{min}, the NYQUIST stability margin, can be obtained only graphically. Its graph versus T_I/T is seen in Fig. 8.24.

A residual system containing an integrator and a first order lag is obtained for instance if a proportional second order lag element is compensated by a *PI* regulator using the pole cancellation technique. The loop transfer function is

$$L(s) = A_P \frac{1+sT_1}{s} \frac{K_P}{(1+sT_1)(1+sT_2)} = \frac{A_P K_P}{s(1+sT_2)} = \frac{K}{s(1+sT)} = \frac{1}{sT_I(1+sT)}, \tag{8.37}$$

which is of form (8.30), and the design formulae above can be applied. In the case of a prescribed damping factor ξ, the gain of the regulator can be calculated. If $\xi = 1$, then $K = 1/4T_2$, and the system has two coinciding real poles. The unit step response of the closed-loop just will not have any overshoot. If $\xi = \sqrt{2}/2 \approx 0.7$, then $K = 1/2T_2$, and the phase margin of the system will be $\varphi_t = 65.53°$. The overshoot of the unit step response of the closed-loop system will be about 5%. As $K = A_P K_P$ is the gain of the whole circuit, the gain of the regulator is obtained by $A_P = K/K_P = 1/2K_P T_2$.

8.3 Empirical Regulator Tuning Methods

Besides the model-based regulator design methods, which provide an outline of the foreseeable properties of the closed-loop control system, several experimental *PID* regulator tuning methods have been suggested in industrial process control, mainly for stable processes. These methods are often used during the installation of the regulator. The recommendations, the "recipes" for regulator parameter tuning, are based on some preliminary measurements executed on the process, or on simulations and practical observations.

Note that these methods are appropriate for fast regulator tuning when putting the control system in operation, but model based design methods provide a better basis for and overview of the behavior of the control system, or for modifying the tuning in case of changes. Usually, experimental methods are considered as the initial settings before introducing theoretically elaborated more extensive methods.

8.3.1 Methods of Ziegler and Nichols

Frequency response method

It is supposed that the technology of the process allows operating the closed-loop control system for a short time on the borderline of stability applying only a proportional regulator. During this experiment the integrating and the differentiating channels of the regulator are switched off ($T_I = \infty$ and $T_D = 0$), then by cautiously increasing A_P the borderline of stability is reached, when sinusoidal oscillations appear. After each change of A_P we have to wait for the new steady state to settle down, which can take a long time. Let $A_{P,cr}$ denote the critical gain and T_{cr} the time period of the constant sinusoidal oscillations. Ziegler and Nichols suggested the following Table 8.2 regulator tuning rules of thumb.

These tuning rules correspond to a damping factor of about $\xi = 0.25$ (which corresponds to a quite high overshoot of about 40%, thus in practice they can be used only for compensating slowly changing disturbances).

Tuning method based on the unit step response

The unit step response of several industrial processes shows an aperiodic characteristic with dead-time (Fig. 8.25). The straight line fitted to the inflection point of

Table 8.2 Regulator tuning according to Ziegler-Nichols (I)

Regulator	T_I	T_D	A_P
P			$0.5A_{P,cr}$
PI	$0.85T_{cr}$		$0.45A_{P,cr}$
PID	$0.5T_{cr}$	$0.125T_{cr}$	$0.6A_{P,cr}$

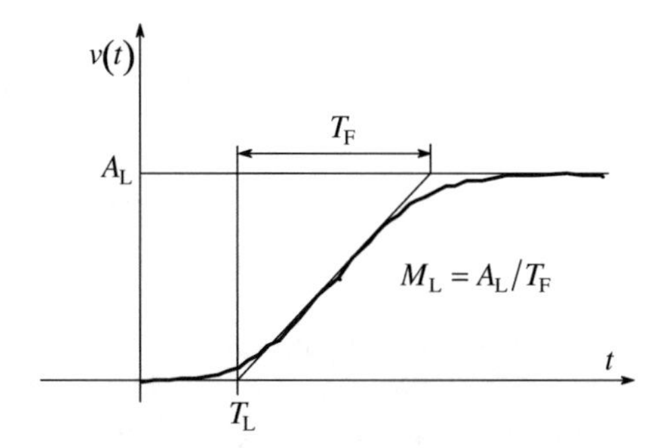

Fig. 8.25 Shape of the measured unit step response of the process

Table 8.3 Regulator tuning according to ZIEGLER-NICHOLS (II)

Regulator	T_I	T_D	A_P
P			$1/T_L M_L$
PI	$3T_L$		$0.9/T_L M_L$
PID	$2T_L$	$0.5T_L$	$1.2/T_L M_L$

the step response determines two quantities indicated in the figure, the so called latent dead-time (T_L) and the latent slope ($M_L = A_L/T_F$). Based on these quantities ZIEGLER and NICHOLS suggested tuning rules summarized in the Table 8.3.

From the above tables it can be seen, that the tuning of the three parameters $\{A_P, T_I, T_D\}$ in both cases is based on two observed values, then the *D*-channel is set as $T_D = T_I/4$. Of course this is a source of further design freedom.

8.3.2 *Method of OPPELT*

Several grapho-analytical methods can be used to fit an approximate first order lag element with dead-time given by (8.38) to the measured unit step response of the process.

$$\hat{P}(s) = \frac{A_L}{1 + sT_F} e^{-sT_L}; \quad A_L = \frac{y_\infty - y_o}{u_\infty - u_o}; \quad T_L = t_1 - t_o \quad \text{and} \quad T_F = t_2 - t_1. \tag{8.38}$$

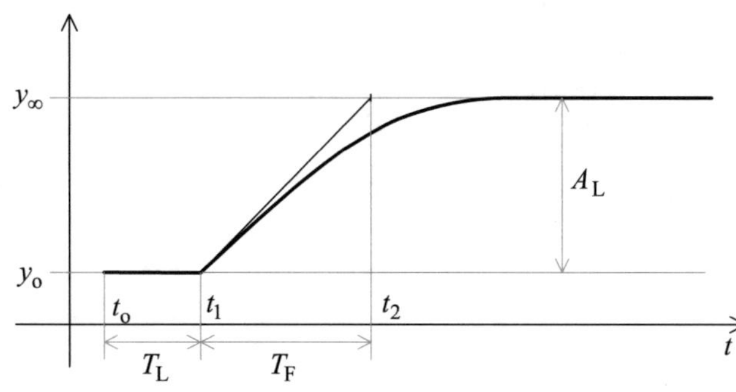

Fig. 8.26 Approximation of an aperiodic process by a first order lag with dead-time

Table 8.4 Regulator tuning according to OPPELT

Regulator	$A_P M_L T_L$	T_I/T_L	T_D/T_L
P	1		
PD	1.2		0.25
PI	0.8	3	
PID	1.2	2	0.42

Let us set manually the nominal operating point, where for an input signal u_o the value of the output signal is y_o. At time t_o let us apply a step input, when the input signal u jumps from u_o to u_∞. The output signal is seen in Fig. 8.26. (The definition of the time constant T_F in the two figures is different, as OPPELT does not suppose an inflection point.)

The tuning parameters according to OPPELT were determined for a damping factor of $\xi = 0.25$ considering the parameters of the approximate model given by (8.38). Therefore similarly to the ZIEGLER-NICHOLS method, in this case also a quite high overshoot can be expected. The proposed tuning values are summarized in the Table 8.4.

8.3.3 Method of CHIEN-HRONES-RESWICK

For the tuning of the regulator parameters, CHIEN, HRONES and RESWICK suggested the values summarized in the Table 8.5.

8.3.4 Method of STREJC

STREJC approximated the process by the model given by the transfer function

Table 8.5 Regulator tuning according to CHIEN-HRONES-RESWICK

Regulator	Fastest aperiodic transient	Fastest oscillating transient with 20% overshoot
P	$A_P = 0.3T_F/T_L$	$A_P = 0.7T_F/T_L$
PI	$A_P = 0.35T_F/T_L$ $T_I = 1.2T_F$	$A_P = 0.6T_F/T_L$ $T_I = 1.0T_F$
PID	$A_P = 0.6T_F/T_L$ $T_I = 1.0T_F$ $T_D = 0.5T_L$	$A_P = 0.95T_F/T_L$ $T_I = 1.35T_F$ $T_D = 0.47T_L$

Table 8.6 Regulator tuning according to STREJC

Regulator	A_P	T_I	T_D
P	$\frac{1}{A_L(n-1)}$		
PI	$\frac{n+2}{4A_L(n-1)}$	$\frac{T(n+2)}{3}$	
PID	$\frac{7n+16}{16A_L(n-2)}$	$\frac{T(7n+16)}{15}$	$\frac{T(n+1)(n+3)}{7n+16}$

$$\hat{P}(s) = \frac{A_{\mathrm{L}}}{(1+sT)^n}. \tag{8.39}$$

On the basis of the parameters of the approximate model, he gave tuning proposals according to the Table 8.6 to set the parameters of the *PID* regulator family.

8.3.5 Relay Method of Åström

In the case of real industrial processes, it is rather difficult to set the critical loop gain corresponding to the borderline of stability of the closed-loop control system. Generally, approaching this range is not permitted because of safety requirements. If we want to use regulator tuning rules based on the critical loop gain, then it is expedient to determine its value with a different method. Åström suggested a method which can well be applied in practice. In a closed-loop system, the *PID* regulator has to be replaced by an amplifier realizing a relay characteristic with hysteresis (see Fig. 8.27).

For the stability analysis of special closed-loop nonlinear systems, the so called describing function method can be applied. The describing function $N(j\omega, a)$ is obtained by harmonic linearization, when the static nonlinear element is excited by a sinusoidal signal of amplitude a, and then the complex division of the basic harmonic of the output signal and the input sinusoidal signal is executed. Generally $N(j\omega, a)$ is a complex function with parameter a. In analyzing the stability, the role of the point $-1+j0$ is replaced by the function $-1/N(j\omega, a)$. The system is at the borderline of stability at the point $(\omega_{\mathrm{cr}}, a_{\mathrm{cr}})$ where the Nyquist diagram of the loop frequency function $L(j\omega)$ intersects the inverse negative describing function $-1/N(j\omega, a)$, that is,

$$L(\omega_{\mathrm{cr}})N(\omega_{\mathrm{cr}}, a_{\mathrm{cr}}) = -1, \quad \text{i.e.} \quad -1/L(\omega_{\mathrm{cr}}) = N(\omega_{\mathrm{cr}}, a_{\mathrm{cr}}). \tag{8.40}$$

Here, a_{cr} is the approximate amplitude of the periodic oscillation in the borderline case of stability. (It is not an entirely accurate value, as the harmonic linearization considers only the first harmonic.) From the time period T_{cr} of the

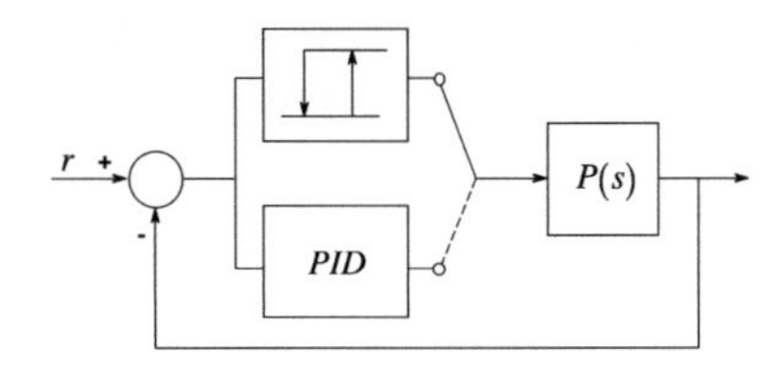

Fig. 8.27 Tuning of a *PID* regulator with the relay method

periodic signal, a good approximation of the angular frequency belonging to the critical point can be calculated by $\omega_{cr} = 2\pi/T_{cr}$.

Let the deadband of the hysteresis be zero, so the regulator is a two-position relay. In this case the input of the process is a rectangular signal, and the process output in steady state is a periodic signal. In the linear case for the critical gain the characteristic equation is written as

$$1 + L(\omega_{cr}) = 1 + A_{P,cr}P(\omega_{cr}) = 0, \quad \text{i.e.} \quad A_{P,cr} = -1/P(\omega_{cr}). \tag{8.41}$$

From (8.40) and (8.41), a simple relationship can be obtained to estimate the critical gain:

$$A_{P,cr} = -1/P(\omega_{cr}) = N(\omega_{cr}, a). \tag{8.42}$$

If steady oscillations of amplitudes $\pm\Delta u$ and $\pm\Delta y$ are measured at the process input and at the process output, respectively, then the critical gain is

$$A_{P,cr} = N(a) = \frac{4\Delta u}{\pi \Delta y}, \tag{8.43}$$

where now $N(a)$, the describing function of the relay, depends only on the amplitude. The most important advantage of this method is that the oscillation of the process output can be gradually set to a still allowed value, namely to $\Delta u = h$, which is one half the height of the relay characteristic, and $\Delta y = a$. If an integrator is serially connected to the relay, then the loop gain belonging to the phase angle of $-270°$ can be determined with this method.

Based on the describing function belonging to the hysteresis characteristic

$$-1/N(a) = -\frac{\pi}{4h}\sqrt{a^2 - g^2} - j\frac{\pi g}{4h} = -\frac{\pi}{4\Delta u}\sqrt{\Delta y^2 - g^2} - j\frac{\pi g}{4\Delta u}, \tag{8.44}$$

which is a straight line parallel to the negative real axis. Here g is the half-width of the hysteresis. Hence it can be easily checked that again the relationship

$$A_{P,cr} = |N(a)| = \frac{4\Delta u}{\pi \Delta y} \tag{8.45}$$

is obtained. Nowadays, in advanced electronic compact regulators, tuning with the relay method is already a built in possibility. (By changing the values h and g of the relay characteristic, further points of the Nyquist diagram also could be analyzed.)

8.3.6 Method of ÅSTRÖM-HÄGGLUND

This method also uses the simple approximation of the process according to (8.38) as a starting point, but as a design parameter it also uses the maximum value M_{max} of the complementary sensitivity function. The corresponding M curve characterizes the distance measured from the point $-1+0j$, thus to some extent the robustness of the regulator could be considered as well. The expression of the *PID* regulator used in this method is given by

$$u(t) = A_P\left[\beta r(t) - y(t) + \frac{1}{T_I}\int_0^t e(\tau)d\tau + T_D\frac{de(t)}{dt}\right] \tag{8.46}$$

where in forming the error signal the reference signal $r(t)$ and the output signal $y(t)$ are taken into account with different weights. (β is the weighting factor of the reference signal.) On the basis of the approximate form (8.38), introduce the following relative parameters:

$$\alpha = A_L\frac{T_L}{T_F} \quad \text{and} \quad \gamma = \frac{T_L}{T_L+T_F}. \tag{8.47}$$

ÅSTRÖM and HÄGGLUND suggested setting the regulator parameters according to the following function:

$$f(\gamma) = a_o\exp\{a_1\gamma + a_2\gamma^2\}. \tag{8.48}$$

Table 8.7 Tuning of the parameters of the *PI* regulator

	$M_{max} = 1.4$			$M_{max} = 2$		
$f(\gamma)$	a_o	a_1	a_2	a_o	a_1	a_2
αA_P	0.29	−2.7	3.7	0.78	−4.1	5.7
T_I/T_L	8.9	−6.6	3.0	8.9	−6.6	3.0
T_I/T_F	0.79	−1.4	2.4	0.79	−1.4	2.4
β	0.81	0.73	1.9	0.44	0.78	−0.45

Table 8.8 Tuning of the parameters of the *PID* regulator

	$M_{max} = 1.4$			$M_{max} = 2$		
$f(\gamma)$	a_o	a_1	a_2	a_o	a_1	a_2
αA_P	3.8	−8.4	7.3	8.4	−9.6	9.8
T_I/T_L	5.2	−2.5	−1.4	3.2	−1.5	0.93
T_I/T_F	0.46	2.8	−2.1	0.28	3.8	−1.6
T_D/T_L	0.89	−0.37	−4.1	0.86	−1.9	−0.44
T_D/T_F	0.077	5.0	−4.8	0.076	3.4	−1.1
β	0.4	0.18	2.8	0.22	0.65	−0.051

The Tables 8.7 and 8.8 give the coefficients a_o, a_1 and a_2 of (8.48) which determine the parameters of the *PI* and *PID* regulators, based on experimental considerations.

Similar tables have been elaborated for integrating processes.

8.4 Handling Amplitude Constraints: "*Anti-Reset Windup*"

Regulator design should take into account the limitations set for the control signal $u(t)$. These limitations may originate from several sources. The limitation may be the property of the actuator structure. Often the actuator is not able to provide an output value higher than a given maximum. For example, a valve in its fully open position provides a maximum flow rate. If it gets a command to transfer a higher value than its maximum, it is not able to execute it, it will be "saturated". The role of a deliberately applied saturation at the process input is to prevent the process from a harmful level of overexcitation which would cause failure in the process.

Whether the restriction occurs because of the properties of the process, or is artificially introduced into the control loop, its effects have to be taken into account. It is expedient to consider the restrictions already in the phase of regulator design, and to design a regulator whose output signal will not reach the limit value. If this is not possible, then the additional phenomena appearing when the restriction occurs have to be handled.

Thus the linear range of the regulator or of the actuator operated by the regulator is finite. The relation of this amplitude limitation with the design goals has already been discussed in Sect. 7.4. During saturation, the closed-loop control system behaves similarly to the open-loop, as the output of the saturation is constant and thus the input of the process is also constant. The process output is settled according to its dynamics. But in the case of regulators which contain an integrating element, another problem may also occur: if the value of the error signal is high, the regulator output may reach the horizontal section of the saturation characteristic. If the integrator works further on, then the input of the saturation characteristic continuous to increase, and a long time has to pass till the sign of the error signal changes and the input signal gets back to the linear section of the characteristic, if this

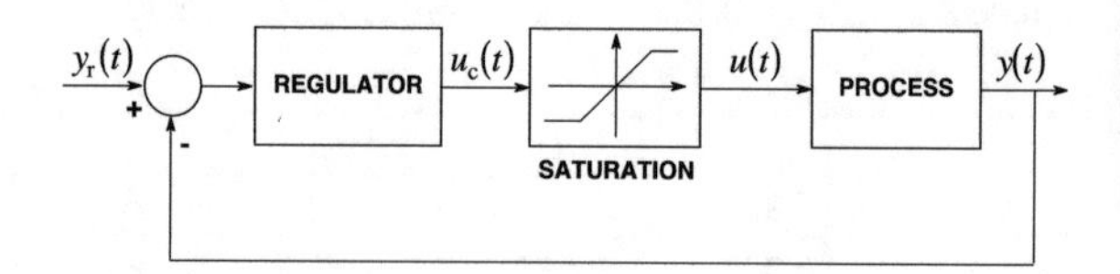

Fig. 8.28 Regulator and actuator with saturation

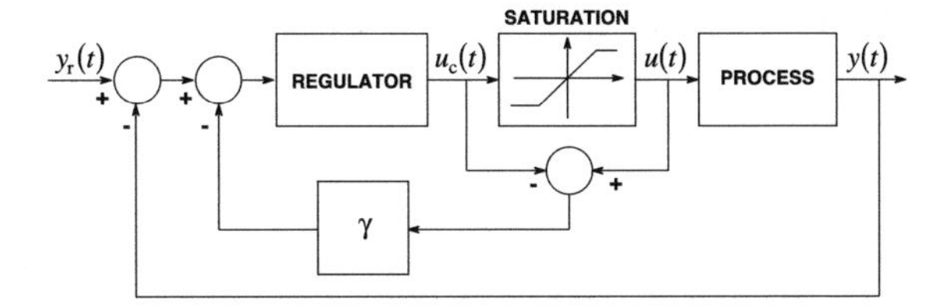

Fig. 8.29 Control system with an extended regulator ensuring the *ARW* effect

happens at all. Therefore the time of the transients will be increased unacceptably, and also steady oscillations may occur which are harmful to the process.

The problem can be solved by the technique of *Anti Reset Windup* (*ARW*) or "*antiwindup*", for short. The main point of this technique is that it uses the model of the static saturation characteristic and with an appropriate feedback it guides the operating point to the crossing point of the linear and the saturated section. The usual saturation characteristic can be described by

$$u(t) = \begin{cases} U_{\max}; & \text{if} \quad u_c(t) > U_{\max} \\ u(t); & \text{if} \quad |u_c(t)| < U_{\max} \\ -U_{\max}; & \text{if} \quad u_c(t) < -U_{\max} \end{cases}, \tag{8.49}$$

which is a more detailed form of (7.39). A closed-loop control system with the saturation is shown in Fig. 8.28.

The *ARW* effect can be reached by the simple feedback illustrated in Fig. 8.29. The extra inner feedback works until the process input is on the horizontal saturation section of the characteristic. This ensures that the regulator output is set to $u(t) = u_c(t)$, belonging to the breakpoint of the characteristic.

In a continuous system this solution can be realized if the signal $u_c(t)$ is available. If the regulator output and the manipulated variable are distinct, then both $u_c(t)$ and $u(t)$ are measurable, and the feedback can be easily realized through the constant element γ. If this is not the case, then a model of the saturated process has to be built. (The realization of such algorithms is much easier in the case of sampled data systems, see Chap. 13.) It has to be ensured an appropriate choice of γ, that the inner feedback acts faster than the dynamics of the process itself.

Another possibility is also available: to reset the integrator component of the regulator when observing saturation. The disadvantage of resetting of the integrator in the case of saturation is that when the regulator comes out of the saturation, there is a mismatch between the state variables of the regulator and those of the process, which results in a deterioration of the control behavior. This can be compensated by a regulator structure where the input of the regulator and the input of the process are similarly restricted, that is, the regulator is put into the feedback path of the saturation. A typical example of this solution is the FOXBORO regulator (Fig. 8.30), which corresponds to a saturated *PI* regulator.

Now, if there is no saturation, then the overall transfer function of a proportional element fed back through a first order lag element with positive feedback results in the transfer function of a *PI* regulator:

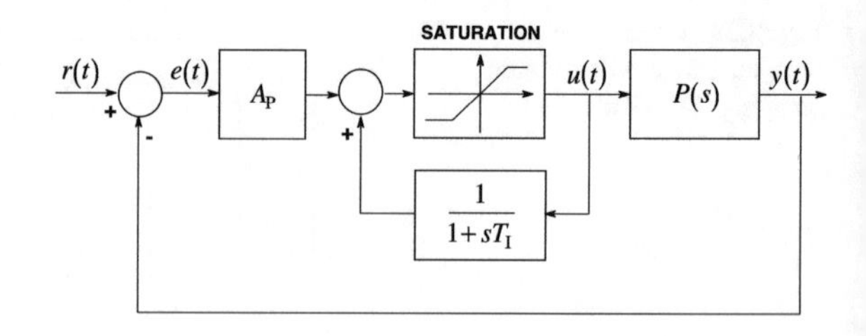

Fig. 8.30 The Foxboro regulator

$$C(s) = A_P \frac{1}{1 - \frac{1}{1+sT_I}} = A_P \frac{1+sT_I}{sT_I}. \tag{8.50}$$

In this structure, windup does not occur.

Most solutions set the integrator output to a given value after leaving the saturation. Several methods have been elaborated (some of them are rather complex) to calculate and set the "reset" value. There is no single procedure which ensures in all cases the appropriate behavior, but the above simple procedures in most cases provide satisfactory operation. Several other methods are known for taking the effect of saturation into account, but these are not discussed here.

8.5 Control of Special Plants

In the sequel, two examples will be shown of the regulator design of special plants, namely for a process containing two integrators, and for compensation of unstable processes. It will also be presented how in some cases the design can be executed analytically, approximating the plant by its dominant pole pair model.

8.5.1 Control of a Double Integrator

Let the transfer function of the process be $P(s) = K/s^2$. The process contains two integrators, in a closed-loop control system with feedback unity and with a

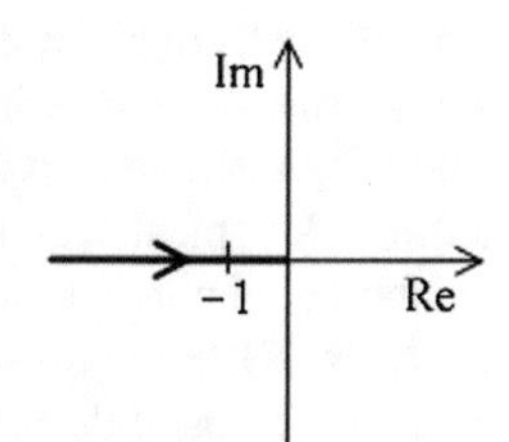

Fig. 8.31 Nyquist diagram of a control system containing two integrators

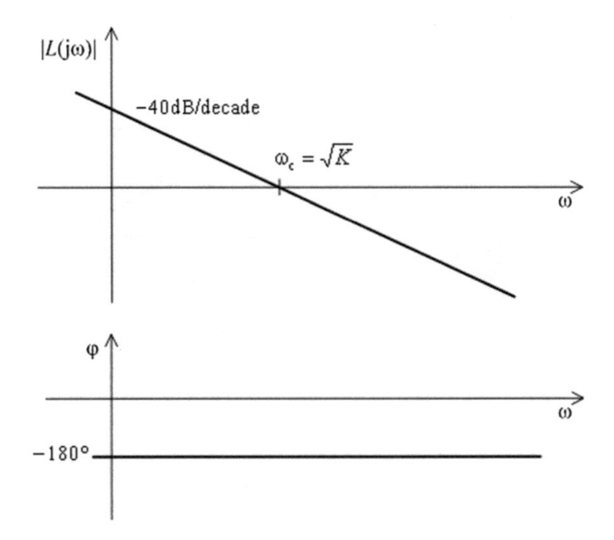

Fig. 8.32 BODE diagram of a control system containing two integrators

proportional regulator working at the borderline of stability. The characteristic equation is $1+K/s^2=0$, or $s^2+K=0$. Its roots $s_{1,2}=\pm j\sqrt{K}$ are located on the imaginary axis. Instead of a good control system this system rather realizes a good oscillator. The NYQUIST diagram of the open-loop is shown in Fig. 8.31. The NYQUIST diagram crosses the point −1, thus the system is at the borderline of stability. The BODE diagram is given in Fig. 8.32.

The quality specifications set for the control system are as follows: stability; the phase margin should be about 60° to ensure an appropriate dynamic response; and for step and ramp reference signals, the tracking error in steady state should be zero (that is, the type number of the control system after the compensation has to remain 2).

These requirements can be fulfilled applying a compensation element which is able to improve the phase conditions. The phase angle of the loop frequency function with the regulator is expressed as $\varphi_L(\omega)=\varphi_C(\omega)+\varphi_P(\omega)=\varphi_t(\omega)-180°$. The regulator has to provide a positive phase margin by $\varphi_t(\omega)=180°+\varphi_L(\omega_c)=\varphi_C(\omega_c)$, because $\varphi_P(\omega)=-180°$. A phase-lead (approximate *PD*) regulator described by the transfer function

$$C(s)=A\frac{1+s\tau}{1+sT};\quad \tau>T$$

guarantees the addition of a positive phase angle (see also Sect. 2.4), as $\varphi_C(\omega)=\arctan(\omega\tau)-\arctan(\omega T)>0$, if $\tau>T$. The higher the ratio τ/T, the higher the values that $\varphi_C(\omega)$ can take. The phase lead element is also called an approximate *PD* element, as from the form

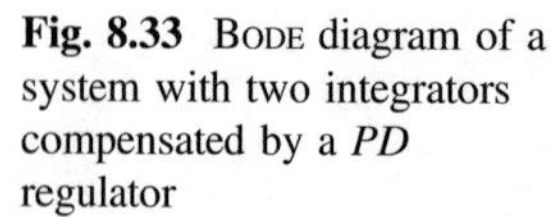

Fig. 8.33 BODE diagram of a system with two integrators compensated by a *PD* regulator

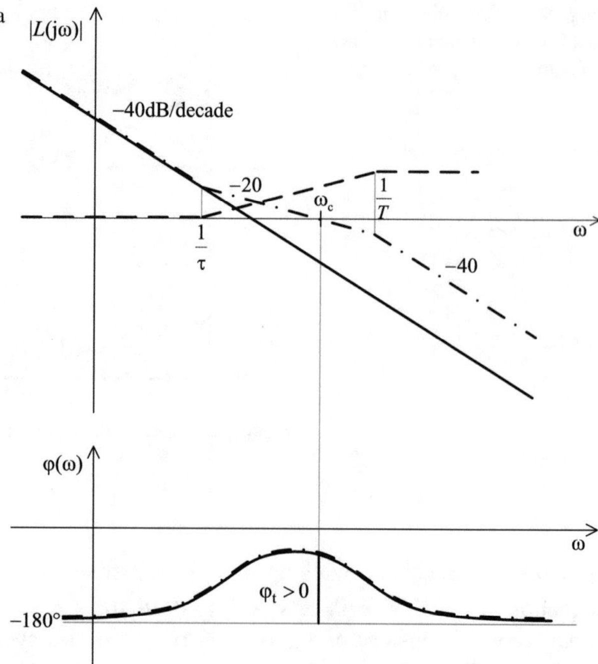

Fig. 8.34 NYQUIST diagram of a system with two integrators compensated by a *PD* regulator

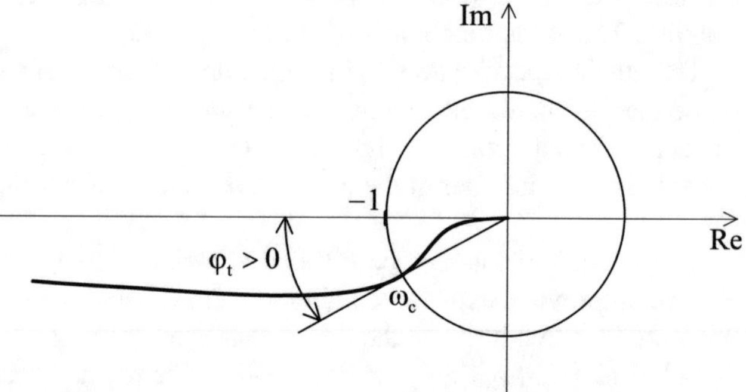

$$C(s) = A\frac{1+s\tau}{1+sT} = A\frac{1+sT+s\tau-sT}{1+sT} = A\left(1+\frac{s(\tau-T)}{1+sT}\right)$$

it can be seen, that $C(s)$ can be obtained as the parallel connection of a proportional and an approximate (that is realizable) differentiating channel. The transfer function of the regulator is often given in the form

$$\tilde{C}_{\mathrm{PD}}(s) = \tilde{A}_{\mathrm{PD}}\frac{1+sT_{\mathrm{D}}}{1+sT}; \quad T_{\mathrm{D}} > T, \tag{8.51}$$

where the notation $\tau = T_{\mathrm{D}}$ is introduced for the differentiating time constant, see also (8.14). Instead of the correct denomination "approximate *PD*", often the

Fig. 8.35 Root locus of a system with two integrators compensated by a *PD* regulator

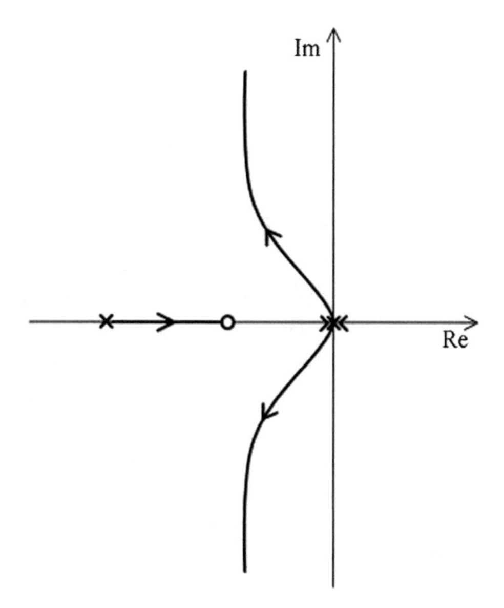

slightly simpler denomination "*PD*" is used. Nevertheless this sloppiness is reasonable, since an accurate *PD* regulator without the pole is non-realizable (its high frequency gain would be infinity).

Figure 8.33 shows that with the *PD* regulator a straight line section of −20 dB/decade slope can be formed on the BODE amplitude-frequency curve. The cut-off frequency is placed on this section. Thus the system will have a positive phase margin; its performance will be fast, as ω_c is shifted toward the higher frequency domain. The gain of the regulator may be chosen to maximize the phase margin. The reachable maximum phase margin depends on the ratio T_D/T. This ratio also influences the maximum value of the control signal appearing at the input of the process at the time instant when the unit step reference signal is switched on. The maximum value of the control signal is $u_{max} = A_P T_D/T$.

As seen from the BODE diagram, the control system is structurally stable: the phase margin is positive for any value of the loop gain.

The NYQUIST diagram of the compensated system is shown in Fig. 8.34. (Note that in the case of processes containing integrators, when there are poles at the origin of the complex plane, it is not necessary to apply the generalized NYQUIST criterion to evaluate the stability from the properties of the conformal mapping of the closed curve shown in Fig. 5.18 surrounding the origin by a circle of infinitesimal radius according to the loop frequency function. It is sufficient to plot the NYQUIST diagram only for the positive frequencies and to check whether when going through the curve from $\omega = 0$ to $\omega = \infty$, the point $-1 + 0j$ is to the left of the curve or not. If it is to the left, then the control system is stable, and the phase margin or the gain margin can be used to measure the distance from the borderline of stability.)

Figure 8.35 gives the root locus of the system. The points of the root locus, which are the roots of the characteristic equation, lie in the left half of the complex plane for all gain values, indicating that the control system is structurally stable.

8.5.2 Control of an Unstable Plant

In state space, the state variables are attached to the poles of the transfer functions of the process and the regulator. These state variables together form the state variables of the open-loop. When an undesired pole of the process is cancelled by a zero of the regulator, actually the corresponding state variable becomes inaccessible from the output or from the input side, namely the system becomes unobservable or uncontrollable (Sect. 3.4). But in spite of the fact that these variables do not appear in the overall transfer function of the loop, they remain parts of the system. To ensure the stability of the control system, not only do the poles of the transfer function have to be at left half side of the complex plane, but so do the unobservable and the uncontrollable poles.

Unstable poles of an unstable process must not be cancelled by the zeros of the regulator. This prohibition can be justified also by the fact that as seen in Chap. 4, the behavior of a closed-loop control system is characterized not only by the overall transfer function between the output and the reference signal, but also by the overall transfer functions between the output signal and the input and the output disturbances, and the overall transfer function between the control and the reference signal. The unstable pole does appear in the transfer function between the output signal and the input disturbance, thus in spite of the compensation the instability of the control system is maintained.

A further consideration is that in real systems, the values of the parameters are not accurate: generally they are obtained by measurements and lie within a range of their possible values. Therefore, an accurate cancellation of an unstable pole is not possible, and the instability is maintained in the control system. This phenomenon can be illustrated by the root locus. Let us consider as an example the control system in Fig. 8.36b. The loop transfer function is given by a proportional system with two lags, where one pole is unstable. From the root locus (Fig. 8.36a) it can be

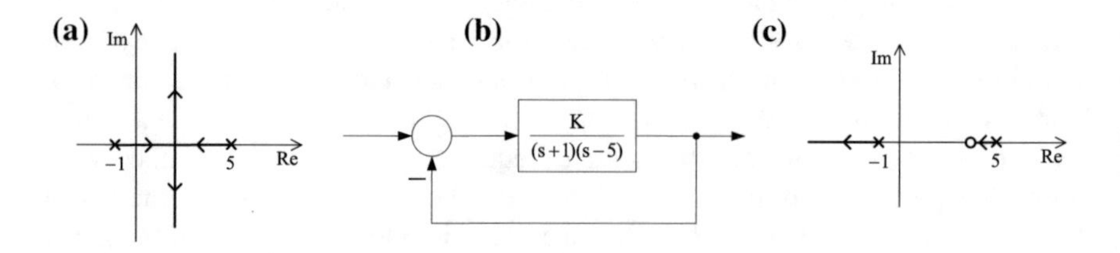

Fig. 8.36 With imperfect zero-pole cancellation the root locus has a branch on the right half of the complex plane

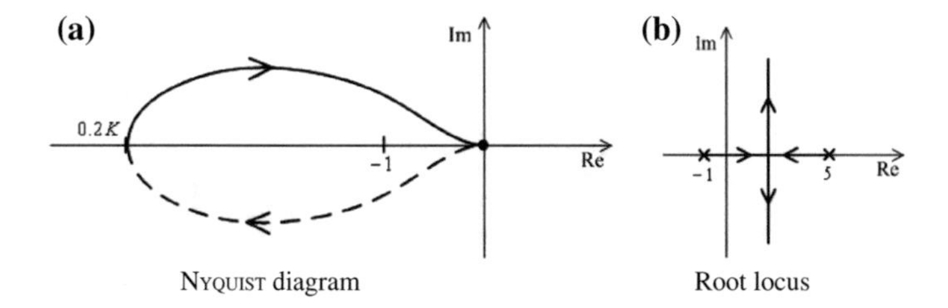

Fig. 8.37 Control of an unstable process with a proportional regulator: the system can not be stabilized!

seen that the closed-loop control system is unstable for all values of the gain. If the unstable pole could be accurately cancelled by the zero of the regulator, the control system would become structurally stable and the root locus would have a single branch on the left half of the complex plane. But as practically perfect cancellation can not be realized, there will remain a branch of the root locus on the right half of the complex plane, and thus the closed-loop control system remains unstable (Fig. 8.36c). (Note that in compensation, a zero by itself can not be realized, it always appears together with a pole.)

When compensating unstable processes, the generalized NYQUIST stability criterion has to be considered to ensure the stable behavior of the closed-loop control system. *PID*-like regulators can be used as compensating elements in such cases as well.

Example 8.1 Let us analyze whether the processes given by the transfer function

$$P_1(s) = \frac{1}{(s+1)(s-5)} = -\frac{0.2}{(1+s)(1-0.2s)} \tag{8.52}$$

and

$$P_2(s) = \frac{1}{(s-1)(s+5)} = -\frac{0.2}{(1-s)(1+0.2s)} \tag{8.53}$$

can be stabilized by the proportional regulator $C(s) = A_P$ or not.

As there is an unstable pole in the open-loop, stable behavior can be ensured if the NYQUIST diagram encircles the point $-1 + j0$ once anticlockwise.

For the first process, the NYQUIST diagram of the open-loop is shown in Fig. 8.37a. As seen, the diagram can encircle the point $-1 + 0j$ only clockwise, thus this process can not be stabilized by a proportional regulator. This property is demonstrated also with the root locus, which contains a pole in the right half-plane at any gain value. (Here, the stabilization of the system can be tried with a *PD*-like compensation, such as $C(s) = A_P(1 + 0.2s)/(1 + 0.02s)$.)

The second process can be stabilized by a proportional regulator, as the direction of encircling by the NYQUIST diagram is counterclockwise, thus choosing an

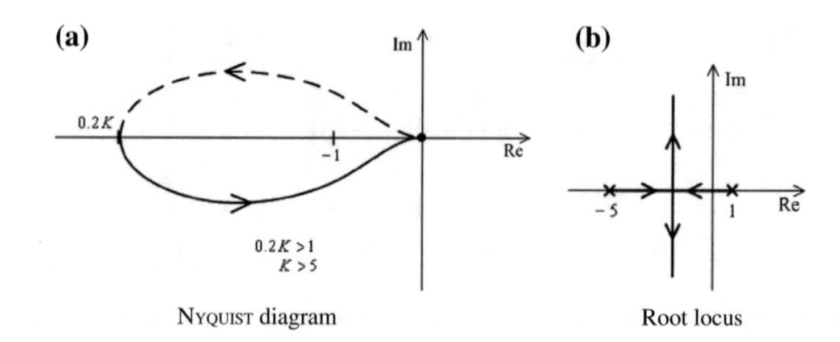

Fig. 8.38 Control of an unstable process with a proportional regulator: the system can be stabilized

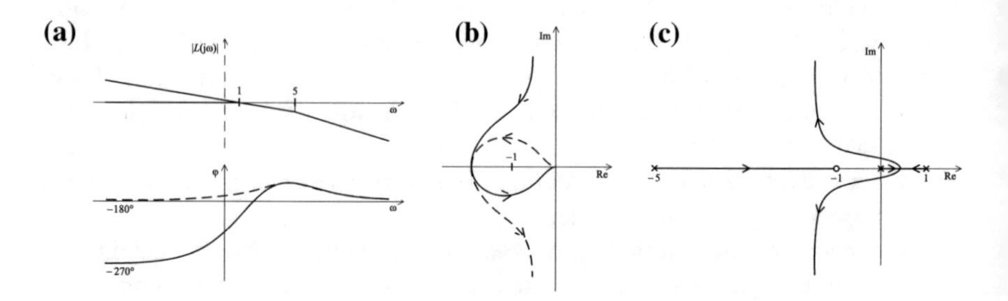

Fig. 8.39 Bode and Nyquist diagram, and the root locus of the unstable process compensated by a *PI* regulator

Fig. 8.40 Unit step response of an unstable process compensated by a *PI* regulator

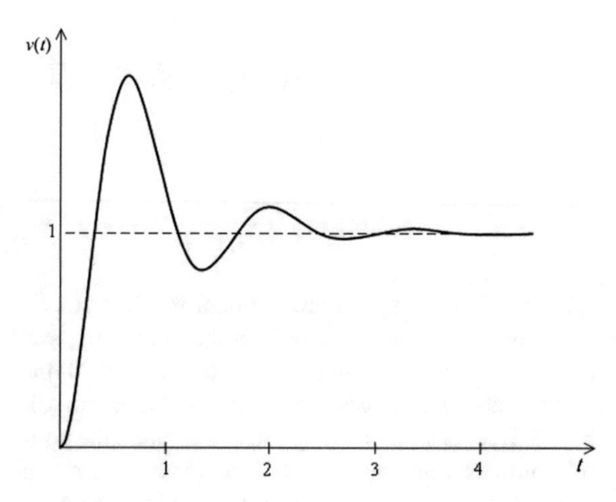

appropriate gain ($K > 5$) the NYQUIST diagram will encircle the point $-1+0j$ once (Fig. 8.38a). This is also demonstrated by the root locus shown in Fig. 8.38b. By increasing the gain, the root locus will get into the left half of the complex plane.

The type-number of the control system is zero, thus it has a static error. The static accuracy can be improved by a *PI* regulator which does not cancel the unstable pole. The transfer function of the regulator is: $C(s) = A_P(1+s)/s$.

The loop transfer function is:

$$L(s) = C(s)P_2(s) = -A_P \frac{1+s}{s} \frac{0.2}{(1-s)(1+0.2s)}. \tag{8.54}$$

The BODE diagram of the original and of the compensated system is shown in Fig. 8.39a, the NYQUIST diagram of the compensated system (whose course can be derived from the BODE diagram) is given in Fig. 8.39b, and the shape of the root locus is provided in Fig. 8.39c. It can be seen that increasing the gain beyond a defined value the closed-loop control system will be stable, the generalized NYQUIST diagram encircles once the point $-1+0j$ in the positive (counterclockwise) direction. The parameter A_P can be designed for maximum phase margin. (The concept of phase margin can be used in this case as well.) Figure 8.40 shows the unit step response of the control system. ■

Example 8.2 The transfer function of an unstable process is:

$$P(s) = \frac{0.5}{(s-0.1)(s+1)(s+5)} = -\frac{1}{(1-10s)(1+s)(1+0.2s)} \tag{8.55}$$

Let us design a regulator which ensures stable behavior, tracking the unit step reference signal without static error, and the initial value of the control signal does not exceed the value of 50.

A qualitatively correct NYQUIST diagram of the open-loop with a proportional regulator is shown in Fig. 8.41. As the loop transfer function has one pole in the right half-plane, the closed-loop will be stable if $-1+0j$ lies within the left side curve in the figure. The condition for that is that the usually interpreted phase margin indicated in the figure be positive.

The asymptotic BODE amplitude-frequency and the phase-frequency diagrams are shown in Fig. 8.42. To reach a more favorable phase margin, *PD* compensation is applied. Thus the section of slope −20 dB/decade is elongated, and the cut-off frequency can be relocated to $\omega_c \approx 1$. To remove the static error, a further *PI* regulator is used. The transfer function of the entire *PID* compensator is

$$C(s) = 10 \frac{1+10s}{10s} \frac{1+s}{1+0.2s}. \tag{8.56}$$

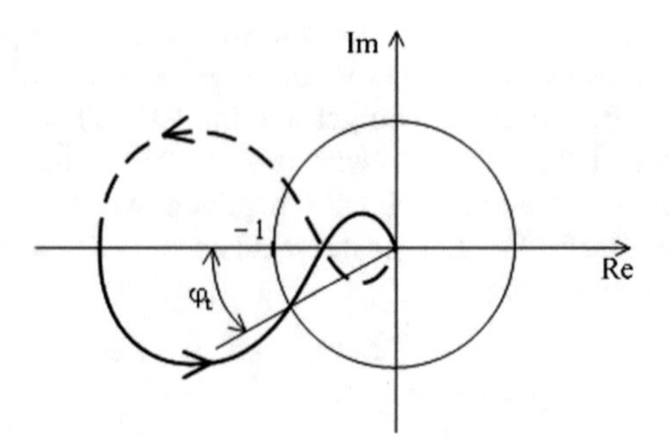

Fig. 8.41 Nyquist diagram of an unstable process compensated by a proportional regulator

The loop transfer function is:

$$L(s) = -\frac{1 + 10s}{s(1 - 10s)(1 + 0.2s)^2}. \tag{8.57}$$

In the Bode amplitude diagram at frequency $\omega = 0.1$, the breakpoint disappears because of the contradictory effects of the zero and the unstable pole, but it remains as a corner-point, where the phase angle asymptotically changes from $-270°$ to $-90°$.

The cut-off frequency and the phase margin of the open-loop are $\omega_c = 0.964$, and $\varphi_t = 56.25$. The initial value of the control signal in the case of a unit step reference signal is just 50. The output and the control signals are shown in Fig. 8.43. ■

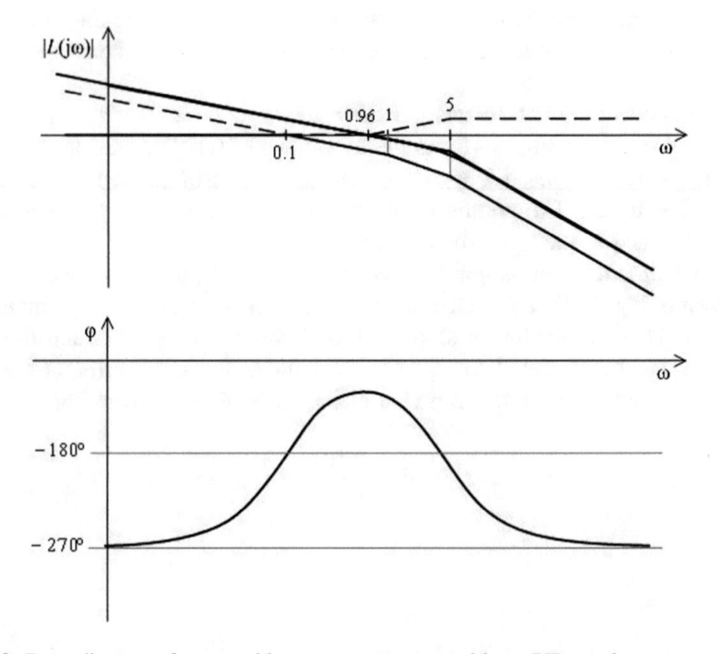

Fig. 8.42 Bode diagram of an unstable process compensated by a *PID* regulator

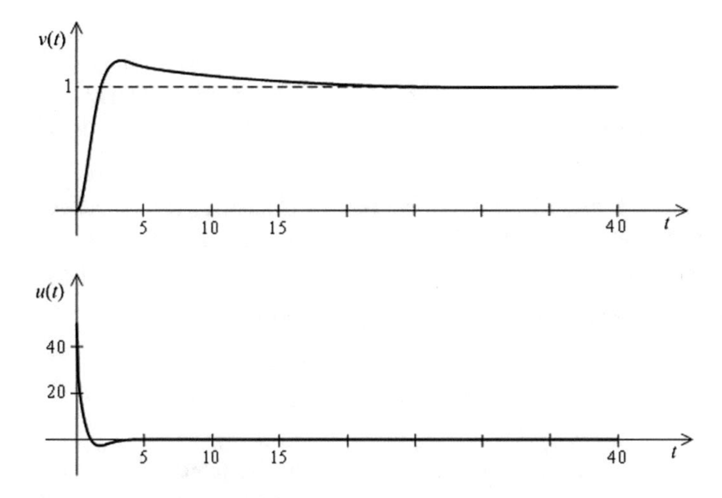

Fig. 8.43 Output and control signals of an unstable process compensated by a *PID* regulator in the case of a unit step reference signal

8.6 Regulator Design Providing a 60° Phase Margin by Pole Cancellation

A regulator is designed so that the process (or its model) meets the quality requirements set for the control system. A common compensation technique is pole cancellation, when the zeros of the transfer function of the regulator are chosen equal to the poles of the process: the unfavorable poles of the process are "cancelled", and instead more favorable poles are introduced. As an example let us consider a proportional process with three time lags. The transfer function of the process is

$$P(s) = \frac{1}{(1+sT_1)(1+sT_2)(1+sT_3)}; \quad T_1 > T_2 > T_3. \tag{8.58}$$

(a) Suppose the prescription for the control system is stable behavior and an overshoot less than 10%. This latter requirement can be fulfilled in the frequency domain by ensuring a phase margin of about 60°.

The requirements can be met by applying a simple proportional regulator: $C(s) = A_P$. The approximate Bode diagram of the open-loop is shown in Fig. 8.44. To ensure stability the cut-off frequency ω_c has to be placed at a straight line section of slope −20 dB/decade. To reach the required phase margin, ω_c is located at the frequency where the phase angle is $\varphi = -120°$. First the loop frequency function is analyzed supposing $A_P = 1$ (dotted line in the figure), then A_P is set to the reciprocal of the amplitude belonging to the phase angle $\varphi = -120°$.

The given requirements can be fulfilled with a proportional regulator. The control system will be slow, as ω_c has to be placed at the straight line section of slope −20 dB/decade, which is in the low frequency domain. The control system is of 0-type, so it tracks the unit step reference signal with a static error, whose value depends on the loop gain.

(b) Suppose the prescription for the control system is stable behavior and an overshoot less than 10%. Furthermore, that the static error be zero for step reference signal.

These prescriptions can be fulfilled by a *PI* regulator.

$$C_{PI}(s) = A_P \frac{1 + sT_I}{sT_I} \tag{8.59}$$

Let us choose the time constant T_I equal to the largest time constant of the process, $T_I = T_1$ (we "cancel" the largest time constant of the process, and "introduce" an integrating effect instead). According to Fig. 8.45 a long straight line section of −20 dB/decade slope is formed in the low frequency domain of the Bode amplitude diagram of the open-loop. Changing the gain A_P of the regulator, the Bode amplitude diagram is shifted parallel until the cut-off frequency is located to ensure the required ~60° phase margin.

With a *PI* regulator the type number will be 1, and besides meeting the prescriptions for stability and dynamic response, the control system also fulfills the static requirements. But as the cut-off frequency can be placed only in the low frequency range, the control system will be slow.

(c) Let the prescription for the control system be stable behavior and an overshoot less than 10%, as well as that the operation of the control system has to be faster.

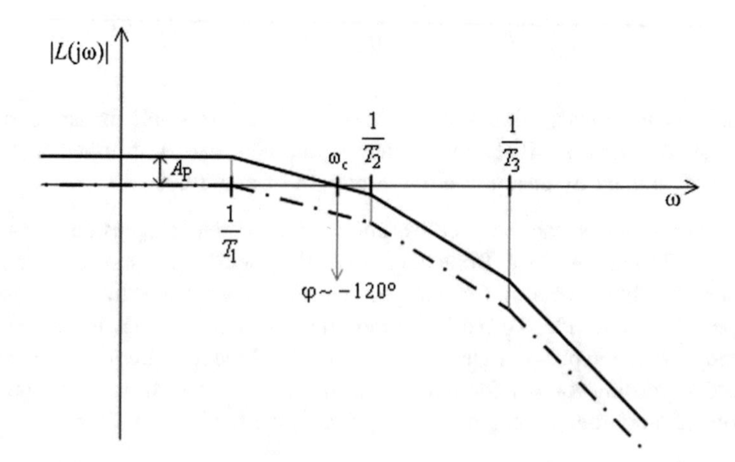

Fig. 8.44 Series compensation with proportional regulator

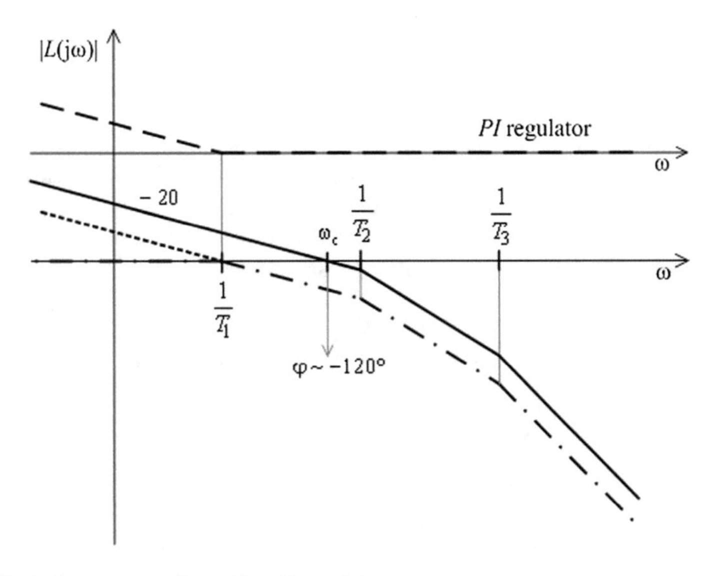

Fig. 8.45 Series compensation with a *PI* regulator

These specifications can be fulfilled by using a *PD* regulator.

$$C_{\mathrm{PD}}(s) = A_{\mathrm{P}} \frac{1 + sT_{\mathrm{D}}}{1 + sT}; \quad T_{\mathrm{D}} > T \tag{8.60}$$

Let us choose the time constant T_{D} equal to the second largest time constant of the process, $T_{\mathrm{D}} = T_2$ (that is, equal to that time constant for which the slope changes from −20 dB/decade to −40 dB/decade at the corresponding breakpoint of the Bode amplitude diagram). The ratio $\eta = T_{\mathrm{D}}/T$ is chosen according to the practical limit of the control signal. (We "cancel" the unfavorable time constant of the process and "introduce" a much smaller time constant instead.) Then changing the gain A_{P} of the regulator, the Bode amplitude diagram is shifted parallel until the cut-off frequency is located to ensure the required ~60° phase margin. The effect of the compensation on the Bode diagram of the open-loop is shown in Fig. 8.46.

The control system will be stable, it has a small overshoot, it will be fast, but as it remains of 0-type, it will have a static error, depending on the loop gain when tracking a unit step reference signal. The acceleration results from the high initial value $u(t = 0) = A_{\mathrm{P}}\eta$ of the control signal.

(d) Let the prescription for the control system be stable behavior, an overshoot less than 10%, fast operation and zero static error for tracking a step reference signal.

These prescriptions can be fulfilled by a *PID* regulator, combining the possibilities of the *PI* and the *PD* regulators.

$$C_{\mathrm{PID}}(s) = A_{\mathrm{P}} \frac{1+sT_{\mathrm{I}}}{sT_{\mathrm{I}}} \frac{1+sT_{\mathrm{D}}}{1+sT} \tag{8.61}$$

Two of the four free parameters are chosen considering the poles of the process. The parameter T_{I} is chosen equal to the largest time constant of the process, and the parameter T_{D} is set equal to the second largest time constant. With this choice $T_{\mathrm{I}} = T_1$ and $T_{\mathrm{D}} = T_2$ the loop transfer function can be simplified, that is the "introduced" zeros "cancel" poles of the process.

$$\begin{aligned} L(s) = C(s)P(s) &= A_{\mathrm{P}} \frac{1+sT_1}{sT_1} \frac{1+sT_2}{1+sT} \frac{1}{(1+sT_1)(1+sT_2)(1+sT_3)} \\ &= \frac{A_{\mathrm{P}}}{sT_1(1+sT)(1+sT_3)} \end{aligned} \tag{8.62}$$

It can be seen that the transfer function of the residual system became simpler, thus the further steps of the design become easier.

The remaining two parameters are chosen considering the prescriptions set for the acceleration and the overexcitation. The parameter T is chosen based on the pole placement ratio. The phase margin (and the overshoot of the step response) can be set with A_{P}. It can be seen in the Bode amplitude diagram of the open-loop that the section of slope -20 dB/decade will be longer because of the choice $T < T_{\mathrm{D}}$. With the gain A_{P} of the regulator, the Bode amplitude diagram is shifted parallel until the cut-off frequency is located to ensure the required $\sim 60°$ phase margin. This is done by checking the frequency where the phase angle is about $-120°$, and A_{P} is then set to the reciprocal of the amplitude corresponding to this

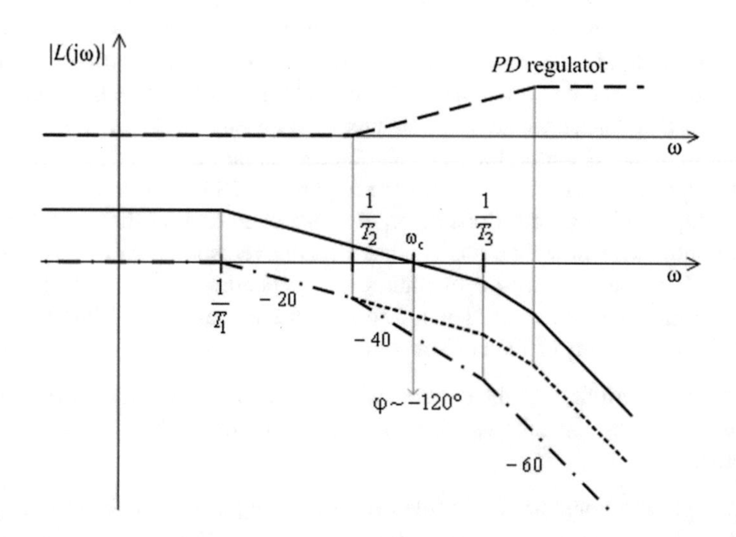

Fig. 8.46 Series compensation with a *PD* regulator

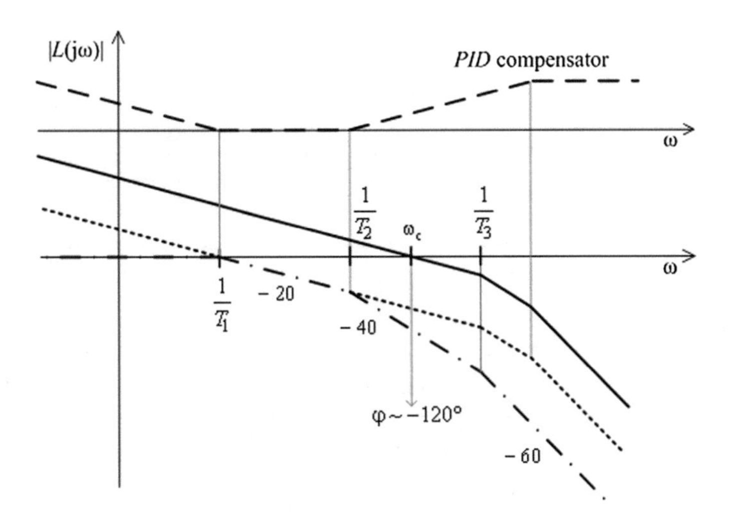

Fig. 8.47 Series compensation with *PID* regulator

Fig. 8.48 In compensation pole cancellation is virtual

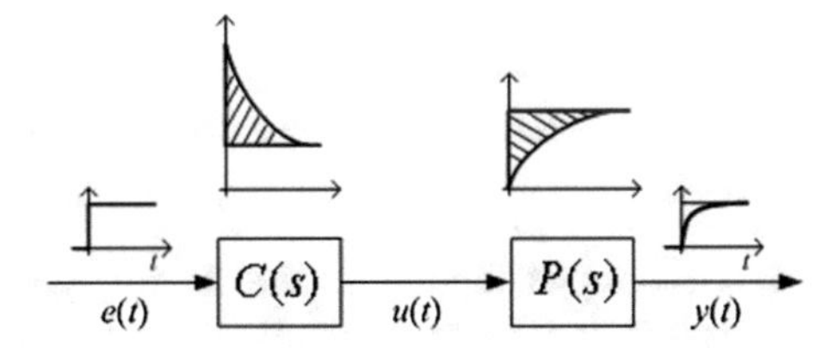

frequency. This frequency will be ω_c, which is located now in the higher frequency domain, therefore the control system will be faster (Fig. 8.47).

But let us observe that the pole cancellation is just formal: in reality the poles do not disappear. The process can not be changed, its poles do exist. Actually the zeros and the poles do not cancel each other, only their effects compensate each other. Figure 8.48 demonstrates that in the case of zero-pole cancellation the overall transfer function of the serially connected regulator and process behaves as if a real pole cancellation has happened, but the effect of the zero of the regulator does appear in the signal $u(t)$. The overexcitation in the control signal depends on the ratio of the zero and the pole of the regulator. The so called *acceleration area* in the control signal decreases the so called *decelerating area* of the process which characterizes the settling time of its unit step response, thus yielding a faster response of the control system.

The main point of the pole cancellation method is that the unfavorable poles of the process are cancelled, and the poles of the regulator ensure a more favorable dynamics for the control system. As no real pole cancellation occurs, it is not necessary to set the zeros of the regulator quite accurately. The tuning can also be refined later, by moving the zeros a bit away from their pole cancellation location.

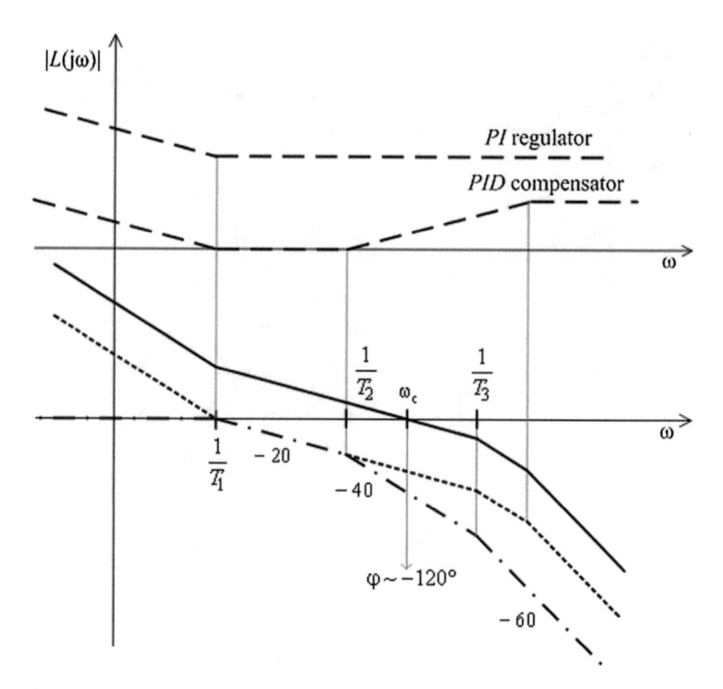

Fig. 8.49 Series *PID* compensation with additional *PI* regulator

(e) Let the prescription for the control system be its stable behavior, an overshoot less than 10%, fast operation, and zero static error for tracking both a step and also a ramp reference signal.

To meet the static requirements the control system has to be of 2-type containing two integrating effects. The low frequency part of the Bode diagram has to be of −40 dB/decade.

Table 8.9 *PID*-like regulators designed for a proportional system with three lags and the characteristic measures of the control systems

	$C(s)$	$L(s)$	e_∞	u_{max}	ω_c
P	7.51	$\frac{7.51}{(1+10s)(1+s)(1+0.2s)}$	0.1174	7.51	0.62
PI	$5.04\frac{1+10s}{10s}$	$\frac{5.04}{10s(1+s)(1+0.2s)}$	0	5.04	0.45
PD	$16.55\frac{1+s}{1+0.2s}$	$\frac{16.55}{(1+10s)(1+0.2s)^2}$	0.057	82.7	1.51
PID	$14.27\frac{1+10s}{10s}\frac{1+s}{1+0.2s}$	$\frac{14.27}{10s(1+0.2s)^2}$	0	71.36	1.33
PI − *PID*	$14.3\frac{1+50s}{50s}\frac{1+10s}{10s}\frac{1+s}{1+0.2s}$	$\frac{14.3(1+50s)}{500s^2(1+0.2s)^2}$	0	71.5	1.34

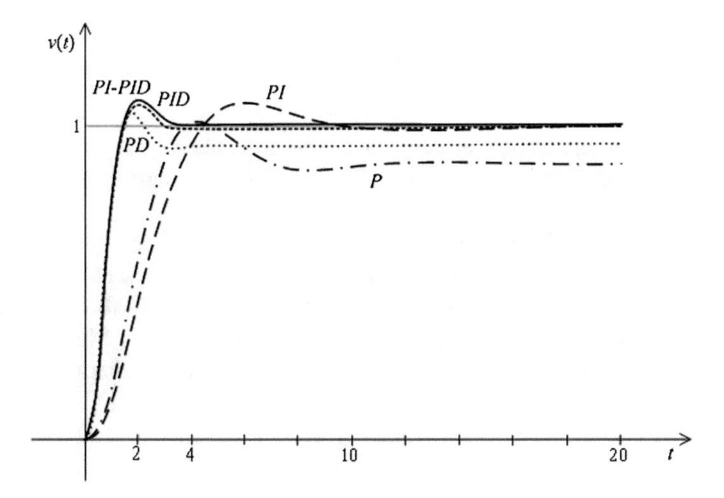

Fig. 8.50 Unit step responses of the compensated control systems

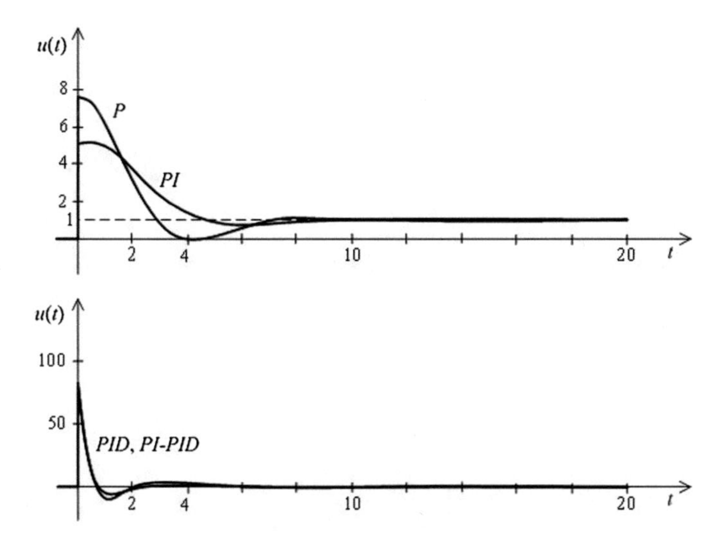

Fig. 8.51 Control signals of the compensated control systems

The previous regulator designed for pole cancellation is extended by a *PI* effect, which forms the Bode diagram of Fig. 8.49. The transfer function of the regulator is

$$C_{\mathrm{PI-PID}}(s) = A_{\mathrm{P}} \frac{1+sT_{\mathrm{I1}}}{sT_{\mathrm{I1}}} \frac{1+sT_{\mathrm{I2}}}{sT_{\mathrm{I2}}} \frac{1+sT_{\mathrm{D}}}{1+sT}, \tag{8.63}$$

where $T_{\mathrm{I1}} > T_{\mathrm{I2}}$. This ratio is selected by the designer, its advisable value is $T_{\mathrm{I1}} \approx 5T_{\mathrm{I2}}$.

This regulator, considering Fig. 6.3, ensures not only a better reference signal tracking, but also a favorable disturbance rejection and less sensitivity to parameter changes.

Example 8.3 Let the time constants of the above process be $T_1 = 10$, $T_2 = 1$, $T_3 = 0.2$. The regulators designed for the above requirements and the characteristics of the control system are given in Table 8.9.

Figure 8.50 shows the unit step responses of the control circuits, whereas Fig. 8.51 presents the corresponding control signals. As can be seen, the static error is zero only in the cases when there is an integrating effect in the regulator. The control system with the *P* and the *PI* regulator is slow, whereas with the *PD*, *PID* and the *PI–PID* regulators it is fast, at the cost of high overexcitation. ■

Chapter 9
Control Systems with State Feedback

In Chap. 3. the description of processes in state-space was investigated. In many cases, this is the kind of description that is primarily available, and not the transfer function of the controlled system. This is the explanation, in part, for why there is a control design methodology directly based on the state-space description. For illustrative purposes, let us consider the state-space representation of an (*LTI*) process to be controlled,

$$\frac{\mathrm{d}\boldsymbol{x}}{\mathrm{d}t} = \dot{\boldsymbol{x}} = \boldsymbol{A}\boldsymbol{x} + \boldsymbol{b}u \qquad y = \boldsymbol{c}^{\mathrm{T}}\boldsymbol{x} \tag{9.1}$$

which corresponds to (3.10) for the case of $d = 0$. This, as was mentioned earlier, does not impair generality, because it is a very rare case when the model contains proportional channel directly affecting the output. The block scheme of (9.1) is shown in Fig. 9.1.

Here u and y are the input and output signals of the process, respectively, and $\boldsymbol{x}$ is the state vector. According to the equivalent transfer function (3.17) we get

$$P(s) = \boldsymbol{c}^{\mathrm{T}}(s\boldsymbol{I} - \boldsymbol{A})^{-1}\boldsymbol{b} = \frac{\mathcal{B}(s)}{\det(s\boldsymbol{I} - \boldsymbol{A})} = \frac{\mathcal{B}(s)}{\mathcal{A}(s)} = \frac{b_1 s^{n-1} + \cdots + b_{n-1}s + b_n}{s^n + a_1 s^{n-1} + \cdots + a_{n-1}s + a_n}. \tag{9.2}$$

Figure 9.2 shows the so-called classical closed control system directly fitting the state-equation description, where r denotes the reference signal. In the closed-loop, the state vector is fed back with the linear proportional vector $\boldsymbol{k}^{\mathrm{T}}$ according to the expression below

$$u = k_{\mathrm{r}} r - \boldsymbol{k}^{\mathrm{T}}\boldsymbol{x} \tag{9.3}$$

L. Keviczky et al., *Control Engineering*, Advanced Textbooks in Control and Signal Processing, https://doi.org/10.1007/978-981-10-8297-9_9

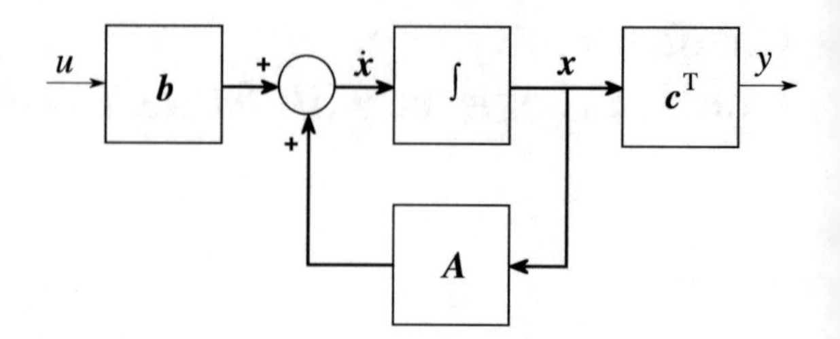

Fig. 9.1 Block scheme of the state-space equation of the *LTI* system

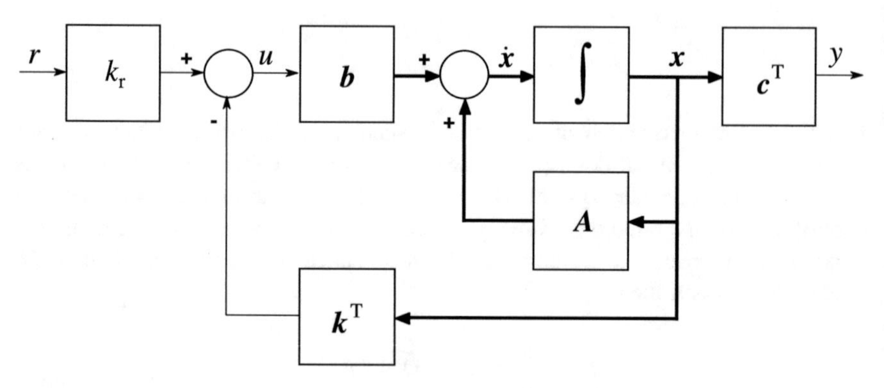

Fig. 9.2 Linear controller with state feedback

Based on Fig. 9.2, the state-equation of the complete closed system can be easily written as

$$\begin{aligned} \frac{\mathrm{d}\boldsymbol{x}}{\mathrm{d}t} &= \left(\boldsymbol{A} - \boldsymbol{b}\boldsymbol{k}^{\mathrm{T}}\right)\boldsymbol{x} + k_{\mathrm{r}}\boldsymbol{b}r \\ y &= \boldsymbol{c}^{\mathrm{T}}\boldsymbol{x} \end{aligned} \tag{9.4}$$

i.e. with the state feedback the dynamics represented by the original system matrix is modified by the dyadic product $\boldsymbol{b}\boldsymbol{k}^{\mathrm{T}}$ to $\left(\boldsymbol{A} - \boldsymbol{b}\boldsymbol{k}^{\mathrm{T}}\right)$.

The transfer function of the closed control loop is

$$\begin{aligned} T_{\mathrm{ry}}(s) &= \frac{Y(s)}{R(s)} = \mathbf{c}^{\mathrm{T}}\left(s\boldsymbol{I} - \boldsymbol{A} + \boldsymbol{b}\boldsymbol{k}^{\mathrm{T}}\right)^{-1}\boldsymbol{b}k_{\mathrm{r}} = \frac{\mathbf{c}^{\mathrm{T}}(s\boldsymbol{I} - \boldsymbol{A})^{-1}\boldsymbol{b}k_{\mathrm{r}}}{1 + \boldsymbol{k}^{\mathrm{T}}(s\boldsymbol{I} - \boldsymbol{A})^{-1}\boldsymbol{b}} \\ &= \frac{k_{\mathrm{r}}}{1 + \boldsymbol{k}^{\mathrm{T}}(s\boldsymbol{I} - \boldsymbol{A})^{-1}\boldsymbol{b}} P(s) = \frac{k_{\mathrm{r}}\mathcal{B}(s)}{\mathcal{A}(s) + \boldsymbol{k}^{\mathrm{T}}\boldsymbol{\Psi}(s)\boldsymbol{b}} \end{aligned} \tag{9.5}$$

which comes from the comparison of the equations valid for the Laplace transforms $\boldsymbol{X}(s) = (s\boldsymbol{I} - \boldsymbol{A})^{-1}\boldsymbol{b}U(s)$ [see (3.12)], $U(s) = k_{\mathrm{r}}R(s) - \boldsymbol{k}^{\mathrm{T}}\boldsymbol{X}(s)$ [see (9.3)] and

$Y(s) = \boldsymbol{c}^{\mathrm{T}}\boldsymbol{X}(s)$ [see (9.1)] using the matrix inversion lemma (for details, see A.9.1 in Appendix A.5). Note that the state feedback leaves the zeros of the process untouched and only the poles of the closed-loop system can be designed by $\boldsymbol{k}^{\mathrm{T}}$.

The so-called calibration factor k_{r} is introduced in order to make the gain of T_{ry} equal to unity $\left(T_{\mathrm{ry}}(0) = 1\right)$. The open-loop is obviously not of integrator type, it cannot provide zero error and unit static transfer gain. It can be assured only if the condition

$$k_{\mathrm{r}} = \frac{-1}{\boldsymbol{c}^{\mathrm{T}}\left(\boldsymbol{A} - \boldsymbol{b}\boldsymbol{k}^{\mathrm{T}}\right)^{-1}\boldsymbol{b}} = \frac{\boldsymbol{k}^{\mathrm{T}}\boldsymbol{A}^{-1}\boldsymbol{b} - 1}{\boldsymbol{c}^{\mathrm{T}}\boldsymbol{A}^{-1}\boldsymbol{b}} \tag{9.6}$$

is fulfilled [see A.9.2 in Appendix A.5.]. The special control loop shown above is called *state-feedback.*

9.1 Pole Placement by State Feedback

The most natural design method for state feedback is the so-called pole placement. In this case the feedback vector $\boldsymbol{k}^{\mathrm{T}}$ has to be chosen to make the characteristic equation of the closed-loop equal to the prescribed (or design) polynomial $\mathcal{R}(s)$, i.e.,

$$\begin{aligned}\mathcal{R}(s) &= s^n + r_1 s^{n-1} + \cdots + r_{n-1}s + r_n = \prod_{i=1}^{n}(s - s_i) = \det\left(s\boldsymbol{I} - \boldsymbol{A} + \boldsymbol{b}\boldsymbol{k}^{\mathrm{T}}\right)\\ &= \mathcal{A}(s) + \boldsymbol{k}^{\mathrm{T}}\boldsymbol{\Psi}(s)\boldsymbol{b}\end{aligned} \tag{9.7}$$

A solution always exists if the process is controllable. (It is reasonable if the order of $\mathcal{R}$ is equal to that of $\mathcal{A}$.) In the exceptional case when the transfer function of the controlled system is known, then the canonical state-equations can be written directly. Based on the controllable canonical form (3.47) the system matrices are

$$\boldsymbol{A}_{\mathrm{c}} = \begin{bmatrix} -a_1 & -a_2 & \ldots & -a_{n-1} & -a_n \\ 1 & 0 & \ldots & 0 & 0 \\ 0 & 1 & & 0 & 0 \\ \vdots & \vdots & \ddots & \vdots & \vdots \\ 0 & 0 & 0 & 1 & 0 \end{bmatrix}; \quad \boldsymbol{c}_{\mathrm{c}}^{\mathrm{T}} = [b_1, b_2, \ldots, b_n]; \quad \text{and} \tag{9.8}$$
$$\boldsymbol{b}_{\mathrm{c}} = [1,0, \ldots 0]^{\mathrm{T}}$$

Considering the special forms of $\boldsymbol{A}_{\mathrm{c}}$ and $\boldsymbol{b}_{\mathrm{c}}$, it can be seen that according to the design equation

$$
\boldsymbol{A}_{\mathrm{c}}-\boldsymbol{b}_{\mathrm{c}}\boldsymbol{k}_{\mathrm{c}}^{\mathrm{T}}=\begin{bmatrix}-a_1 & -a_2 & \ldots & -a_{n-1} & -a_n\\ 1 & 0 & \ldots & 0 & 0\\ 0 & 1 & \ldots & 0 & 0\\ \vdots & \vdots & \ddots & \vdots & \vdots\\ 0 & 0 & 0 & 1 & 0\end{bmatrix}-\begin{bmatrix}1\\0\\0\\ \vdots\\0\end{bmatrix}\boldsymbol{k}_{\mathrm{c}}^{\mathrm{T}}
$$

$$
=\begin{bmatrix}-r_1 & -r_2 & \ldots & -r_{n-1} & -r_n\\ 1 & 0 & \ldots & 0 & 0\\ 0 & 1 & \ldots & 0 & 0\\ \vdots & \vdots & \ddots & \vdots & \vdots\\ 0 & 0 & 0 & 1 & 0\end{bmatrix} \tag{9.9}
$$

the choice

$$
\boldsymbol{k}^{\mathrm{T}}=\boldsymbol{k}_{\mathrm{c}}^{\mathrm{T}}=[r_1-a_1, r_2-a_2, \ldots, r_n-a_n] \tag{9.10}
$$

ensures the satisfaction of the characteristic equation (9.7), i.e., the prescribed poles. The choice of the calibration factor can be determined by simple calculations

$$
k_{\mathrm{r}}=\frac{a_n+(r_n-a_n)}{b_n}=\frac{r_n}{b_n} \tag{9.11}
$$

Based on Eqs. (9.4) and (9.6) it can be seen that in the case of state feedback pole placement, the transfer function turns out to be

$$
T_{\mathrm{ry}}(s)=\frac{k_{\mathrm{r}}\mathcal{B}(s)}{\mathcal{R}(s)} \tag{9.12}
$$

as was shown at (9.5).

Example 9.1 Consider an unstable process with transfer function

$$
P(s)=\frac{-8}{(s+2)(s-4)}=\frac{1}{(1+0.5s)(1-0.25s)}=\frac{-8}{s^2-2s-8}=\frac{-8}{\mathcal{A}(s)}
$$

where $\mathcal{A}(s)=(s+2)(s-4)=s^2-2s-8=s^2+a_1s+a_2$. To stabilize the process we should mirror the right half-plane unstable pole $p_2^{\mathrm{c}}=4$ into the left plane, i.e. $p_2^{\mathrm{c}}=-4$ is to be obtained. This can be arranged by the choice of the polynomial $\mathcal{R}(s)=(s+2)(s+4)=s^2+6s+8=s^2+r_1s+r_2$. So the necessary stabilizing feedback vector is

$$
\mathbf{k}^{\mathrm{T}}=[r_1-a_1 \quad r_2-a_2]=[6-(-2) \quad 8-(-8)]=[8 \quad 16]
$$

■

The most frequent case of state feedback is when rather than the transfer function, the state-space form of the control system is given. In relation with

Eq. (3.67) it has already been discussed that all controllable systems can be described in controllable canonical form using the transformation matrix $\boldsymbol{T}_{\mathrm{c}} = \boldsymbol{M}_{\mathrm{c}}^{\mathrm{c}}(\boldsymbol{M}_{\mathrm{c}})^{-1}$. This linear transformation also refers to the feedback vector

$$\begin{aligned} \boldsymbol{k}^{\mathrm{T}} &= \boldsymbol{k}_{\mathrm{c}}^{\mathrm{T}}\boldsymbol{T}_{\mathrm{c}} = \boldsymbol{k}_{\mathrm{c}}^{\mathrm{T}}\boldsymbol{M}_{\mathrm{c}}^{\mathrm{c}}\boldsymbol{M}_{\mathrm{c}}^{-1} \\ \boldsymbol{k}^{\mathrm{T}} &= \boldsymbol{b}_{\mathrm{c}}^{\mathrm{T}}\boldsymbol{M}_{\mathrm{c}}^{-1}\mathcal{R}(\boldsymbol{A}) = [0, 0, \ldots, 1]\boldsymbol{M}_{\mathrm{c}}^{-1}\mathcal{R}(\boldsymbol{A}) \end{aligned} \tag{9.13}$$

The design relating to the controllable canonical form (9.10), together with the linear transformation relationship corresponding to the first row of the non-controllable form (9.13) is called BASS-GURA algorithm. The algorithm in the second row of (9.13) is called ACKERMANN method after its developer (see the details in the A.9.3 of Appendix A.5).

In the BASS-GURA algorithm, the inverse of the controllability matrix $\boldsymbol{M}_{\mathrm{c}}$ has to be determined by the general system matrices $\boldsymbol{A}$ and $\boldsymbol{b}$, on the one hand, and the controllability matrix $\boldsymbol{M}_{\mathrm{c}}^{\mathrm{c}}$ of the controllable canonical form [see (3.61)], on the other. Since this latter term depends only on the coefficients a_i in the denominator of the process transfer function, then the denominator has to be calculated: $\mathcal{A}(s) = \det(s\boldsymbol{I} - \boldsymbol{A})$. Since $[0, 0, \ldots, 1]\boldsymbol{M}_{\mathrm{c}}^{-1}$ is the last row of the inverse of the controllability matrix, and besides this $\mathcal{R}(\boldsymbol{A})$ has to be also calculated, in the ACKERMANN method it is not necessary to calculate $\mathcal{A}(s)$.

It can be easily seen that state feedback formally corresponds to a serial compensation $R_{\mathrm{s}} = \mathcal{A}(s)/\mathcal{R}(s)$ (Fig. 9.3a). The real operation and effect of state feedback can be easily understood by the equivalent block schemes using the transfer functions shown in Fig. 9.3. The "controller" $R_{\mathrm{f}}(s)$ of the closed-loop is in the feedback line (Fig. 9.3b). The transfer function of the closed-loop (9.12) is

$$T_{\mathrm{ry}}(s) = \frac{k_{\mathrm{r}}\mathcal{B}(s)}{\mathcal{R}(s)} = \frac{k_{\mathrm{r}}\mathcal{B}(s)}{\mathcal{A}(s) + \mathcal{B}(s)} = \frac{k_{\mathrm{r}}P(s)}{1 + K_{\mathrm{k}}(s)P(s)} = \frac{k_{\mathrm{r}}\mathcal{A}(s)}{\mathcal{R}(s)}\frac{\mathcal{B}(s)}{\mathcal{A}(s)} = k_{\mathrm{r}}R_{\mathrm{s}}(s)P(s) \tag{9.14}$$

where

$$R_{\mathrm{f}} = K_{\mathrm{k}}(s) = \frac{\mathcal{K}(s)}{\mathcal{B}(s)} = \frac{\mathcal{R}(s) - \mathcal{A}(s)}{\mathcal{B}(s)} = \frac{\boldsymbol{k}^{\mathrm{T}}(s\boldsymbol{I} - \boldsymbol{A})^{-1}\boldsymbol{b}}{\mathbf{c}^{\mathrm{T}}(s\boldsymbol{I} - \boldsymbol{A})^{-1}\boldsymbol{b}} \tag{9.15}$$

and the calibration factor is

$$k_{\mathrm{r}} = \frac{\boldsymbol{k}^{\mathrm{T}}\boldsymbol{A}^{-1}\boldsymbol{b} - 1}{\boldsymbol{c}^{\mathrm{T}}\boldsymbol{A}^{-1}\boldsymbol{b}} = \frac{1 + K_{\mathrm{k}}(0)P(0)}{P(0)}. \tag{9.16}$$

Based on the block schemes of Fig. 9.3 it can be stated that the state-feedback also stabilizes the unstable terms, since due to the effect of the polynomial $\mathcal{K}(s) = \mathcal{R}(s) - \mathcal{A}(s)$, there is a pole allocation for any process, so by choosing a

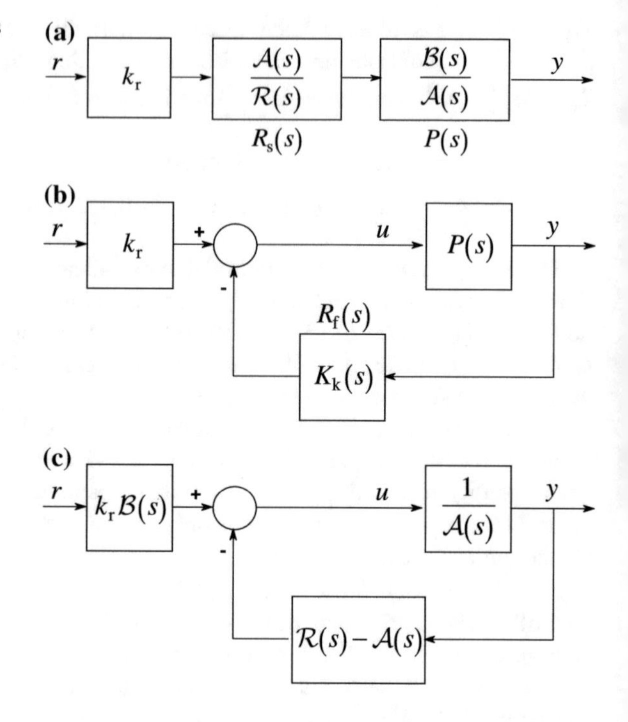

Fig. 9.3 Equivalent schemes of the state feedback design by transfer functions and polynomials

stable $\mathcal{R}(s)$, the stabilization is achieved. The feedback polynomial $\mathcal{K}(s)$ corresponds formally to $\boldsymbol{k}^{\mathrm{T}}$. The fact that the numerator $\mathcal{B}(s)$ of the process is present in the denominator of $K_{\mathrm{k}}(s)$ requires special consideration. It is used to be said in these cases that the controller can be applied only to minimum-phase (inverse stable) processes, where the roots of $\mathcal{B}(s)$ are stable. As a consequence of the special character of the state feedback, however, here $\mathcal{B}(s)$ is not replaced by its model $\widehat{\mathcal{B}}(s)$, but the method itself realizes the exact $1/\mathcal{B}(s)$.

Further methods have been developed for the calculation of the pole placement state feedback vector $\boldsymbol{k}^{\mathrm{T}}$. From among these, the so-called MAYNE-MURDOCH method is briefly shown here, on the basis of which useful statements can be made. In the BASS-GURA and ACKERMANN methods the controllable canonical form has a special role. A similarly important canonical form is the diagonal form. Let the diagonal form $\mathbf{A}_{\mathrm{d}} = \mathbf{diag}[\lambda_1, \ldots, \lambda_n]$ be built with the eigenvalues λ_i, i.e. the roots of $\mathcal{A}(s)$, and let the roots of the design polynomial $\mathcal{R}(s)$ be the prescribed values of $\{\mu_1, \ldots, \mu_n\}$. Assuming that the eigenvalues are single, the MAYNE-MURDOCH method gives the following closed form expression for the product $k_i^{\mathrm{d}} b_i^{\mathrm{d}}$,

$$k_i^{\mathrm{d}} b_i^{\mathrm{d}} = \frac{\Pi_{j=1}^{n} \left(\lambda_i - \mu_j\right)}{\Pi_{\substack{j=1 \\ i \neq j}}^{n} \left(\lambda_i - \lambda_j\right)} \qquad i = 1, \ldots, n \tag{9.17}$$

from which k_i^{d} can be easily determined. Here the coefficient b_i^{d} is an element of the parameter vector $\boldsymbol{b}^{\mathrm{d}} = \left[b_1^{\mathrm{d}}, \ldots, b_n^{\mathrm{d}}\right]^{\mathrm{T}} = [\beta_1, \ldots, \beta_n]^{\mathrm{T}}$ of the diagonal form [see also (3.38)]. The most interesting consequence of (9.17) is that it clearly shows that the absolute value of the feedback gain k_i^{d} required by the pole placement increases directly proportionally to the "moving" distance between the poles of the open- and closed-loop.

9.2 Observer Based State Feedback

The method of state-feedback shown in the previous section requires the direct measurement of the state vector of the state-equation describing the process. Only very rarely can this be fulfilled: generally only in the case of low order dynamics (e.g., in mechanical systems measuring the values of the distance, velocity and acceleration). Thus the usefulness of the method depends on the possible measurement or estimation of the state vector. To determine the state vector the so-called observer principle has been developed. This method requires the knowledge of the system matrices $\boldsymbol{A}$, $\boldsymbol{b}$ and $\boldsymbol{c}^{\mathrm{T}}$, by means of which an exact model of the process is realized and using the same excitation that is applied for the original process, this model (observer) provides estimated values $\hat{\boldsymbol{x}}$ and $\hat{y}$ of the variables $\boldsymbol{x}$ and y. The state-feedback is realized by using $\hat{\boldsymbol{x}}$. The principle is shown in Fig. 9.4.

More strictly the estimated values $\hat{\boldsymbol{A}}, \hat{\boldsymbol{b}}$ and $\hat{\boldsymbol{c}}^{\mathrm{T}}$ in the observer should have been used instead of $\boldsymbol{A}$, $\boldsymbol{b}$ and $\boldsymbol{c}^{\mathrm{T}}$. The speciality of the observer, however, is that it applies not only a parallel model, but it calculates an error $\varepsilon = y - \hat{y}$ from the deviation of the original and estimated output values of the process, and has a feedback via a proportional feedback vector $\boldsymbol{l}$ to the input of the integrator of the observer. This feedback is in operation until the error exists, i.e., until the output of the process and the observer become equal. This operation can tolerate a rather large error in the knowledge of the system matrices.

It can be seen in the figure that now the state-feedback is

$$u = k_{\mathrm{r}} r - \boldsymbol{k}^{\mathrm{T}} \hat{\boldsymbol{x}} \tag{9.18}$$

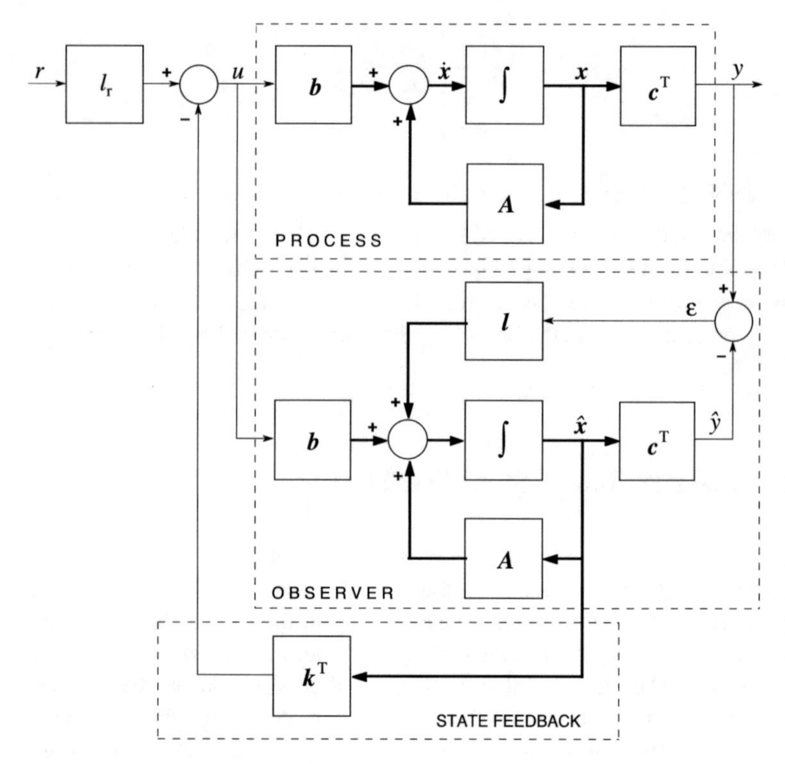

Fig. 9.4 Observer based state-feedback

thus simply $\hat{\boldsymbol{x}}$ is used instead of $\boldsymbol{x}$. Through a long and very complex deduction, whose details will not be discussed here, we get the overall closed-loop transfer function in the form

$$T_{\mathrm{ry}}(s) = \frac{k_{\mathrm{r}}P(s)}{1+\boldsymbol{k}^{\mathrm{T}}(s\boldsymbol{I}-\boldsymbol{A})^{-1}\boldsymbol{b}} = \frac{k_{\mathrm{r}}\mathcal{B}(s)}{\mathcal{R}(s)}, \tag{9.19}$$

which, perhaps surprisingly, is exactly equal to (9.12), i.e., to the case of state-feedback without an observer. (A detailed proof can be seen in A.9.5 of Appendix A.5.) This means that the tracking property of the closed-loop does not depend on the choice of the vector $\boldsymbol{l}$. (The theoretical explanation for this phenomenon is that the observer is the non-controllable part of the whole closed-loop.) The feedback "controller" introduced in Fig. 9.3 can also be determined now as

$$R_{\mathrm{f}} = \boldsymbol{k}^{\mathrm{T}}\left(s\boldsymbol{I}-\boldsymbol{A}+\boldsymbol{b}\boldsymbol{k}^{\mathrm{T}}+\boldsymbol{l}\boldsymbol{c}^{\mathrm{T}}\right)^{-1}\boldsymbol{l} = \frac{\boldsymbol{k}^{\mathrm{T}}\left(s\boldsymbol{I}-\boldsymbol{A}+\boldsymbol{b}\boldsymbol{k}^{\mathrm{T}}\right)^{-1}\boldsymbol{l}}{1+\boldsymbol{c}^{\mathrm{T}}\left(s\boldsymbol{I}-\boldsymbol{A}+\boldsymbol{b}\boldsymbol{k}^{\mathrm{T}}\right)^{-1}\boldsymbol{l}} \tag{9.20}$$

which has a more complex form than in (9.15).

To investigate the operation of the observer, let us define a new state vector error as

$$\tilde{\boldsymbol{x}} = \boldsymbol{x} - \hat{\boldsymbol{x}} \tag{9.21}$$

which can also be written as

$$\frac{\mathrm{d}\tilde{\boldsymbol{x}}}{\mathrm{d}t} = \left(\boldsymbol{A} - \boldsymbol{l}\boldsymbol{c}^{\mathrm{T}}\right)\tilde{\boldsymbol{x}} \tag{9.22}$$

which is very similar to (9.4) without excitation. For the design of observers, a method very similar to what was used in the case of the state-feedback, is applied, where by the choice of $\boldsymbol{l}$ our goal is to ensure the dynamics of (9.21) by the second characteristic polynomial

$$\det\left(s\boldsymbol{I} - \boldsymbol{A} + \boldsymbol{l}\boldsymbol{c}^{\mathrm{T}}\right) = \mathcal{F}(s) = s^n + f_1 s^{n-1} + \cdots + f_{n-1}s + f_n \tag{9.23}$$

A solution always exists if the process is observable. (It is reasonable to assume that the order of $\mathcal{F}$ is equal to that of $\mathcal{A}$.) It is an exceptional case when the transfer function of the process to be controlled is known, by means of which the canonical state-equations can be directly written. Based on the observable canonical form of (3.53), the system matrices are

$$\boldsymbol{A}_{\mathrm{o}} = \begin{bmatrix} -a_1 & 1 & 0 & \dots & 0 \\ -a_2 & 0 & 1 & \dots & 0 \\ \vdots & \vdots & \vdots & \ddots & \vdots \\ -a_{n-1} & 0 & 0 & \dots & 1 \\ -a_n & 0 & 0 & \dots & 0 \end{bmatrix}; \quad \boldsymbol{c}_{\mathrm{o}}^{\mathrm{T}} = [1, 0, \dots, 0]; \quad \boldsymbol{b}_{\mathrm{o}} = [b_1, b_2, \dots, b_n]^{\mathrm{T}} \tag{9.24}$$

Considering the special form of $\boldsymbol{A}_{\mathrm{o}}$ and $\boldsymbol{c}_{\mathrm{o}}^{\mathrm{T}}$ it can be easily seen, that according to the design equation

$$\boldsymbol{A}_{\mathrm{o}} - \boldsymbol{l}_{\mathrm{o}}\boldsymbol{c}_{\mathrm{o}}^{\mathrm{T}} = \begin{bmatrix} -a_1 & 1 & 0 & \dots & 0 \\ -a_2 & 0 & 1 & \dots & 0 \\ \vdots & \vdots & \vdots & \ddots & \vdots \\ -a_{n-1} & 0 & 0 & \dots & 1 \\ -a_n & 0 & 0 & \dots & 0 \end{bmatrix} - \boldsymbol{l}_{\mathrm{o}}[1, 0, \dots, 0] =$$

$$= \begin{bmatrix} -f_1 & 1 & 0 & \dots & 0 \\ -f_2 & 0 & 1 & \dots & 0 \\ \vdots & \vdots & \vdots & \ddots & \vdots \\ -f_{n-1} & 0 & 0 & \dots & 1 \\ -f_n & 0 & 0 & \dots & 0 \end{bmatrix}, \tag{9.25}$$

the choice

$$\boldsymbol{l} = \boldsymbol{l}_\mathrm{o} = [f_1 - a_1, f_2 - a_2, \ldots, f_n - a_n]^\mathrm{T} \tag{9.26}$$

ensures the satisfaction of the characteristic equation of (9.23), i.e. the prescribed poles.

The general case now is that the state-space form of the process to be controlled is given instead of its transfer function. Referring to Eq. (3.79), it has been discussed that all observable systems can be written in observable canonical form by using the transformation matrix $\boldsymbol{T}_\mathrm{o} = \left(\boldsymbol{M}_\mathrm{o}^\mathrm{o}\right)^{-1}\boldsymbol{M}_\mathrm{o}$. This similarity transformation has an effect also on the feedback vector

$$\boldsymbol{l} = (\boldsymbol{T}_\mathrm{o})^{-1}\boldsymbol{l}_\mathrm{o} = \boldsymbol{M}_\mathrm{o}^{-1}\boldsymbol{M}_\mathrm{o}^\mathrm{o}\boldsymbol{l}_\mathrm{o} \tag{9.27}$$

To calculate (9.27) the inverse of the observability matrix $\boldsymbol{M}_\mathrm{o}$ is required using the system matrices $\boldsymbol{A}$ and $\boldsymbol{c}^\mathrm{T}$. Similarly the observability matrix $\boldsymbol{M}_\mathrm{o}^\mathrm{o}$ of the observable canonical form has to be formed [see (3.73)]. Since this latter one depends only on the coefficients a_i in the denominator of the transfer function of the process, so the denominator has to be calculated: $\mathcal{A}(s) = \det(s\boldsymbol{I} - \boldsymbol{A})$. This method of calculating the observer vector is called the Ackermann method, after its developer.

There is an interesting similarity in the design methods of the dynamics of the observer and the state-feedback, often called duality, i.e., they correspond to each other under the conditions: $\boldsymbol{A} \leftrightarrow \boldsymbol{A}^\mathrm{T}$, $\boldsymbol{b} \leftrightarrow \boldsymbol{c}^\mathrm{T}$, $\boldsymbol{k} \leftrightarrow \boldsymbol{l}^\mathrm{T}$, $\boldsymbol{M}_\mathrm{c}^\mathrm{c} \leftrightarrow \left(\boldsymbol{M}_\mathrm{o}^\mathrm{o}\right)^\mathrm{T}$.

Based on the equations of the error (9.21) and the process (9.1), the joint equations of the state-feedback and the observer are

$$\begin{aligned} \frac{\mathrm{d}}{\mathrm{d}t}\begin{bmatrix} \boldsymbol{x} \\ \tilde{\boldsymbol{x}} \end{bmatrix} &= \begin{bmatrix} \boldsymbol{A} - \boldsymbol{b}\boldsymbol{k}^\mathrm{T} & \boldsymbol{b}\boldsymbol{k}^\mathrm{T} \\ \boldsymbol{0} & \boldsymbol{A} - \boldsymbol{l}\boldsymbol{c}^\mathrm{T} \end{bmatrix}\begin{bmatrix} \boldsymbol{x} \\ \tilde{\boldsymbol{x}} \end{bmatrix} + \begin{bmatrix} k_\mathrm{r}\boldsymbol{b} \\ \boldsymbol{0} \end{bmatrix} r \\ e &= y - \hat{y} = \boldsymbol{c}^\mathrm{T}\tilde{\boldsymbol{x}} \end{aligned} \tag{9.28}$$

Since the system matrix of the right hand side is block diagonal, the characteristic equation of the closed-loop is

$$\det\left(s\boldsymbol{I} - \boldsymbol{A} + \boldsymbol{b}\boldsymbol{k}^\mathrm{T}\right)\det\left(s\boldsymbol{I} - \boldsymbol{A} + \boldsymbol{l}\boldsymbol{c}^\mathrm{T}\right) = \mathcal{R}(s)\mathcal{F}(s) \tag{9.29}$$

Thus the polynomial is the product of two terms: the first term relates to the state-feedback, the other one to the observer. It is important to note, that $\mathcal{F}(s)$, in spite of (9.29), does not appear in the transfer function $T_\mathrm{ry}(s)$ of the closed-loop of (9.5). This interesting fact can be explained by the re-definition of the whole system given in the block diagram of Fig. 9.4, applying appropriate transfer functions.

Equation (9.29) of the observer based state-feedback, according to which the state-feedback and the characteristic equation of the observer are independent, is called the *separation principle*.

9.3 Observer Based State Feedback Using Equivalent Transfer Functions

The block scheme containing transfer functions has already been applied in the Fig. 9.3. A further generalized form of the approach used there can also be applied, which is shown in Fig. 9.5.

It follows from Fig. 9.5 that the resulting equivalent serial compensator is now again

$$R_s = \frac{1}{1+R_f P} = \frac{1}{1+K_k P} = \frac{\mathcal{A}(s)}{\mathcal{A}(s)+\mathcal{K}(s)} = \frac{\mathcal{A}(s)}{\mathcal{R}(s)} \tag{9.30}$$

It must be stated that R_s is a fictitious term: it is used only for demonstrating the final signal formation, i.e., $k_r R_s P$ ensures the same T_{ry} as (9.14). If the pole cancellation represented by R_s is intended to be performed by a serial compensator, then it cannot be applied to unstable processes, since the unstable zeros and poles cannot be eliminated by cancellation. The signal $\bar{x}$ (which is not the same as $\boldsymbol{x}$) introduced in Fig. 9.4 represents that finally both the state-feedback and the observer are *SISO* subsystems which can be performed by transfer functions, i.e., it is always possible to find equivalent representations for the input and output. Applying this approach and based on Fig. 9.4, the block scheme using transfer functions can be drawn as shown in Fig. 9.6.

After a long transformation procedure and block manipulations the block scheme of Fig. 9.6 can be traced back to the very simple, unit feedback closed-loop shown

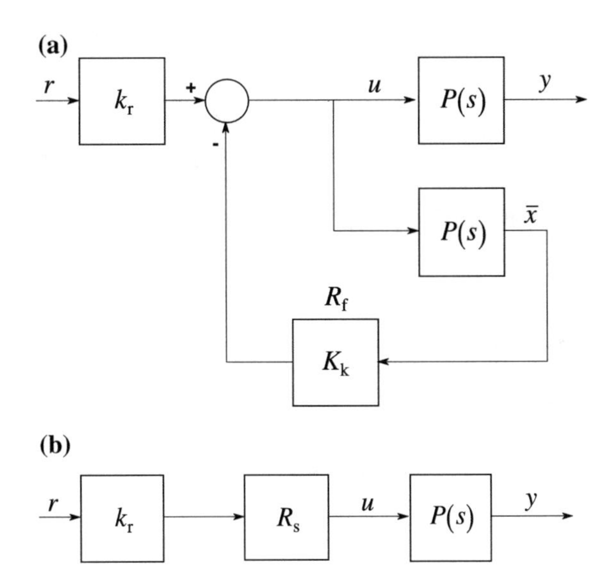

Fig. 9.5 The further equivalent schemes of the state feedback with transfer functions

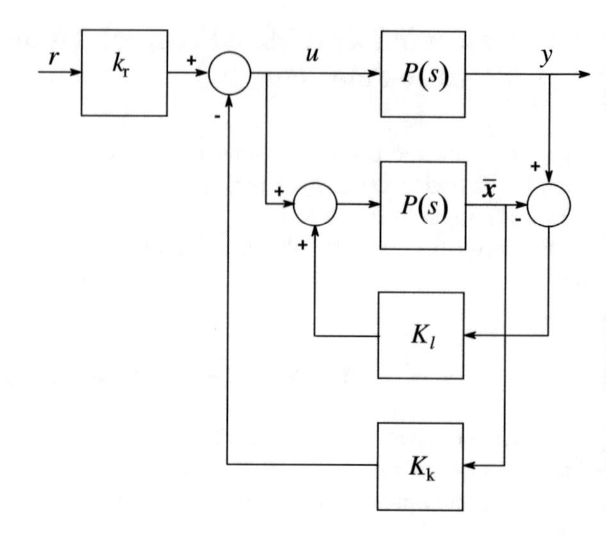

Fig. 9.6 State-feedback and observer using transfer functions

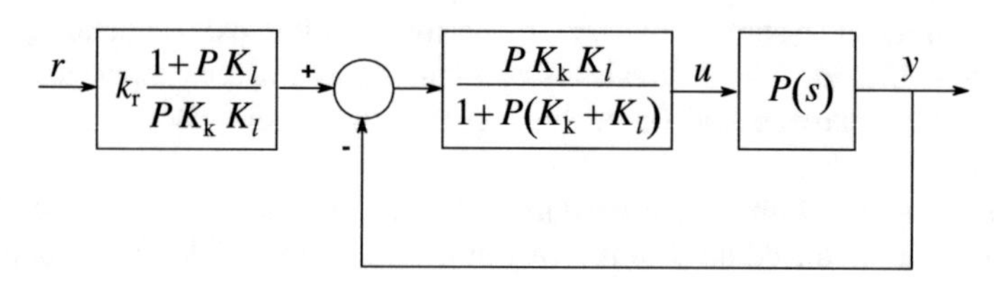

Fig. 9.7 The reduced block scheme of the state-feedback and observer

in Fig. 9.7. Here the relationship (9.15) defining K_k is also used, and K_l is introduced in a similar way

$$K_k(s) = \frac{\mathcal{K}(s)}{\mathcal{B}(s)}; \quad K_l(s) = \frac{\mathcal{L}(s)}{\mathcal{B}(s)}, \tag{9.31}$$

where the design polynomial equations

$$\mathcal{K}(s) = \mathcal{R}(s) - \mathcal{A}(s) \quad \text{and} \quad \mathcal{L}(s) = \mathcal{F}(s) - \mathcal{A}(s) \tag{9.32}$$

result from the conditions of the two kinds of pole placements.

It is easily seen that the resulting transfer function of the inner closed-loop

$$\frac{P^2 K_k K_l}{1 + P(K_k + K_l) + P^2 K_k K_l} = \frac{PK_k}{1 + PK_k} \frac{PK_l}{1 + PK_l} = \frac{\mathcal{K}}{\mathcal{A} + \mathcal{K}} \frac{\mathcal{L}}{\mathcal{A} + \mathcal{L}} = \frac{\mathcal{K}}{\mathcal{R}} \frac{\mathcal{L}}{\mathcal{F}} \tag{9.33}$$

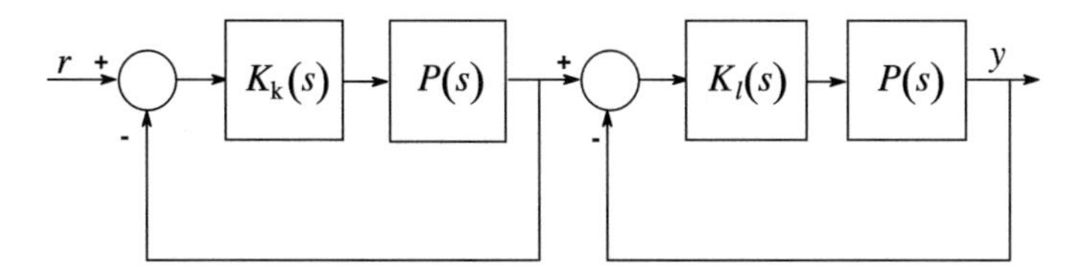

Fig. 9.8 Equivalent observer block schemes of the inner system

has a special form, but its denominator completely corresponds to the characteristic equation (9.29), i.e., represents two serially connected independent closed-loops (see Fig. 9.8). This fact is called the separation principle of the state-feedback and the observer. To ensure stability, both loops must be stable. This can be arranged by proper pole placement design.

At the same time, the transfer function of the whole system is

$$T_{\mathrm{ry}}(s) = k_{\mathrm{r}} \frac{1+PK_{\mathrm{k}}}{PK_l K_{\mathrm{k}}} \frac{PK_l}{1+PK_l} \frac{PK_{\mathrm{k}}}{1+PK_{\mathrm{k}}} = \frac{k_{\mathrm{r}} P}{1+PK_l} = \frac{k_{\mathrm{r}} \frac{\mathcal{B}}{\mathcal{A}}}{1+\frac{\mathcal{B}}{\mathcal{A}}\frac{\mathcal{K}}{\mathcal{B}}} = \frac{k_{\mathrm{r}} \mathcal{B}}{\mathcal{A}+\mathcal{K}} = \frac{k_{\mathrm{r}} \mathcal{B}(s)}{\mathcal{R}(s)}, \tag{9.34}$$

which is completely the same as (9.19). As expected, the poles of the observer do not appear in T_{ry}. The inner character of the whole system can be better seen from the final block scheme shown in Fig. 9.9 for the tracking properties.

This simple structure is not valid for the disturbance rejection capabilities of the closed-loop. This can be simply seen if the sensitivity function of the closed-loop is constructed,

$$\frac{1}{1+\dfrac{P^2 K_{\mathrm{k}} K_l}{1+P(K_{\mathrm{k}}+K_l)}} = \frac{1+P(K_{\mathrm{k}}+K_l)}{1+P(K_{\mathrm{k}}+K_l)+P^2 K_{\mathrm{k}} K_l} = \left(1+\frac{\mathcal{L}}{\mathcal{R}}\right)\left(1-\frac{\mathcal{L}}{\mathcal{F}}\right), \tag{9.35}$$

which shows that both $\mathcal{R}$ and $\mathcal{F}$ appear in the transfer function of the disturbance rejection according to (9.29). Equation (9.35) has a special form, since formally it is the product of the output noise rejection transfer functions of two serially connected closed-loops, while it is known, that the tracking properties are indeed the result of a product of the transfer functions, but this phenomenon is not valid for the

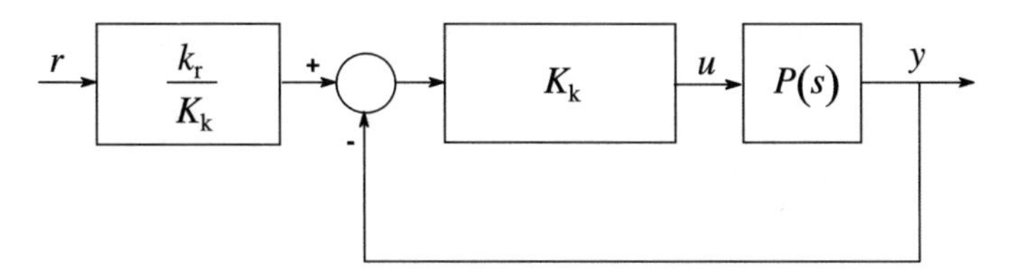

Fig. 9.9 The reduced block scheme of the state-feedback and the observer for the tracking properties

sensitivity functions. Note that the resulting noise rejection properties are not independent of the tracking ones, therefore the joint application of the state-feedback and the observer is not appropriate to realize an actual *TDOF* control loop.

9.4 Two-Step Design Methods Using State Feedback

It has been already seen in the discussion of the state-feedback based control that the most advantageous features of that method are:

- the applicability of the method does not depend on whether the process is stable or not
- the tracking property does not depend on the applied observer, thus it can be directly designed
- the method is not very sensitive to the exact knowledge of the parameter matrices of the state-equation.

(This last feature is usually demonstrated by experimental and simulation examples, but it can be proved that the error, using an observer, can be reduced by the $[1+K_l(s)P(s)]$ part of the original one, compared to the modeling error obtained by the simple parallel model of the state-equation of the process, thus being like that which would be obtained via a closed-loop $1/[1+K_l(s)P(s)]$. So it can be reduced by the feedback $K_l(s)$ of the observer in a specific frequency region. If the model of the process is applied, which is quite conventional practice, then both loops of the Fig. 9.8 must be robust stable.)

The unfavorable (unwanted) features are:

- the state feedback is basically a zero-type control, therefore the remaining error can be eliminated by the calibration factor, which, in the case of using a process model, never provides a precise result
- the state feedback can not change the zeros of the process
- the disturbance rejection property can not be designed directly.

Mostly because of these latter features, usually further steps are applied to the state-feedback based control systems. The necessity of the calibration factor can be eliminated in the simplest way by using a cascade integrating controller, as shown in Fig. 9.10.

Instead of (9.4), the joint state-equation of the closed-loop can be written as

$$\dot{\boldsymbol{x}}^*(t) = \begin{bmatrix} \dot{\boldsymbol{x}}(t) \\ \dot{\delta}(t) \end{bmatrix} = \begin{bmatrix} \boldsymbol{A} & \boldsymbol{0} \\ \boldsymbol{c}^{\mathrm{T}} & 0 \end{bmatrix} \begin{bmatrix} \boldsymbol{x}(t) \\ \delta(t) \end{bmatrix} + \begin{bmatrix} \boldsymbol{b} \\ 0 \end{bmatrix} u(t) + \begin{bmatrix} \boldsymbol{0} \\ -1 \end{bmatrix} r(t) \\ = \left(\boldsymbol{A}^* - \boldsymbol{b}^* \boldsymbol{k}_*^{\mathrm{T}}\right) \boldsymbol{x}^*(t) + \boldsymbol{v}^* r(t) \tag{9.36}$$

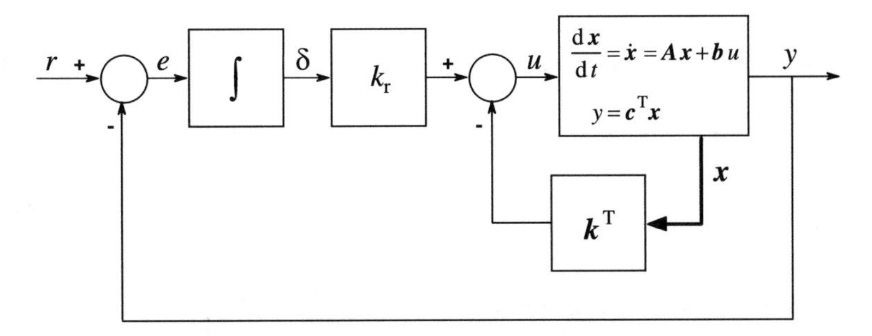

Fig. 9.10 Joint state-feedback and integrating controller

by introducing the new state variable $\delta(t)$, which is the integral of the error $e(t) = r(t) - y(t)$ in the outer loop. In this extended state-equation, the notation

$$\boldsymbol{A}^* = \begin{bmatrix} \boldsymbol{A} & \boldsymbol{0} \\ \boldsymbol{c}^{\mathrm{T}} & 0 \end{bmatrix}; \quad \boldsymbol{b}^* = \begin{bmatrix} \boldsymbol{b} \\ 0 \end{bmatrix}; \quad \boldsymbol{v}^* = \begin{bmatrix} \boldsymbol{0} \\ -1 \end{bmatrix} \tag{9.37}$$

and the new extended feedback equation

$$u(t) = -\begin{bmatrix} \boldsymbol{k}^{\mathrm{T}} & k_{\mathrm{r}} \end{bmatrix} \begin{bmatrix} \boldsymbol{x}(t) \\ \delta(t) \end{bmatrix} = -\boldsymbol{k}_*^{\mathrm{T}} \boldsymbol{x}^*(t) = k_{\mathrm{r}} \int_0^t e(\tau)\,\mathrm{d}\tau - \boldsymbol{k}^{\mathrm{T}} \boldsymbol{x}(t) \tag{9.38}$$

are employed. Equation (9.38) clearly shows the integrating effect. The term $\boldsymbol{k}^{\mathrm{T}}\boldsymbol{x}(t)$, however, can be considered as a generalization of the differentiating effect.

Thus the closed control loop including an integrator can be formulated by a state-equation of order greater by one, where besides the coefficient $\boldsymbol{k}^{\mathrm{T}}$, now k_{r} has to be also determined. To design the extended system, the characteristic polynomial $\mathcal{R}^*(s)$ of order $(n+1)$ has to be required, and then the design Eq. (9.10) of the ACKERMANN method can be directly applied here too. If the process is not presented in the transfer function form, then first the general state-equation has to be transformed into the controllable canonical form, as was already shown in (9.13).

Note that the extended task can not be solved sequentially, i.e., in such a way that first the $\boldsymbol{k}^{\mathrm{T}}$ relating to $\mathcal{R}(s)$ is determined, then k_{r} based on $\mathcal{R}^*(s) = \mathcal{R}(s)(s - s_{n+1})$ is calculated. The task must be solved in one step for $\boldsymbol{k}_*^{\mathrm{T}}$ by $\mathcal{R}^*(s)$.

An integrating effect can also be included by the design of the state-feedback for a modified process $P^*(s) = P(s)/s$ instead of the transfer function $P(s)$. Note that the two state feedback vectors, obtained for the previous case and for this approach, are not equal!

Obviously beside the *I*-controller, other—higher order—controllers can be also applied, but the pole placement is not always automatically given by the ACKERMANN method, and can result in complicated systems of non-linear equations.

In the case of observer based state-feedback, at the feedback of the observer error, not only zero-type, but one-type or higher-type controllers can also be applied by the methods shown above.

The untouched zeros of the process can be modified by a serial compensator

$$K_s(s) = G_s(s)\frac{\mathcal{N}(s)}{\mathcal{B}_+(s)} \tag{9.39}$$

too, where the numerator of the process $\mathcal{B}(s) = \mathcal{B}_+(s)\mathcal{B}_-(s)$ is assumed according to the method applied in the Chap. 7. Here $\mathcal{B}_+$ is stable, $\mathcal{B}_-$, however, contains the unstable zeros. For realizability, $\mathcal{N}(s)/\mathcal{B}_+(s)$ must be proper, thus only as many zeros can be placed in the transfer function of the closed-loop as many stable zeros are in the process. Finally the resulting transfer function has the form

$$T_{ry}(s) = \frac{\mathcal{N}(s)}{\mathcal{R}(s)} k_r G_s(s)\mathcal{B}_-(s) \tag{9.40}$$

where the effect of the invariant $\mathcal{B}_-(s)$ can be optimally attenuated by the filter $G_s(s)$. In many cases, however, the simple, but not optimal, choice $G_s(s) = 1$ is used.

An acceptable design of the disturbance rejection feature can be reached by the application of the YOULA-parameterized controller in the outer cascade loop. It can be done because by the state-feedback any process, even an unstable one, can be stabilized. The qualitative control of the unstable processes has two steps in general. In the first step the process is stabilized by the controller, then the required qualitative goals can be reached by a second outer control loop or even in *TDOF* structures.

The state-feedback based stabilizing controller can only be applied to processes without dead-time. If the process has considerable time-delay, then one possibility is to approach the dead-time by rational fractions [see Sect. 2.5]. The other solution is to use computer based sampled data control [see Chap. 15].

9.5 The *LQ* Controller

The method shown in the previous sections of this chapter could perform arbitrary (stabilizing) pole placement by the so-called state feedback from the state vector of the process. By this state feedback technique further optimization tasks can also be solved. The goal of this task is to optimally control the *LTI* process (9.1) by the minimization of a complex optimality criterion

$$I = \frac{1}{2}\int_{0}^{\infty}\left[\boldsymbol{x}^{\mathrm{T}}(t)\boldsymbol{W}_{\mathrm{x}}\boldsymbol{x}(t) + W_{\mathrm{u}}u^2(t)\right]\mathrm{d}t. \tag{9.41}$$

Here $\boldsymbol{W}_{\mathrm{x}}$ is a real symmetrical positive semi-definite matrix weighting the state vector, and W_{u} is a positive constant weighting the excitation. The solution minimizing the criterion is a state-feedback

$$u(t) = -\boldsymbol{k}_{\mathrm{LQ}}^{\mathrm{T}}\boldsymbol{x}(t) \tag{9.42}$$

[see (9.3)], where the feedback vector $\boldsymbol{k}_{\mathrm{LQ}}^{\mathrm{T}}$ has the form

$$\boldsymbol{k}_{\mathrm{LQ}}^{\mathrm{T}} = \frac{1}{W_{\mathrm{u}}}\boldsymbol{b}^{\mathrm{T}}\boldsymbol{P}. \tag{9.43}$$

Here the symmetrical positive semi-definite matrix $\boldsymbol{P}$ comes from the solution of the algebraic Riccati matrix equation

$$\boldsymbol{P}\boldsymbol{A} + \boldsymbol{A}^{\mathrm{T}}\boldsymbol{P} - \frac{1}{W_{\mathrm{u}}}\boldsymbol{P}\boldsymbol{b}\boldsymbol{b}^{\mathrm{T}}\boldsymbol{P} = -\boldsymbol{W}_{\mathrm{x}}. \tag{9.44}$$

Since this Riccati equation is non-linear in $\boldsymbol{P}$, it has no explicit algebraic solution. The *CAD* systems frequently used in the control technique, however, generally provide several numerical algorithms for the solution of this equation. This controller is called *Linear Quadratic* (*LQ*) controller. This stands for: *linear regulator—quadratic criterion.*

The state-equation of the *LQ* controller based closed-loop is

$$\frac{\mathrm{d}\boldsymbol{x}}{\mathrm{d}t} = \left(\boldsymbol{A} - \boldsymbol{b}\boldsymbol{k}_{\mathrm{LQ}}^{\mathrm{T}}\right)\boldsymbol{x}; \quad \overline{\boldsymbol{A}} = \boldsymbol{A} - \boldsymbol{b}\boldsymbol{k}_{\mathrm{LQ}}^{\mathrm{T}}. \tag{9.45}$$

The details of the *LQ* based method are given in A.9.6 of Appendix A.5. (The above controller is very simple, but its derivation is quite time consuming.)

If the transfer function of the process is known, then the controllable canonical form can be easily given. For special $\boldsymbol{A}_{\mathrm{c}}$ and $\boldsymbol{b}_{\mathrm{c}}$, Eq. (9.10) gives the classical state feedback design algorithm. In the *LQ* method the feedback vector $\boldsymbol{k}_{\mathrm{LQ}}^{\mathrm{T}}$ is obtained from the design (from the solution of the Riccati equation). So turning back the derivation of (9.10) the characteristic polynomial $\mathcal{R}(s)$ of the resulting closed-loop system can be given by its coefficients as

$$[r_1, r_2, \ldots, r_n]^{\mathrm{T}} = \boldsymbol{k}_{\mathrm{LQ}}^{\mathrm{T}} + [a_1, a_2, \ldots, a_n]^{\mathrm{T}}. \tag{9.46}$$

It is also possible to employ an observer for constructing the state vector in *LQ* control.

In engineering practice it is simpler to solve the stabilizing task by pole allocation state-feedback, since there the prescribed poles are directly known. It is evident, however, that in this case the quality of the transient processes are less known. The *LQ* controller, beside the stabilization, also makes it possible to design even the quality of the transient processes, but it needs long term practice to determine the proper weighting matrix $\boldsymbol{W}_{\mathrm{x}}$ and weighting factor W_{u}, usually through a trial-and-error method.

A simpler version of the *LQ* controller is when, instead of the states, only the square of the output is weighted, similarly to the input, i.e., instead of (9.41) the criterion

$$I = \frac{1}{2}\int_{0}^{\infty} \left[W_{\mathrm{y}} y^2(t) + W_{\mathrm{u}} u^2(t)\right] \mathrm{d}t \tag{9.47}$$

is used. This task (in the case of $d = 0$), after some identical manipulations, can be traced back to the original *LQ* controller

$$W_{\mathrm{y}} y^2 = y W_{\mathrm{y}} y = \boldsymbol{x}^{\mathrm{T}} \boldsymbol{c} W_{\mathrm{y}} \boldsymbol{c}^{\mathrm{T}} \boldsymbol{x} = \boldsymbol{x}^{\mathrm{T}} \left(\boldsymbol{c} W_{\mathrm{y}} \boldsymbol{c}^{\mathrm{T}}\right) \boldsymbol{x} = \boldsymbol{x}^{\mathrm{T}} \left(W_{\mathrm{y}} \boldsymbol{c}\boldsymbol{c}^{\mathrm{T}}\right) \boldsymbol{x} \tag{9.48}$$

by a special choice of the weighting matrix like

$$\boldsymbol{W}_{\mathrm{x}} = W_{\mathrm{y}} \boldsymbol{c}\boldsymbol{c}^{\mathrm{T}}. \tag{9.49}$$

Observe that the state-feedback $\boldsymbol{k}_{\mathrm{LQ}}^{\mathrm{T}}$ leaves the process zeros untouched.

Chapter 10
General Polynomial Method for Controller Design

It was shown in Sect. 7.1 that the YOULA-parameterization can well be used for the design of optimal controllers in the case of stable processes. The only disadvantage of the general method is that it cannot be applied to unstable processes, so different kind of parameterization is required. Let us find the controller $C(s)$ in the form of rational function.

$$C(s) = \frac{\mathcal{Y}(s)}{\mathcal{X}(s)} = \frac{\mathcal{Y}}{\mathcal{X}}. \tag{10.1}$$

Let the prescribed stable characteristic polynomial of the closed-loop be denoted by $\mathcal{R}(s)$, i.e., the characteristic equation is given by $\mathcal{R}(s) = 0$. Similarly to state feedback, here the design of the stability and performance is also carried out via prescribed poles (pole-placement). Let the transfer characteristics of the delay free process be

$$P(s) = \frac{\mathcal{B}(s)}{\mathcal{A}(s)} = \frac{\mathcal{B}}{\mathcal{A}}. \tag{10.2}$$

The characteristic equation expressing the design goal is

$$\mathcal{A}(s)\mathcal{X}(s) + \mathcal{B}(s)\mathcal{Y}(s) = \mathcal{A}\mathcal{X} + \mathcal{B}\mathcal{Y} = \mathcal{R} = \mathcal{R}(s) \tag{10.3}$$

where $\mathcal{A}$, $\mathcal{B}$ and $\mathcal{R}$ are known polynomials, the unknown parameters to be determined are in polynomials $\mathcal{X}$ and $\mathcal{Y}$. Equation (10.3) is called a DIOPHANTINE *equation* (*DE*). Since it is not assumed that the process is stable, the resulting controller is therefore also called a *stabilizing controller*.

This *DE* has solution if, and only if, all common factors of $\mathcal{A}$ and $\mathcal{B}$ are also the common factors of $\mathcal{R}$. If $\mathcal{A}$ and $\mathcal{B}$ are relative prime (i.e., they have no polynomial common factor), this *DE* always has a solution for any $\mathcal{R}$, and the number of the solutions is infinity. If a pair $\mathcal{X}_o$, $\mathcal{Y}_o$ fulfills the equation, then the pair

L. Keviczky et al., *Control Engineering*, Advanced Textbooks in Control and Signal Processing, https://doi.org/10.1007/978-981-10-8297-9_10

$$\mathcal{X} = \mathcal{X}_o + \mathcal{D}\mathcal{B}; \quad \mathcal{Y} = \mathcal{Y}_o - \mathcal{D}\mathcal{A} \tag{10.4}$$

is also a solution of this *DE*, where $\mathcal{D}$ is an arbitrary polynomial. If the process polynomials are relative prime (if $\mathcal{A} \neq 0$) then there is always a solution $\mathcal{X}_o$, $\mathcal{Y}_o$ of this *DE* such that either $\mathcal{Y}_o = 0$ or $\deg\{\mathcal{Y}_o\} < \deg\{\mathcal{A}\}$. This latter solution $\mathcal{X}_o$, $\mathcal{Y}_o$ is called the minimal one, because there is no other solution $\mathcal{X}$, $\mathcal{Y}$ whose polynomials have degree less than the degree of $\mathcal{Y}_o$.

Since there are an infinite number of solutions of this *DE*, there exists a special one satisfying the assumption

$$\deg\{\mathcal{X}\} < \deg\{\mathcal{B}\}. \tag{10.5}$$

Similarly, there exists a solution for which

$$\deg\{\mathcal{Y}\} < \deg\{\mathcal{A}\}. \tag{10.6}$$

Both assumptions are fulfilled at the same time (simultaneously), if

$$\deg\{\mathcal{A}\} + \deg\{\mathcal{B}\} \geq \deg\{\mathcal{R}\}. \tag{10.7}$$

In the case of (10.6), the DE has a special minimum order solution if

$$\deg\{\mathcal{X}\} = \deg\{\mathcal{Y}\}. \tag{10.8}$$

If (10.6) is not valid, then there exists a solution when $\mathcal{X}$ or $\mathcal{Y}$ is minimal.

In practice, two basic cases can be distinguished:

(a) Let $\mathcal{R}(s)$ be an arbitrary polynomial of order $\deg\{\mathcal{R}\} = 2\deg\{\mathcal{A}\} - 1$. In this case the solution of the *DE* can be sought by controller polynomials of order $\deg\{\mathcal{X}\} = \deg\{\mathcal{A}\} - 1$ and $\deg\{\mathcal{Y}\} = \deg\{\mathcal{A}\} - 1$. Consequently the controller will be proper.

(b) Let $\mathcal{R}(s)$ be an arbitrary polynomial of order $\deg\{\mathcal{R}\} = 2\deg\{\mathcal{A}\}$. In this case the solution of the *DE* can be sought by controller polynomials of order $\deg\{\mathcal{X}\} = \deg\{\mathcal{A}\}$ and $\deg\{\mathcal{Y}\} = \deg\{\mathcal{A}\} - 1$. Consequently the controller will be strictly proper.

Therefore for a process of degree n usually a stabilizing regulator of degree $(n-1)$ is searched, because in this case *DE* always has solution. It can be seen from (10.4), that the order of $\mathcal{Y}$ can be less than the order of $\mathcal{A}$. Theoretically $\mathcal{X}$ could be of lower order than $\mathcal{B}$, but in this case the obtained controller cannot be realized. That is why the stabilizing controller is sought as a transfer function of order $(n-1)$.

It seems to be a reasonable choice if the order of $\mathcal{R}$ is equal to the order of $\mathcal{A}$. In a fortunate case it is possible to find a stabilizing controller of corresponding order and pole excess to a process having a pole excess bigger than one. This procedure,

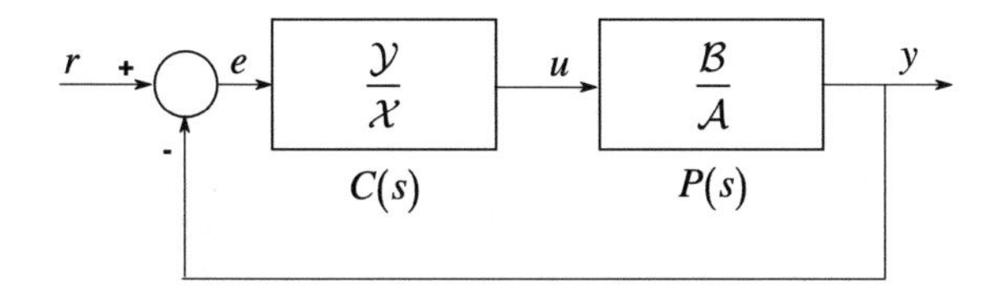

Fig. 10.1 One-degree-of-freedom (*ODOF*) stabilized closed-loop controller

however, cannot be performed in a systematic way and according to (10.7) the solution is not surely minimal.

Equation (10.4) is also valid for the rational function $D = \mathcal{G}/\mathcal{D}$. In this case besides $\mathcal{R}$, $\mathcal{D}$ also appears in the denominator of the overall transfer function of the closed-loop. The form (10.4) parameterizes all stabilizing controllers by D. The parameter D is called the YOULA-KUČERA parameter.

It can be seen easily that the transfer function of the one-degree-of-freedom (*ODOF*) stabilized closed-loop shown in Fig. 10.1 is

$$T = \frac{\mathcal{B}\mathcal{Y}}{\mathcal{A}\mathcal{X}+\mathcal{B}\mathcal{Y}} = \frac{\mathcal{Y}}{\mathcal{R}}\mathcal{B} = R_n'\mathcal{B}. \tag{10.9}$$

Equation (10.9) shows that stabilization is achieved, but the numerator of the process and the polynomial $\mathcal{Y}$ resulting from the solution of the *DE* appears in the numerator of the overall transfer function. Note that none of them can be directly influenced, so the numerator of the transfer function of the closed-loop cannot be designed. (See the similarities with the results obtained for state-feedback.)

In spite of the not completely preferable design possibilities, a *TDOF* control loop can be constructed where the reference signal tracking, at least, can be designed. This system is shown in Fig. 10.2a. An equivalent block scheme is presented in Fig. 10.2b which can be directly compared to the generic *TDOF* (*GTDOF*) control loop obtained by a YOULA parameterized controller for stable processes according to Fig. 7.10a. The controller is obviously different now.

The transfer function of the control loop shown in Fig. 10.2 is

$$T_r = R_r\mathcal{B}. \tag{10.10}$$

Here $\mathcal{Y}$ already does not appear, only the numerator of the process, and R_r is independent of R_n, thus it is really a *TDOF* control. The noise-rejection behavior can be computed from T

$$S = 1 - T = 1 - R_n'\mathcal{B} = 1 - \frac{\mathcal{Y}}{\mathcal{R}}\mathcal{B}. \tag{10.11}$$

It has been already seen in the discussion of the YOULA-parameterized controller, that in the numerator of the transfer function of the process only the stable zeros can

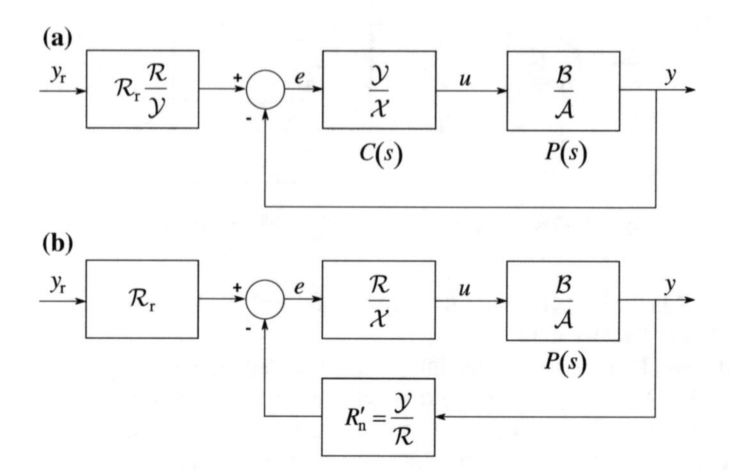

Fig. 10.2 *TDOF* stabilized closed-loop

be cancelled. This method can be extended to the stable poles of the denominator in the design method using the *DE*. Assume that the transfer function of the process is

$$P(s) = P_+(s)P_-(s) \text{ or, for short, } P = P_+P_-, \tag{10.12}$$

where P_+ is stable, its inverse is also stable (*SIS*: *Stable Inverse Stable*). P_- is unstable, and its inverse is also unstable (*UIU*: *Unstable Inverse Unstable*). Thus a practical factorization is

$$P = \frac{\mathcal{B}}{\mathcal{A}} = \frac{\mathcal{B}_+\mathcal{B}_-}{\mathcal{A}_+\mathcal{A}_-} = \left(\frac{\mathcal{B}_+}{\mathcal{A}_+}\right)\left(\frac{\mathcal{B}_-}{\mathcal{A}_-}\right) = P_+P_-. \tag{10.13}$$

Here $\mathcal{A}_+$ contains the stable poles of the process and $\mathcal{A}_-$ does contains the unstable ones. Similarly $\mathcal{B}_+$ contains the stable zeros and $\mathcal{B}_-$ the unstable zeros. The *DE* has to be constructed to make possible the cancellation of the stable roots $\mathcal{B}_+$ and $\mathcal{A}_+$. In order to define the design procedure in a completely general way, predefined polynomials $\mathcal{Y}_d$ and $\mathcal{X}_d$ are introduced in the numerator and denominator of the controller. The following design *DE* can be written for this most general case:

$$\begin{array}{c} (\mathcal{A}_+\mathcal{A}_-)(\mathcal{B}_+\mathcal{X}_d\mathcal{X}) + (\mathcal{B}_+\mathcal{B}_-)(\mathcal{A}_+\mathcal{Y}_d\mathcal{Y}) = \mathcal{R} = \mathcal{A}_+\mathcal{B}_+\mathcal{R} \\ \mathcal{A} \qquad\quad \mathcal{X} \quad + \quad \mathcal{B} \qquad\quad \mathcal{Y} \quad = \mathcal{R}' \end{array} \tag{10.14}$$

A lower order *DE* can be obtained by simplifying with the reducing factors

$$\begin{gathered}(\mathcal{A}_{-}\mathcal{X}_{\mathrm{d}})\mathcal{X}' + (\mathcal{B}_{-}\mathcal{Y}_{\mathrm{d}})\mathcal{Y}' = \mathcal{R} \\ \mathcal{A}'\mathcal{X}' + \mathcal{B}'\mathcal{Y}' = \mathcal{R}\end{gathered} \tag{10.15}$$

where $\mathcal{A}' = \mathcal{A}_{-}\mathcal{X}_{\mathrm{d}}$ and $\mathcal{B}' = \mathcal{B}_{-}\mathcal{Y}_{\mathrm{d}}$ are known, and the controller is obtained as

$$C = \frac{\mathcal{Y}}{\mathcal{X}} = \frac{\mathcal{A}_{+}\mathcal{Y}_{\mathrm{d}}\mathcal{Y}'}{\mathcal{B}_{+}\mathcal{X}_{\mathrm{d}}\mathcal{X}'}. \tag{10.16}$$

It is evident, that the stabilizing controller cancelled only the stable zeros and poles, and introduced the desired polynomials $\mathcal{Y}_{\mathrm{d}}$ and $\mathcal{X}_{\mathrm{d}}$ into the numerator and denominator. The Youla regulator is an integrating one, if a unit gain concerning the reference model is ensured: $R_{\mathrm{n}}(\omega = 0) = R_{\mathrm{n}}(s = 0) = 1$. This cannot be automatically ensured for the stabilizing controller resulting from a *DE*. It can be guaranteed only if $\mathcal{X}_{\mathrm{d}}$ brings a pole $s = 0$ into the denominator.

Since now $\mathcal{Y}_{\mathrm{d}}$ can be considered as the numerator of the reference model, and $\mathcal{R}$, however, as the denominator, it follows that in the general case, the corrected reference model is

$$R_{\mathrm{n}}' = \frac{\mathcal{Y}_{\mathrm{d}}}{\mathcal{R}}, \tag{10.17}$$

which depends only on us, so it can completely be designed.

Equivalent block schemes of the general stabilized control loop are shown in Fig. 10.3.

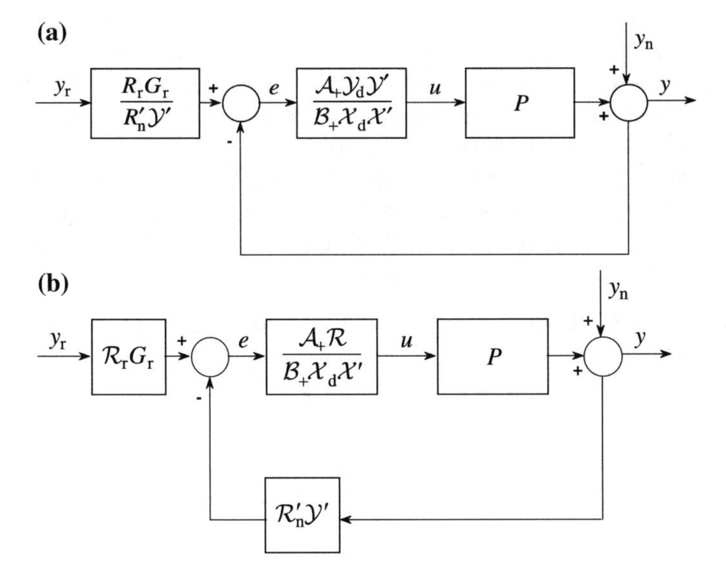

Fig. 10.3 *TDOF* general stabilized closed-loop

It can be easily checked that the transfer function of the whole loop is

$$T_{\mathrm{r}} = P_{\mathrm{r}} G_{\mathrm{r}} \mathcal{B}_{-} \tag{10.18}$$

and the sensitivity function of the closed-loop is

$$S = 1 - P'_{\mathrm{w}} \mathcal{Y}' \mathcal{B}_{-}. \tag{10.19}$$

So the transfer characteristics of the whole closed-loop control is

$$y = T_{\mathrm{r}} y_{\mathrm{r}} + S y_{\mathrm{n}} = R_{\mathrm{r}} G_{\mathrm{r}} \mathcal{B}_{-} y_{\mathrm{r}} + \left(1 - R'_{\mathrm{n}} \mathcal{Y}' \mathcal{B}_{-}\right) y_{\mathrm{n}}. \tag{10.20}$$

It is evident that the filter G_{r} can be freely chosen, and can be optimized to attenuate the effect of $\mathcal{B}_{-}$. Unfortunately, the same is not valid for the optimal design concerning the disturbance rejection, because there, $\mathcal{Y}'$ results from the modified *DE* (10.15), so it cannot be freely chosen, therefore the attenuation of the effect of $\mathcal{Y}'$ cannot be easily solved, as has been seen in the YOULA-parameterization for the tracking problem (10.20).

The form of the resulting stabilizing controller shown in (10.16) can be further simplified:

$$C = \frac{\mathcal{A}_{+} \mathcal{Y}_{\mathrm{d}} \mathcal{Y}'}{\mathcal{B}_{+} \mathcal{X}_{\mathrm{d}} \mathcal{X}'} = \frac{\left(\frac{\mathcal{Y}_{\mathrm{d}}}{\mathcal{R}}\right) \mathcal{Y}' \mathcal{A}}{\mathcal{B}_{+} \left(1 - \frac{\mathcal{Y}_{\mathrm{d}}}{\mathcal{R}} \mathcal{Y}' \mathcal{B}_{-}\right)} = \frac{P'_{\mathrm{w}} \mathcal{Y}'}{1 - P'_{\mathrm{w}} \mathcal{Y}' \mathcal{B}_{-}} \frac{\mathcal{A}}{\mathcal{B}_{+}}, \tag{10.21}$$

which is very similar to the form of the optimal YOULA regulator (7.14). Observe that though only the stable factors $\mathcal{A}_{+}$ and $\mathcal{B}_{+}$ are cancelled, formally the controller cancels the whole denominator of the process.

If the feature obtained for the noise-rejection in (10.20) cannot be accepted, an outer cascade control loop has to be applied, which can already be designed by the YOULA-parameterization, since the system has already been stabilized by the inner loop. This two-step method was discussed in detail in the chapter on the control loops applying state-feedback [see Sect. 9.4].

The stabilizing controller obtained by the *DE* can be applied only to delay free processes. If the process has significant dead-time, then there is the possibility of approximating the delay by a rational function [see Sect. 2.5]. The other possibility is to use sampled data control system [see Chap. 14].

Example 10.1 Let the controlled system be a first order ($n = 1$) unstable process

$$P = \frac{\mathcal{B}}{\mathcal{A}} = \frac{0.5}{1 - 0.5s} = \frac{-1}{s - 2}, \tag{10.22}$$

whose pole $p = 2$ is on the right half of the complex plane. Find the controller $C = \mathcal{Y}/\mathcal{X}$ that stabilizes the process by prescribing the characteristic polynomial

$\mathcal{R}(s) = s + 2 = 0$. The controller is sought in the form of order $n - 1 = 0$, which can be ensured by the structure

$$C = \frac{\mathcal{Y}}{\mathcal{X}} = \frac{K}{1} = K, \tag{10.23}$$

i.e., by a proportional controller. Based on (10.3) one can write

$$\begin{aligned} \mathcal{A}\mathcal{X} + \mathcal{B}\mathcal{Y} &= \mathcal{R} \\ (s - 2) - K &= s + 2 \end{aligned} \tag{10.24}$$

where $C = K = -4$ is obtained for the controller. It can be checked by simple computation, that the transfer function of the closed-loop is

$$T = \frac{4}{s+2} = \frac{2}{1 + 0.5s}, \tag{10.25}$$

thus the unstable pole can be mirrored about the imaginary axis, and by this means, the system is stabilized. The static gain of the closed-loop system is not unity, because the controller is proportional and not an integrating one. To reach better quality in performance, it is reasonable to use a further outer cascade control loop, as was seen with the state feedback controllers. Based on (10.4), the resulting stabilizing controllers $C(s)$ and $T(s)$ are given for different parameters $D(s) = \mathcal{G}/\mathcal{D}$ in Table 10.1. ■

The first row of the Table 10.1 contains the first solution obtained in (10.23) and (10.25). It is well seen, that only the first controller can be realized, so the other solutions have only theoretical importance. For higher order processes the expressions are more complicated, but even for these cases it is reasonable to summarize the different order solutions in tables and choose the lowest order realizable controller. In the same way it is also reasonable to give the solutions being lower order than the $(n - 1)$ order controller.

Example 10.2 Let the controlled system be a first order $(n = 1)$ stable process

$$P = \frac{\mathcal{B}}{\mathcal{A}} = \frac{1}{1 + 10s} = \frac{0.1}{s + 0.1}, \tag{10.26}$$

Table 10.1 .

$D(s) = \mathcal{G}/\mathcal{D}$	$C(s)$	$T(s)$
0	-4	$\frac{4}{s+2}$
1	$\frac{s-6}{2}$	$-\frac{s-6}{s+2}$
$1+s$	$\frac{s^2-s-6}{s+2}$	$-\frac{s^2-s-6}{s+2}$
$\frac{s+2}{s+1}$	$\frac{s^2-4s-8}{2s+3}$	$-\frac{s^2-4s-8}{(s+1)(s+2)}$

which we would like to speed up. Assuming an *ODOF* system, the design goal is expressed by the reference model

$$R_r = R_n = \frac{1}{1+2s} = \frac{0.5}{s+0.5}. \tag{10.27}$$

The YOULA-regulator

$$C_{opt} = C_{id} = \frac{R_n P_+^{-1}}{1-R_n} = \left(1 - \frac{1}{1+2s}\right)^{-1} \frac{1+10s}{1+2s} = \frac{1+10s}{2s} \tag{10.28}$$

is now an integrating one, so the transfer function of the closed-loop is

$$T(s) = \frac{1}{1+2s}. \tag{10.29}$$

For the *DE* design, based on (10.27), the characteristic equation is $\mathcal{R}(s) = s + 0.5 = 0$. As in the previous example, the controller is again sought in a form of $n-1=0$ degree, thus the proportional controller (10.23) is employed. The Eq. (10.3) now becomes

$$\begin{gathered} \mathcal{A}\mathcal{X} + \mathcal{B}\mathcal{Y} = \mathcal{R} \\ (s+0.1) + 0.1K = s + 0.5 \end{gathered} \tag{10.30}$$

where $C = K = 4$ is obtained for the regulator. It can be easily checked that the transfer function of the closed system is

$$T = \frac{0.4}{s+0.5} = \frac{0.8}{1+2s}. \tag{10.31}$$

The prescribed pole -0.5 is successfully placed, but the control loop has zero-type, therefore for the gain of T the value 0.8 is obtained. The above two examples well represent the practice of how the YOULA-parameterization can be reasonably applied to stable processes, while for stabilizing unstable processes the application of *DE* or the state-feedback discussed in Chap. 9 can provide a solution. ■

Chapter 11
Sampled Data Control Systems

In practice, most of the results delivered by control theory are realized by digital computers equipped with appropriate real-time facilities. Depending on how many control loops are implemented for a given application, the digital controller can be realized by various devices ranging from single-chip microcontrollers via single board controllers to PLCs or industrial PCs. The reliable network technology available today allows system developers to implement the controllers in a distributed topology, as well.

The digital realization of control algorithms also reflects the contemporary available computing technology. The control devices applied in industry integrate a number of open-loop and closed-loop control components as a single compact digital unit.

A simple scheme of a sampled data control system is shown in Fig. 11.1. Most of the processes control engineers deal with are continuous in nature. In this chapter it will be assumed that the control signal applied to the process (control input), as well as the process variable (output signal), are both continuous-time (CT) signals. While the input and output signals of a process are assumed to be continuous (called 'analog' in practice), the digital processing assumes the data is available in discrete-time (DT) form as a sequence of numbers. Consequently, the analog world represented by physical signals should be interfaced to the world of data used by digital computations. These interfaces are the *sampler* transforming the analog signals to discrete ones, and the *holder* transforming the discrete signals to analog ones. In practice, these interfaces are typically implemented by *analog to digital* (A/D) and *digital to analog* (D/A) converters, respectively. As far as Fig. 11.2 is concerned, observe that a real-time clock governs the operation of the digital computer to control the sampling and holding in a synchronized way. To distinguish the analog and discrete versions of the signals involved in sampled data control systems, the following notation will be used:

L. Keviczky et al., *Control Engineering*, Advanced Textbooks in Control and Signal Processing, https://doi.org/10.1007/978-981-10-8297-9_11

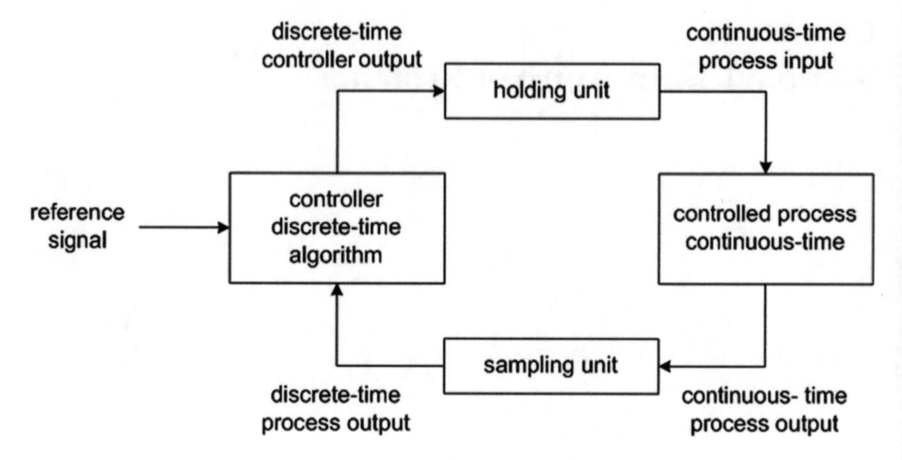

Fig. 11.1 Schematic diagram of closed-loop sampled data control systems

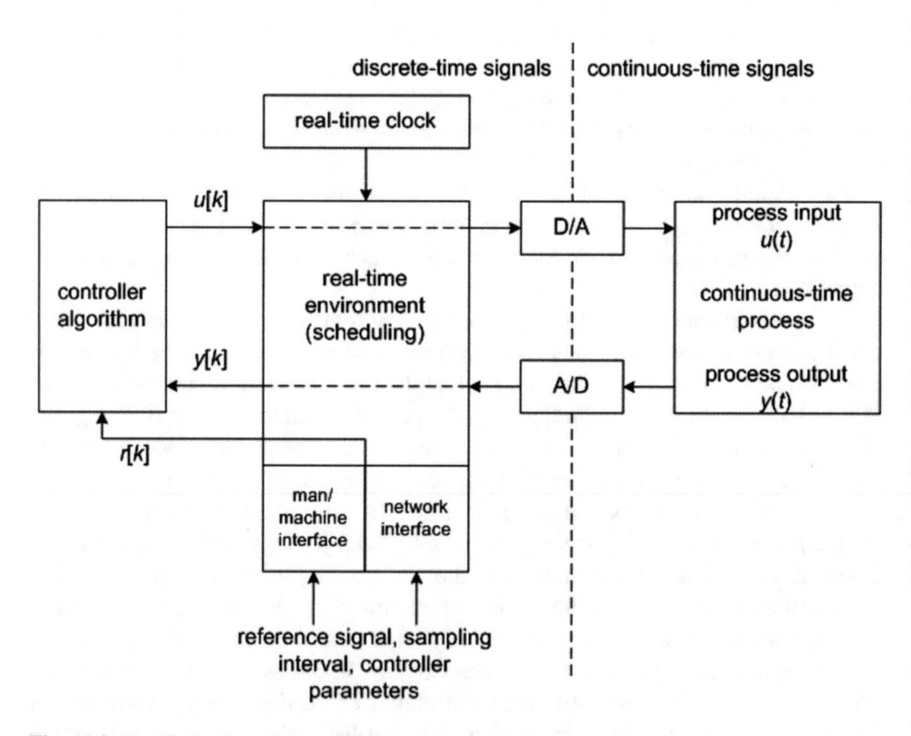

Fig. 11.2 Detailed set-up of sampled data control systems

- $x(t)$: CT signal
- $x[k] = x(kT_s)$: DT signal, where T_s denotes the sampling time and kT_s with $k = 0, 1, 2, \ldots$ assigns the sampling instants.

Comparing the sampled data control system in Fig. 11.2 to a CT closed-loop control system it can be seen that the control signal $u(t)$ is produced by a CT regulator $C(s)$, while the sequence of the control signal $u[k]$ is produced by a digital control algorithm running in a real-time digital environment. At each sampling instant the digital regulator carries out the following actions:

- Receive the sampled process variable and transmit the digitized data to the sampled data control algorithm.
- Receive the set value previously adjusted at a Man-Machine Interface (e.g., typed in) or delivered by a communication network (as a result of a calculation performed at a higher hierarchy level).
- Realize the digital control algorithm to calculate the digital control input $u[k]$ and send this digital value to the D/A converter.

The progress of the digital technologies surrounding the digital controller (intelligent sensors and actuators, advanced Man-Machine Interface devices, a wide range of cheap, though powerful and reliable network technologies) indicate that digital controllers will dominate the field over CT controllers in the future. A comparison of the continuous and digital control technologies can be summarized in favor of digital controllers as follows:

- The digital technology applied is more reliable and cheaper.
- Its flexibility is superior considering both the implementation and the variety of the control algorithms.
- Possible modifications and/or extensions are far easier to accomplish.
- Accuracy is kept constant over a long period of time.
- There are easy ways to deliver the set point value for the controller, to overwrite the controller's parameters, as well as to monitor the controller's operation.

There are, however, a few issues requiring special care:

- Between two samples the control system is left to operate in open-loop.
- The sampling rate should carefully be selected to be in harmony with the dynamics of the process and to comply with the capabilities of the real-time environment (performance and the number representation employed).
- The output of the digital controller (control signal) must be interpolated from a digital sequence to a CT function, thus the waveform of the control signal is restricted.
- Sampling introduces additional difficulties for the design (dead-time and unwanted dynamics).

11.1 Sampling

From the operation of a CT process information can be collected by observing the CT input/output signals of the process. In case this information is elaborated by a digital device, these CT signals will be represented by their samples. Assuming equidistant sampling (constant sampling time), an important result of systems theory helps us to decide whether a CT signal can be reconstructed from its samples or not. Namely, for band limited CT signals, SHANNON's sampling theorem requires that at least two samples must be available from the highest frequency component of the band-limited CT signal for a reconstruction. Concerning the importance of SHANNON's sampling theorem, we refer to the fact that this theorem constitutes the basis for the usability of a flexible, programmable, digital environment in CT signal processing (visualization, analysis in the frequency domain, etc.).

As was mentioned earlier, sampling is physically performed by A/D and D/A converters governed by the real-time clock of the controller. The A/D converter is driven by an analog signal $x(t)$ and produces a sequence $x[k]$ in a coded digital form. The operation of an A/D converter is commonly symbolized by a periodically closed switch (Fig. 11.3).

For various applications an A/D converter is selected according to several of the parameters attached to the converters. The speed of the conversion is characterized by the conversion time, which can be even less than $1\,\mu s$. Further parameters are the noise rejection and the resolution specifying the number of bits used for the digitized code (typically a value from 8 to 16).

Using again the notation introduced earlier, a sampling according to $f[k] = f(kT_s)$ is also called mathematical sampling. The sequence $f[k]$ can be derived by *impulse modulation* according to

$$f^*(t) = \sum_{n=-\infty}^{\infty} f(nT_s)\delta(t - nT_s), \tag{11.1}$$

where $f^*(t)$ is a sequence of DIRAC-impulses, and $f[k]$ is a sequence of numbers made out from the area of these DIRAC-impulses. Figure 11.4 shows the impulse modulation supposing $f(nT_s) \equiv 0$, $(n < 0)$. Applying mathematical sampling, fundamental analysis can be carried out in the frequency domain. The analysis also points out the necessity of appropriate sampling according to the sampling theorem.

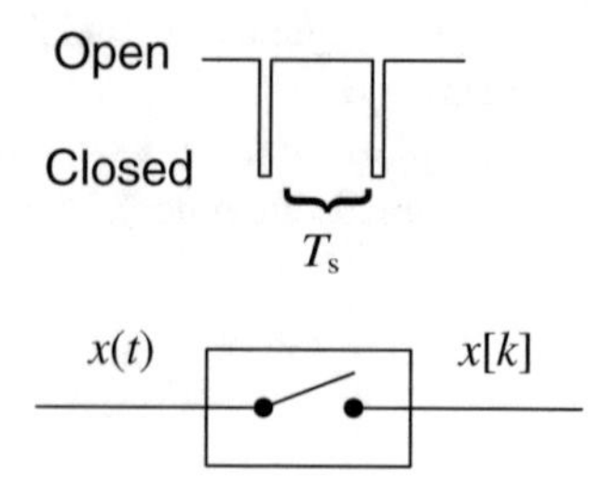

Fig. 11.3 Periodically controlled switch symbolizing the operation of the A/D converter

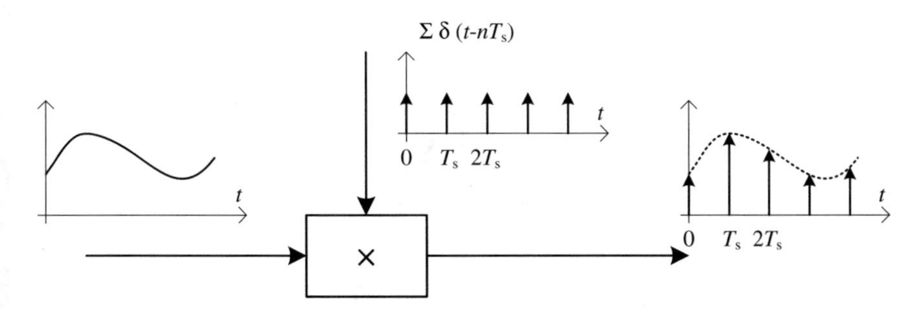

Fig. 11.4 Impulse modulation

Consider a CT signal $f(t)$ with a frequency spectrum of $F(j\omega)$. The theorem of signals and systems specifies the frequency spectrum of $f^*(t)$ to be

$$F^*(j\omega) = \frac{1}{T_s}\sum_{k=-\infty}^{\infty} F(j\omega + jk\omega_s), \tag{11.2}$$

where $\omega_s = 2\pi/T_s$ is the sampling radian frequency.

It is seen that the Fourier transform of the sampled signal is a sum of the side frequency components exhibited by $F(j\omega + jk\omega_s)$. The frequency folding phenomenon is avoided if $\omega_s \geq 2\omega_{max}$ (or $\omega_s/2 \geq \omega_{max}$) holds, where ω_{max} is the maximum frequency of the band limited $f(t)$ CT signal. The frequency $\omega_s/2$ is of great importance for the spectrum of the sampled CT signal, and is also called the Nyquist *frequency*. The phenomenon is presented by the Figs. 11.5, 11.6 and 11.7, where the spectrum of the CT signal, the spectrum of an appropriately sampled DT signal, as well as the spectrum of the inappropriately sampled signal are shown, respectively. Figure 11.8 explains how a non-existing (*alias*) signal may appear as a consequence of a slow sampling rate. In this example the slow sampling rate applied to the high frequency sinusoidal signal produces samples, which can be interpreted as samples of a non-existing, low frequency sinusoidal component. In a closed-loop control system, control actions to compensate this low frequency component are unnecessary and would only induce an additional disturbance in the closed-loop. To avoid aliasing, a low pass filter (also called an anti-aliasing filter) must be placed between the measured output signal and the A/D converter. The fundamental component of the spectrum of the DT signal is

$$F^*(j\omega) = \frac{1}{T_s}F(j\omega). \tag{11.3}$$

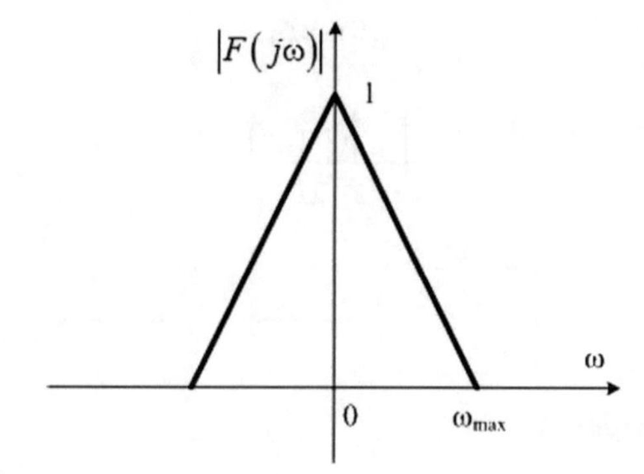

Fig. 11.5 Spectrum of a *CT* signal

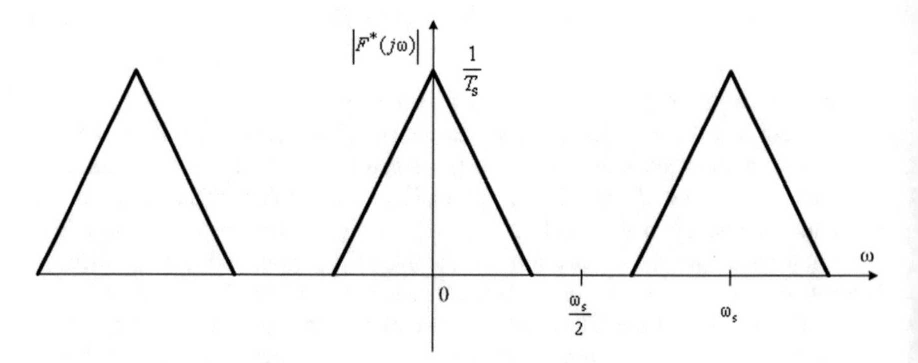

Fig. 11.6 Spectrum of a DT signal assuming an appropriate sampling rate

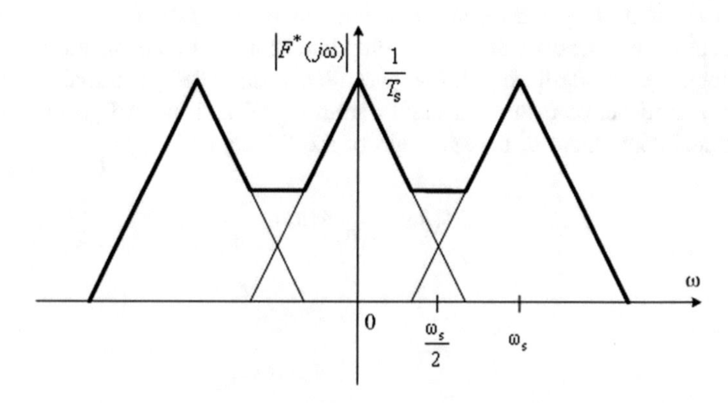

Fig. 11.7 Spectrum of a DT signal assuming an inappropriately low sampling rate

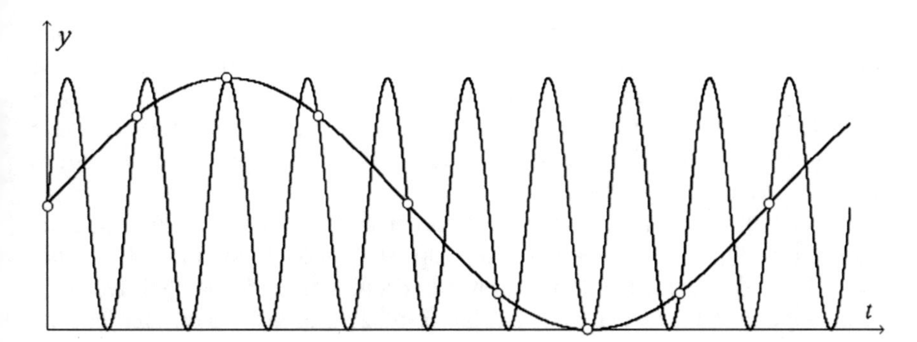

Fig. 11.8 Appearance of the non-existing component according to the low sampling rate

11.2 Holding

Signals in a sampled data control loop form a hybrid system simultaneously presenting CT and DT signals. From a systems engineering point of view a, CT process and the resulting DT signal processing in the controller should be interfaced to each other. Sampling performs the continuous-discrete transformation. However, a functional unit to perform the discrete-continuous transformation should inevitably be inserted into the loop to ensure closed-loop operation. The holding unit is also a controlled element to receive and decode a digital input signal, as well as to produce a CT approximation between two samples. As far as the nature of the approximation is concerned (constant, linear, quadratic, etc.), there is no general requirement. However, the easy realization of the constant approximation (*zero-order holding*) has become a practical standard. Applying a zero-order holding (ZOH) unit results in a staircase waveform for the CT output (see Fig. 11.9). Note that **MATLAB**® offers the application of the *stairs* function to perform a ZOH discrete-continuous transformation.

Fig. 11.9 Application of a ZOH unit

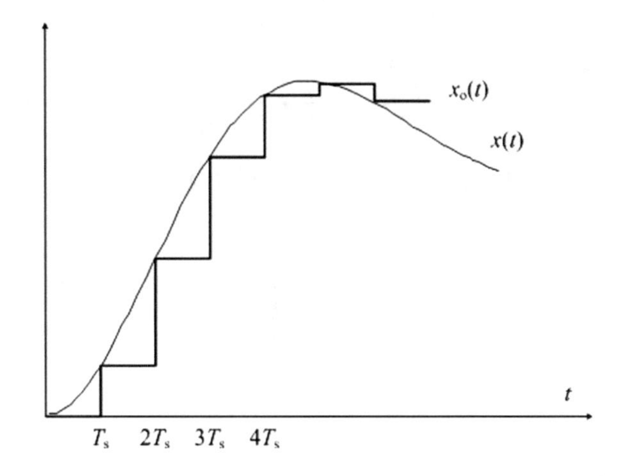

As mentioned earlier, the signal dynamics between two sampling instants is not limited to be a constant, but it could follow a first- or second-order course. In the case of first-order holding, the output of the holding unit is determined by a straight line defined by the last two sampled data (see Fig. 11.10). As in practice zero order holding satisfies all the requirements, holding with higher-order approximations will not be considered in the sequel.

The mathematical description of the ZOH unit should comply with the operation shown by Fig. 11.11. Consider the impulse response of the ZOH unit: $1(t) - 1(t - T_s)$. As the unit step response is produced by an integrator driven by a Dirac impulse and the delay by T_s can be taken into account by a transfer function of e^{-sT_s}, Fig. 11.12 allows seeing that the transfer function of the ZOH unit is

$$W_{\mathrm{ZOH}}(s) = \frac{1}{s} - \frac{e^{-sT_s}}{s} = \frac{1 - e^{-sT_s}}{s} \tag{11.4}$$

Once its transfer function is known, the frequency function of the ZOH unit can easily be determined (see A.11.3 of Appendix A.11.1). Some fundamental frequency domain properties of the ZOH unit, however, can be discovered using the following Taylor series approximation in the low-frequency domain:

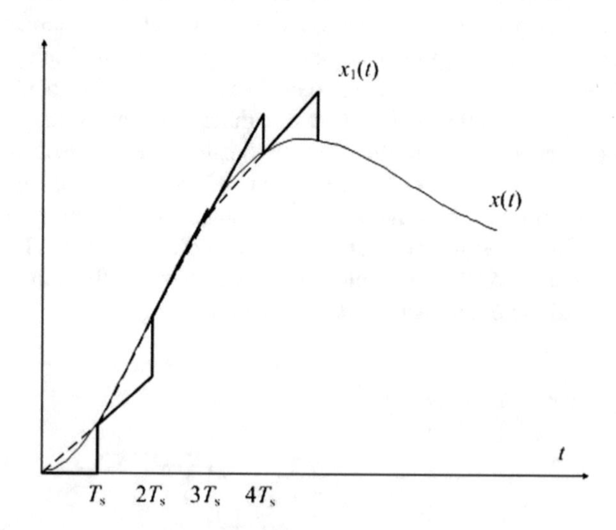

Fig. 11.10 Application of a first order holding unit

Fig. 11.11 Impulse response of the ZOH unit

δ(t)

zero order hold

1

t

T_s

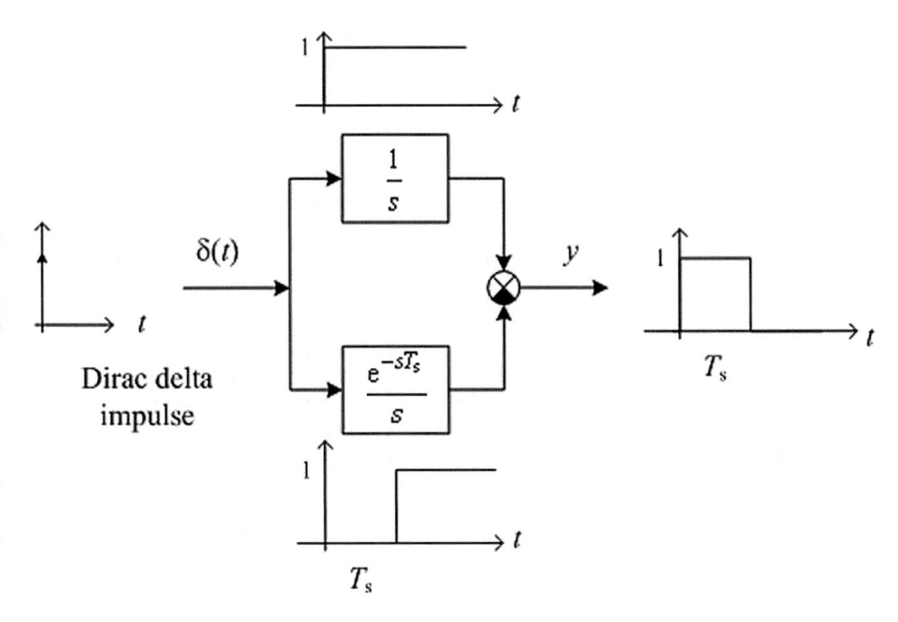

Fig. 11.12 Impulse response components of the ZOH unit

$$W_{\mathrm{ZOH}}(s)=\frac{1-e^{-sT_s}}{s}\cong\frac{1-\left(1-sT_s+\frac{s^2T_s^2}{2}-\cdots\right)}{s}\approx T_s\left(1-\frac{sT_s}{2}+\cdots\right)$$
$$\approx T_s e^{-sT_s/2} \tag{11.5}$$

According to the discussion of the spectral properties of sampled data systems in Sect. 11.1, the frequency function of a CT system built up as a series connection of a ZOH unit and a CT process given by a transfer function $H(s)$ can well be approximated as

$$H_{\mathrm{d}}(j\omega)=\frac{1}{T_s}\tilde{H}(j\omega)\approx\frac{1}{T_s}T_s e^{-j\omega T_s/2}H(j\omega)=e^{-j\omega T_s/2}H(j\omega), \tag{11.6}$$

where $\tilde{H}$ is the approximate transfer function of the joint ZOH unit and the given CT system. The low-frequency approximation indicates that the ZOH unit inserts a delay by $T_s/2$ into the loop. As pointed out earlier, the appearance of any delay in the loop is an unwanted effect in the control loop both considering the stability of the closed-loop system and the quality of its transient response. It is to be emphasized, however, that the application of a holding unit is an absolutely inevitable element in the control loop to interface the DT and CT signal domains.

The physical arrangement in Fig. 11.13 verifies the validity of the above approximation. Samples of a sinusoidal CT signal drive a ZOH unit. The output of

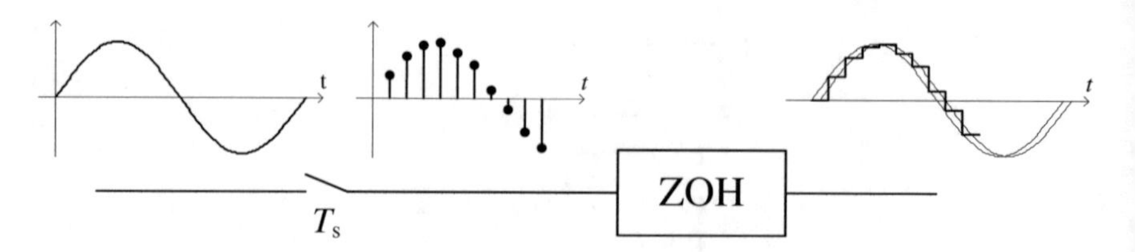

Fig. 11.13 Scheme to demonstrate the delay of the ZOH unit

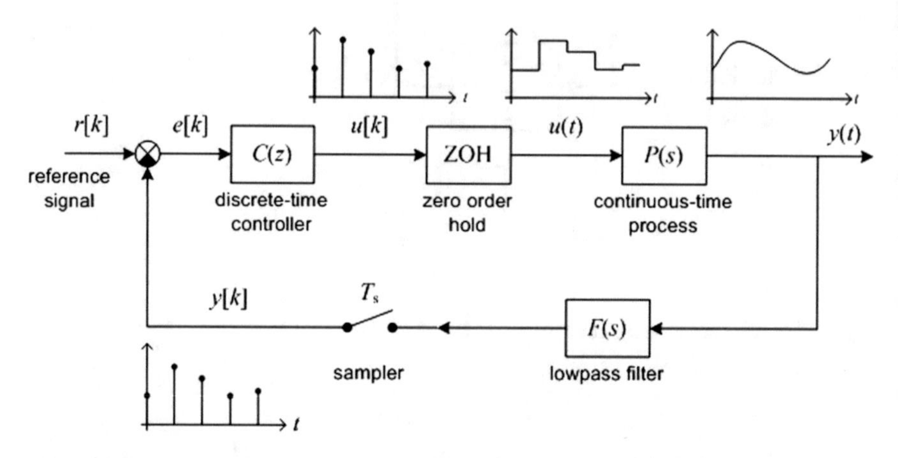

Fig. 11.14 Detailed block diagram of a closed-loop sampled data control system

the ZOH unit is a staircase signal whose fundamental harmonic component has also been drawn together with the original sinusoidal CT signal. The exact phase delay between these two sinusoidal signals depends on the sampling time. However, the approximation according to a delay of $T_s/2$ is rather convincing.

Selecting the sampling time is certainly a new aspect of DT design in comparison with simple CT systems. This issue will be discussed later on. However, selecting a sampling frequency ω_s, a Nyquist frequency of $\omega_N = \omega_s/2$ is also selected, which will suggest the application of a proper low-pass filter to avoid frequency folding. That is, the low-pass filter should pass a highest frequency component ω_{max} to satisfy the $\omega_N \geq \omega_{max}$ condition. A detailed block diagram of a closed-loop sampled data control system is drawn in Fig. 11.14. The time-domain behavior of the signals involved are also shown.

11.3 Description of Discrete-Time Signals, the z-Transformation and the Inverse z-Transformation

The *z-transformation* is a widely used way to describe DT signals and systems. Given a sequence $f[k]$ $(k = 0, 1, 2, \ldots)$ of data, its z-transform is defined by the infinite series

$$\mathcal{Z}\{f[k]\} = \sum_{k=0}^{\infty} z^{-k} f[k] = f[0] + z^{-1} f[1] + z^{-2} f[2] + \cdots, \tag{11.7}$$

where z is the complex valued operator of the transformation. Note that $f[k]$ is assumed to be a positive-time-function, i.e., $f[k] \equiv 0$, $(k<0)$. Though $\mathcal{Z}\{f[k]\}$ is a function of z^{-1}, in practice the notation $F(z) = \mathcal{Z}\{f[k]\}$ is used for z-transforms.

The *region of convergence* (*ROC*) in the complex plane is given by a circle of radius R_1. The infinite series in (11.7) is assumed to be convergent outside of the circle given by R_1, i.e., $F(z) = f[0] + z^{-1} f[1] + z^{-2} f[2] + \cdots$ is convergent for $|z| > R_1$.

Discussing CT closed-loop control systems, the application of the LAPLACE transformation turned out to be a very useful tool both for analysis and design, provided that inverse LAPLACE transformation capabilities are also available. In a similar way inverse z-transformation techniques are to be developed for DT system analysis and design. An analytical expression for the inverse z-transformation is given by the following integral:

$$f[k] = \mathcal{Z}^{-1}\{F(z)\} = \frac{1}{2\pi j} \oint_{R_2} F(z) z^{k-1} dz, \tag{11.8}$$

where the integration runs along the circle R_2 around the origin in the complex plane with a radius allowing all the poles of $F(z)z^{k-1}$ to be within the circle. The inversion integral above is of theoretical importance. The proof of (11.8) is given in A.11.2 of Appendix A.5.

11.3.1 Basic Properties of the z-Transformation

Consider a DT signal $f[k] (k = 0, 1, 2, \ldots)$ with the z-transform of $F(z) = \mathcal{Z}\{f[k]\}$.

Multiplication by a constant coefficient

If c is a constant coefficient, then the z-transform of $g[k] = cf[k] (k = 0, 1, 2, \ldots)$ is

$$\mathcal{Z}\{g[k]\} = \sum_{k=0}^{\infty} z^{-k} g[k] = cf[0] + z^{-1}cf[1] + z^{-2}cf[2] + \cdots = c\sum_{k=0}^{\infty} z^{-k} g[0]$$
$$= cF(z). \tag{11.9}$$

Linearity

If c_1 and c_2 are constant coefficients, we have

$$\mathcal{Z}\{c_1 f_1[k] + c_2 f_2[k]\} = \sum_{k=0}^{\infty} z^{-k}(c_1 f_1[k] + c_2 f_2[k]) = c_1 F_1(z) + c_2 F_2(z), \tag{11.10}$$

which is considered as the linearity property for the z-transform: the z-transform of the linear combination of two signals is equal to the linear combination of the z-transforms of the signals involved.

Shift in the time-domain

Find the z-transform of $f[k-n]$, where $f[k-n]$ is derived from a DT signal $f[k]$ by a *delay* of n steps, assuming that n is a positive integer. For the delayed signal

$$\mathcal{Z}\{f[k-n]\} = z^{-n} F(z) \tag{11.11}$$

holds, as introducing $m = k - n$ and taking note of the fact that $f[m] \equiv 0$, $(m<0)$,

$$\mathcal{Z}\{f[k-n]\} = \sum_{k=0}^{\infty} f[k-n] z^{-k} = z^{-n} \sum_{k=0}^{\infty} f[k-n] z^{-(k-n)}$$
$$= z^{-n} \sum_{m=-n}^{\infty} f[m] z^{-m} = z^{-n} \sum_{m=0}^{\infty} f[m] z^{-m} = z^{-n} F(z)$$

follows. In a similar way, assuming n is a negative integer, a bit more involved relation can be derived for *advanced* signals:

$$\mathcal{Z}\{f[k+n]\} = \sum_{k=0}^{\infty} f[k+n] z^{-k} = z^{n} \sum_{k=0}^{\infty} f[k+n] z^{-(k+n)}$$
$$= z^{n} \left\{ \sum_{k=0}^{\infty} f[k+n] z^{-(k+n)} + \sum_{k=0}^{n-1} f[k] z^{-k} - \sum_{k=0}^{n-1} f[k] z^{-k} \right\}$$
$$= z^{n} \left\{ \sum_{k=0}^{\infty} f[k] z^{-k} - \sum_{k=0}^{n-1} f[k] z^{-k} \right\} = z^{n} \left\{ F(z) - \sum_{k=0}^{n-1} f[k] z^{-k} \right\}$$
$$= z^{n} F(z) - z^{n} f[0] - z^{n-1} f[1] - \cdots - z f[n-1].$$

Multiplication by a^k

Given $F(z)$ as the z-transform of $f[k]$ the z-transform of the DT signal $a^k f[k]$ is

$$\mathcal{Z}\{a^k f[k]\} = \sum_{k=0}^{\infty} a^k f[k] z^{-k} = \sum_{k=0}^{\infty} f[k]\left(a^{-1}z\right)^{-k} = F\left(a^{-1}z\right). \tag{11.12}$$

Observe that the above derivation even allows a to be complex. Also, introducing $a = e^{-b}$, the multiplication rule can be expressed in the following form:

$$\mathcal{Z}\{e^{-bk} f[k]\} = \sum_{k=0}^{\infty} e^{-bk} f[k] z^{-k} = \sum_{k=0}^{\infty} f[k]\left(e^{b}z\right)^{-k} = F\left(e^{b}z\right). \tag{11.13}$$

11.3.2 The z-Transformation of Elementary Time Series

In the sequel, the z-transforms of a few elementary DT *signals* will be derived.

Unit impulse: $f[k] \equiv 1, \quad \mathrm{k} = 0$, otherwise $f[k] \equiv 0$

$$\mathcal{Z}\{f[k]\} = \sum_{k=0}^{\infty} z^{-k} f[k] = 1 + \sum_{k=1}^{\infty} z^{-k} 0 = 1. \tag{11.14}$$

Unit step: $f[k] \equiv 1 \ (k = 0, 1, 2, \ldots)$ and $f[k] \equiv 0, \ (k<0)$

Apply the relation valid for the sum of a geometric series provided the *ROC* $|z| > 1$:

$$\mathcal{Z}\{f[k]\} = \sum_{k=0}^{\infty} z^{-k} f[k] = \sum_{k=0}^{\infty} z^{-k} 1 = \frac{1}{1 - z^{-1}} = \frac{z}{z - 1}. \tag{11.15}$$

Unit ramp: $f[k] \equiv kT_s \ (k = 0, 1, 2, \ldots)$

Again, provided the *ROC* is $|z| > 1$, we have

$$\begin{aligned}\mathcal{Z}\{f[k]\} &= \sum_{k=0}^{\infty} z^{-k} f[k] = \sum_{k=0}^{\infty} z^{-k} kT_s = T_s z^{-1}\left(1 + 2z^{-1} + 3z^{-2} + \cdots\right) \\ &= T_s z^{-1} \frac{1}{\left(1 - z^{-1}\right)^2} = \frac{T_s z}{(z - 1)^2}\end{aligned} \tag{11.16}$$

Power function: $f[k] \equiv a^k \ (k = 0, 1, 2, \ldots)$ where a is a complex constant. Provided the *ROC* is $|z| > |a|$, $\mathcal{Z}\{f[k]\}$ turns out to be

$$\mathcal{Z}\{f[k]\} = \sum_{k=0}^{\infty} z^{-k} f[k] = \sum_{k=0}^{\infty} z^{-k} a^k = \frac{1}{1 - az^{-1}} = \frac{z}{z - a}. \tag{11.17}$$

Applying the rule developed earlier for the multiplication by a^k, the above relation can be derived in a straightforward way:

$$\mathcal{Z}\{a^k\} = \mathcal{Z}\{a^k 1[k]\} = \frac{z}{z-1}\bigg|_{z:=a^{-1}z} = \frac{a^{-1}z}{a^{-1}z-1} = \frac{1}{1-az^{-1}}.$$

Exponential function: $f[k] \equiv e^{-akT_s}$ $(k = 0, 1, 2, \ldots)$

Using the relation derived for the power function, assuming the *ROC* is $|z| > |e^{-aT_s}|$,

$$\mathcal{Z}\{f[k]\} = \sum_{k=0}^{\infty} z^{-k} f[k] = \sum_{k=0}^{\infty} z^{-k} e^{-akT_s} = \frac{1}{1 - e^{-aT_s} z^{-1}} = \frac{z}{z - e^{-aT_s}}. \quad (11.18)$$

Sinusoidal function: $f[k] \equiv \sin(\omega k T_s)$ $(k = 0, 1, 2, \ldots)$

Applying the relation obtained earlier for the exponential function with

$$\sin(\omega k T_s) = \frac{1}{2j}\left(e^{j\omega k T_s} - e^{-j\omega k T_s}\right) \quad (11.19)$$

and assuming the *ROC* is $|z| > 1$, the following relation can be derived:

$$Z\{f[k]\} = \sum_{k=0}^{\infty} z^{-k} \sin(\omega k T_s) = \frac{z\sin(\omega T_s)}{z^2 - 2z\cos(\omega T_s) + 1}. \quad (11.20)$$

The results derived so far, together with some extensions are summarized in Table 11.1.

Table 11.1 LAPLACE and z-transforms of some functions

$F(s)$	$f(t)$	$f[k] = f(kT_s)$	$F(z)$
1	$\delta(t)$	$\delta[k]$	1
$\frac{1}{s}$	$1(t)$	$1[k]$	$\frac{z}{z-1}$
$\frac{1}{s+a}$	e^{-at}	e^{-akT_s}	$\frac{z}{z-e^{-aT_s}}$
$\frac{1}{s^2}$	t	kT_s	$\frac{T_s z}{(z-1)^2}$
$\frac{1}{(s+a)^2}$	te^{-at}	$kT_s e^{-akT_s}$	$\frac{T_s e^{-aT_s} z}{(z-e^{-aT_s})^2}$
$\frac{1}{s^3}$	t^2	$(kT_s)^2$	$\frac{T_s^2 z(z+1)}{(z-1)^3}$
$\frac{b-a}{(s+a)(s+b)}$	$e^{-at} - e^{-bt}$	$e^{-akT_s} - e^{-bkT_s}$	$\frac{(e^{-aT_s} - e^{-bT_s})z}{(z-e^{-aT_s})(z-e^{-bT_s})}$
$\frac{\omega}{s^2+\omega^2}$	$\sin(\omega t)$	$\sin(\omega k T_s)$	$\frac{z\sin(\omega T_s)}{z^2 - 2z\cos(\omega T_s) + 1}$
$\frac{s}{s^2+\omega^2}$	$\cos(\omega t)$	$\cos(\omega k T_s)$	$\frac{z^2 - z\cos(\omega T_s)}{z^2 - 2z\cos(\omega T_s) + 1}$

As far as the application of the this table is concerned an important comment should be added here. Table 11.1 contains four columns. However, an unambiguous mapping exists only for the pairs of $f(t) \leftrightarrow F(s)$ and $f[k] \leftrightarrow F(z)$. More precisely, given a CT signal $f(t)$, the table always indicates a z-transform $F(z)$, however, given a z-transform by $F(z)$, only a CT signal can be obtained, whose samples are only defined at the sampling instants. In such a way a number of CT signals can be derived with diverse intersampling behavior.

11.3.3 The Inverse z-Transformation

Assume that $F(z)$ is the z-transform of the DT signal $f[k]$. Here we ask the question of how to find the DT signal $f[k]$ once $F(z)$ is given. This task is called the inverse z-transformation. In theory, the answer to our question has already been formulated by

$$f[k] = \mathcal{Z}^{-1}\{F(z)\} = \frac{1}{2\pi j} \oint_{R_2} F(z) z^{k-1} dz. \tag{11.21}$$

In practice one of the following three options are used:

Polynomial division

Assume $F(z) = 10z/[(z-1)(z-0.2)]$ then divide the polynomial in the numerator by the polynomial in the denominator and read the samples of $f[k]$ $(k = 0, 1, 2, \ldots)$ as the coefficients obtained along the division:

$$(10z) : (z^2 - 1.2z + 0.2) = f[0] + z^{-1} f[1] + z^{-2} f[2] + \cdots$$

Thus for the first few samples $f[0] = 0, f[1] = 10, f[2] = 12, f[3] = 12.4$ are obtained. The method is not quite efficient. In addition there are *CAD* tools leading to numerical results in a far more efficient way.

Calculating the DT *impulse response*

Consider $F(z)$ as an impulse response function (see later on the pulse transfer function) between a DT input signal with a z-transform of $U(z)$ and a response with a z-transform of $Y(z)$:

$$Y(z) = F(z) U(z). \tag{11.22}$$

Applying a unit impulse input with $U(z) = 1$, then for the z-transform of the response $Y(z) = F(z)$. The calculated samples obviously match the values obtained by polynomial division.

Partial Fractional Expansion (PFE)

The key point here is to decompose $F(z)$ to a sum of components existing in tables of z-transform pairs. Observe the structure of $F(z)$ in the rightmost column of Table 11.1. To ensure the appearance of z in the numerator set up the *PFE* form as follows:

$$\frac{F(z)}{z} = \frac{c_o}{z} + \frac{c_1}{z - p_1} + \frac{c_2}{z - p_2} + \cdots + \frac{c_n}{z - p_n}. \tag{11.23}$$

In this decomposition simple real poles have been assumed. As an example, consider

$$\frac{F(z)}{z} = \frac{12.5}{z-1} - \frac{12.5}{z-0.2} \quad \text{or} \quad F(z) = \frac{12.5z}{z-1} - \frac{12.5z}{z-0.2},$$

which lead to

$$f[k] = \mathcal{Z}^{-1}\{F(z)\} = \mathcal{Z}^{-1}\left\{\frac{12.5z}{z-1} - \frac{12.5z}{z-0.2}\right\} = 12.5\left(1 - 0.2^k\right),\ k = 0, 1, 2, \ldots$$

Observe that unlike the first two methods, the *PFE* delivers $f[k]$ in analytical form, so its value can be easily computed for arbitrary $k \geq 0$.

Handling complex poles

Setting up the *PFE* form becomes more involved in the case of complex poles of $F(z)$ even for a simple complex pole pair.

Example 11.1 Consider

$$F(z) = \frac{z^3 + 1}{z^3 - z^2 - z - 2}.$$

Following the *PFE* procedure,

$$\frac{F(z)}{z} = -\frac{0.5}{z} + \frac{0.643}{z-2} + \frac{c}{z-p} - \frac{\bar{c}}{z-\bar{p}},$$

is obtained, where $c = 0.429 + j0.0825$ and $p = -0.5 - j0.866$, and $\bar{c}$ and $\bar{p}$ denote their complex conjugate values, respectively. Evaluating the last two terms as one second-order component, two sinusoidal components can be reconstructed based on the last two rows in Table 11.1. Combining these two sinusoidal components into one single sinusoidal form yields

$$\begin{aligned} f[k] &= -0.5\delta[k] + 0.643(2)^k + c(p)^k + \bar{c}(\bar{p})^k \\ &= -0.5\delta[k] + 0.643(2)^k + \cos\left(\frac{4\pi}{3} + 10.89^\circ\right) \end{aligned}$$

by using the trigonometric identity

$$c(p)^k + \bar{c}(\bar{p})^k = 2C\sigma^k \cos(\Omega k + \Theta),$$

where $c = Ce^{j\Theta}$ and $p = \sigma e^{j\Omega}$. ■

Handling multiple poles

Example 11.2 The discussion concerning repeated poles will be reduced to a numerical example. Consider

$$F(z) = \frac{6z^3 + 2z^2 - z}{z^3 - z^2 - z + 1}.$$

The structure of the *PFE* is governed by the poles of $F(z)$. In this case $p_{1,2} = 1$ and $p_3 = -1$, i.e., $p_{1,2} = 1$ is a double pole. Accordingly,

$$\frac{F(z)}{z} = \frac{5.25}{z-1} + \frac{3.5}{(z-1)^2} + \frac{0.75}{z+1}.$$

Observe that in the *PFE* form the repeated pole shows up in single *and* double forms, as well. The inverse z-transformation then leads to

$$f[k] = 5.25(1)^k + 3.5k + 0.75(-1)^k, \quad k = 0, 1, 2, \ldots.$$ ■

General method for single poles

If the LAPLACE-transform of a *CT* signal has the form $F(s) = \mathcal{F}_\mathrm{z}(s)/\mathcal{F}_\mathrm{p}(s)$, where $\mathcal{F}_\mathrm{z}(s)$ and $\mathcal{F}_\mathrm{p}(s)$ are the numerator and denominator of $F(s)$, respectively, then the z-transform of the sampled signal is given by

$$F(z) = \sum_{i=1}^{n} \frac{\mathcal{F}_\mathrm{z}(p_i)}{\mathcal{F}'_\mathrm{p}(p_i)} \frac{z}{z - e^{p_i T_\mathrm{s}}} = \sum_{i=1}^{n} \frac{\mathcal{F}_\mathrm{z}(p_i)}{\mathcal{F}'_\mathrm{p}(p_i)} \frac{1}{1 - e^{p_i T_\mathrm{s}} z^{-1}}. \tag{11.24}$$

For single poles the simple ‘cover up’ technique to determine the residues can be used

$$\frac{1}{\mathcal{F}'(p_i)} = \lim_{s \to p_i} (s - p_i) \frac{1}{\mathcal{F}_\mathrm{p}(s)} \tag{11.25}$$

to ease the evaluation of $F(z)$. The application of this method assumes that the denominator is available in factored form. One component of the sum can be calculated by covering up $s - p_i$ and substituting $s = p_i$ in the rest.

11.3.4 Initial and Final Value Theorems

Once $F(z)$ is given, the initial and final value theorems provide tools to determine $f[0]$ and $\lim_{k\to\infty} f[k]$ in a direct way, without performing the inverse z-transformation.

Initial value theorem $(k \to 0)$

Starting from the definition of the z-transform

$$\mathcal{Z}\{f[k]\} = \sum_{k=0}^{\infty} z^{-k} f[k] = f[0] + z^{-1} f[1] + z^{-2} f[2] + \cdots$$

it can be seen that $f[0]$ can be easily determined as $z \to \infty$:

$$f[0] = \lim_{z\to\infty} F(z). \tag{11.26}$$

Final value theorem $(k \to \infty)$

The basic idea is to separate the 'last' element of the sequence $f[0]$ by subtracting the original and the delayed sequences:

$$\begin{aligned}\lim_{k\to\infty} f[k] &= \lim_{z\to 1}\{F(z) - z^{-1}F(z)\}\\ &= (f[0] + f[1] + f[2] + \cdots) - (f[-1] + f[0] + f[1] + \cdots).\end{aligned}$$

Assuming positive-time-functions $(f[k] = 0,\ k < 0)$ results in

$$\lim_{k\to\infty} f[k] = \lim_{z\to 1}\{(1 - z^{-1})F(z)\}. \tag{11.27}$$

The final value theorem can only be applied if the signal has a steady state value.

11.4 Description of Sampled Data Systems in the Discrete-Time and in the Operator and Frequency Domain

Discussing CT systems, the need for an abstract system description for closed-loop analysis and design was seen. Considering sampled data systems, the value of the control input and that of the samples of the output only change at the sampling instants. It seems to be reasonable then to create a mathematical model to describe the system behavior only at the sampling instants. The process of describing the behavior of a sampled CT system will be referred to as discretization. As a starting point, the *state-space* description will be discussed, then the methods of the *impulse response function* (pulse transfer function) and the *difference equation* will be derived. This triple is in complete harmony with the triple used for CT systems (state-space, transfer function, differential equation).

11.4.1 The State-Space Model

Assume the *LTI* state-space model of the CT system to be discretized along with a sampling time of T_s (see 3.10):

$$\begin{aligned}\dot{\boldsymbol{x}}(t) &= \boldsymbol{A}\boldsymbol{x}(t) + \boldsymbol{b}u(t)\\ y(t) &= \boldsymbol{c}^{\mathrm{T}}\boldsymbol{x}(t) + du(t)\end{aligned} \tag{11.28}$$

The solution of the state equation as discussed earlier for CT systems (3.2.1) with the initial time t_o and the initial state vector $\boldsymbol{x}(t_o)$ is

$$\boldsymbol{x}(t) = e^{\boldsymbol{A}(t-t_o)}\boldsymbol{x}(t_o) + \int_{t_o}^{t} e^{\boldsymbol{A}(t-\tau)}\boldsymbol{b}u(\tau)\mathrm{d}\tau = e^{\boldsymbol{A}(t-t_o)}\boldsymbol{x}(t_o) + \left[\int_{t_o}^{t} e^{\boldsymbol{A}(t-\tau)}u(\tau)\mathrm{d}\tau\right]\boldsymbol{b}. \tag{11.29}$$

Assuming that the discretized model will contain a ZOH unit (in other words the continuous control input will be a staircase function) perform the integration from kT_s to $(k+1)T_s$:

$$\begin{aligned}\boldsymbol{x}(kT_s + T_s) &= e^{\boldsymbol{A}T_s}\boldsymbol{x}(kT_s) + \int_{kT_s}^{kT_s+T_s} e^{\boldsymbol{A}(kT_s+T_s-\tau)}\boldsymbol{b}u(\tau)\mathrm{d}\tau\\ &= e^{\boldsymbol{A}T_s}\boldsymbol{x}(kT_s) + \boldsymbol{b}u(kT_s)\int_{kT_s}^{kT_s+T_s} e^{\boldsymbol{A}(kT_s+T_s-\tau)}\mathrm{d}\tau\\ &= e^{\boldsymbol{A}T_s}\boldsymbol{x}(kT_s) + \left[u(kT_s)\int_{0}^{T_s} e^{\boldsymbol{A}\lambda}\mathrm{d}\lambda\right]\boldsymbol{b}\end{aligned}$$

The above manipulations used the fact that $u(\tau) = \text{constant}$ within each sampling period $([kT_s \le \tau < (k+1)T_s])$, that is, $u(\tau) = u(kT_s)$. Furthermore, to simplify the evaluation of the integral, $\lambda = kT_s + T_s - \tau$ has been introduced. Using the standard notations $\boldsymbol{x}(kT_s + T_s) = \boldsymbol{x}[k+1]$, $\boldsymbol{x}(kT_s) = \boldsymbol{x}[k]$ and $u(kT_s) = u[k]$, the above equation can be rewritten as

$$\boldsymbol{x}[k+1] = e^{\boldsymbol{A}T_s}\boldsymbol{x}[k] + \left[\int_0^{T_s} e^{\boldsymbol{A}\lambda}\mathrm{d}\lambda\right]\boldsymbol{b}u[k]. \tag{11.30}$$

Introducing the parameter matrices for the discretized model

$$\boldsymbol{F} = e^{\boldsymbol{A}T_s} \quad \text{and} \quad \boldsymbol{g} = \int_0^{T_s} e^{\boldsymbol{A}\lambda}\mathrm{d}\lambda\,\boldsymbol{b} \tag{11.31}$$

the following discretized state-model is obtained: $\boldsymbol{x}[k+1] = \boldsymbol{F}\boldsymbol{x}[k] + \boldsymbol{g}u[k]$. Note that if $\boldsymbol{A}$ is invertible, then the integration in the expression for $\boldsymbol{g}$ can be carried out and this leads to the closed-form formula $\boldsymbol{g} = \boldsymbol{A}^{-1}\left(e^{\boldsymbol{A}T_s} - \boldsymbol{I}\right)\boldsymbol{b}$. The output equation of the CT state-model of $y(t) = \boldsymbol{c}^{\mathrm{T}}\boldsymbol{x}(t) + du(t)$ can simply be sampled by substituting $t = kT_s$:

$$y(kT_s) = \boldsymbol{c}^{\mathrm{T}}\boldsymbol{x}(kT_s) + du(kT_s)$$

or, to emphasize the DT nature of the model,

$$y[k] = \boldsymbol{c}^{\mathrm{T}}\boldsymbol{x}[k] + du[k]. \tag{11.32}$$

The DT state-model can be summarized as follows. The DT *state difference equation*

$$\boldsymbol{x}[k+1] = \boldsymbol{F}\boldsymbol{x}[k] + \boldsymbol{g}u[k] \tag{11.33}$$

and the DT *output equation* according to (11.32) form the DT state-model for $k = 0, 1, 2, \ldots$. Comparing the DT state-model with the CT state-model, (11.31) shows how to derive the parameter matrices of the DT state difference equation from the parameter matrices of the CT state-model, while the parameters of the output equation ($\boldsymbol{c}^{\mathrm{T}}$ and d) are identical for the DT and CT state-models. The matrix $\boldsymbol{F}$ will be called the state transition matrix, in particular assuming zero excitation, it governs the transition between $\boldsymbol{x}[k]$ and $\boldsymbol{x}[k+1]$ according to

$$\boldsymbol{x}[k+1] = \boldsymbol{F}\boldsymbol{x}[k]. \tag{11.34}$$

Example 11.3 To study the nature of the operations while discretizing a system and the influence of the sampling time on these operations, consider a second order example, namely a double integrator. Select the state variables as shown in Fig. 11.15.

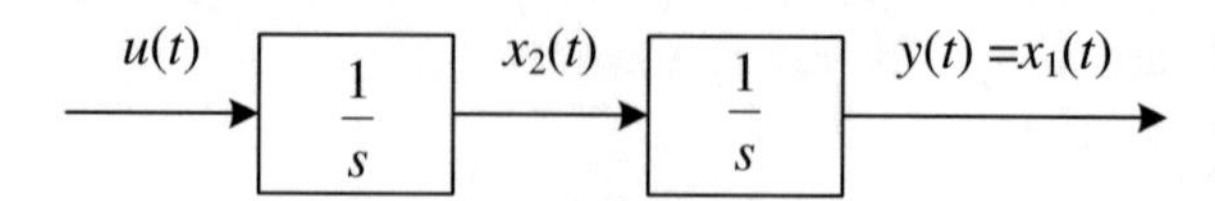

Fig. 11.15 CT double integrator

The CT state equations:

$$\dot{x}_1(t) = x_2(t)$$
$$\dot{x}_2(t) = u(t)$$

The CT state-model:

$$\dot{\boldsymbol{x}}(t) = \begin{bmatrix} \dot{x}_1(t) \\ \dot{x}_2(t) \end{bmatrix} = \begin{bmatrix} 0 & 1 \\ 0 & 0 \end{bmatrix} \begin{bmatrix} x_1(t) \\ x_2(t) \end{bmatrix} + \begin{bmatrix} 0 \\ 1 \end{bmatrix} u(t) = \boldsymbol{A}\boldsymbol{x}(t) + \boldsymbol{b}u(t),$$

hence

$$\boldsymbol{F} = e^{\boldsymbol{A}T_s} = e^{\begin{bmatrix} 0 & 1 \\ 0 & 0 \end{bmatrix} T_s}.$$

To determine the above matrix exponential evaluate the infinite series (see 3.20 and 3.26)

$$e^{\boldsymbol{A}T_s} = \boldsymbol{I} + \boldsymbol{A}T_s + \frac{1}{2}\boldsymbol{A}^2 T_s^2 + \cdots$$

note that $\boldsymbol{A}^k = 0\,(k \geq 2)$, thus

$$e^{\boldsymbol{A}T_s} = \boldsymbol{I} + \boldsymbol{A}T_s + \frac{1}{2}\boldsymbol{A}^2 T_s^2 + \cdots = \boldsymbol{I} + \boldsymbol{A}T_s = \begin{bmatrix} 1 & 0 \\ 0 & 1 \end{bmatrix} + \begin{bmatrix} 0 & T_s \\ 0 & 0 \end{bmatrix} = \begin{bmatrix} 1 & T_s \\ 0 & 1 \end{bmatrix}$$

and

$$\boldsymbol{g} = \int_0^{T_s} e^{\boldsymbol{A}\lambda} \mathrm{d}\lambda\, \boldsymbol{b} = \int_0^{T_s} \begin{bmatrix} 1 & \lambda \\ 0 & 1 \end{bmatrix} \mathrm{d}\lambda \begin{bmatrix} 0 \\ 1 \end{bmatrix} = \int_0^{T_s} \begin{bmatrix} \lambda \\ 1 \end{bmatrix} \mathrm{d}\lambda = \begin{bmatrix} T_s^2/2 \\ T_s \end{bmatrix}.$$

In the case of higher order systems the application of *CAD* tools is advised. ■

Given the initial state $\boldsymbol{x}[0]$, the solution of the state difference equation is well known from the theory of "Signals and Systems":

$$\boldsymbol{x}[k] = \boldsymbol{F}^k \boldsymbol{x}[0] + \sum_{m=0}^{k-1} \boldsymbol{F}^{k-m-1} \boldsymbol{g} u[m], \tag{11.35}$$

where the first term depends on the initial value of the state vector, while the second is a weighted sum of the input samples at $0, 1, \ldots, (k-1)$. It can be seen that in the above solution the $\boldsymbol{F}$ state transition matrix plays a key role. It will be shown later on that $\boldsymbol{F}$ has a fundamental role in determining other important system properties, like stability, as well.

11.4.2 Input-Output Models Based on the Shift Operator

The input-output models will be presented according to the operators used to describe the relation between the input and output samples. First an expressive modeling approach based on the application of the shift operator will be shown. This approach directly supports the discussion using the concept of difference equations. Just as the LAPLACE-transform played a fundamental role in the description of CT systems, the application of z-transforms will be useful for DT systems.

Discussing CT systems, it has been shown that an input-output model given by a differential equation or by a transfer function can be transformed into an infinite number of input-output equivalent state-models. On the contrary, input-output equivalent state-models always exhibit one single input-output model. This property is valid for DT systems, as well. To present this property, first introduce the *shift operator* according to

$$q\boldsymbol{x}[k] = \boldsymbol{x}[k+1] \quad (k = \ldots, -1, 0, 1, \ldots). \tag{11.36}$$

The function of the q operator is to advance a scalar- or vector-valued DT sequence $\boldsymbol{x}[k]$ by one single step. The repeated application of the shift operator leads to advancing the sequence by several steps. E.g., for m steps,

$$q^m\boldsymbol{x}[k] = \boldsymbol{x}[k+m] \quad (k=\ldots, -1, 0, 1, \ldots) \tag{11.37}$$

is to be applied. A delay can be realized by using the inverse of the operator q:

$$q^{-1}\boldsymbol{x}[k] = \boldsymbol{x}[k-1] \quad (k = \ldots, -1, 0, 1, \ldots) \tag{11.38}$$

or

$$q^{-m}\boldsymbol{x}[k] = \boldsymbol{x}[k-m] \quad (k=\ldots, -1, 0, 1, \ldots).$$

In the sequel, the operator q^{-1} will be referred to as the delay or shift operator. Apply q to the DT state-model:

$$\begin{aligned} \boldsymbol{x}[k+1] &= q\boldsymbol{x}[k] = \boldsymbol{F}\boldsymbol{x}[k] + \boldsymbol{g}u[k] \\ y[k] &= \boldsymbol{c}^{\mathrm{T}}\boldsymbol{x}[k] + du[k] \end{aligned} \tag{11.39}$$

Solving for $\boldsymbol{x}[k]$

$$\boldsymbol{x}[k] = (q\boldsymbol{I} - \boldsymbol{F})^{-1}\boldsymbol{g}u[k] \tag{11.40}$$

allows rewriting the output equation as

$$y[k] = \mathbf{c}^{\mathrm{T}}(q\mathbf{I} - \mathbf{F})^{-1}\mathbf{g}u[k] + du[k] = \left[\mathbf{c}^{\mathrm{T}}(q\mathbf{I} - \mathbf{F})^{-1}\mathbf{g} + d\right]u[k]. \tag{11.41}$$

The dependence of the output sequence on the input sequence can be expressed by

$$y[k] = G(q)u[k] = \left[\mathbf{c}^{\mathrm{T}}(q\mathbf{I} - \mathbf{F})^{-1}\mathbf{g} + d\right]u[k], \tag{11.42}$$

where $G(q)$, the *transfer function operator*, has been introduced. Analyzing $G(q)$, it can be seen that it is a rational function

$$G(q) = \mathbf{c}^{\mathrm{T}}(q\mathbf{I} - \mathbf{F})^{-1}\mathbf{g} + d = \mathbf{c}^{\mathrm{T}}\frac{\mathbf{adj}(q\mathbf{I} - \mathbf{F})}{\det(q\mathbf{I} - \mathbf{F})}\mathbf{g} + d = \frac{\mathcal{B}(q)}{\mathcal{A}(q)}, \tag{11.43}$$

where $\det(q\mathbf{I} - \mathbf{F})$ and $\mathbf{c}^{\mathrm{T}}\mathbf{adj}(q\mathbf{I} - \mathbf{F})\mathbf{g}$ are polynomials in the operator q with real coefficients:

$$\begin{aligned} \mathcal{A}(q) &= \det(q\mathbf{I} - \mathbf{F}) = q^n + a_1 q^{n-1} + \cdots + a_n \\ \mathcal{B}(q) &= b_0 q^{n_{\mathrm{B}}} + b_1 q^{n_{\mathrm{B}}-1} + \cdots + b_{n_{\mathrm{B}}} \end{aligned} \tag{11.44}$$

As far as the degrees of the introduced polynomials $\mathcal{A}(q)$ and $\mathcal{B}(q)$ are concerned, n is the number of the state variables, while n_B is subject to the constraint $n_{\mathrm{B}} \le n$. For $d = 0$, which holds for most systems, (11.43) is a strictly proper rational function, so it is reasonable to rewrite (11.44) for $n_{\mathrm{B}} = n - 1$ as

$$\begin{aligned} \mathcal{A}(q) &= \det(q\mathbf{I} - \mathbf{F}) = q^n + a_1 q^{n-1} + \cdots + a_n \\ &= q^n\left(1 + a_1 q^{-1} + \cdots + a_n q^{-n}\right) = q^n \mathcal{A}\left(q^{-1}\right) \\ \mathcal{B}(q) &= b_1 q^{n-1} + b_2 q^{n-2} + \cdots + b_n \\ &= q^n\left(b_1 q^{-1} + b_2 q^{-2} + \cdots + b_n q^{-n}\right) = q^n \mathcal{B}\left(q^{-1}\right) \end{aligned} \tag{11.45}$$

Similarly to the terminology introduced for CT systems, the roots of $\mathcal{A}(q) = 0$ will be called the DT poles and the roots of $\mathcal{B}(q) = 0$ will be called the DT zeros.

Based on the introduced polynomials and utilizing the time domain interpretation of the shift operator, another model of DT systems, namely the difference equation model, can be derived. As discussed earlier,

$$y[k] = G(q)u[k] = \frac{\mathcal{B}(q)}{\mathcal{A}(q)}u[k], \tag{11.46}$$

which can be written as

$$\mathcal{A}(q)y[k] = \mathcal{B}(q)u[k]. \tag{11.47}$$

Further substitution gives

$$\left(q^n + a_1 q^{n-1} + \cdots + a_n\right)y[k] = \left(b_1 q^{n-1} + b_2 q^{n-2} + \cdots + b_n\right)u[k] \tag{11.48}$$

and finally

$$y[k+n] + a_1 y[k+n-1] + \cdots + a_n y[k] = b_1 u[k+n-1] + b_1 u[k+n-2] + \cdots + b_n u[k].$$

Divide both sides of (11.48) by q^n:

$$\left(1 + a_1 q^{-1} + \cdots + a_n q^{-n}\right)y[k] = \left(b_1 q^{-1} + b_2 q^{-2} + \cdots + b_n q^{-n}\right)u[k] \tag{11.49}$$

and rearrange it

$$y[k] = b_1 u[k-1] + b_2 u[k-2] + \cdots + b_n u[k-n] - a_1 y[k-1] - \cdots - a_n y[k-n] \tag{11.50}$$

which is the input-output difference equation. Another interpretation suggests considering (11.50) as a *recursive formula* to calculate the output sample. It can be seen that to derive the recursive formula, the polynomials of (11.50) have been used in their form arranged by q^{-1}. Thus the pulse transfer function operator can equivalently be defined as follows:

$$G\left(q^{-1}\right) = \frac{\mathcal{B}(q^{-1})}{\mathcal{A}(q^{-1})} = \frac{\mathcal{B}(q^{-1})}{1 + \widetilde{\mathcal{A}}(q^{-1})}, \tag{11.51}$$

which is also called the filter form, for based on (11.51) the recursive formula by (11.50) can always be derived:

$$\begin{aligned} y[k] &= \boldsymbol{B}\left(q^{-1}\right)u[k] - \tilde{\boldsymbol{A}}\left(q^{-1}\right)y[k] \\ &= b_1 u[k-1] + b_1 u[k-2] + \cdots + b_n u[k-n] - a_1 y[k-1] - \cdots - a_n y[k-n] \end{aligned} \tag{11.52}$$

which is linear-in-the-parameters $\{a_i; b_i\}$ and ready to be implemented in a computing environment.

Example 11.4 Find the pulse transfer function operator $G(q)$ for the double integrator discussed earlier in Example 11.3. Apply $T_s = 1$:

$$G(q) = \mathbf{c}^{\mathrm{T}}(q\mathbf{I} - \mathbf{F})^{-1}\mathbf{g} + d = [1 \quad 0]\begin{bmatrix} q-1 & -1 \\ 0 & q-1 \end{bmatrix}^{-1}\begin{bmatrix} 0.5 \\ 1 \end{bmatrix} = \frac{0.5(q+1)}{(q-1)^2}$$
$$= \frac{0.5q+0.5}{q^2-2q+1}.$$

Both poles of the system are equal to $p_{\mathrm{d}_{1,2}} = 1$, while the single zero is $z_{\mathrm{d}} = -1$.

■

11.4.3 Modeling Based on the z-Transformation

The z-transform of a DT sequence $f[k]$, $(k = 0, 1, 2, \ldots)$ is defined as follows:

$$\mathcal{Z}\{f[k]\} = \sum_{k=0}^{\infty} z^{-k} f[k] = f[0] + z^{-1}f[1] + z^{-2}f[2] + \cdots, \tag{11.53}$$

It can be shown that the complex variable of the z-transform (z) is in a close relationship with the complex variable of the LAPLACE-transform (s). As a guiding principle, a discretization is looked for where the values attached to the sequence of impulses are identical to those obtained from sampling. The theory of hybrid systems calls this principle *impulse invariance*. Consider the mathematically sampled form of a CT signal $f(t)$:

$$f^*(t) = \sum_{m=0}^{\infty} f(mT_{\mathrm{s}})\delta(t - mT_{\mathrm{s}}), \tag{11.54}$$

then find the LAPLACE-transform of $f^*(t)$:

$$\mathcal{L}\{f^*(t)\} = \mathcal{L}\left\{\sum_{m=0}^{\infty} f(mT_{\mathrm{s}})\delta(t - mT_{\mathrm{s}})\right\} = f(0) + f(T_{\mathrm{s}})e^{-sT_{\mathrm{s}}} + f(2T_{\mathrm{s}})e^{-2sT_{\mathrm{s}}} + \cdots$$

Note that $f^*(t)$ is a DT signal. To comply with the impulse invariance principle

$$\mathcal{L}\{f^*(t)\} = \mathcal{Z}\{f[k]\} \tag{11.55}$$

is obtained, which leads to

$$f(0) + f(T_{\mathrm{s}})e^{-sT_{\mathrm{s}}} + f(2T_{\mathrm{s}})e^{-2sT_{\mathrm{s}}} + \cdots = f[0] + z^{-1}f[1] + z^{-2}f[2] + \cdots,$$

implying that

$$z = e^{sT_s}. \tag{11.56}$$

To emphasize the importance of the relation $z = e^{sT_s}$ find the stability region for DT systems. For CT systems, the left half-plane of the $s = \sigma + j\omega$ complex plane turned out to be the stability region concerning the poles of the CT system. The $z = e^{sT_s}$ maps this left half-plane into the unit disc in the complex z-plane. To see this, observe that $z = e^{sT_s}$ maps the borderline of the stability of the s-plane $s = j\omega\,(-\infty < \omega < \infty)$ to the unit circle. As the frequency increases the region of stability is to the left of the borderline in the s-plane, so will happen in the z plane. More exactly, sampling will map the fundamental band $-\pi/T_s < \omega < \pi/T_s$ in the s half-plane into the unit disc (bands outside the fundamental band repeat themselves). The two lines parallel to the negative real axis in the s-plane at $\omega = \pm j\pi/T_s$ will be transformed to one single line (the negative real axis of the z-plane). Poles according to the limit of the SHANNON sampling law will be transformed to the negative real axis of the z-plane.

The simple basic building blocks for DT models can be determined by approximating the CT operators while acting on the sequence of sampled data. Start with simple differentiation.

Backward difference:

Based on

$$\frac{\mathrm{d}y}{\mathrm{d}t} \approx \frac{y[k] - y[k-1]}{T_s} = \frac{1 - z^{-1}}{T_s} y[k] \tag{11.57}$$

the operator $H_D(s) = s$ for differentiation has a DT equivalent by,

$$G_D(z) = \frac{z-1}{T_s z} = \frac{1 - z^{-1}}{T_s}. \tag{11.58}$$

Forward difference:

$$\frac{\mathrm{d}y}{\mathrm{d}t} \approx \frac{y[k+1] - y[k]}{T_s} = \frac{z-1}{T_s} y[k] = \frac{1 - z^{-1}}{T_s z^{-1}} y[k], \tag{11.59}$$

suggests using

$$G_D(z) = \frac{z-1}{T_s} = \frac{1 - z^{-1}}{T_s z^{-1}}. \tag{11.60}$$

as the DT equivalent of the CT differentiation.

Consider now the simple integration by $H_{\mathrm{I}}(s) = 1/s$.

Right hand rectangular rule:

$$y(t) = \int_0^t u(\tau)\mathrm{d}\tau \approx \sum_{i=0}^{k} u(i)T_{\mathrm{s}} = T_{\mathrm{s}} \sum_{i=0}^{k} u(i) = y[k] = y[k-1] + T_{\mathrm{s}}u[k], \quad (11.61)$$

allows setting up the following recursive formula:

$$y[k] - y[k-1] = \left(1 - z^{-1}\right)y[k] = T_{\mathrm{s}}u[k]. \quad (11.62)$$

Consequently,

$$G_{\mathrm{I}}(z) = \frac{T_{\mathrm{s}}}{1 - z^{-1}} = \frac{T_{\mathrm{s}}z}{z - 1}. \quad (11.63)$$

Left hand rectangular rule:

$$y(t) = \int_0^t u(\tau)\mathrm{d}\tau \approx \sum_{i=0}^{k} u(i)T_{\mathrm{s}} = T_{\mathrm{s}} \sum_{i=0}^{k} u(i) = y[k] = y[k-1] + T_{\mathrm{s}}u[k-1], \quad (11.64)$$

leads to

$$y[k+1] - y[k] = (z - 1)y[k] = T_{\mathrm{s}}u[k]. \quad (11.65)$$

Consequently,

$$G_{\mathrm{I}}(z) = \frac{T_{\mathrm{s}}}{z - 1} = \frac{T_{\mathrm{s}}z^{-1}}{1 - z^{-1}}. \quad (11.66)$$

To recapture the methodology used for discussing CT systems, LAPLACE-transforms were effectively used by introducing various transfer functions. For DT systems the z-transforms may play a similarly important role once the notion of *pulse transfer function* is introduced as the ratio of the z-transforms of two signals. A natural requirement here is that both the input and the output signal should be DT sequences. This means that the ZOH unit should also be taken into account. Figure 11.16 shows the proper arrangement driven by the DT input $u[k]$ and generating the DT output signal $y[k]$.

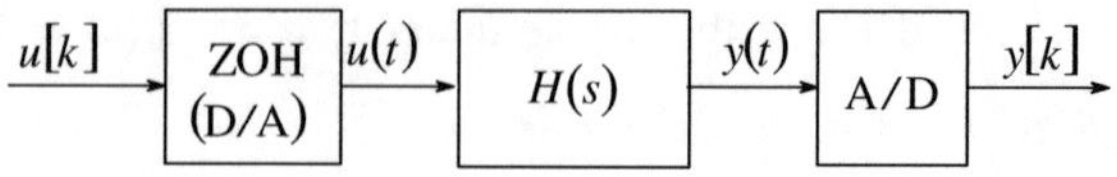

Fig. 11.16 A CT block extended by a ZOH and a sampler

The *pulse transfer function* of the discretized system (assuming zero initial conditions) is

$$G(z) = \frac{\mathcal{Z}\{y[k]\}}{\mathcal{Z}\{u[k]\}}. \tag{11.67}$$

For an arbitrary input the sampled response can be calculated by

$$y[k] = \mathcal{Z}\left\{y(t)|_{t=kT_s}\right\} = \mathcal{Z}\left\{\mathcal{L}^{-1}[Y(s)]_{t=kT_s}\right\} = \mathcal{Z}\left\{\mathcal{L}^{-1}[U(s)H(s)]_{t=kT_s}\right\}. \tag{11.68}$$

It can be seen that the above expression is input-dependent, in other words instead of a general DT model, only a DT model valid for a given class of inputs can be formulated. Aim at setting up a *Step Response Equivalent* (*SRE*) DT model valid for a unit step sequence of $u[k]$ resulting in a CT step response $u(t)$ input:

$$y[k] = \mathcal{Z}\left\{\mathcal{L}^{-1}\left[\frac{H(s)}{s}\right]_{t=kT_s}\right\}. \tag{11.69}$$

In this case the CT and DT models will exhibit identical outputs at the sampling instants. Observe that the LAPLACE-transform yields the step response of the CT system:

$$\mathcal{L}^{-1}\left\{\frac{H(s)}{s}\right\} = v(t), \tag{11.70}$$

whose samples determine the DT step response

$$v[k] = \mathcal{L}^{-1}\left[\frac{H(s)}{s}\right]_{t=kT_s}, \tag{11.71}$$

thus the DT model can be written as

$$G(z) = \frac{\mathcal{Z}\{v[k]\}}{\mathcal{Z}\{u[k]\}} = \frac{\mathcal{Z}\{v[k]\}}{\frac{z}{z-1}} = \frac{z-1}{z}\mathcal{Z}\{v[k]\} = \left(1-z^{-1}\right)\mathcal{Z}\{v[k]\}. \tag{11.72}$$

Note that the above relation can be found in some textbooks as

$$G(z) = \left(1 - z^{-1}\right)\mathcal{Z}\left\{\frac{H(s)}{s}\right\}. \tag{11.73}$$

The above form is rather expressive, however, it is not correct in the mathematical sense, since a LAPLACE-transform obviously has no z-transform. The correct procedure is to perform an inverse LAPLACE-transformation, then sampling the CT signal, a DT signal is obtained by substituting $t = kT_s$. In principle, the *SRE* transformation matches the case of using a ZOH unit, i.e., between the sampling instants there are always step-like excitations.

In a similar way, *Ramp Response Equivalent* (*RRE*) DT models can also be derived:

$$y[k] = \mathcal{Z}\left\{\mathcal{L}^{-1}\left[\frac{H(s)}{s^2}\right]_{t=kT_s}\right\}. \tag{11.74}$$

In principle, the *RRE* transformation matches the case of using a first order holding unit in the closed-loop. In this arrangement, there are always ramp-like excitations between two samples.

Example 11.5 Apply the results obtained so far for setting up a DT model discretizing a double integrator. The unit step response of the double integrator is

$$v(t) = \frac{t^2}{2} \quad (\mathrm{t} \geq 0), \tag{11.75}$$

or in a sampled form

$$v[k] = \frac{(kT_s)^2}{2} \quad (\mathrm{k} \geq 0), \tag{11.76}$$

z-transformation gives

$$\mathcal{Z}\{v[k]\} = \frac{T_s^2}{2} z(z+1)/(z-1)^3, \tag{11.77}$$

and finally

$$G(z) = \left(1 - z^{-1}\right)\mathcal{Z}\{v[k]\} = \frac{z-1}{z}\mathcal{Z}\{v[k]\} = \frac{z-1}{z}\frac{T_s^2}{2}\frac{z(z+1)}{(z-1)^3} = \frac{T_s^2}{2}\frac{(z+1)}{(z-1)^2}. \tag{11.78}$$

Specifically, $G(z) = (z+1)/2(z-1)^2 = (0.5z+0.5)/(z^2 - 2z + 1)$ is obtained for $T_s = 1$. ■

Observe the formal coincidence of the forms derived by the pulse transfer operator $G(q)$ and by the z-transforms. For the relations of the z-transformation the interpretation of multiplication by z is advancing a sequence by one sample and the interpretation of multiplication by z^{-1} is delaying a sequence by one sample, in general it can be stated that $G(q)$ can be obtained by substituting $z = q$ in $G(z)$. Despite the formal matching keep it in mind that the shift operator and the variable of the z-transforms are different notions!

Repeating the derivation used for the shift operator it can be shown that $G(z) = \dfrac{\mathcal{B}(z)}{\mathcal{A}(z)} = \boldsymbol{c}^{\mathrm{T}} \dfrac{\mathbf{adj}(z\boldsymbol{I} - \boldsymbol{F})}{\det(z\boldsymbol{I} - \boldsymbol{F})} \boldsymbol{g} + d$. From this the characteristic equation is found to be $\det(z\boldsymbol{I} - \boldsymbol{F}) = 0$. For the stability of the DT model then $|z_i| < 1$ $(i = 1, 2, .., n)$ applies.

Example 11.6 Derive an analytical relation to determine the *SRE* DT model of the first order lag given by $H(s) = K/(1 + sT)$. Use the simplified expression of $G(z) = (1 - z^{-1})\mathcal{Z}\{H(s)/s\}$

$$\begin{aligned} G(z) &= (1 - z^{-1})\mathcal{Z}\left\{\frac{H(s)}{s}\right\} = (1 - z^{-1})\mathcal{Z}\left\{\frac{K}{s(1 + sT)}\right\} = (1 - z^{-1})\mathcal{Z}\left\{\frac{K}{s} - \frac{KT}{1 + sT}\right\} \\ &= (1 - z^{-1})\frac{Kz}{z - 1} - (1 - z^{-1})\frac{Kz}{z - e^{-T_s/T}} = K\left(1 - \frac{z - 1}{z - e^{-T_s/T}}\right) = K\frac{1 - e^{-T_s/T}}{z - e^{-T_s/T}} \\ &= \frac{K(1 - e^{-T_s/T})z^{-1}}{1 - e^{-T_s/T}z^{-1}} = \frac{b_1 z^{-1}}{1 + a_1 z^{-1}} \end{aligned} \tag{11.79}$$

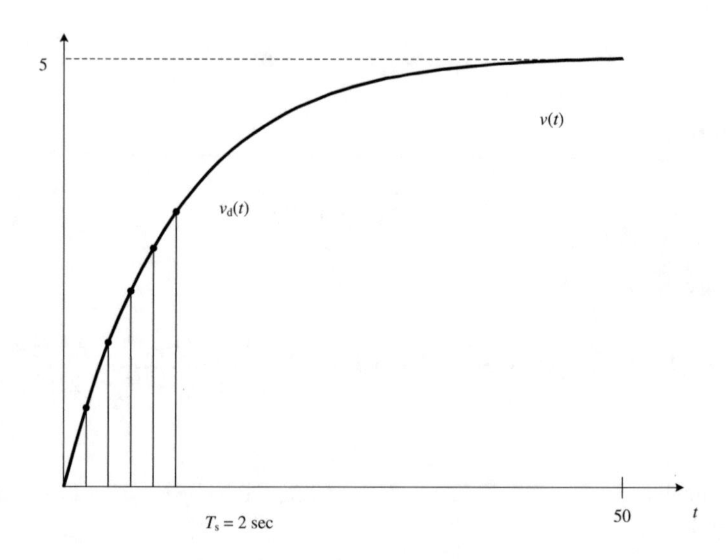

Fig. 11.17 CT and *SRE* DT step responses of a first order lag

Figure 11.17 shows the step response for $K = 5$ and $T = 10$. Check the initial and final value of the step response using the initial value and final theorems:

$$y[0] = \lim_{z\to\infty} Y(z) = \lim_{z\to\infty}\{U(z)G(z)\} = \lim_{z\to\infty}\left\{\frac{z}{z-1}\frac{K\left(1-e^{-T_s/T}\right)}{z-e^{-T_s/T}}\right\} = 0$$

and

$$\begin{aligned}\lim_{k\to\infty} y[k] &= \lim_{z\to 1}\left\{\left(1-z^{-1}\right)Y(z)\right\} = \lim_{z\to 1}\left\{\left(1-z^{-1}\right)U(z)G(z)\right\}\\ &= \lim_{z\to 1}\left\{\left(1-z^{-1}\right)\frac{z}{z-1}\frac{K\left(1-e^{-T_s/T}\right)}{z-e^{-T_s/T}}\right\} = K\end{aligned}$$

It may be of interest to analyze the dependence of the coefficients b_1 and a_1 on T and T_s, respectively. Introducing the relative sampling rate as $x = T_s/T$ we get

$$\begin{aligned} b_1 &= K(1-e^{-x})\\ a_1 &= -e^{-x}\end{aligned} \tag{11.80}$$

It is easy to check the steady-state gain: $G(z = 1) = K$. The pole of $G(z)$ turns out to be $p_1^{d} = -a_1 = e^{-x}$. ■

Example 11.7 To derive the DT *SRE* model of a second order CT process is a far more tiresome procedure, though this relationship is frequently needed. For a process with a complex pole pair $p_{1,2}^{c} = a \pm jb$, the well-known expressions using the damping factor of ξ and the natural frequency ω_o are $a = -\xi\omega_o$ and $b = \omega_o\sqrt{1-\xi^2}$. The DT pair of complex conjugates takes the following form:

$$\left(z-p_1^{d}\right)\left(z-\bar{p}_1^{d}\right) = z^2 - \left[2e^{2aT_s}\cos(bT_s)\right]z + e^{2aT_s}, \tag{11.81}$$

■

11.4.4 Analysis of DT Systems in the Frequency Domain

The analysis of systems driven by sinusoidal signals forms a fundamental method of system analysis. Consider the response of a stable DT model by

$$G(z) = \frac{b_1 z^{n-1} + b_2 z^{n-2} + \cdots + b_n}{z^n + a_1 z^{n-1} + a_2 z^{n-2} + \cdots + a_n} = \frac{\mathcal{B}(z)}{\mathcal{A}(z)} \tag{11.82}$$

driven by a sequence of sinusoidal samples

$$u[k] = K\cos(\omega_o k)$$

From Table 11.1,

$$U(z) = \mathcal{Z}\{u[k]\} = K\frac{z^2 - z\cos(\omega_o T_s)}{z^2 + 2z\cos(\omega_o T_s) + 1},$$

and the z-transform of the output is

$$Y(z) = U(z)G(z) = K\frac{z^2 - z\cos(\omega_o T_s)}{z^2 + 2z\cos(\omega_o T_s) + 1}G(z) = K\frac{z^2 - z\cos(\omega_o T_s)}{z^2 + 2z\cos(\omega_o T_s) + 1}\frac{\mathcal{B}(z)}{\mathcal{A}(z)}$$

Further manipulations give

$$\frac{Y(z)}{z} = K\frac{z - \cos(\omega_o T_s)}{(z - e^{j\omega_o T_s})(z - e^{-j\omega_o T_s})}\frac{\mathcal{B}(z)}{\mathcal{A}(z)}$$

and the *PFE* turns out to be

$$\frac{Y(z)}{z} = \frac{\mathcal{V}(z)}{\mathcal{A}(z)} + \frac{c}{z - e^{j\omega_o T_s}} + \frac{\bar{c}}{z - e^{-j\omega_o T_s}}$$

where $\mathcal{V}(z)$ is a polynomial with degree less than n, while the coefficient c equals

$$c = \left[\left(z - e^{j\omega_o T_s}\right)\frac{Y(z)}{z}\right]_{z=e^{j\omega_o}} = \cdots = \frac{K}{2}G\left(e^{j\omega_o T_s}\right).$$

As $\bar{c}$ is the complex conjugate of c,

$$Y(z) = \frac{z\mathcal{V}(z)}{\mathcal{A}(z)} + \frac{K}{2}\left[\frac{zG(e^{j\omega_o T_s})}{z - e^{j\omega_o T_s}} + \frac{z\bar{G}(e^{j\omega_o T_s})}{z - e^{j\omega_o T_s}}\right]$$

is the sum of a transient $y_{tr}[k]$ and a steady-state component $y_{ss}[k]$:

$$y[k] = y_{tr}[k] + y_{ss}[k]. \tag{11.83}$$

Stability requires $\lim_{k\to\infty} y_{tr}[k] = 0$ and the steady-state component is

$$y_{ss}[k] = K\left|G\left(e^{j\omega_o T_s}\right)\right|\cos(\omega_o T_s k + \Theta), \tag{11.84}$$

where

$$G\left(e^{j\omega_o T_s}\right) = \left|G\left(e^{j\omega_o T_s}\right)\right| e^{j\Theta}. \tag{11.85}$$

The sinusoidal excitation causes the steady-state response of a stable DT system to be sinusoidal with the same frequency but an amplitude of

$$G(z)|_{z=e^{j\omega_o T_s}} = G\left(e^{j\omega_o T_s}\right) = \left|G\left(e^{j\omega_o T_s}\right)\right| e^{j\Theta}. \tag{11.86}$$

The above result is in harmony with the result obtained earlier for CT systems. The results are obviously different, however, the DT and CT systems share identical properties.

Based on the discussion of DT systems so far, it is rather general that in the expression of the $G(z)$ pulse transfer function and that of the transfer function operator $G(q)$ involve the variables z^{-1} and q^{-1}, rather than z and q, respectively. To derive the filter forms $G(z^{-1})$ and $G(q^{-1})$ can be derived from $G(z)$ and $G(q)$ by dividing both their numerator and the denominator by the highest power term. These forms are

$$\begin{aligned} G\left(z^{-1}\right) &= \frac{\mathcal{B}(z^{-1})}{\mathcal{A}(z^{-1})} = \frac{b_1 z^{-1} + b_2 z^{-2} + \cdots + b_n z^{-n}}{1 + a_1 z^{-1} + a_2 z^{-2} + \cdots + a_n z^{-n}} \\ &= \frac{\left(b_1 + b_2 z^{-1} + \cdots + b_n z^{-(n-1)}\right) z^{-1}}{1 + a_1 z^{-1} + a_2 z^{-2} + \cdots + a_n z^{-n}} \end{aligned} \tag{11.87}$$

or

$$\begin{aligned} G\left(q^{-1}\right) &= \frac{\mathcal{B}(q^{-1})}{\mathcal{A}(q^{-1})} = \frac{b_1 q^{-1} + b_2 q^{-2} + \cdots + b_n q^{-n}}{1 + a_1 q^{-1} + a_2 q^{-2} + \cdots + a_n q^{-n}} \\ &= \frac{\left(b_1 + b_2 q^{-1} + \cdots + b_n q^{-(n-1)}\right) q^{-1}}{1 + a_1 q^{-1} + a_2 q^{-2} + \cdots + a_n q^{-n}} \end{aligned} \tag{11.88}$$

Note that if the transfer function of the CT system is strictly proper, than the *SRE* transformation results in transfer functions $G(z)$ and $G(q)$ with pole excess one. The CT time-delay e^{-sT_d} will be represented by

$$z^{-d} \quad ; \quad q^{-d} \tag{11.89}$$

where

$$d = \text{int}\left\{\frac{T_d}{T_s}\right\} \tag{11.90}$$

with $d = \text{int}\{\ldots\}$ meaning the separation of the integer part. z^{-d} corresponds to the general delay relationship by (11.11). The z^{-1} and q^{-1} terms in the expression of

$G(z^{-1})$ and $G(q^{-1})$, respectively, are not part of the z^{-d} and q^{-d} terms representing the time-delay. Thus for a process containing an actual time-delay it is reasonable to use the pulse transfer function

$$G(z^{-1}) = \frac{b_1 + b_2 z^{-1} + \cdots + b_n z^{-(n-1)}}{1 + a_1 z^{-1} + a_2 z^{-2} + \cdots + a_n z^{-n}} z^{-(d+1)}, \tag{11.91}$$

which is the *SRE* transform of the general CT system given by

$$H(s) = k \frac{\prod_{i=1}^{m} \left(s - z_i^{\mathrm{c}}\right)}{\prod_{i=1}^{n} \left(s - p_i^{\mathrm{c}}\right)} e^{-sT_{\mathrm{d}}}; \quad m<n. \tag{11.92}$$

In the CT state-model form the gain of the direct constant channel is zero $(d = 0)$.

11.4.5 *Transformation of Zeros*

The allocation of poles and zeros of an *nth* order $G(z)$ pulse transfer function is an interesting issue. Assume the *SRE* transformation is used for the discretization. Based on the example elaborated earlier for a first order lag, it can be seen that the transformation of the poles follows $p_i^{\mathrm{d}} = e^{-x_i}$, where $x_i = T_{\mathrm{s}}/T_i$ and the pole of the CT system is $-1/T_i$. (The relation is valid for complex conjugate poles as well, however, it is reasonable to handle those poles as complex conjugate pairs. Anyway, the transformation relationship will be far more involved.) The exponential relationship will map stable CT poles to stable DT poles, for the complete CT stability region (the left half-plane) is transformed into the stable DT region (the unit disc). Also, unstable CT poles (in the right half-plane) will be transformed into unstable DT poles (outside the unit disc). The transformation of the zeros, however, does not follow this pattern. Note first that $G(z)$ always has $(n-1)$ zeros. Provided that the CT system has only m stable zeros, then m out of the $(n-1)$ DT zeros are transformed approximately by $z_i^{\mathrm{d}} \approx e^{-x_i}$ (the exact relationship is far more involved). The rest of the zeros—there are $(n-m-1)$ of them—follow a different law, as those zeros have no CT counterparts. Typically, these (unmatched) zeros are located on the real axis. If the pole excess is larger than one, even for minimum phase systems (having stable zeros), the unstable zeros (out of the unmatched zeros) should be accounted for. In short, even minimum-phase CT systems may turn into non-minimum phase DT systems after the *SRE* transformation.

Similarly, an important observation is that a fractional time-delay (a delay which is not an integer multiple of the sampling time) may result in unstable zeros. As unstable zeros play an important role in the DT controller design procedure, it should be noted that nonminimum-phase CT systems represent a very special class. However, DT nonminimum-phase systems are rather common.

11.5 Structural Properties of State Equations

The state-space description of a sampled data system consists of the state difference equation

$$\boldsymbol{x}[k+1] = \boldsymbol{F}\boldsymbol{x}[k] + \boldsymbol{g}u[k] \tag{11.93}$$

and

$$y[k] = \boldsymbol{c}^{\mathrm{T}}\boldsymbol{x}[k] + du[k]. \tag{11.94}$$

Introducing a shift element the realization of the system is shown in Fig. 11.18.

The state difference equation describes the development of the states of the system, whereas the output equation gives how the output depends on the state variables and incidentally directly on the input. The pulse transfer function of the system (supposing zero initial conditions) is given by

$$G(z) = \frac{\mathcal{Z}\{y[k]\}}{\mathcal{Z}\{u[k]\}} = \boldsymbol{c}^{\mathrm{T}}(z\boldsymbol{I} - \boldsymbol{F})^{-1}\boldsymbol{g} + d. \tag{11.95}$$

Similarly to CT systems, an infinite number of state space representations can be derived from a given pulse transfer function, which provide the same output signal for a given input signal (therefore these representations are called input-output equivalent descriptions). To verify this statement, in the sequel several different input-output equivalent state space descriptions will be derived from a given pulse transfer function.

Example 11.8 Consider the pulse transfer function

$$G(z) = \frac{b'_0z^3 + b'_1z^2 + b'_2z + b'_3}{z^3 + a_1z^2 + a_2z + a_3} = \frac{b_1z^2 + b_2z + b_3}{z^3 + a_1z^2 + a_2z + a_3} + d, \tag{11.96}$$

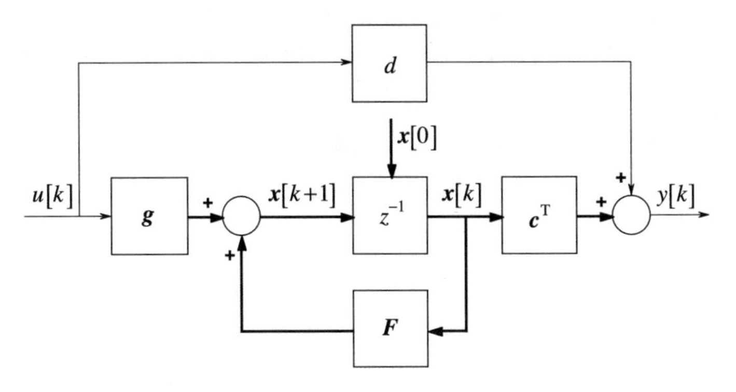

Fig. 11.18 Block-diagram representing the state equation of a discrete system

Transform it to the form

$$G(z) = \frac{b_1 z^{-1} + b_2 z^{-2} + b_3 z^{-3}}{1 + a_1 z^{-1} + a_2 z^{-2} + a_3 z^{-3}} + d. \tag{11.97}$$

Then

$$Y(z) = \frac{b_1 z^{-1} + b_2 z^{-2} + b_3 z^{-3}}{1 + a_1 z^{-1} + a_2 z^{-2} + a_3 z^{-3}} U(z) + dU(z). \tag{11.98}$$

Observe that the value of d is non-zero only if the degrees of the numerator and of the denominator of $G(z)$ are identical (the degree of the numerator can be at most the same as that of the denominator).

Let us realize first the relationship

$$Y(z) = \left[\left(b_1 z^{-1} + b_2 z^{-2} + b_3 z^{-3}\right) / \left(1 + a_1 z^{-1} + a_2 z^{-2} + a_3 z^{-3}\right)\right] U(z),$$

then the value of $dU(z)$ has to be added to the output. By cross multiplication

$$Y(z) + a_1 z^{-1} Y(z) + a_2 z^{-2} Y(z) + a_3 z^{-3} Y(z) = b_1 z^{-1} U(z) + b_2 z^{-2} U(z) + b_3 z^{-3} U(z) \tag{11.99}$$

is obtained. Rearranging the equation $Y(z)$ can be expressed as

$$Y(z) = [b_1 U(z) - a_1 Y(z)] z^{-1} + [b_2 U(z) - a_2 Y(z)] z^{-2} + [b_3 U(z) - a_3 Y(z)] z^{-3}. \tag{11.100}$$

The realization will provide a block-scheme, which besides the constants and the summation elements, contains shift blocks with pulse transfer function of z^{-1}, realizing a one step delay. As seen from the equation above, the signal $[b_1 U(z) - a_1 Y(z)]$ has to pass through one delay element, the signal $[b_2 U(z) - a_2 Y(z)]$ has to pass through two, whereas signal $[b_3 U(z) - a_3 Y(z)]$ has to go through three. Based on these considerations the block-diagram showing the realization is drawn in Fig. 11.19. The state variables can be chosen at the outputs of the shift elements. Thus the state equation can be written as

$$\begin{aligned} x_1[k+1] &= -a_3 x_3[k] + b_3 u[k] \\ x_2[k+1] &= x_1[k] - a_2 x_3[k] + b_2 u[k] \\ x_3[k+1] &= x_2[k] - a_1 x_3[k] + b_1 u[k] \\ y[k] &= x_3[k] + du[k] \end{aligned} \tag{11.101}$$

In the figure the time domain and z domain notations appear at the same time, thus the equations of the given realization can be directly "read": if the output of a shift element is $x_i[k]$, then its input is $x_i[k+1]$. It is easily seen in the figure that a

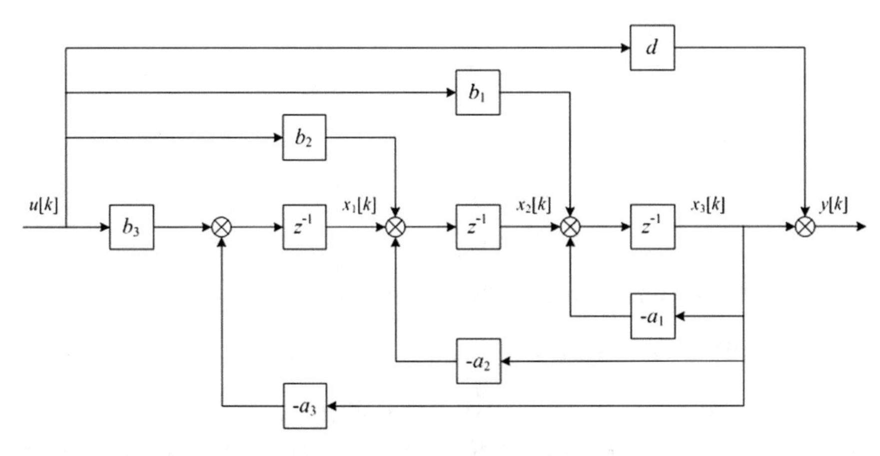

Fig. 11.19 Block-scheme of the observable canonical form

direct channel between $u[k]$ and $y[k]$ without dynamics exists only if $d \neq 0$. This means that an abrupt change in the input signal causes an immediate change in the output signal only if the degrees of the numerator and of the denominator of the pulse transfer function are identical. If this condition does not hold, then a change in the input signal affects the output signal only in later steps. Finally, arranging the equations derived above according to the state space form yields

$$\boldsymbol{x}[k+1] = \begin{bmatrix} x_1[k+1] \\ x_2[k+1] \\ x_3[k+1] \end{bmatrix} = \begin{bmatrix} 0 & 0 & -a_3 \\ 1 & 0 & -a_2 \\ 0 & 1 & -a_1 \end{bmatrix} \boldsymbol{x}[k] + \begin{bmatrix} b_3 \\ b_2 \\ b_1 \end{bmatrix} u[k] = \bar{\boldsymbol{F}}_{\mathrm{o}} \boldsymbol{x}[k] + \bar{\mathbf{g}}_{\mathrm{o}} u[k]$$

$$y[k] = \begin{bmatrix} 0 & 0 & 1 \end{bmatrix} \begin{bmatrix} x_1[k] \\ x_2[k] \\ x_3[k] \end{bmatrix} + du[k] = \bar{\mathbf{c}}_{\mathrm{o}}^{\mathrm{T}} \boldsymbol{x}[k] + du[k] \tag{11.102}$$

Note that the derived state space description gives the so called *observable canonical form* (see the explanation of this phrase around the formula (3.55) in the CT case).

Derive now another realization of the subsystem by

$$Y(z) = \left[\left(b_1 z^{-1} + b_2 z^{-2} + b_3 z^{-3} \right) / \left(1 + a_1 z^{-1} + a_2 z^{-2} + a_3 z^{-3} \right) \right] U(z)$$

(the realization of the dynamic subsystem will be extended by the $du[k]$ element afterwards). The method is the following: after cross multiplication an intermediate variable is defined; first this variable is created from the input signals, then the output is built using the intermediate $v[k]$ variable. Now the pulse transfer function is arranged as

$$\frac{Y(z)}{b_1 z^{-1} + b_2 z^{-2} + b_3 z^{-3}} = \frac{U(z)}{1 + a_1 z^{-1} + a_2 z^{-2} + a_3 z^{-3}} = V(z). \tag{11.103}$$

Let us consider first the equation

$$\frac{U(z)}{1 + a_1 z^{-1} + a_2 z^{-2} + a_3 z^{-3}} = V(z), \tag{11.104}$$

which in the time domain provides the recursive relationship

$$v[k] = u[k] - a_1 v[k-1] - a_2 v[k-2] - a_3 v[k-3]. \tag{11.105}$$

Regarding the realization let us start with three serially connected shift elements driven by the input $u[k]$ (Fig. 11.20). Coming back to the equation $V(z) = Y(z)/(b_1 z^{-1} + b_2 z^{-2} + b_3 z^{-3})$ the difference equation in the time domain describing the output signal can be written as $y[k] = b_1 v[k-1] + b_2 v[k-2] + b_3 v[k-3]$, thus the whole realization can be drawn as given in Fig. 11.21. Based on Fig. 11.21 the following equations can be written:

$$\begin{aligned} x_1[k+1] &= -a_1 x_1[k] - a_2 x_2[k] - a_3 x_3[k] + u[k] \\ x_2[k+1] &= x_1[k] \\ x_3[k+1] &= x_2[k] \\ y[k] &= b_1 x_1[k] + b_2 x_2[k] + b_3 x_3[k] + du[k] \end{aligned} \tag{11.106}$$

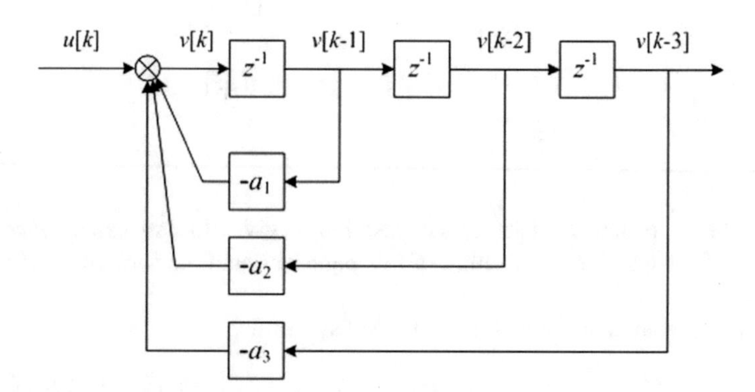

Fig. 11.20 Creating an intermediate variable

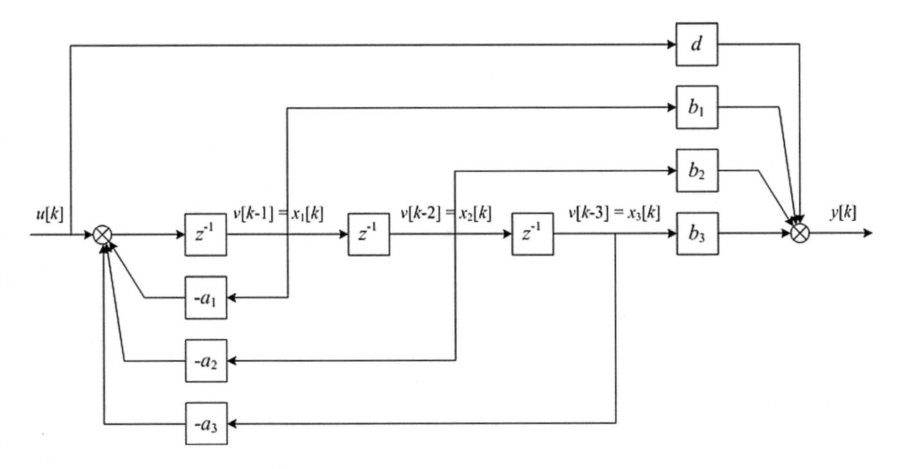

Fig. 11.21 Realization of the controllable canonical form

Thus the state space model is

$$\boldsymbol{x}[k+1] = \begin{bmatrix} x_1[k+1] \\ x_2[k+1] \\ x_3[k+1] \end{bmatrix} = \begin{bmatrix} -a_1 & -a_2 & -a_3 \\ 1 & 0 & 0 \\ 0 & 1 & 0 \end{bmatrix} \boldsymbol{x}[k] + \begin{bmatrix} 1 \\ 0 \\ 0 \end{bmatrix} u[k] = \boldsymbol{F}_c \boldsymbol{x}[k] + \boldsymbol{g}_c u[k]$$

$$y[k] = [b_1 \quad b_2 \quad b_3] \begin{bmatrix} x_1[k] \\ x_2[k] \\ x_3[k] \end{bmatrix} + du[k] = \mathbf{c}_c^T \boldsymbol{x}[k] + du[k] \tag{11.107}$$

This model is called the *controllable canonical form* (see the explanation of this phrase for the formula (3.46) in the CT case). ■

The solution of the DT state space system in the form $\boldsymbol{x}[k] = (q\boldsymbol{I} - \boldsymbol{F})^{-1}\boldsymbol{g}u[k]$ directly shows that in the case of a diagonal state transition matrix $\boldsymbol{F}$ the matrix inversion can be avoided; therefore transforming the pulse transfer function to a form which results in a diagonal representation will lead to an advantageous structure. Generally let $G(z)$ be

$$G(z) = \frac{b_1 z^{-1} + b_2 z^{-2} + \cdots + b_{n-1} z^{-(n-1)}}{1 + a_1 z^{-1} + a_2 z^{-2} + \cdots + a_n z^{-n}} + d = \sum_{i=1}^{n} \frac{\beta_i z^{-1}}{1 + \alpha_i z^{-1}} + d, \tag{11.108}$$

then introducing the state variables by

$$x_i[k+1] = \beta_i u[k] - \alpha_i x_i[k] \tag{11.109}$$

the state difference equation will be

$$\boldsymbol{x}[k+1] = \begin{bmatrix} x_1[k+1] \\ x_2[k+1] \\ \vdots \\ x_n[k+1] \end{bmatrix} = \begin{bmatrix} -\alpha_1 & 0 & 0 & 0 \\ 0 & -\alpha_2 & 0 & 0 \\ 0 & 0 & \ddots & 0 \\ 0 & 0 & 0 & -\alpha_n \end{bmatrix} \begin{bmatrix} x_1[k] \\ x_2[k] \\ \vdots \\ x_n[k] \end{bmatrix} + \begin{bmatrix} \beta_1 \\ \beta_2 \\ \vdots \\ \beta_n \end{bmatrix} u[k]$$
$$= \boldsymbol{F}_{\mathrm{d}}\boldsymbol{x}[k] + \boldsymbol{g}_{\mathrm{d}}u[k] \tag{11.110}$$

and the output equation is given by

$$y[k] = [\,1 \quad 1 \quad \ldots \quad 1\,]\boldsymbol{x}[k] + du[k] = \boldsymbol{c}_{\mathrm{d}}^{\mathrm{T}}\boldsymbol{x}[k] + du[k]. \tag{11.111}$$

It was supposed that the poles of $G(z)$ are single and real. The derived form is also called *parallel canonical form*. The realization is shown in Fig. 11.22.

It can be easily seen that besides $\boldsymbol{F}_{\mathrm{d}} = \mathbf{diag}[-\alpha_1, -\alpha_2, \ldots, -\alpha_n]$ the state space model with

$$\begin{aligned} \boldsymbol{x}[k+1] &= \boldsymbol{F}_{\mathrm{d}}\boldsymbol{x}[k] + \boldsymbol{g}_{\mathrm{d}}u[k] \\ y[k] &= \boldsymbol{c}_{\mathrm{d}}^{\mathrm{T}}\boldsymbol{x}[k] + du[k] \end{aligned} \tag{11.112}$$

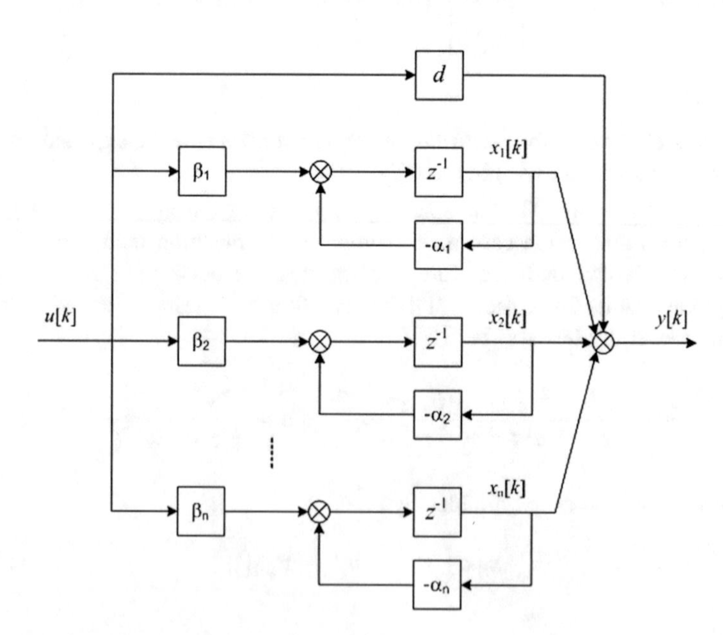

Fig. 11.22 Block-diagram of the parallel canonical form

gives a parallel canonical form, where $g_i c_i = g_i = \beta_i$ is fulfilled. (Comparing this with the CT forms (3.38), (3.39) it can be seen that in the DT form, simply $\gamma_i = 1$ was chosen.)

(In the state space forms—following the conventions—the gain of the direct constant channel was denoted by d, whose value is the same both for the CT and the DT case. As in the DT case its value generally is zero, it was not so disturbing, that the same letter was used also for the discrete dead-time.)

Chapter 12
Sampled Data Controller Design for Stable Discrete-Time Processes

Comparing Chaps. 7 and 9 it is worth distinguishing stable and unstable processes at the design step of a CT controller. For stable processes precise mathematical methods—based on or derived from YOULA-parameterization—are available. Even the conventional *PID* regulator—under the usual tuning—corresponds to a rough approach of the YOULA-controller. The unstable processes require different methods, where the most important task is to stabilize the process. For this task, in Chap. 9, a general polynomial method based on state-feedback and observer was presented, and in Chap. 10 another method based on DIOPHANTINE-equations (*DE*) was presented. For sampled data systems, a similar logic will be followed to design the controllers. The main difference is that the establishment of the different signal-forming items realizing the different methods for DT control loops in computer based control systems is significantly simpler.

In this section the transfer function operators $G(\ldots)$ are functions of z and z^{-1} or q and q^{-1} depending on the character of the DT description. The design methods presented here are practically polynomial (algebraic) methods, so they can be applied to all models discussed for the sampled systems, only the chosen model form has to be applied correspondingly.

12.1 The YOULA Controller for Sampled Data Systems

In Chap. 7 a general control parameterization method and a design method based on that was shown. The so-called YOULA-parameterization method was suggested for the design of one- or two degree of freedom (*ODOF*, *TDOF*) control loops. The advantage of the method is that the design of the closed-loop is assigned to two reference models, namely the reference signal tracking behavior may be assigned to R_{r}, the disturbance rejection to R_{n}, and the design of the controller can be given in a

L. Keviczky et al., *Control Engineering*, Advanced Textbooks in Control and Signal Processing, https://doi.org/10.1007/978-981-10-8297-9_12

relatively simple closed form. The disadvantage of the method is that it can be applied only to stable processes.

The presentation of the YOULA-parameterization, its relationship to the *IMC* principle, the optimality and the best reachable control, were discussed in a very general way, so—in many cases—it is enough to replace the transfer functions with the pulse transfer functions, and all the relations are valid here, too. Therefore the general statements of Chap. 7 are not repeated here, instead the differences and deviations are emphasized. Consider the DT process

$$\begin{aligned} G(z^{-1}) &= G_+(z^{-1})\bar{G}_-(z^{-1}) = G_+(z^{-1})G_-(z^{-1})z^{-d} \quad \text{or} \\ G &= G_+\bar{G}_- = G_+G_-z^{-d}, \end{aligned} \tag{12.1}$$

where G_+ is stable, its inverse is also stable and realizable (*ISR*). The inverse of $\overline{G}_-$ is unstable (*Inverse Unstable*: *IU*) and non-realizable (*IUNR*). In general the inverse of the delay z^{-d} is unrealizable, because it would mean an ideal predictor.

The optimal controller obtained for the general case [see (7.14)] is

$$C_{\mathrm{opt}} = \frac{R_{\mathrm{n}}K_{\mathrm{n}}}{1 - R_{\mathrm{n}}K_{\mathrm{n}}G} = \frac{Q_{\mathrm{opt}}}{1 - Q_{\mathrm{opt}}G} = \frac{R_{\mathrm{n}}G_{\mathrm{n}}G_+^{-1}}{1 - R_{\mathrm{n}}G_{\mathrm{n}}G_-z^{-d}} = R_{\mathrm{n}}G_{\mathrm{n}}C'_{\mathrm{opt}}, \tag{12.2}$$

where the optimal YOULA parameter is

$$Q_{\mathrm{opt}} = R_{\mathrm{n}}G_{\mathrm{n}}G_+^{-1} = R_{\mathrm{n}}K_{\mathrm{n}} \quad \text{and} \quad K_{\mathrm{n}} = G_{\mathrm{n}}G_+^{-1} \tag{12.3}$$

and

$$Q_{\mathrm{r}} = R_{\mathrm{r}}G_{\mathrm{r}}G_+^{-1} = R_{\mathrm{r}}K_{\mathrm{r}} \quad \text{and} \quad K_{\mathrm{r}} = G_{\mathrm{r}}G_+^{-1}. \tag{12.4}$$

For sampled systems the equivalent optimal control system corresponding to the generalized *IMC* principle is completely the same as seen in Fig. 7.9, whose simplified version is shown in Fig. 12.1.

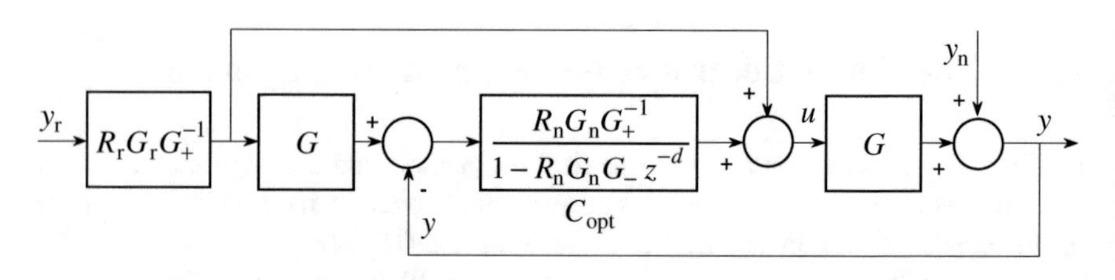

Fig. 12.1 Optimal sampled-time control system based on the generalized *IMC* principle

The most important signals of the *TDOF* closed-control loop are

$$
\begin{aligned}
u_{\mathrm{opt}} &= R_{\mathrm{r}} G_{\mathrm{r}} G_{+}^{-1} y_{\mathrm{r}} - R_{\mathrm{n}} G_{\mathrm{n}} G_{+}^{-1} y_{\mathrm{n}} \\
e_{\mathrm{opt}} &= \left(1 - R_{\mathrm{r}} G_{\mathrm{r}} G_{-} z^{-d}\right) y_{\mathrm{r}} - \left(1 - R_{\mathrm{n}} G_{\mathrm{n}} G_{-} z^{-d}\right) y_{\mathrm{n}} = \left(1 - T_{\mathrm{r}}^{\mathrm{opt}}\right) y_{\mathrm{r}} - S_{\mathrm{n}}^{\mathrm{opt}} y_{\mathrm{n}} \\
y_{\mathrm{opt}} &= R_{\mathrm{r}} G_{\mathrm{r}} G_{-} z^{-d} y_{\mathrm{r}} + \left(1 - R_{\mathrm{n}} G_{\mathrm{n}} G_{-} z^{-d}\right) y_{\mathrm{n}} = T_{\mathrm{r}}^{\mathrm{opt}} y_{\mathrm{r}} + \left(1 - T_{\mathrm{n}}^{\mathrm{opt}}\right) y_{\mathrm{n}} = T_{\mathrm{r}}^{\mathrm{opt}} y_{\mathrm{r}} + S_{\mathrm{n}}^{\mathrm{opt}} y_{\mathrm{n}}
\end{aligned}
\tag{12.5}
$$

where the equalities $T_{\mathrm{r}}^{\mathrm{opt}} = R_{\mathrm{r}} G_{\mathrm{r}} G_{-} z^{-d}$ and $T_{\mathrm{n}}^{\mathrm{opt}} = R_{\mathrm{n}} G_{\mathrm{n}} G_{-} z^{-d}$ are valid. The further equivalent forms of the best reachable optimal control systems are shown in Fig. 12.2. (These figures are for illustration only, their realizability has to be investigated in each case!).

As mentioned earlier, the theory of the optimality of G_{r} and G_{n} will not be discussed here. The choice $G_{\mathrm{r}} = G_{\mathrm{n}} = 1$ employed in this simple case leaves the invariant process factor G_{-} unchanged, so it appears unchanged in the signals of the system.

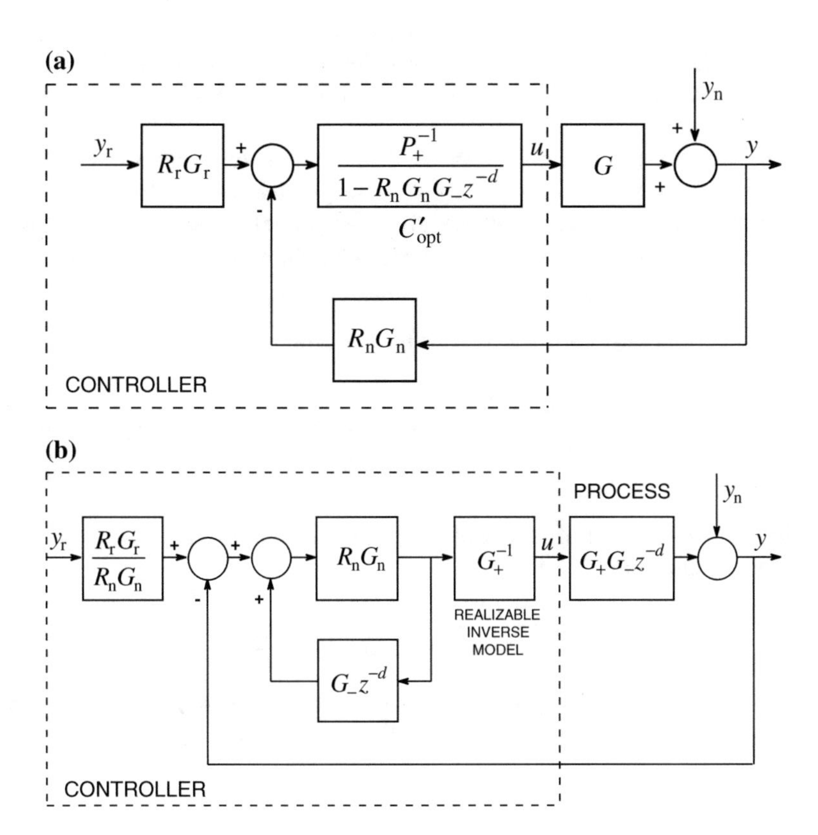

Fig. 12.2 The equivalent forms of the best reachable optimal sampled data control loop

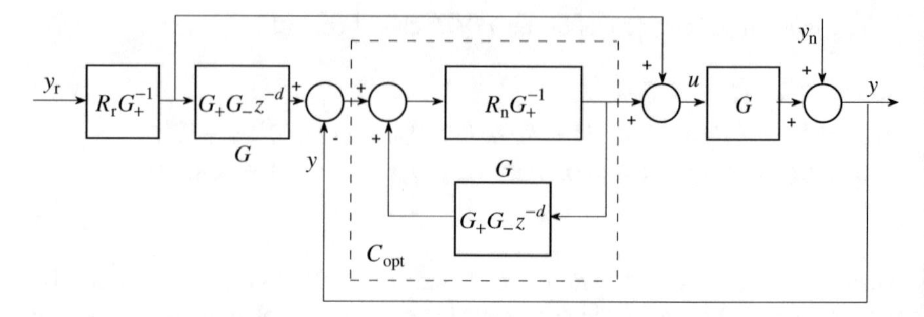

Fig. 12.3 YOULA-parameterized realizable sampled control loop with the choice of $G_r = G_n = 1$

$$
\begin{aligned}
u &= R_r G_+^{-1} y_r - R_n G_+^{-1} y_n \\
e &= \left(1 - R_r G_- z^{-d}\right) y_r - \left(1 - R_n G_- z^{-d}\right) y_n = (1 - T_r) y_r - S_n y_n \\
y &= R_r G_- z^{-d} y_r + \left(1 - R_n G_- z^{-d}\right) y_n = T_r y_r + (1 - T_n) y_n = T_r y_r + S_n y_n
\end{aligned}
\tag{12.6}
$$

so the realizability of the transfer functions $R_r G_+^{-1}$, $R_n G_+^{-1}$, and $R_n G_-$ has to be ensured, respectively. It can be clearly seen that the realizability can be simply handled by the appropriate choice of the order and pole excess of the reference models R_r and R_n. A realizable but not optimal control system is shown in Fig. 12.3.

In the case of sampled data systems it is worth noting, that using the *SRE* transformation it is always true for the delay-free part $(G_+ G_-)$ in the pulse transfer function of the process—independently from the pole excess of the CT process—that the pole excess is equal to one. So the realizability of the items $R_r G_+^{-1}$ and $R_n G_+^{-1}$ is ensured even for first order reference models R_r and R_n.

Example 12.1 Let the controlled system be a first order process with delay

$$
G = \frac{0.2z^{-1}}{1 - 0.8z^{-1}} z^{-3} = \frac{0.2z^{-4}}{1 - 0.8z^{-1}} \quad \text{i.e.} \quad G_+ = \frac{0.2z^{-1}}{1 - 0.8z^{-1}} \quad \text{and} \quad G_- = 1
\tag{12.7}
$$

and the goal of the control is to make it faster. Let the tracking and disturbance rejection reference models be

$$
R_r = \frac{0.8z^{-1}}{1 - 0.2z^{-1}} \quad \text{and} \quad R_n = \frac{0.5z^{-1}}{1 - 0.5z^{-1}}.
\tag{12.8}
$$

Since $G_- = 1$ there is nothing to be compensated optimally, i.e., $G_r = 1$ and $G_n = 1$ can be chosen. The optimal controller is

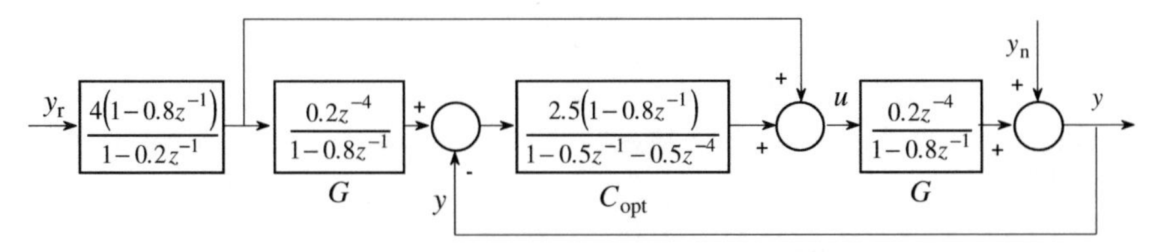

Fig. 12.4 The optimal control loop of Example 12.1

$$
\begin{aligned}
C_{\text{opt}} &= \frac{R_n G_n G_+^{-1}}{1 - R_n G_n G_- z^{-d}} = \frac{1}{1 - R_n z^{-d}} R_n G_+^{-1} \\
&= \frac{1}{1 - \frac{0.5z^{-1}}{1-0.5z^{-1}} z^{-3}} \frac{0.5z^{-1}}{1 - 0.5z^{-1}} \frac{1 - 0.8z^{-1}}{0.2z^{-1}} = \frac{2.5(1 - 0.8z^{-1})}{1 - 0.5z^{-1} - 0.5z^{-4}}
\end{aligned}
\tag{12.9}
$$

and the serial compensation has the form

$$
R_r G_+^{-1} = \frac{0.8z^{-1}}{1 - 0.2z^{-1}} \frac{1 - 0.8z^{-1}}{0.2z^{-1}} = \frac{4(1 - 0.8z^{-1})}{1 - 0.2z^{-1}}
\tag{12.10}
$$

so the optimal *TDOF* control loop has the scheme shown in Fig. 12.4 (see Fig. 12.2b). Note that $C_{\text{opt}}(z = 1) = \infty$, i.e., the controller has an integrating character, which comes from the condition $R_n(z = 1) = 1$.

It can be easily checked that the output of the closed-loop is

$$
\begin{aligned}
y_{\text{opt}} &= R_r z^{-d} y_r + \left(1 - R_n z^{-d}\right) y_n = \frac{0.8z^{-1}}{1 - 0.2z^{-1}} z^{-3} y_r + \left(1 - \frac{0.5z^{-1}}{1 - 0.5z^{-1}} z^{-3}\right) y_n \\
&= \frac{0.8z^{-4}}{1 - 0.2z^{-1}} y_r + \left(1 - \frac{0.5z^{-4}}{1 - 0.5z^{-1}}\right) y_n
\end{aligned}
\tag{12.11}
$$

which completely corresponds to the designed *TDOF* control system. ■

12.2 The SMITH Controller for Sampled Data System

Let us consider a simple process with delay based on (12.1) in the sampled data control system

$$
\begin{aligned}
G(z^{-1}) &= G_+(z^{-1}) \bar{G}_-(z^{-1}) = G_+(z^{-1}) G_-(z^{-1}) z^{-d} \quad \text{or} \\
G &= G_+ \bar{G}_- = G_+ G_- z^{-d},
\end{aligned}
\tag{12.12}
$$

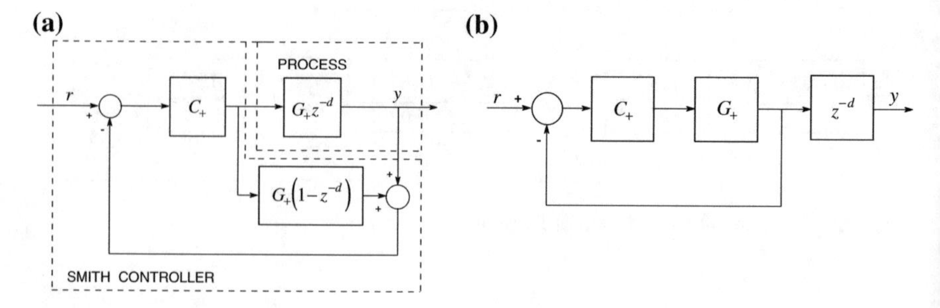

Fig. 12.5 The scheme of the sampled data SMITH controller

where G_+ is stable. For the DT process of (12.12) the SMITH-predictor principle discussed in Chap. 7.1. is shown by the control system given in Fig. 12.5a. Since this control loop is equivalent with the scheme given in Fig. 12.5b, the goal of the control is clearly seen, to separate the original closed-loop containing the delay to a delay-free closed-loop and the delay appears serially connected. So the controller C_+ for the process G_+ can also be designed by a conventional method (not taking the delay into account).

Figure 12.5a can be redrawn for the equivalent forms (a) and (b) of Fig. 12.6 by simple block-manipulations.

The *IMC* structure of Fig. 12.6a clearly shows that the SMITH controller is a YOULA-parameterized special controller with YOULA parameter

$$Q_+ = \frac{C_+}{1+C_+G_+} = \frac{C_+G_+}{1+C_+G_+}G_+^{-1} = \frac{L_+}{1+L_+}G_+^{-1} = R_+G_+^{-1}, \qquad (12.13)$$

if the controller C_+ stabilizes the delay-free part G_+ of the process. Here $L_+ = C_+G_+$ is the loop transfer function of the closed-loop shown in Fig. 12.5b, furthermore the complementary sensitivity function

$$T_+ = R_+ = \frac{L_+}{1+L_+} \qquad (12.14)$$

will be the reference model R_+.

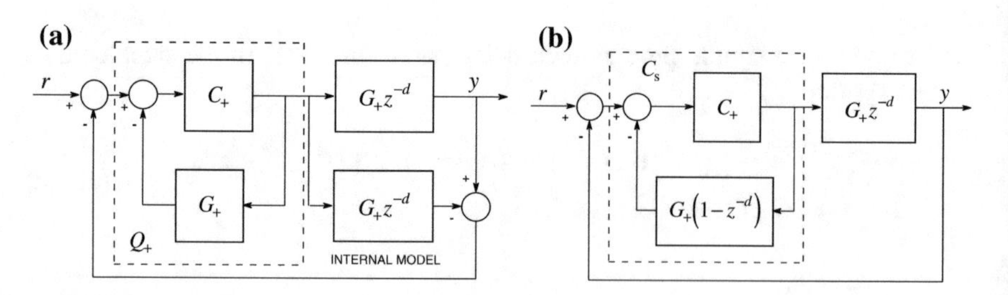

Fig. 12.6 Schemes of the equivalent sampled data SMITH controller

Figure 12.6b shows the equivalent complete closed-loop, where the *YP* sampled data serial controller is

$$C_s = \frac{Q_+}{1 - Q_+ G_+ z^{-d}} = \frac{C_+}{1 + C_+ G_+ (1 - z^{-d})} = C_+ K_S \tag{12.15}$$

the form of which, at the same time, suggests the realization of the inner closed-loop representing the mode of the realizability. Here K_S represents the serial transfer function by means of which the SMITH controller modifies the effect of the original controller C_+. Thus

$$K_S = \frac{1}{1 + C_+ G_+ (1 - z^{-d})} = \frac{1}{1 + L_+ (1 - z^{-d})}. \tag{12.16}$$

Contrary to the CT systems, the realization of the sampled-time SMITH controller does not involve any difficulty in practice, since C_S can be easily realized in part or completely by computer aided systems (see the statements in the previous section about the linear DT filters).

12.3 The TRUXAL-GUILLEMIN Regulator for Sampled Data Systems

The TRUXAL-GUILLEMIN method can be applied to the design of the controller of *ODOF* sampled data control systems. According to this method, the prescribed design goal has to be formulated for the transfer function of the closed system, which is a process with delay

$$T = \frac{CG}{1 + CG} = \frac{CG_+ z^{-d}}{1 + CG_+ z^{-d}} = R_n z^{-d}, \tag{12.17}$$

where it is assumed that in the formula (12.1) of the DT process $G_- = 1$. Now, based on this condition the following simple algebraic equation is obtained for C

$$CG_+ = R_n + CG_+ z^{-d} R_n. \tag{12.18}$$

From this the controller can be chosen according to

$$C = \frac{R_n}{1 - R_n z^{-d}} (G_+)^{-1} = C_{TG}. \tag{12.19}$$

Observe that this form is equal to the basic case ($G_n = 1$, $G_- = 1$) of the sampled data YOULA controller (12.2). The controller can be realized according to

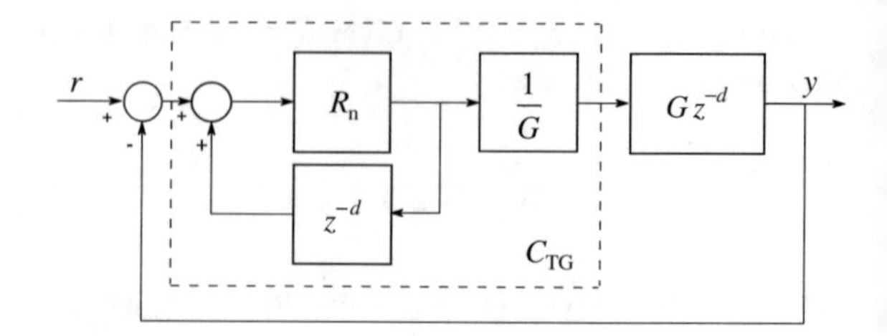

Fig. 12.7 The realization of the Truxal-Guillemin regulator

Fig. 12.7, but there is no problem even with the complete realization of the formula (12.19) in computer aided control systems.

Thus R_n corresponds to one of the reference models of the Youla method. For the *ODOF* case $R_n = R_r$. Let the reference model and the process be given in the forms $R_n = \mathcal{B}_n/\mathcal{A}_n$ and $G = \mathcal{B}/\mathcal{A}$, respectively. In this case the polynomial form of the controller is

$$C_{TG} = \frac{\mathcal{B}_n}{\mathcal{A}_n - \mathcal{B}_n} \frac{\mathcal{A}}{\mathcal{B}}. \tag{12.20}$$

The controller is realizable if the pole excess of R_n is greater than or equal to that of the process. It was seen in Chap. 11 for the DT case that the pole excess of the pulse transfer function of the process is one (in practice generally for zero order hold, thus in the case of unit step equivalent (*SRE*) transformation). Thus the controller (12.20) can be realized, in general, because R_n is usually chosen to ensure the necessary pole excess. If R_n has unit gain ($R_n(1) = 1$), then the controller is of 1-type.

12.4 Design of Regulators Providing Finite Settling Time

In the case of DT systems it is possible to design a controller which is able to track exactly the unit step reference signal within finite steps, or make the error signal zero in finite steps. This controller is called a *Dead-Beat* (*DB*) controller which provides finite settling time. Let us assume that in an *ODOF* control system the process is a relative prime $G = \mathcal{B}/\mathcal{A}$, and C_{DB} is the "*deadbeat*" controller to be designed. Assuming a unit step reference signal, its z-transform is $R(z) = z/(z-1) = 1/(1-z^{-1})$. The dead-beat control requires that the z-transform of the error must be a finite order polynomial $\mathcal{P}_e(z)$, i.e.,

$$E(z) = S(z)R(z) = \frac{1}{1 + C_{DB}G} R(z) = \left[1 - \frac{C_{DB}G}{1 + C_{DB}G}\right] R(z) = \mathcal{P}_e(z). \tag{12.21}$$

It follows from this that both the sensitivity function and the complementary sensitivity transfer functions must also be finite order polynomials, i.e., the polynomials

$$S = \left(1 - z^{-1}\right)\mathcal{P}_e\left(z^{-1}\right) \quad ; \quad T = \frac{C_{DB}G}{1 + C_{DB}G} = 1 - \left(1 - z^{-1}\right)\mathcal{P}_e(z) = \mathcal{P}_y(z) \tag{12.22}$$

have finite order. Similarly it has to be required that the complementary sensitivity transfer function referring to the output of the controller must be a finite order polynomial

$$\frac{C_{DB}}{1 + C_{DB}G} = \mathcal{P}_u(z). \tag{12.23}$$

This kind of transfer function is usually said to be a *finite impulse response* (*FIR*) type, also known as a *moving-average filter*. Based on the above, we may write

$$T = \frac{C_{DB}G}{1 + C_{DB}G} = \mathcal{P}_y(z) = \mathcal{P}_u(z)\frac{\mathcal{B}}{\mathcal{A}}, \tag{12.24}$$

whence the condition for the dead-beat control is

$$\frac{\mathcal{B}(z)}{\mathcal{A}(z)} = \frac{\mathcal{P}_y(z)}{\mathcal{P}_u(z)} = G, \tag{12.25}$$

which—in the case of relative prime process polynomials—can be fulfilled if

$$\mathcal{P}_y(z) = \mathcal{M}(z)\mathcal{B}(z) \quad \text{and} \quad \mathcal{P}_u(z) = \mathcal{M}(z)\mathcal{A}(z). \tag{12.26}$$

Since in steady state the error is zero, the condition $\mathcal{P}_y(1) = 1$ must be fulfilled. As a consequence the gain of the design polynomial $\mathcal{M}(z)$ must be

$$\mathcal{M}(1) = \frac{1}{\mathcal{B}(1)}. \tag{12.27}$$

Finally, based on (12.23), (12.24) and (12.26), the controller has the form

$$C_{DB} = \frac{\mathcal{P}_u}{1 - \mathcal{P}_u G} = \frac{\mathcal{M}\mathcal{A}}{1 - \mathcal{M}\mathcal{B}}. \tag{12.28}$$

Thus the most important step of the design of a dead-beat control is the choice of the design polynomial $\mathcal{M}(z)$. The dead-beat behavior for the input and the output of the process is given by (12.26). It is worth investigating the forms of the signals on the basis of (12.21) and (12.28):

$$\mathcal{P}_{\mathrm{e}}(z) = \frac{z(1 - \mathcal{MB})}{z - 1} = \frac{1 - \mathcal{MB}}{1 - z^{-1}} = \mathcal{N}. \tag{12.29}$$

Here, (12.27) is taken into account, according to which the factor $(1 - \mathcal{MB})$ has always the root $z = 1$, since $1 - \mathcal{M}(1)\mathcal{B}(1) = 0$.

Equation (12.28) has also the forms

$$C_{\mathrm{DB}} = \frac{\mathcal{MA}}{1 - \mathcal{MB}} = \frac{\mathcal{P}_{\mathrm{u}}}{1 - \mathcal{P}_{\mathrm{y}}} = \frac{\mathcal{P}_{\mathrm{y}}}{1 - \mathcal{P}_{\mathrm{y}}}\frac{\mathcal{A}}{\mathcal{B}} = \frac{R_{\mathrm{n}}}{1 - R_{\mathrm{n}}}\frac{\mathcal{A}}{\mathcal{B}}, \tag{12.30}$$

where the substitution $R_{\mathrm{n}} = \mathcal{P}_{\mathrm{y}}$ is applied. Thus the same form is obtained as the TRUXAL-GUILLEMIN regulator (12.19) or the basic case of the YOULA regulator (7.9). The significant difference is that now R_{n} is a *FIR* filter, thus it is a polynomial and (12.29) must also be fulfilled.

Let us summarize the applied restrictions concerning the design of a dead-beat controller:

- it is assumed that the process to be controlled is stable
- the reference signal of the closed-loop is assumed to be unit step
- the dead-beat behavior is valid only at the sampling instants.

Note that if the above conditions are not fulfilled, the dead-beat controller design can still be performed in certain cases (e.g., by polynomial or state-space techniques), but it can not be made by the simple and clear design methods to be presented next.

Example 12.2 The method is presented for a second order CT process with dead-time. Let the transfer function of the CT process be

$$P(s) = \frac{e^{-s}}{(1 + 10s)(1 + 5s)}. \tag{12.31}$$

The first step is to discretize the CT process by a zero-order hold term under the sampling time $T_{\mathrm{s}} = 1$ s

$$G(z) = \frac{\mathcal{B}(z)}{\mathcal{A}(z)} = \frac{0.0091(z + 0.9048)}{(z - 0.9048)(z - 0.8187)z} \tag{12.32}$$

(*SRE* transformation). Notice that the factor z represents the delay supposing a sampling time of $T_{\mathrm{d}} = 1$ s, since $T_{\mathrm{s}} = T_{\mathrm{d}}$. The effect of the delay can be better seen from the form

$$G(z) = \frac{0.0091(z+0.9048)}{(z-0.9048)(z-0.8187)} z^{-1}. \tag{12.33}$$

Regarding the goal of the design, the following considerations can be made. Without a delay, the effect of the input signal $u[0] \neq 0$, appearing at the sampling time $k = 0$ at the input of the zero order hold, will appear in the output—at the earliest—at the sampling time $k = 1$ due to the order of the process. Consequently, if the delay is $T_\text{d} = 1$ s, then the effect of the input $u[0] \neq 0$ will appear—at the earliest—at the sampling time $k = 2$. As a consequence, the best tracking control that can be constructed for the unit step reference signal $y_\text{r}[k] = 1[k]$ turns out to be, in discrete form, $y[k] = 1[k-2]$. Thinking in terms of the pulse transfer function, in this example the condition

$$T = \frac{C(z)G(z)}{1+C(z)G(z)} = \mathcal{P}_\text{y}(z) = z^{-2}$$

is required for the transfer function of the closed system (12.24), from which the controller is:

$$C(z) = \frac{\mathcal{P}_\text{y}}{1-\mathcal{P}_\text{y}} \frac{\mathcal{A}}{\mathcal{B}} = \frac{z^{-2}}{(1-z^{-2})} \frac{1}{G(z)} = \frac{1}{G(z)(z^2-1)} = C_\text{DB}.$$

Expressing the controller in terms of the polynomials of the pulse transfer function of the process:

$$C(z) = \frac{1}{G(z)(z^2-1)} = \frac{\mathcal{A}(z)}{\mathcal{B}(z)(z^2-1)} = \frac{109.9(z^3-1.7236z^2+0.7408z)}{z^3+0.9048z^2-z-0.9048}. \tag{12.34}$$

It can be clearly seen that for a unit step reference signal, the steady-state output can be expected to be error-free, since the loop transfer function $L(z) = C(z)G(z)$ has a pole at $z = 1$, or in other words the controller contains an integrator. The

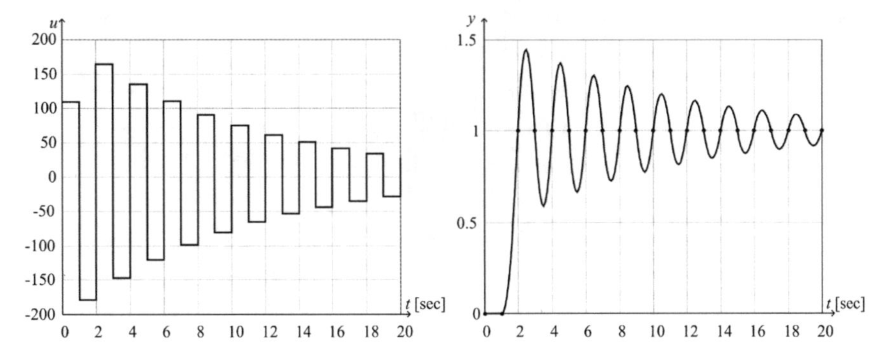

Fig. 12.8 Dead-beat control (2 steps)

behavior of the closed-loop is shown in Fig. 12.8. Considering the discrete-time instants, the result is the same as is expected for the output, but in the case of sampled-time systems the quality of the closed control loop is determined by the continuous output signal, which, however, shows an unacceptable oscillation. Furthermore the actuator signal does not have a dead-beat character: its changes have extreme dynamics, because the condition (12.26) is not fulfilled.

Investigate the hidden oscillations between the sampling times, which is called in the English language literature "*intersampling ripples*". It can be seen in the time diagrams that the oscillation of the CT output is caused by the oscillation of the step-like inputs generated by the zero-order hold term. These values are produced by the regulator due to the fact that it has a so-called slightly (or under) damped pole $p_1 = -0.9048$. The relevance of this qualification can be explained in two ways. Strictly investigating the effect of the specific pole, consider the pulse transfer function

$$G_1(z) = \frac{z+1}{z+0.9048} \tag{12.35}$$

and its step response is shown in Fig. 12.9.

Based on the figure it can be asked, considering only the pulse transfer function itself of the controller, where are those poles which will produce a stable, well attenuated step response, i.e., they do not generate an oscillating output from a unit step input. In the case of CT systems, the regions of the well- and under-damped poles are separated by lines belonging to constant attenuations (damping factors) in the complex frequency domain of s. These lines make a constant angle φ (which depends on the damping) with the negative real axis: $\cos(\varphi) = \xi$. Now the mapping of these lines have to be found in the z-plane. In the s-plane the points $s = \sigma + j\omega$ are on the lines of constant damping with the condition $\omega = \sigma\sqrt{1-\xi^2}\Big/\xi$. For a given damping, the mapping of $z = e^{sT_s}$ to $z = e^{\sigma + j\sigma T_s\sqrt{1-\xi^2}/\xi}$ can be calculated

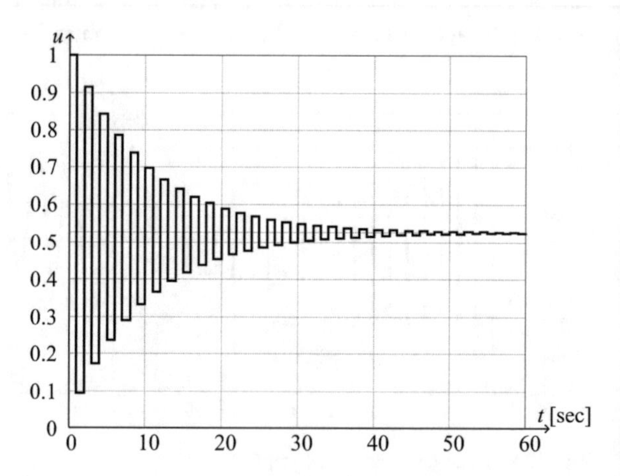

Fig. 12.9 The unwanted dynamics of the actuator signal

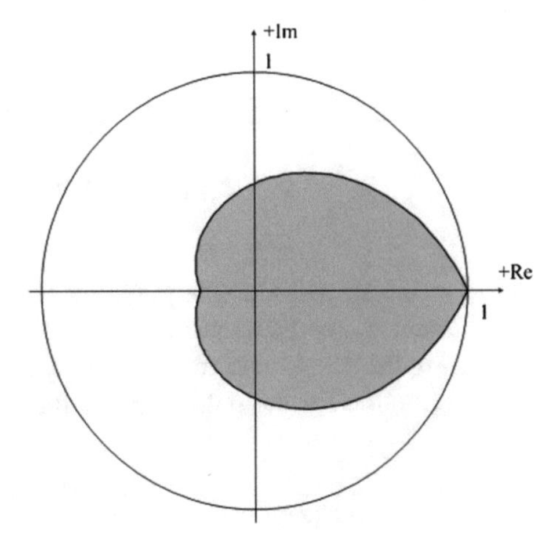

Fig. 12.10 The region of the well damped poles in the z-plane

and drawn for different values of σ. For example, the curve in the z-plane corresponding to the constant line $\xi = 0.4$ can be seen in Fig. 12.10. This well damped ($\xi > 0.4$) region shown in the figure is also called the "heart form curve" in the literature. The formula $\xi = 1 \Big/ \sqrt{1 + \pi^2 \Big/ [\ln(|p_1|)]^2}$ is obtained after some long calculations showing how ξ depends on a root p_1 falling on the negative real axis.

Based on the previous investigations it can be stated that the oscillation is generated by the controller itself, because it can be clearly seen from

$$C(z) = \frac{\mathcal{A}(z)}{\mathcal{B}(z)(z^2 - 1)} \tag{12.36}$$

that $C(z)$ has the roots of the polynomial $\mathcal{B}(z)$ as its poles (in this case $\mathcal{B}(z)$ is of first order, i.e., it has only one root). The oscillating effect of the roots depends on their position relative to the heart-shaped figure belonging to a given damping. In the present case the root of $\mathcal{B}(z)$ is $z = -0.9048$, which is outside of the well damped region. In order to avoid oscillation the direct compensation of the slightly damped roots of $\mathcal{B}(z)$, i.e., simply saying their cancellation with the corresponding poles of the controller, has to be avoided. Separate the roots of $\mathcal{B}(z)$ in such a way that $\mathcal{B}_+(z)$ contains the well damped roots of $\mathcal{B}(z)$ (they are inside of the heart-shaped region) and $\mathcal{B}_-(z)$ contains the slightly damped roots (outside of the heart-shaped region)

$$\mathcal{B}(z) = \mathcal{B}_+(z)\mathcal{B}_-(z) \tag{12.37}$$

and the condition $\mathcal{B}_-(z)|_{z=1} = \mathcal{B}_-(1) = 1$ must be fulfilled. For the example

$$\mathcal{B}(z) = \mathcal{B}_+(z)\mathcal{B}_-(z) = 0.01733(0.525z + 0.475), \tag{12.38}$$

where $\mathcal{B}_+(z) = 0.01733$ (the polynomial $\mathcal{B}_+(z)$ has no well damped pole) and $\mathcal{B}_-(z) = 0.525z + 0.475$ (the polynomial $\mathcal{B}(z)$ has one slightly damped pole). Next let m_+ and m_- be the degrees of the polynomials $\mathcal{B}_+(z)$ and $\mathcal{B}_-(z)$, respectively.

During the design process, take the applied separation into account, and modify our expectation for the pulse transfer function of the closed discrete-time system:

$$T = \frac{C(z)G(z)}{1 + C(z)G(z)} = \mathcal{P}_y(z) = \mathcal{B}_-(z)z^{-2}z^{-m_-} = \mathcal{B}_-(z)z^{-m_- -2}, \tag{12.39}$$

so the assumption (12.26) is also fulfilled. From the above condition the controller becomes:

$$C(z) = \frac{\mathcal{A}(z)}{\mathcal{B}_+(z)[z^3 - \mathcal{B}_-(z)]} = \frac{57.7(z^3 - 1.7236z^2 + 0.7408z)}{z^3 - 0.525z - 0.475}. \tag{12.40}$$

From the above form it is easily seen why the condition $\mathcal{B}_-(z)|_{z=1} = \mathcal{B}_-(1) = 1$ has to be assumed during the separation of the polynomial $\mathcal{B}(z)$. For in this case, due to the fact that $[z^3 - \mathcal{B}_-(z)]_{z=1} = 0$, $z = 1$ is still the pole of the loop transfer function $L(z)$, i.e., it is of 1-type. The results obtained with the modified controller are illustrated in Fig. 12.11. The closed system is slowed down, the settling time increased from 2 to 3 s, in other words, from 2 steps to 3 steps, but, at the same time, the moderate dynamics of the actuator signal can be observed. The oscillation is completely eliminated, but the magnitude of the initial value of the actuator signal can not be considered acceptable for any kind of application.

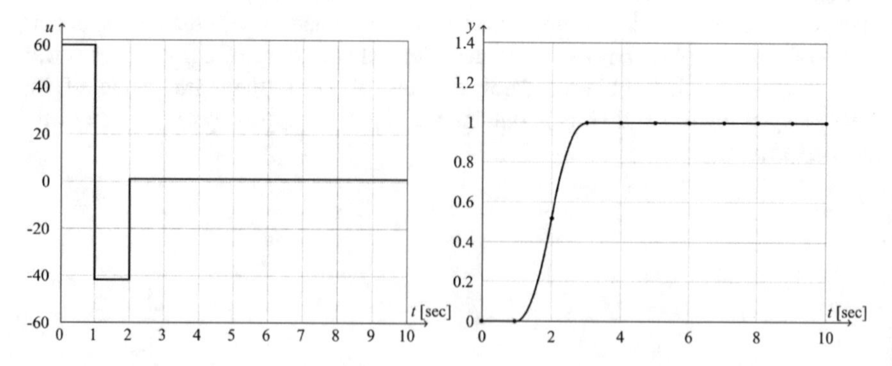

Fig. 12.11 Dead-beat controller (3 steps)

The slowing down can be performed in many different ways, from which such a solution is shown next, where the dead-beat character is kept, but the settling time is further increased depending on the degree of the design polynomial. Let the slowing design polynomial be

$$T(z) = \mathcal{P}_{\mathrm{y}}(z) = 0.2z^2 + 0.3z + 0.5, \tag{12.41}$$

of degree $m_{\mathrm{T}} = 2$, and the condition $T(z)|_{z=1} = T(1) = 1$ is used in the specification of the closed-loop pulse transfer function according to

$$\frac{C(z)G(z)}{1 + C(z)G(z)} = \mathcal{P}_{\mathrm{y}}(z)\mathcal{B}_-(z)z^{-2-m_- - m_{\mathrm{T}}} = \mathcal{P}_{\mathrm{y}}(z)\mathcal{B}_-(z)z^{-5}. \tag{12.42}$$

The controller becomes

$$\begin{aligned} C(z) &= \frac{\mathcal{A}(z)\mathcal{P}_{\mathrm{y}}(z)}{\mathcal{B}_+(z)\left[z^5 - \mathcal{B}_-(z)\mathcal{P}_{\mathrm{y}}(z)\right]} \\ &= \frac{11.54\left(z^5 - 0.2235z^4 + 0.6555z^3 - 3.198z^2 + 1.852z\right)}{z^5 - 0.105z^3 - 0.2525z^2 - 0.405z - 0.2375} \end{aligned}$$

The operation of the closed-loop in the time domain is seen in Fig. 12.12.

In connection with the controller of (12.30) it has already been mentioned that the design equation of the dead-beat controller actually corresponds to the basic case (7.9) of the Youla regulator. For comparison with the general case (12.2), consider the formula (12.1) of the DT process according to the above separation (12.37),

$$G = G_+ G_- z^{-d} = \frac{\mathcal{B}_+ \mathcal{B}_-}{\mathcal{A}} z^{-2}, \tag{12.43}$$

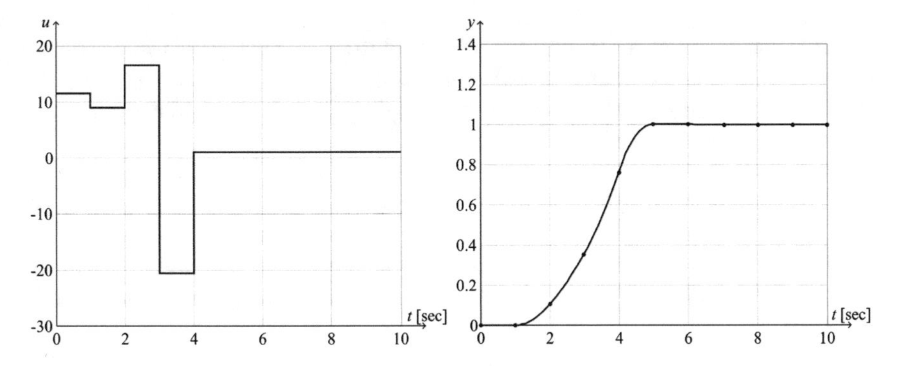

Fig. 12.12 Dead-beat controller (5 steps)

where

$$\mathcal{B}_+(z) = 0.01733 \quad \text{and} \quad \mathcal{B}_-(z) = 0.525z + 0.475 \tag{12.44}$$

Here $\mathcal{B}_+$ is the factor having acceptable inverse form, $\mathcal{B}_-$ is the factor having non-acceptable inverse form in the numerator of the pulse transfer function. Notice that the inverse of $\mathcal{B}$ is not unstable now, but is underdamped, therefore it is also considered an unwanted factor. The form of the optimal controller obtained for the general case according to (12.2) with the choice $G_r = G_n = 1$ is

$$\begin{aligned} C_{opt} &= \frac{R_n G_+^{-1}}{1 - R_n G_- z^{-d}} = \frac{\mathcal{P}_y G_+^{-1}}{1 - \mathcal{P}_y G_- z^{-d}} = \frac{\mathcal{P}_y \mathcal{A}}{\mathcal{B}_+\left(1 - \mathcal{P}_y \mathcal{B}_- z^{-d}\right)} \\ &= \frac{z^{-2}\mathcal{A}}{\mathcal{B}_+\left(1 - z^{-2}\mathcal{B}_- z^{-1}\right)}, \end{aligned} \tag{12.45}$$

which is completely the same as the controller of (12.40),

$$C(z) = \frac{\mathcal{A}(z)}{\mathcal{B}_+(z)\left[z^3 - \mathcal{B}_-(z)\right]} = \frac{57.7\left(z^3 - 1.7236z^2 + 0.7408z\right)}{z^3 - 0.525z - 0.475}. \tag{12.46}$$

It is confirmed again that the Youla-controller is generally valid for stable processes.

■

12.5 Predictive Controllers

Assume that the pulse transfer function of a control system in an *ODOF* loop is

$$G\left(z^{-1}\right) = \frac{\mathcal{B}\left(z^{-1}\right)}{\mathcal{A}\left(z^{-1}\right)} z^{-d} = G_+\left(z^{-1}\right) G_-\left(z^{-1}\right) \quad ; \quad G_-\left(z^{-1}\right) = z^{-d}, \tag{12.47}$$

which corresponds to a CT process with dead-time. A relationship is sought by which the value of the output signal at the sampling time $k + d$ can be estimated from the information available up to the sampling time k. To achieve this let us introduce a special polynomial equations

$$1 = \mathcal{A}\mathcal{F} + \mathcal{P}z^{-d} \tag{12.48}$$

whose solution is unambiguous, seeking $\mathcal{F}$ of degree $(d - 1)$ and $\mathcal{P}$ of degree $(n - 1)$, if $\mathcal{A}$ has degree n. Equation (12.48) is a special form of the *DE* discussed in Chap. 10. Using equivalent rewriting $G(z^{-1})$ can be decomposed as

$$\begin{aligned} G &= \frac{\mathcal{B}}{\mathcal{F}} z^{-d} = \frac{\mathcal{BAF} + \mathcal{BP} z^{-d}}{\mathcal{A}} z^{-d} = \left(\mathcal{BF} + \frac{\mathcal{BP} z^{-d}}{\mathcal{A}} \right) z^{-d} \\ &= \mathcal{BF} z^{-d} + \mathcal{P} \left(\frac{\mathcal{B}}{\mathcal{A}} z^{-d} \right) z^{-d} = \mathcal{BF} z^{-d} + \mathcal{P} G z^{-d} \end{aligned} \tag{12.49}$$

Apply both sides to the series of the input signal $u[k]$

$$\begin{aligned} y[k] &= \mathcal{BF} z^{-d} u[k] + \mathcal{P} z^{-d} G u[k] = \mathcal{BF} u[k-d] + \mathcal{P} z^{-d} y[k] \\ &= \mathcal{BF} u[k-d] + \mathcal{P} y[k-d] = y[k|k-d] \end{aligned} \tag{12.50}$$

The equation can also be written for sampling time $k+d$

$$y[k+d] = \mathcal{BF} u[k] + \mathcal{P} y[k] = y[k+d|k], \tag{12.51}$$

where $y[k+d|k]$ is the estimate or prediction of the series $y[k]$ for the sampling time $k+d$. Notice that the prediction is error-free and it uses only the information available at the time instant k concerning both the input and output. Both polynomials $\mathcal{BF}$ and, $\mathcal{P}$ are functions of z^{-1} and in (12.51) their coefficients weight only the current and the previous values of the signals u and y. Based on the d-step predictor a special, so-called predictive controller can be constructed. If the goal is to track the output of a reference model R_r, then the equation of the controller is

$$R_r y_r[k] = y[k+d|k] = \mathcal{BF} u[k] + \mathcal{P} y[k] \tag{12.52}$$

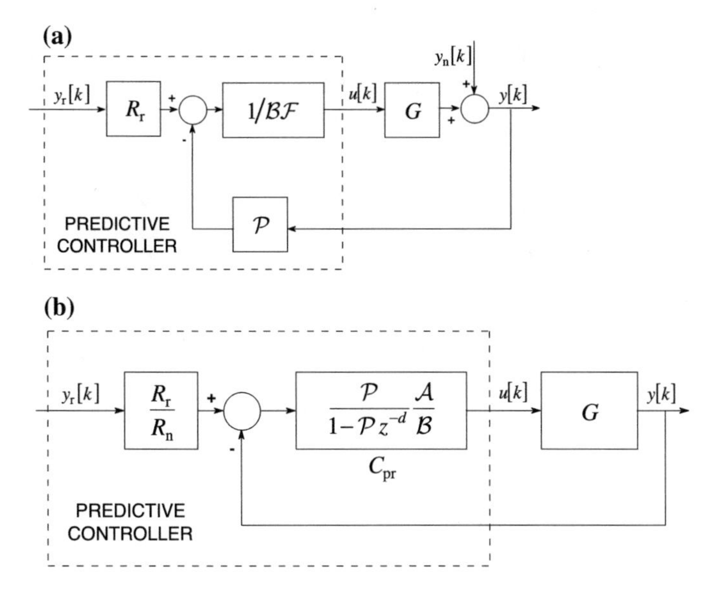

Fig. 12.13 Forms of predictive controllers

from which the input signal is

$$u[k] = \frac{R_r y_r[k] - \mathcal{P} y[k]}{\mathcal{B}\mathcal{F}} \tag{12.53}$$

The direct mapping of (12.53) can be seen in Fig. 12.13a. The equivalent block scheme of the Fig. 12.13b, however, shows the conventional closed-loop control.

Thus the predictive controller has the form

$$C_{pr} = \frac{\mathcal{P}}{1 - \mathcal{P} z^{-d}} \frac{\mathcal{A}}{\mathcal{B}} = \frac{\mathcal{P}}{\mathcal{B}\mathcal{F}}, \tag{12.54}$$

i.e. the reference model for the disturbance rejection is now

$$R_n = \mathcal{P}, \tag{12.55}$$

which does not depend on the designer, but comes from (12.48). The predictive controller is formally equal to a YOULA regulator where $G_- = z^{-d}$. The transfer characteristic of the complete control loop is

$$y = R_r z^{-d} y_r + \left(1 - R_n z^{-d}\right) y_n = R_r z^{-d} y_r + \left(1 - \mathcal{P} z^{-d}\right) y_n \tag{12.56}$$

by means of which the predictive controller completely solves the problem for reference signal tracking, but R_n can not be designed. The d-step predictor of (12.51) is linear in its parameters, therefore it is easy to apply in parameter estimation (identification) techniques to determine the parameters of the controller. From the above control design principle, a new, widely applied computer controlled method has been developed, which is called *Model Predictive Control* (*MPC*).

For the noise rejection behavior of a closed system, a method is introduced which penalizes the change or variance of the input. So the dynamics of the closed-loop, though in a restricted way, can be acceptable by the proper choice of the penalty weights.

Example 12.3 Let the controlled system be a first order process with delay $d = 2$

$$G(z^{-1}) = \frac{\mathcal{B}(z^{-1})}{\mathcal{A}(z^{-1})} = \frac{0.7z^{-1} - 1.0z^{-2}}{1 - 1.5z^{-1} + 0.2z^{-2}} z^{-1} = \frac{0.7 - 1.0z^{-1}}{1 - 1.5z^{-1} + 0.2z^{-2}} z^{-2} \tag{12.57}$$

Compute the d-step ahead predictor by solving the DIOPHANTINE equation

$$1 = \mathcal{A}\mathcal{F} + \mathcal{P} z^{-d} \tag{12.58}$$

where $n = 2$ and the polynomials $\mathcal{F}$ and $\mathcal{P}$ are of degrees $d - 1$ and $n - 1$ respectively. The equation is

$$1=\left(1-1.5z^{-1}+0.2z^{-2}\right)\left(1+f_1z^{-1}\right)+\left(p_{\text{o}}+p_1z^{-1}\right)z^{-2} \tag{12.59}$$

The solution is: $f_1 = 1.5$, $p_{\text{o}} = 2.05$ and $p_1 = -0.3$. So the predictive regulator is given by

$$C_{\text{pr}}=\frac{\mathcal{P}}{1-\mathcal{P}z^{-d}}\frac{\mathcal{A}}{\mathcal{B}}=\frac{\mathcal{P}}{\mathcal{B}\mathcal{F}}=\frac{p_{\text{o}}+p_1z^{-1}}{(0.7-1.0z^{-1})(1+f_1z^{-1})}=\frac{2.05-0.3z^{-1}}{0.7+0.05z^{-1}-1.5z^{-2}} \tag{12.60}$$

■

12.6 The Best Reachable Discrete-Time Control

12.6.1 General Theory

The decomposition of the control error discussed in Sect. 7.5 is completely valid for DT systems, so all relationships can be applied in unchanged form.

It is worth noting that in the DT case the fastest reachable first order reference model can be easily determined under the amplitude restriction of the output of the controller

$$|u(t)|=U_{\text{max}} \tag{12.61}$$

if the YP controller is applied. Let the first order reference model with unit gain be

$$R_{\text{n}}=\frac{(1+a_{\text{n}})z^{-1}}{1+a_{\text{n}}z^{-1}}=\frac{(1+a_{\text{n}})}{z+a_{\text{n}}}. \tag{12.62}$$

Let the first (so-called leading) coefficient in the numerator of the pulse transfer function of the process be b_1. Then the following restriction

$$\frac{1+a_{\text{n}}}{b_1}\leq U_{\text{max}} \tag{12.63}$$

must be fulfilled by the first biggest jump of the step response series, from which the maximum value of the coefficient a_{n} of the reference model is

$$a_{\text{n}}\leq b_1\,U_{\text{max}}-1. \tag{12.64}$$

12.6.2 Empirical Rules

It has already been seen in the recent discussion of the best reachable control, that the basic restriction derives from the saturation of the actuator or from the process dynamics itself. It can even be supplemented with the noise of the measurements. This noise may derive from the physical operation of the sensor, but also from the electronics, and in DT control from the A/D converter. The measurement noise usually appears in high frequency regions, therefore the uncertainty caused by them restricts the high frequency gain A_∞ of the controller. For simplicity, assume the A/D and D/A converters are those generally used, with 12 bits, which corresponds to 4096 levels. Thus a conversion error or measurement error of 1 bit, by being increased 4096 times, reaches the whole signal region.

In practice measurement changes greater than 5% are not allowed. This means that the high frequency gain of the controller must be smaller than 200, i.e., $A_\infty < 200$.

Chapter 13
Design of Conventional Sampled Data Regulators

It was seen in Chap. 12 that—similarly to the continuous-time case—there are exact theoretical methods for the design of discrete controllers in the case of stable processes. On the basis of these methods the structure of the optimal controller and the optimal values of its parameters can be determined for the most varied cases. For sampled data control systems it can not be stated that already long before the elaboration of these theoretical methods a class of controllers had been evolved and widespread in practice, which still have decisive importance in the control of industrial processes, as these control algorithms have been developed simultaneously with the applications of computer control systems. Considering the development it was more typical that at the beginning sampled data controllers just copied the conventional continuous controllers. The probable reason for that was that the continuous controllers in their history of several decades had gained high reliability and recognition in practice. Whereas in the framework of computer control the realization of the higher order complex controllers described in Chap. 12 is simple, to date the discretized versions of the conventional controllers are still what is mainly used in practical operating control systems. Therefore this controller family is discussed here in a separate chapter. As the sampled data controllers are realized in software, it is common to call them DT (discrete-time) algorithms.

Several methods are available for the design of sampled data control systems. The great number of methods in practical use is explained by the fact that a discrete-time control algorithm is realized by a program, and not by electronic equipment containing operational amplifiers and passive elements. This provides great flexibility for the designer. The hybrid nature of the design problem—let us think of the necessity of the simultaneous consideration of continuous and discrete signals—also provides the possibility for the application of various design concepts.

Besides the variety of methods it has to be emphasized that—whatever strategy is chosen—the discrete-time controller can be interpreted as a signal processing unit which determines $u[k]$, the current value of the control signal (the input of the process) based on the sampled current and previous filtered values of the output signal

L. Keviczky et al., *Control Engineering*, Advanced Textbooks in Control and Signal Processing, https://doi.org/10.1007/978-981-10-8297-9_13

$$y[k],\ y[k-1],\ y[k-2],\ldots \tag{13.1}$$

and the previous values of the control signal

$$u[k-1],\ u[k-2],\ u[k-3],\ldots \tag{13.2}$$

calculated by the control algorithm. That is, the digital controller realizes the following mapping:

$$\{y[k],\ y[k-1],\ y[k-2],\ldots;u[k-1],\ u[k-2],\ u[k-3],\ldots \Rightarrow u[k]\} \tag{13.3}$$

The simplest case of this mapping is when a serial controller is realized given by its pulse transfer function $C(z)$. In the case of a serial controller the input of the controller is the error signal

$$e[k] = r[k] - y[k] \tag{13.4}$$

where $r[k]$ is the reference signal. $u[k]$ is the output of the controller

$$C(z) = \frac{U(z)}{E(z)}, \tag{13.5}$$

which provides the input of the process.

To demonstrate how this mapping works, let us consider a second order controller:

$$C(z) = \frac{U(z)}{E(z)} = \frac{q_o z^2 + q_1 z + q_2}{z^2 + r_1 z + r_2} = \frac{q_o + q_1 z^{-1} + q_2 z^{-2}}{1 + r_1 z^{-1} + r_2 z^{-2}}, \tag{13.6}$$

with the recursive expansion shown in Chap. 11

$$u[k] = q_o e[k] + q_1 e[k-1] + q_2 e[k-2] - r_1 u[k-1] - r_2 u[k-2], \tag{13.7}$$

which gives the algorithm for the realization. In the above example the degree of the numerator of the controller was deliberately chosen to be equal to the degree of the denominator, because in this case the controller reacts immediately, without any delay to eliminate the error, assuming that a steady state characterized by zero error values $e[j] = 0\ (j<k)$ is followed by an error $e[k] \neq 0$. It has to be emphasized that the pulse transfer function is a common, but not the only representation of the digital control algorithms.

Summarizing the steps of the realization of a digital controller, they are:

- Sampling
- Calculation of the input signal $u[k]$ with the knowledge of $y[k]$ (running the control algorithm)

- Supplying the holding element with the calculated input signal
- Shifting $u[k]$, $y[k]$ and/or $e[k]$ to prepare the next step:

$$u[k-2] := u[k-1] \quad u[k-1] := u[k] \quad e[k-2] := e[k-1] \quad e[k-1] := e[k].$$

Depending on the complexity of the digital control algorithm, as well as on the chosen form of representing the numbers and the computing capacity, typically the time required for the execution of the operations above can be neglected compared to the sampling time. But in the case of fast sampling it may occur that the calculation time of the control algorithm is comparable to the sampling time T_s. In this case, the mapping has to be modified, as the calculated input signal $u[k]$ can not be considered as a sample belonging to the sampling instant k, as its value becomes available only later. Then the steps of the digital control algorithm are modified taking into account that after the sampling the algorithm can only be started to calculate a single step:

- Sampling
- Starting the calculation of $u[k]$ with the knowledge of $y[k]$ (starting the run of the control algorithm)
- Supplying the holding element with the most recent available calculated input signal
- Preparing the next step.

As an example let us suppose that the calculation time of the algorithm is less than the sampling time T_s, but is close to it. Then the mapping is

$$\{y[k], y[k-1], y[k-2], \ldots; u[k], u[k-1], u[k-2], \ldots \Rightarrow u[k+1]\} \qquad (13.8)$$

Consequently the $u[k]$ belonging to the time instant k can be generated according to the mapping

$$\{y[k-1], y[k-2], y[k-3], \ldots; u[k-1], u[k-2], u[k-3], \ldots \Rightarrow u[k]\}. \qquad (13.9)$$

This also means that a digital controller designed for a dead-time process with time-delay T_d has to be redesigned for an increased time-delay of $T_d + T_s$ so that the delay of the calculations can be taken into account.

13.1 Design Methods for the Discrete-Time *PID* Regulator Family

Creating the sampled version of the ideal *PID* regulator according to expression (8.1) let us approximate the integrating element with the right hand rectangle rule and the differentiating effect by backward difference. Then the pulse transfer function is

$$C_{\mathrm{PID}}(z) = A_{\mathrm{P}}\left(1 + \frac{1}{T_{\mathrm{I}}}\frac{T_{\mathrm{s}}z}{z-1} + T_{\mathrm{D}}\frac{z-1}{T_{\mathrm{s}}z}\right) = A_{\mathrm{P}} + \frac{A_{\mathrm{P}}}{T_{\mathrm{I}}}\frac{T_{\mathrm{s}}z}{z-1} + A_{\mathrm{P}}T_{\mathrm{D}}\frac{z-1}{T_{s}z}. \tag{13.10}$$

Transforming the right hand side of the equation to get a common denominator, the pulse transfer function is obtained in the form of a second order discrete-time (DT) filter,

$$C_{\mathrm{PID}}(z) = \frac{q_{\mathrm{o}} + q_1 z^{-1} + q_2 z^{-2}}{1 - z^{-1}}, \tag{13.11}$$

where

$$q_{\mathrm{o}} = A_{\mathrm{P}}\left(1 + \frac{T_{\mathrm{s}}}{T_{\mathrm{I}}} + \frac{T_{\mathrm{D}}}{T_{\mathrm{s}}}\right);\ q_1 = -A_{\mathrm{P}}\left(1 + \frac{2T_{\mathrm{D}}}{T_{\mathrm{s}}}\right) \text{ and } q_2 = A_{\mathrm{P}}\frac{T_{\mathrm{D}}}{T_{\mathrm{s}}} \tag{13.12}$$

In Sect. 8.4, when discussing the traditional continuous-time (CT) regulators, we already dealt with the effect of the constraints and with the extension of the regulator structure with *ARW*. In the case of CT regulators, there is generally no opportunity to introduce a signal into the inner structure of the regulator. Therefore the feedback from the saturation is realized at the place where the error is measured, as generally this point is accessible (see Fig. 8.29). But when realizing sampled data regulators, also the details are in our hand, thus the feedback can be led directly to the input of the integrator, as shown in Fig. 13.1.

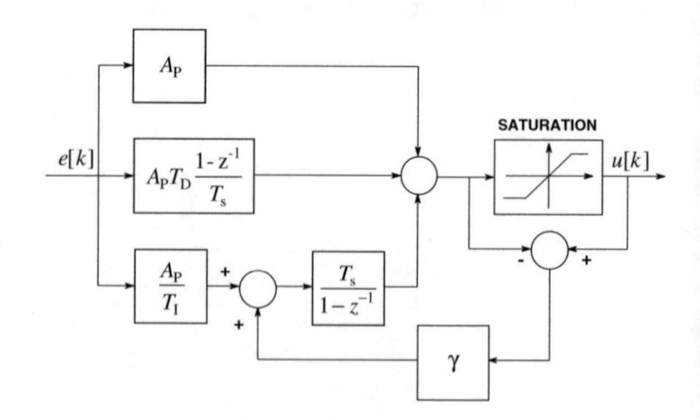

Fig. 13.1 Digital *PID* regulator extended by the *ARW* effect

For the realization of the discrete-time version of the approximate *PID* regulator given by (8.4) the following *SRE* form can be used:

$$\widehat{C}_{\mathrm{PID}}(z) = A_{\mathrm{P}} + \frac{A_{\mathrm{P}}T_{\mathrm{s}}}{T_{\mathrm{I}}}\frac{z^{-1}}{1-z^{-1}} + \frac{A_{\mathrm{P}}T_{\mathrm{D}}}{T_{\mathrm{s}}}\frac{1-z^{-1}}{1-e^{-T_{\mathrm{s}}/T_{z}-1}}. \tag{13.13}$$

After some conversions, also this regulator shows the form of a second order DT filter:

$$\widehat{C}_{\mathrm{PID}}(z) = \frac{q_{\mathrm{o}}+q_{1}z^{-1}+q_{2}z^{-2}}{(1-z^{-1})(1-e^{-T_{\mathrm{s}}/T}z^{-1})} = \frac{q_{\mathrm{o}}+q_{1}z^{-1}+q_{2}z^{-2}}{1+r_{1}z^{-1}+r_{2}z^{-2}} \tag{13.14}$$

where

$$q_{\mathrm{o}} = A_{\mathrm{P}}\left(1+\frac{T_{\mathrm{D}}}{T}\right);\; q_{1} = -A_{\mathrm{P}}\left(1+e^{-T_{\mathrm{s}}/T}-\frac{T_{\mathrm{s}}}{T_{\mathrm{I}}}+\frac{2T_{\mathrm{D}}}{T}\right) \tag{13.15}$$

$$q_{2} = A_{\mathrm{P}}\left\{e^{-T_{\mathrm{s}}/T}\left(1-\frac{T_{\mathrm{s}}}{T_{\mathrm{I}}}\right)+\frac{T_{\mathrm{D}}}{T}\right\} \tag{13.16}$$

$$r_{1} = -\left(1+e^{-T_{\mathrm{s}}/T}\right);\; r_{2} = e^{-T_{\mathrm{s}}/T} \tag{13.17}$$

The *ARW* extension can be realized as shown in Fig. 13.1.

The discrete-time form of the continuous-time approximate *PID* regulator according to (8.8) is given by the following DT pulse transfer function:

$$\widehat{C}_{\mathrm{PID}}(z) = K_{\mathrm{C}}\frac{\left(z-z_{1}^{\mathrm{cd}}\right)\left(z-z_{2}^{\mathrm{cd}}\right)}{(z-1)\left(z-e^{-T_{\mathrm{s}}/T}\right)} = K_{\mathrm{C}}\frac{\left(1-z_{1}^{\mathrm{cd}}z^{-1}\right)\left(1-z_{2}^{\mathrm{cd}}z^{-1}\right)}{(1-z^{-1})\left(1-e^{-T_{\mathrm{s}}/T}z^{-1}\right)} = \tilde{C}_{\mathrm{PID}}\left(z^{-1}\right) \tag{13.18}$$

(The superscript ‘cd’ refers to the discrete-time form of the regulator.)

13.1.1 Tuning of Sampled Data PI Regulators

From (13.18) the pulse transfer function of a digital *PI* regulator is given by

$$C_{\mathrm{PI}}(z) = K_{\mathrm{C}}\frac{z-z_{1}^{\mathrm{cd}}}{z-1} = K_{\mathrm{C}}\frac{1-z_{1}^{\mathrm{cd}}z^{-1}}{1-z^{-1}} = C_{\mathrm{PI}}\left(z^{-1}\right) \tag{13.19}$$

The discrete-time *PI* regulator replaces the pole p_{1}^{pd} (which generally is the smallest, belonging to the largest time constant) of the DT pulse transfer function corresponding to the CT process with the pole $z=1$ (placing it at frequency $\omega=0$). The design of the *PI* regulator takes into consideration the pulse transfer function

$$C_{\mathrm{PI}}(z) = K_{\mathrm{C}}\frac{z - e^{-T_s/T_1}}{z-1} = K_{\mathrm{C}}\frac{1 - e^{-T_s/T_1}z^{-1}}{1-z^{-1}} = C_{\mathrm{PI}}\left(z^{-1}\right) \tag{13.20}$$

with the choice of $z_1^{\mathrm{cd}} = p_1^{\mathrm{pd}} = e^{-T_s/T_1}$, where $T_1 = \max\{T_i\}$. The gain K_{C} of the regulator has to be set on the basis of the frequency function of the sampled CT process, where the approximate relationship between the frequency functions of the discrete- and the continuous-time processes is $P_{\mathrm{d}}(j\omega) \approx e^{-j\omega T_s/2}P(j\omega)$. This means that the amplitude-frequency function remains unchanged: $a_s(j\omega) \approx a(j\omega)$, whereas the phase-frequency function is changed unfavorably: $\varphi_s(j\omega) = \varphi(j\omega) - \omega T_s/2$. This effect can be taken into account very simply by prescribing a value for the phase margin stricter than the original φ_{to} required for the CT system, as $\varphi_{\mathrm{to}}^{\mathrm{s}} = \varphi_{\mathrm{to}} + \omega T_s/2$.

The discrete-time and the continuous-time *PI* compensating control algorithms have the same effect and their tuning procedures are also similar.

13.1.2 Tuning of Sampled Data PD Regulators

According to (13.18) the pulse transfer function of a digital *PD* regulator is

$$\tilde{C}_{\mathrm{PD}}(z) = K_{\mathrm{C}}\frac{z - z_1^{\mathrm{cd}}}{z - p^{\mathrm{cd}}} = K_{\mathrm{C}}\frac{1 - z_1^{\mathrm{cd}}z^{-1}}{1 - p^{\mathrm{cd}}z^{-1}} = \tilde{C}_{\mathrm{PD}}\left(z^{-1}\right), \tag{13.21}$$

which is the *SRE* counterpart of the continuous-time *PD* regulator of (8.14). The digital *PD* regulator given by (13.21) replaces the pole p_2^{pd} of the pulse transfer function of the process by p^{cd}, which belongs to a higher frequency, thus accelerating the control system.

The initial and the final values of the unit step response of the *PD* regulator (13.21) are

$$t = 0; \quad \lim_{z\to\infty}\frac{z}{z-1}K_{\mathrm{C}}\frac{z - z_1^{\mathrm{cd}}}{z - p^{\mathrm{cd}}} = K_{\mathrm{C}}; \text{ and} \tag{13.22}$$

$$t = \infty; \quad \lim_{z\to 1}\left(\frac{z-1}{z}\right)\frac{z}{z-1}K_{\mathrm{C}}\frac{z - z_1^{\mathrm{cd}}}{z - p^{\mathrm{cd}}} = K_{\mathrm{C}}\frac{1 - z_1^{\mathrm{cd}}}{1 - p^{\mathrm{cd}}}. \tag{13.23}$$

Thus the overexcitation applied for the acceleration is

$$\eta = \frac{K_{\mathrm{C}}}{K_{\mathrm{C}}\frac{1 - z_1^{\mathrm{cd}}}{1 - p^{\mathrm{cd}}}} = \frac{1 - p^{\mathrm{cd}}}{1 - z_1^{\mathrm{cd}}}. \tag{13.24}$$

When designing the regulator for the acceleration, first $z_1^{\mathrm{cd}} = p_2^{\mathrm{pd}} = e^{-T_s/T_2}$ is chosen. Then the value of p^{cd} is calculated from the allowed highest overexcitation $\eta_{\max}$:

$$p^{\mathrm{cd}} \geq 1 - \left(1 - e^{-T_s/T_2}\right)\eta_{\max}. \tag{13.25}$$

In the case of stable processes, with an appropriately chosen sampling time, $0 \leq p^{\mathrm{cd}} \leq 1$ and $0 \leq z_1^{\mathrm{cd}} = p_2^{\mathrm{pd}} \leq 1$. Thus the inequality

$$p^{\mathrm{cd}} \geq \max\left\{0, 1 - \left(1 - e^{-T_s/T_2}\right)\eta_{\max}\right\} \tag{13.26}$$

has to be fulfilled. The time constant T which belongs to the borderline case $p^{\mathrm{cd}} = e^{-T_s/T}$ is given by

$$T = \frac{-T_s}{\ln\left[1 - \left(1 - e^{-T_s/T_2}\right)\eta_{\max}\right]}. \tag{13.27}$$

In the ideal case, $p^{\mathrm{cd}} = 0$, and the ideal *PD* regulator is expressed by the pulse transfer function

$$C_{\mathrm{PD}}(z) = K_{\mathrm{C}}\frac{z - z_1^{\mathrm{cd}}}{z} = K_{\mathrm{C}}\left(1 - z_1^{\mathrm{cd}}z^{-1}\right) = C_{\mathrm{PD}}\left(z^{-1}\right). \tag{13.28}$$

The overexcitation in the case of an ideal *PD* regulator is $\eta = 1/\left(1 - z_1^{\mathrm{cd}}\right)$. The equivalent continuous-time *PD* regulator corresponding to the ideal discrete-time *PD* regulator in the low frequency domain, supposing $z_1^{\mathrm{cd}} = p_2^{\mathrm{pd}} = e^{-T_s/T_2}$, is obtained by

$$C_{\mathrm{PD}}(s) \approx K_{\mathrm{C}}\left(1 - p_2^{\mathrm{pd}}\right)(1 + sT_2)e^{-s(T_{\mathrm{d2}} - T_s)} = K_{\mathrm{C}}\left(1 - e^{-T_s/T_2}\right)(1 + sT_2)e^{-sT_{\mathrm{d2}}}, \tag{13.29}$$

where, according to the experimental formula of TUSCHÁK,

$$T_{\mathrm{d2}} = \frac{T_s}{1 + \sqrt[3]{e^{-T_s/T_2}}}. \tag{13.30}$$

If $T_s \geq 3T_2$, then in (13.27)

$$1 - e^{-T_s/T_2} \approx \frac{T_s}{T_2} \quad \text{and} \quad T_{\mathrm{d2}} \approx \frac{T_s}{T_2}. \tag{13.31}$$

Finally the transfer function of the approximate continuous *PD* regulator which is valid in the low frequency domain is

$$\tilde{\tilde{C}}_{\mathrm{PD}}(s) = K_{\mathrm{C}}\frac{T_s}{T_2}(1 + sT_2)e^{-sT_s/2}. \tag{13.32}$$

The effect of the discrete-time *PD* compensation in the considered frequency domain is as if a *CT* compensating element had changed the time constant T_2 of the process to a dead-time of value $T_s/2$, and in the meantime it results in an overexcitation of value T_2/T_s [more accurately, according to (13.24)]. That is, each single discrete-time *PD* element replaces one pole of the pulse transfer function of the process by a dead-time of value $T_s/2$ with an overexcitation as if the corner frequency of the pole had been placed to $\omega = 1/T_s$. (This rule is called the TUSCHÁK effect.)

A discrete *PD* regulator provides more favorable overexcitation relations than a continuous *PD* regulator. This is because the initial overexcitation is maintained during the entire sampling interval. Thus even with a smaller amplitude, sufficient energy can be transmitted to the process. That is, the overexcitation accelerates the process via the excess energy provided by the regulator.

The design of the gain of the regulator is executed according to the classical method prescribing a given phase margin. The value of the prescribed phase margin φ_{to} formulated for the original CT system is now modified to a value of $\varphi_{to}^{s} = \varphi_{to} + \omega T_s/2$.

13.1.3 Tuning of Sampled Data PID Regulators

The most frequently used form of discrete-time *PID* regulator is given by (13.18). Actually this is a combination of a discrete-time *PI* regulator described by (13.19) and a discrete-time *PD* regulator described by (13.21) forming their serial connection. Thus the design process provided for the two previous cases has to be repeated for their combination. The pulse transfer function of the regulator is given by

$$\hat{C}_{PID}(z) = K_C \frac{\left(z - z_1^{cd}\right)\left(z - z_2^{cd}\right)}{(z-1)\left(z - e^{-T_s/T}\right)} = K_C \frac{\left(1 - z_1^{cd} z^{-1}\right)\left(1 - z_2^{cd} z^{-1}\right)}{(1 - z^{-1})\left(1 - e^{-T_s/T} z^{-1}\right)} = \tilde{C}_{PID}\left(z^{-1}\right), \tag{13.33}$$

where the parameters are tuned to be $z_1^{cd} = p_1^{pd} = e^{-T_s/T_1}$ and $z_2^{cd} = p_2^{pd} = e^{-T_s/T_2}$. The initial and final values of the unit step response of the *PID* regulator given by (13.18) are

$$t = 0; \quad \lim_{z \to \infty} \frac{z}{z-1} K_C \frac{\left(z - z_1^{cd}\right)\left(z - z_2^{cd}\right)}{(z-1)\left(z - e^{-T_s/T}\right)} = K_C; \tag{13.34}$$

$$t = \infty; \quad \frac{1}{P(0)}. \tag{13.35}$$

Thus the value of the overexcitation ensuring the acceleration is

$$\eta = \frac{K_C}{P(0)}. \tag{13.36}$$

It seems formally that K_{C} is the gain of the discrete-time *PID* regulator, although this is not the case. K_{C} is a multiplier factor. If the gain of $\widehat{C}_{\mathrm{PID}}$ is normalized to one, then in the formula

$$\widehat{C}_{\mathrm{PID}}(z)=k_{\mathrm{C}}\frac{\left(1-e^{-T_{\mathrm{s}}/T}\right)}{\left(1-z_1^{\mathrm{cd}}\right)\left(1-z_2^{\mathrm{cd}}\right)}\frac{\left(z-z_1^{\mathrm{cd}}\right)\left(z-z_2^{\mathrm{cd}}\right)}{(z-1)\left(z-e^{-T_{\mathrm{s}}/T}\right)},\tag{13.37}$$

the actual gain of the regulator is k_{C}. Comparing the two formulas yields

$$K_{\mathrm{C}}=k_{\mathrm{C}}\frac{\left(1-e^{-T_{\mathrm{s}}/T}\right)}{\left(1-z_1^{\mathrm{cd}}\right)\left(1-z_2^{\mathrm{cd}}\right)},\tag{13.38}$$

and so

$$\eta=\frac{k_{\mathrm{C}}}{P(0)}\frac{\left(1-e^{-T_{\mathrm{s}}/T}\right)}{\left(1-z_1^{\mathrm{cd}}\right)\left(1-z_2^{\mathrm{cd}}\right)}=\frac{k_{\mathrm{C}}}{P(0)}\frac{\left(1-e^{-T_{\mathrm{s}}/T}\right)}{\left(1-e^{-T_{\mathrm{s}}/T_1}\right)\left(1-e^{-T_{\mathrm{s}}/T_2}\right)}.\tag{13.39}$$

Now p_1^{cd} and T can be obtained from the allowed largest overexcitation value $\eta_{\max}$:

$$p_1^{\mathrm{cd}}\geq 1-\frac{\eta_{\max}P(0)}{k_{\mathrm{C}}}\left(1-e^{-T_{\mathrm{s}}/T_1}\right)\left(1-e^{-T_{\mathrm{s}}/T_2}\right).\tag{13.40}$$

Also in this case we have to take care with the considerations related to (13.26). The time constant T corresponding to the borderline case $p_1^{\mathrm{cd}}=e^{-T_s/T}$ is given by

$$T=\frac{-T_s}{\ln\left[1-\frac{\eta_{\max}P(0)}{k_{\mathrm{C}}}\left(1-e^{-T_{\mathrm{s}}/T_1}\right)\left(1-e^{-T_{\mathrm{s}}/T_2}\right)\right]}.\tag{13.41}$$

Also the discrete *PID* regulator produces better overexcitation relations than the continuous regulator. This is a consequence of the fact that the initial overexcitation value is maintained through the entire sampling interval. Thus even with a smaller amplitude, sufficient energy can be transmitted to the process. That is, the overexcitation accelerates the process via the energy excess provided by the regulator.

In this case, as before, the design of the gain of the regulator is executed according to the classical method prescribing a given phase margin. The value of the prescribed phase margin φ_{to} formulated for the original CT system is now modified to be $\varphi_{\mathrm{to}}^{\mathrm{s}}=\varphi_{\mathrm{to}}+\omega_{\mathrm{c}}T_{\mathrm{s}}/2$.

If the formula (13.10) of the discrete *PID* regulator is used, then regulator tuning means the selection of the four parameters $\{T_{\mathrm{s}},A,T_{\mathrm{I}},T_{\mathrm{D}}\}$ if all three channels are used. If (13.13) is employed, then five parameters $\{T_{\mathrm{s}},A,T_{\mathrm{I}},T_{\mathrm{D}},T\}$ have to be determined. If the realization is based on formula (13.18), then the parameters

$\{T_s, K_C, z_1^{cd}, z_2^{cd}, T\}$ have to be tuned. The parameters $\{T_s, A, T_I, T_D, T\}$ and $\{T_s, K_C, z_1^{cd}, z_2^{cd}, T\}$ can be transformed to each other. In man-machine relations generally the parameters of the CT regulator are used, whereas for programming the DT parameters are used.

As a practical rule of thumb for the choice of the sampling time, it is suggested to take 7–10 samples during the CT step response of the process. Thus the actual sampling time depends on the dynamics of the process. Typical values for the sampling time are, for flow rate control: 1–3 s, level control: 5–10 s, pressure control: 1–5 s, temperature control: 10–20 s.

13.2 Other Design Methods

In the sequel, mainly design questions will be dealt with, but if required also some realization aspects will be considered.

Comparing DT systems with the structure of CT control systems, besides the process and the regulator there are further conversion units which serve to interconnect the discrete and continuous parts of the system. It was seen that the sampling provides the samples in a natural way for the control algorithm, which runs in the discrete sampling instants and calculates the control signal; the holding element supplements the CT process with a preceding dynamic CT element. In Chap. 11 it was seen that in the case of a zero order holding element this dynamics can be characterized on the one hand by the transfer function

$$W_{\mathrm{ZOH}} = \frac{1 - e^{-sT_s}}{s}$$

and on the other hand its effect corresponds approximately to a time-delay element with dead-time of one half of the sampling time, which can be easily taken into account in the frequency domain.

The following straightforward question may arise: why is the application of the methods presented for the synthesis of CT systems not sufficient for the design of DT systems? It will be seen that DT equivalents of the design concepts applied for CT systems can be elaborated, but a mechanical copying can not be followed in this case either. For demonstration, let us analyze the concept of structural stability well known from the theory of CT control systems.

Example 13.1 The transfer function of a CT process is

$$P(s) = \frac{1}{(1+5s)(1+10s)}. \tag{13.42}$$

The process is sampled with sampling time $T_s = 1$ [s] and a DT control system is realized using a serial proportional (*P*) regulator and a zero order holding element. The *SRE* equivalent model of the CT process together with the holding element is

$$P_{\mathrm{d}}(z)=\frac{0.0091z+0.0082}{z^2-1.7236z+0.7408}=\frac{0.0091(z+0.9048)}{(z-0.9048)(z-0.8187)}.$$

The process gain is 1, the gain of the proportional regulator is equal to the loop gain, $A=K$, and the characteristic equation of the closed-loop system is

$$1+KP_{\mathrm{d}}(z)=0, \tag{13.43}$$

or in polynomial form

$$z^2+(0.0091K-1.7236)+0.0082K+0.7408=0. \tag{13.44}$$

The root locus starts from the poles of the open-loop for $K=0$ also in the discrete-time case, as for $K=0$ the poles of the closed-loop are the same as those of the open-loop ($p_1=0.9054$, $p_2=0.8182$ in the z-plane). By increasing K the poles of the closed-loop approach more and more the unit circle, finally at $K=31.6$ the absolute value of the closed-loop poles is 1, $p_{1,2}=0.718\pm j0.696$ and $|p_{1,2}|=1$. For the range of $K>31.6$ the closed-loop poles get to the outside of the unit circle, thus the closed-loop control system becomes unstable. This result is unexpected at first sight, as the corresponding CT control system with loop transfer function

$$L(s)=KP(s)=\frac{K}{(1+5s)(1+10s)} \tag{13.45}$$

is structurally stable. Thus the property of structural stability is not transferred from the continuous system to the sampled data system. This can be explained by the fact that the holding element introduces a "virtual" dead-time into the control system, which excludes the possibility of structural stability. ■

All the methods discussed in Chap. 12. practically worked with rational functions of the variable z, namely with their numerators and denominators (which are polynomials). The discrete-time versions of the usual *PID* regulators were discussed in Sect. 13.1. The general polynomial method and the state feedback regulator shown for CT systems will be given in Chaps. 14 and 15 for the DT case. It is also expedient to deal with design methods which extend the CT frequency domain regulator design methods to the discrete case. In the sequel these further design approaches will also be presented.

These design methods can be discussed in the following three possible ways:

(a) *Design of an intermediate continuous-time regulator and its discretization*

The design of the continuous regulator is executed on the basis of the transfer function of the continuous process extended by the transfer function of the holding element. The designed CT regulator is only an intermediate step of the design, as finally a discrete-time algorithm described by the shift operator has to be obtained.

(b) *Design of a discrete-time regulator based on the discrete-time process model*

With this method the whole design process is executed in the discrete-time domain. The sampled CT model including the holding element is transformed to a discrete-time process model and a discrete-time regulator is designed based on this model.

(c) *Direct design of a discrete-time regulator based on the continuous-time process model*

The main point of the method is that based on the low frequency behavior of the CT process and considering the quality specifications for the closed-loop control system, the discrete-time regulator is formed by serial connection of discrete *PI* and *PD* elements whose parameters are tuned suited to the breakpoints of the frequency diagram of the continuous system.

13.2.1 Design of an Intermediate Continuous-Time Regulator and its Discretization

Repeating the basic concept of the design, here a CT regulator is designed for the CT process, then its discrete-time equivalent is determined. Based on the discrete form of the regulator, a program is written which realizes the control algorithm. The regulator is designed for the CT plant whose transfer function is obtained by serial connection of the original plant with transfer function $P(s)$ with the zero order holding element whose transfer function is $W_{\mathrm{ZOH}} = (1 - e^{-sT_s})/s$, thus obtaining

$$P_e(s) = W_{\mathrm{ZOH}} P(s) = \frac{1 - e^{-sT_s}}{s} P(s). \tag{13.46}$$

As this resulting transfer function is transcendental, an approximation has to be used. The *one-step approximation* considers the zero order holding element as an element with pure dead-time whose dead-time value is one-half of the sampling time. With this conversion the transfer function remains transcendental, but this form can be handled with the methods of continuous regulator design. With the *two-step approximation*, first the discrete form of the serially connected zero order holding element with the process is determined by the *SRE* transformation, then this form—which is already not transcendental—is transformed back to the CT domain, thus yielding a CT transfer function. The method of the design of an intermediate CT regulator is applied in practice when high frequency sampling is possible, and in this case there is no significant difference in the accuracy of the different discretization methods. Summarizing the possible methods, the steps of the method using one-step approximation for the discretization are

$$P_e(s) \approx e^{-sT_s/2} P(s) \Rightarrow C(s) \Rightarrow C_{\mathrm{d}}(z) \Rightarrow C_{\mathrm{d}}(q), \tag{13.47}$$

whereas the discretization in two-steps consists of the following steps:

$$P_{\rm d}(z) = \left(1 - z^{-1}\right)\mathcal{Z}\left\{\frac{P(s)}{s}\right\} \Rightarrow P_{\rm d}(s) \Rightarrow C(s) \Rightarrow C_{\rm d}(z) \Rightarrow C_{\rm d}(q). \tag{13.48}$$

In the conversion above, a complex variable w can be introduced which is analogous to the complex variable s. With this notation, (13.48) can be rewritten in the following form:

$$P_{\rm d}(z) = \left(1 - z^{-1}\right)\mathcal{Z}\left\{\frac{P(s)}{s}\right\} \Rightarrow P_{\rm d}(w) \Rightarrow C(w) \Rightarrow C_{\rm d}(z) \Rightarrow C_{\rm d}(q). \tag{13.49}$$

Here the notation $P_{\rm d}({\rm w})$ indicates that this CT transfer function, which includes also the effect of the holding element, is different from the transfer function of the CT process.

It can be seen that the discretization $C(s) \Rightarrow C_{\rm d}(z)$ is an element of both schemes, thus first this conversion will be discussed. $C_{\rm d}(z) \Rightarrow C_{\rm d}(q)$ means just a formal substitution: $C_{\rm d}(q) = C_{\rm d}(z)|_{z=q}$.

Let us suppose that $C(s)$, the transfer function of the CT regulator which is the basis of the discrete realization, has been designed. The discretization is executed by searching for the *equivalent discrete-time model*, or by *numerical integration*.

Creation of the equivalent discrete-time model

In this case, an equivalent discrete-time realization of the transfer function $C(s)$ is sought. The viewpoint of this search has to be specified. The search for an equivalent is in the sense that for a given input (for example a unit step, ramp or sinusoidal signal) the output of the DT system should be the same as the output of the CT system at the sampling instants. Possible criteria are: identical unit step response, identical impulse response, identical frequency transfer, identical poles-zeros, etc. It has to be emphasized that if the equivalence holds for one criterion, it is not guaranteed to hold for the others. It can be thought that with scarcer sampling the conditions are not improved. Here the problems of quantization and finite word length will not be discussed.

Unit step response equivalent (*SRE*) discretization:

The sampled values of the output of the continuous regulator are:

$$u[kT_{\rm s}] = u[k] = \mathcal{L}^{-1}\left\{\frac{1}{s}C(s)\right\}\Bigg|_{t=kT_{\rm s}}. \tag{13.50}$$

The z-transform of the sampled output signal of the regulator is:

$$U(z) = \frac{1}{1 - z^{-1}}C_{\rm d}(z), \tag{13.51}$$

and hence

$$C_d(z) = \left(1 - z^{-1}\right) \mathcal{Z}\left\{ \mathcal{L}^{-1}\left[\frac{1}{s} C(s)\right]_{t=kT_s} \right\}, \tag{13.52}$$

as also obtained previously.

Pulse response equivalent discretization:

The aim is to ensure the condition $u[k] = u(t)|_{t=kT_s}$. In this case, the following relationship is obtained for the discretization of the regulator:

$$C_d(z) = \mathcal{Z}\{u[k]\} = \mathcal{Z}\left\{ \mathcal{L}^{-1}[C(s)]_{t=kT_s} \right\} = C(z) \tag{13.53}$$

Discretization ensuring equivalent pole-zero mapping:

In this case each pole and each zero is transformed according to the mapping $z = e^{sT_s}$, and the static gain is kept at the same value. In this case a specific problem may arise: what to do if the system has no zero? It can be observed that the frequency function of a dynamical CT system which has no zeros tends to zero when the frequency tends to infinity. For DT systems the largest frequency is $\omega_{max} = \pi/T_s$, here the value of z is

$$z = e^{sT_s}\Big|_{s=j\omega_{max}} = e^{j\pi} = -1. \tag{13.54}$$

Therefore the nonexistent CT zero has to be mapped to the DT zero $z = -1$, that is in the numerator of the discretized model there will appear a factor $(z+1)$.

Discretization by numerical integration

To derive the discretization algorithms let us review some basic relations of numerical integration. Suppose the task is the integration of a continuous function $f(t)$. Let us denote the result of the integration by $i(t)$ (and let us use now the shift operator q).

$$i(t) = \int_{\tau=0}^{t} f(\tau)\mathrm{d}\tau \tag{13.55}$$

On the basis of the samples $f[k]$ of the function $f(t)$ available in the sampling instants T_s, the value of the integral in the sampling points $t = kT_s$ can be approximated in different ways. With right side rectangles, it is

$$i[k] = i[k-1] + T_s f[k]; \quad \frac{1}{s} \Rightarrow \frac{qT_s}{q-1}. \tag{13.56}$$

With left side rectangles:

$$i[k] = i[k-1] + T_s f[k-1]; \quad \frac{1}{s} \Rightarrow \frac{T_s}{q-1}. \tag{13.57}$$

With trapezoidal approximation:

$$i[k] = i[k-1] + T_s \frac{f[k-1] + f[k]}{2}; \quad \frac{1}{s} \Rightarrow \frac{T_s}{2}\frac{q+1}{q-1}. \tag{13.58}$$

Bilinear transformation:

A bilinear transformation is applied if a continuous transfer function $C(s)$ is discretized using numerical integration based on the trapezoid rule, resulting in the rational function of the shift operator; namely a formal substitution of the s variable is applied. Let us examine what kind of difference the formal substitution according to (13.58) yields in the frequency functions of the CT and the corresponding DT model. At a given frequency ω, the frequency function of the continuous model is $C(j\omega) = C(s)|_{s=j\omega}$. Realizing the discretization by

$$s = \frac{2}{T_s}\frac{q-1}{q+1}, \tag{13.59}$$

in the discretized model the frequency ω is expressed by

$$\left.\frac{2}{T_s}\frac{z-1}{z+1}\right|_{z=e^{j\omega T_s}} = \frac{2}{T_s}\frac{e^{j\omega T_s}-1}{e^{j\omega T_s}+1} = \frac{2}{T_s}\frac{e^{j\omega T_s/2}-e^{-j\omega T_s/2}}{e^{j\omega T_s/2}+e^{-j\omega T_s/2}} = j\frac{2}{T_s}\mathrm{tg}\left(\frac{\omega T_s}{2}\right). \tag{13.60}$$

The meaning of this expression is that there is a difference between the frequency functions of the continuous and the discretized systems. The difference depends on the frequency, as the rational function evaluated at a given frequency by substituting $s = j\omega$ and the evaluation of the same rational function by substituting

$$s = j\frac{2}{T_s}\mathrm{tg}\left(\frac{\omega T_s}{2}\right) \tag{13.61}$$

provide different results. The extent of the *frequency distortion* resulting from the relationship $\frac{2}{T_s}\mathrm{tg}\left(\frac{\omega T_s}{2}\right)$ in the low frequency domain is not significant because of the approximation

$$\frac{2}{T_s}\mathrm{tg}\left(\frac{\omega T_s}{2}\right) \approx \frac{2}{T_s}\frac{\omega T_s}{2} = \omega. \tag{13.62}$$

But close and closer to the sampling frequency, the deviation increases. At the same time it can be seen that introducing the "frequency scaling" of the discretization according to

$$s = \frac{\omega_1}{\frac{2}{T_s}\operatorname{tg}\left(\frac{\omega_1 T_s}{2}\right)} \frac{2}{T_s} \frac{q-1}{q+1} \tag{13.63}$$

makes the approximation errorless at a unique frequency ω_1 that can be arbitrarily chosen by the designer.

Delta transformation:
The delta transformation is based on the application of the numerical integration technique using left side rectangles. The differential operator resulting from (13.57) expressed as

$$\delta = \frac{q-1}{T_s} \approx s \tag{13.64}$$

is called the delta operator. Applying it to the DT signal series $f[kT_s]$ the following relationship is obtained:

$$\delta f[k] = \delta f[kT_s] = \frac{f(kT_s + T_s) - f(kT_s)}{T_s} = \frac{f[k+1] - f[k]}{T_s}. \tag{13.65}$$

The transfer function of a system expressed by the delta operator (referred to as its delta transform) can be derived formally both from the CT and the DT forms. This is obvious, as the both substitutions according to (13.64) and $q = \delta T_s + 1$ can be executed, but it has to be mentioned that while one substitution is approximate, the other one leads to exact results.

Starting from the DT form and using the substitution $q = \delta T_s + 1$ it can be seen that the transfer function expressed with the delta operator is also a rational function:

$$G(q) = \frac{\mathcal{B}(q)}{\mathcal{A}(q)} = \frac{\mathcal{B}(\delta T_s + 1)}{\mathcal{A}(\delta T_s + 1)} = \frac{\bar{\mathcal{B}}(\delta)}{\bar{\mathcal{A}}(\delta)} = \bar{G}(\delta). \tag{13.66}$$

The delta transform can be also derived from the CT transfer function $H(s)$, then $G(\delta) = H(s)|_{s=\delta}$, thus this maintains the structure of $H(s)$ (e.g. the pole excess is the same).

The theoretical importance of the delta transformation is that it creates a direct relationship between the CT and DT systems. More precisely, for a CT system given by the transfer function $H(s)$, the following relationship holds:

$$\lim_{T_s \to 0} G(\delta) = H(s). \tag{13.67}$$

Recently the delta transformation has been frequently used in practical applications. The reason for this is related simple to Eq. (13.67), as with a small sampling time, both the poles and the zeros of the delta transform and those of the continuous transfer function are close to each other.

As an example let us consider the continuous process given by the transfer function

$$P(s) = \frac{1}{(1+5s)(1+10s)}. \tag{13.68}$$

The *SRE* pulse transfer function for sampling time $T_s = 1$ [s] is

$$P_d(q) = \frac{0.0091(q+0.9048)}{(q-0.9048)(q-0.8187)}. \tag{13.69}$$

The information related to the poles—because of mapping $z = e^{sT_s}$—appears in a small deviation from 1, which is numerically unfavorable. The situation is even worse for smaller sampling times, e.g., in the case of $T_s = 0.1$ [s], the pulse transfer function is

$$P_d(q) = \frac{9.9006(q+0.99)10^{-5}}{(q-0.99)(q-0.9802)}. \tag{13.70}$$

Applying the relationship $q = \delta T_s + 1$ to (13.69), after some mathematical manipulations the following delta transform is obtained for the pulse transfer function:

$$G(\delta) = 1.0043\frac{1+0.525\delta}{(1+10.5042\delta)(1+5.5157\delta)}.$$

It can be seen that the poles and the gain are close to the continuous poles and gain. They are still closer for an even smaller sampling time. Therefore the delta transform has better numerical properties than the original pulse transfer form.

The development of hardware platforms realizing digital regulators makes possible the employment of smaller sampling times, therefore the use of the delta transform in discretization has gained in importance. In the case of fast sampling the accuracy provided by the delta transformation (13.64) is appropriate and it provides a good method which can replace the application of the bilinear transformation (where frequency distortion has to be handled).

Discretization of the continuous PID regulators

A *PID* regulator offers a good control solution for a wide range of processes, thus in control engineering practice they are the most frequently used regulators. The term *PID* refers to the fact that the regulator creates the control signal from the error signal as the sum of the outputs of three parallel channels. Furthermore, both for the continuous and the discrete cases, there are different variants of the *PID* regulators. In Chap. 12, in the discussion of the design of the discrete-time *PID* regulator family, only the *SRE* transformation was used. Considering the DT variants, there are several discretized forms of a given continuous-time *PID* regulator, depending on the discretization technique used for the CT-DT transformation. In the sequel, some discretization techniques will be presented as examples. Similarly, further forms can be derived. Different analog and digital realizations of *PID* regulators generate the control signal from the error signal according to the following equation:

$$u(t) = A\left[e(t) + \frac{1}{T_{\mathrm{I}}}\int\limits_{\tau=0}^{t} e(\tau)\mathrm{d}\tau + T_{\mathrm{D}}\frac{\mathrm{d}e(t)}{\mathrm{d}t}\right]. \tag{13.71}$$

where the error signal is the difference of the output signal from the reference signal:

$$e(t) = y_{\mathrm{r}} - y(t). \tag{13.72}$$

Supposing a differentiation based on two points and an integration according to the right side rectangle, the operation of the individual channels is directly described as follows.

P-channel:

$$u_{\mathrm{P}}[k] = Ae[k] \tag{13.73}$$

I-channel:

$$u_{\mathrm{I}}[k] = u_{\mathrm{I}}[k-1] + \frac{AT_{\mathrm{s}}}{T_{\mathrm{I}}}e[k] \tag{13.74}$$

D-channel:

$$U_{\mathrm{D}}(s) = AsT_{\mathrm{D}}E(s) \Rightarrow u_{\mathrm{D}}[k] = \frac{AT_{\mathrm{D}}}{T_{\mathrm{s}}}(e[k] - e[k-1]). \tag{13.75}$$

The control signal, which is the input of the process, is given by

$$\begin{aligned} u[k] &= u_{\mathrm{P}}[k] + u_{\mathrm{I}}[k] + u_{\mathrm{D}}[k] = \left[A + \frac{AT_{\mathrm{s}}}{T_{\mathrm{I}}(1-q^{-1})} + \frac{AT_{\mathrm{D}}}{T_{\mathrm{s}}}(q-1)\right]e[k] \\ &= A\left[1 + \frac{T_{\mathrm{s}}q}{T_{\mathrm{I}}(q-1)} + \frac{T_{\mathrm{D}}(q-1)}{T_{\mathrm{s}}q}\right]e[k] \end{aligned} \tag{13.76}$$

Using another numerical integration method, first let us express the *PID* compensation with the LAPLACE transforms:

$$U(s) = A\left[1 + \frac{1}{sT_{\mathrm{I}}} + sT_{\mathrm{D}}\right]E(s). \tag{13.77}$$

Discretize the integrator according to the bilinear transformation:

$$
\begin{aligned}
u[k] &= A\left[1+\frac{T_s}{2T_I}\frac{1+q^{-1}}{1-q^{-1}}+\frac{T_D}{T_s}\left(1-q^{-1}\right)\right]e[k] \\
&= A\left[\left(1+\frac{T_s}{2T_I}\right)+\frac{T_s}{2T_I}\frac{2q^{-1}}{1-q^{-1}}+\frac{T_D}{T_s}\left(1-q^{-1}\right)\right]e[k] \\
&= \left[K_P+K_I\frac{q^{-1}}{1-q^{-1}}+K_D\left(1-q^{-1}\right)\right]e[k]
\end{aligned}
\tag{13.78}
$$

where

$$
K_P = A+\frac{K_I}{2};\; K_I=\frac{AT_s}{T_I} \text{ and } K_D=\frac{AT_D}{T_s}. \tag{13.79}
$$

From the above relationships the discrete-time *PID* regulator using the shift operator is given by

$$
\left(1-q^{-1}\right)u[k] = \left[K_P\left(1-q^{-1}\right)+K_I q^{-1}+K_D\left(1-q^{-1}\right)^2\right]e[k], \tag{13.80}
$$

and after some rearrangement the following relationship is obtained:

$$
u[k] = u[k-1]+(K_P+K_D)e[k]-(K_P-K_I+2K_D)e[k-1]+K_D e[k-2]. \tag{13.81}
$$

This form of the discrete *PID* regulator is called the *position algorithm*, whereas expressing the control increment provides the so-called *velocity algorithm*:

$$
u[k]-u[k-1] = (K_P+K_D)e[k]-(K_P-K_I+2K_D)e[k-1]+K_D[k-2]. \tag{13.82}
$$

In a noisy measurement environment, the practical realization of the *D*-channel certainly can not be done according to $T_D \mathrm{d}e(t)/\mathrm{d}t$ appearing in the theoretical equation, as differentiating a noisy output signal by means of this channel would result in a control signal with causelessly high amplitude. Note that in the case of sampled data systems the application of a low pass filter after the measurement of the output signal would reduce the effect of the high frequency noise. The general solution, already discussed in the previous chapters, is the filtering of the differentiating effect by a first order lag element. In the LAPLACE domain

$$
U(s) = A\left[1+\frac{1}{sT_I}+\frac{sT_D}{1+sT_D/N}\right]E(s),
$$

where the usual value of N is between 5 and 20. Higher values of N allow the differentiating effect at higher frequency ranges. The discretized forms of the proportional (P), the integrating (I), and the differentiating (D) channels can also be generated independently of each other, following simple considerations, without using the formalism of numerical integration. Thus the following relationships can be given:

P-channel:

$$u_{\mathrm{P}}[k] = Ae[k] \tag{13.83}$$

I-channel:

$$u_{\mathrm{I}}[k] = u_{\mathrm{I}}[k-1] + \frac{AT_{\mathrm{s}}}{T_{\mathrm{I}}} e[k] \tag{13.84}$$

D-channel:

$$U_{\mathrm{D}}(s) = A\frac{sT_{\mathrm{D}}}{1+sT_{\mathrm{D}}/N} E(s) \Rightarrow U_{\mathrm{D}}(s) + \frac{sT_{\mathrm{D}}}{N} U_{\mathrm{D}}(s) = AT_{\mathrm{D}} sE(s) \tag{13.85}$$

The differentiation can be executed considering two consecutive samples by

$$\frac{\mathrm{d}f(t)}{\mathrm{d}t} \Rightarrow \frac{f[k]-f[k-1]}{T_{\mathrm{s}}}. \tag{13.86}$$

Then,

$$u_{\mathrm{D}}[k] + \frac{T_{\mathrm{D}}}{NT_{\mathrm{s}}}(u_{\mathrm{D}}[k] - u_{\mathrm{D}}[k-1]) = \frac{AT_{\mathrm{D}}}{T_{\mathrm{s}}}(e[k] - e[k-1]) \tag{13.87}$$

and

$$u_{\mathrm{D}}[k] = \frac{T_{\mathrm{D}}}{T_{\mathrm{D}} + NT_{\mathrm{s}}} u_{\mathrm{D}}[k-1] + \frac{AT_{\mathrm{D}}N}{T_{\mathrm{D}} + NT_{\mathrm{s}}}(e[k] - e[k-1]). \tag{13.88}$$

The control signal is obtained as the sum of the outputs of the three parallel channels:

$$u[k] = u_{\mathrm{P}}[k] + u_{\mathrm{I}}[k] + u_{\mathrm{D}}[k]. \tag{13.89}$$

When applying a parallel PID realization, the operation of the individual channels becomes transparent. A further advantage of this method is that special considerations can be easily taken into account. For instance if we do not want to include the effect of the changes appearing in the reference signal in the operation of the differentiating channel, then this can be directly realized according to the following relationship:

$$u_{\mathrm{D}}[k]=\frac{T_{\mathrm{D}}}{T_{\mathrm{D}}+NT_{\mathrm{s}}}u_{\mathrm{D}}[k-1]-\frac{AT_{\mathrm{D}}N}{T_{\mathrm{D}}+NT_{\mathrm{s}}}(y[k]-y[k-1]). \tag{13.90}$$

Similarly the handling of integrator windup can be realized by observing directly the variable $u_{\mathrm{I}}[k]$ and limiting its value.

Of course the above relations can be also formulated using the shift operator. Then for the *I*-channel and for the *D*-channel the following relationships are obtained:

$$u_{\mathrm{I}}[k]=q^{-1}u_{\mathrm{I}}[k-1]+\frac{AT_{\mathrm{s}}}{T_{\mathrm{I}}}e[k]\Rightarrow u_{\mathrm{I}}[k]=\frac{AT_{\mathrm{s}}}{T_{\mathrm{I}}(1-q^{-1})}e[k], \tag{13.91}$$

$$u_{\mathrm{D}}[k]=\frac{(1-q^{-1})AT_{\mathrm{D}}N}{\left(1-\frac{q^{-1}T_{\mathrm{D}}}{T_{\mathrm{D}}+NT_{\mathrm{s}}}\right)(T_{\mathrm{D}}+NT_{\mathrm{s}})}e[k]=\frac{(1-q^{-1})AT_{\mathrm{D}}N}{T_{\mathrm{D}}+NT_{\mathrm{s}}-q^{-1}T_{\mathrm{D}}}e[k] \tag{13.92}$$

By adding the three components, the resulting equation is

$$\begin{aligned}u[k]&=u_{\mathrm{P}}[k]+u_{\mathrm{I}}[k]+u_{\mathrm{D}}[k]=A\left(1+\frac{T_{\mathrm{s}}}{T_{\mathrm{I}}(1-q^{-1})}+\frac{(1-q^{-1})T_{\mathrm{D}}N}{T_{\mathrm{D}}+NT_{\mathrm{s}}-q^{-1}T_{\mathrm{D}}}\right)e[k]\\&=A\left\{\frac{T_{\mathrm{I}}(1-q^{-1})(T_{\mathrm{D}}+NT_{\mathrm{s}}-q^{-1}T_{\mathrm{D}})+T_{\mathrm{s}}(T_{\mathrm{D}}+NT_{\mathrm{s}}-q^{-1}T_{\mathrm{D}})+T_{\mathrm{D}}T_{\mathrm{I}}N(1-q^{-1})}{T_{\mathrm{I}}(1-q^{-1})(T_{\mathrm{D}}+NT_{\mathrm{s}}-q^{-1}T_{\mathrm{D}})}\right\}e[k]\end{aligned} \tag{13.93}$$

Observe that after some rearrangements the general form of the discrete *PID* regulator can be written as a function of the shift operator in the following form:

$$u[k]=\frac{b_{\mathrm{o}}+b_1q^{-1}+b_2q^{-2}}{1+a_1q^{-1}+a_2q^{-2}}e[k]. \tag{13.94}$$

Hence the recursive relationship (difference equation) providing the realization algorithm is

$$u[k]=b_{\mathrm{o}}e[k]+b_1e[k-1]+b_2e[k-2]-a_1u[k-1]-a_2u[k-2]. \tag{13.95}$$

Let us summarize the above considerations. The regulator design methods based on the presented discretization methods can be divided into three groups:

Regulator design with the method of bilinear transformation (w-transformation)

The steps of the design are the following:

(1) Calculate the w-transform of the discrete process model $P_d(z) = (1 - z^{-1})\mathcal{Z}\{P(s)/s\}$:

$$P(w) = P_d(z)|_{z=(1+wT_s/2)/(1-wT_s/2)} \tag{13.96}$$

(2) Design a continuous $C(w)$ regulator for the continuous-time $P(w)$ process with one of the design methods. This design typically can be executed based on the frequency function of the $P(w)$ process, calculated as $P(j\nu) = P(w)|_{w=j\nu}$.

(3) Determine the discrete-time regulator using the inverse w-transformation:

$$C(z) = C(w)|_{w=2(1-z^{-1})/T_s(1+z^{-1})} \tag{13.97}$$

(4) Checking the behavior of the control system is the essential final step of the design. In this case it will also be seen whether the frequency distortion, giving by $\nu = \frac{2}{T_s}\mathrm{tg}\left(\frac{\omega T_s}{2}\right)$, causes significant deterioration of the designed behavior of the closed-loop system or not. If so (this may occur with relatively low frequency sampling), then in the w-transformation, a *prewarping* (rescaling) can be applied, by $w' = \frac{\omega_c}{\mathrm{tg}(\omega_c T_s/2)}\frac{1-z^{-1}}{1+z^{-1}}$, to avoid the frequency distortion in the surroundings of the cut-off frequency ω_c. Note that the step of the CT regulator design in the w-transformation method, that is, the determination of $C(w)$, is a non-conventional task, as $P(w)$ is a non-minimum-phase transfer function.

Regulator design with the method of the delta (δ) transformation

In practice the discretization of a CT regulator using the delta transformation is executed as follows: we sketch a block-diagram realization of the CT regulator built of integrators, constants and summation elements, then the CT integrators are replaced by the delta integrators $\frac{1}{\delta} = \frac{T_s}{q-1}$. In the regulator realization in each step the outputs of the integrators are calculated according to

$$i[k] = i[k-1] + T_s f[k-1]. \tag{13.98}$$

Regulator design using the discretized PID regulators

With any of the presented methods the position or the velocity algorithm directly provides the control algorithm.

As a summary it can be stated, that the method of the design of an intermediate CT regulator is justified mainly in the case of small sampling time, when the conditions for the application are numerically favorable. Another advantage is that the regulator design methods used in the case of CT systems can be directly applied. But this is also the disadvantage of the method: using it we do not go beyond the limits of the CT design techniques.

13.2.2 Design of Discrete-Time Regulators Using Discrete-Time Process Models

In this design the hybrid system (containing CT and DT parts) is transformed to a discrete-time equivalent system and the regulator design is executed in the DT domain. The first step is the conversion of the CT process together with the holding element to a discrete-time process model. A zero order holding element is supposed, then the *SRE* discrete model of the CT process is

$$P_{\mathrm{d}}(z) = \left(1 - z^{-1}\right)\mathcal{Z}\left\{\frac{P(s)}{s}\right\}. \tag{13.99}$$

The aim of the design is to determine the pulse transfer function $C(z)$ of the series regulator, which results in the overall pulse transfer of the closed-loop control system

$$T(z) = \frac{C(z)P_{\mathrm{d}}(z)}{1 + C(z)P_{\mathrm{d}}(z)} \text{ or more generally, } T(z) = \frac{C(z)G(z)}{1 + C(z)G(z)}, \tag{13.100}$$

ensuring the prescribed design specifications. Chapters. 12, 14 and 15 deal with these methods, therefore here we do not go into the details of these design methods.

13.2.3 Design of Discrete-Time Regulators Using Continuous-Time Process Models

The basis of this method is the observation that in the low frequency domain, sampled data systems can be well approximated by CT systems. As seen previously in the frequency domain the typical steps of the regulator design of CT systems are the following: to ensure the static accuracy *PI* elements are designed as serial parts of the regulator; then to improve the servo and disturbance rejection properties of the closed-loop system, approximate *PD* (in reality phase-lead) elements are added to the regulator, which increases the cut-off frequency; finally the gain is tuned by setting the phase margin to a prescribed value. In the case of processes which have the characteristic of a low pass filter, the corner frequencies of the *PI* and the approximate *PD* elements are placed in the low or the middle frequency domain.

The design of a discrete-time regulator uses a similar method. The serial *PI* and *PD* elements here are also chosen according to the frequency function, practically the approximate Bode diagram of the CT process, but after having chosen the characteristic (*PI* or *PD*) and the breakpoint frequencies of these serial regulator elements, not the CT, but *the discrete-time serial regulator elements are determined directly*. The calculation of the gain is again the final step of regulator design, setting it to ensure the prescribed phase margin.

Summarizing the steps of the design:

(1) Introducing *PI* and *PD* regulators according to the frequency function of the continuous process
(2) Giving the discrete-time forms of the *PI* and *PD* regulators introduced as serial elements in the first step and calculating the loop transfer function
(3) Determining the gain of the regulator to ensure the prescribed phase margin
(4) Checking the value of the cut-off frequency
(5) Checking the performance of the closed-loop, analyzing the course of the output and the control signal (static and dynamic response, intersampling behavior).

Comparing this design procedure with that for a continuous regulator, the second, third and fifth steps require special considerations.

Creating the discrete forms of PI and PD regulators

The discretized forms are generated according to pole/zero equivalence.

PI element:

The continuous-time transfer function is $(1+sT_{\mathrm{I}})/s$. Further characteristics are summarized in the table below.

	Continuous-time	Discrete-time
Transfer function	$\frac{1+sT_{\mathrm{I}}}{s}$	$k_1 \frac{z-z_1}{z-p_1}$
Zero	$z_{\mathrm{PI}} = -1/T_{\mathrm{I}}$	$z_1 = e^{z_{\mathrm{PI}}T_{\mathrm{s}}} = e^{-T_{\mathrm{s}}/T_{\mathrm{I}}}$
Pole	$p_{\mathrm{PI}} = 0$	$p_1 = e^{p_{\mathrm{PI}}T_{\mathrm{s}}} = 1$

The discrete-time equivalent of the *PI* element is $\left(z - e^{-T_{\mathrm{s}}/T_{\mathrm{I}}}\right)/(z-1)$. The gain is not indicated, as the overall loop gain is determined at the end of the design procedure.

PD element:

The continuous-time transfer function is $(1+s\tau)/(1+sT)$. Further characteristics are summarized in the table below.

	Continuous-time	Discrete-time
Transfer function	$\frac{1+s\tau}{1+sT}$	$k_2 \frac{z-z_2}{z-p_2}$
Zero	$z_{\mathrm{PD}} = -1/\tau$	$z_2 = e^{z_{\mathrm{PD}}T_{\mathrm{s}}} = e^{-T_{\mathrm{s}}/T}$
Pole	$p_{\mathrm{PD}} = -1/T$	$p_2 = e^{p_{\mathrm{PD}}T_{\mathrm{s}}} = e^{-T_{\mathrm{s}}/T}$

The discrete-time equivalent of the *PD* element is: $\left(z - e^{-T_{\mathrm{s}}/\tau}\right)/\left(z - e^{-T_{\mathrm{s}}/T}\right)$. The gain is not indicated, as the overall loop gain is determined at the end of the design procedure. Now the value of T which in the continuous case used to be the fifth, tenth or twentieth part of τ, now can be arbitrarily small, as in the extreme case the amplitude-frequency function of the element $1+s\tau$ can not tend to infinity when the frequency is increasing because of the finite sampling frequency. Therefore in practice we can consider a value of T which fulfills the condition $e^{-T_{\mathrm{s}}/T} \approx 0$; the corresponding pulse transfer function of the discrete *PD* element is $\left(z - e^{-T_s/\tau}\right)/z$, and here the only parameter is determined by the chosen breakpoint frequency.

The discrete-time regulator may contain serially connected *PI* and *PD* elements resulting in the transfer function

$$C(z) = AC_{\mathrm{PI}}(z)C_{\mathrm{PD}}(z) = A\frac{\left(z - e^{-T_s/T}\right)\left(z - e^{-T_s/\tau}\right)}{(z-1)z} \tag{13.101}$$

Its usual name is a *PIPD* regulator. Note that it is possible to use more *PI* and *PD* elements.

The third step of the design procedure is the choice of the gain A of the regulator. The structure of the loop is now fixed: the only free parameter is the loop gain $K = AP_{\mathrm{d}}(1)$. The loop transfer function $L(z) = C(z)P_{\mathrm{d}}(z)$ is then investigated, where K determines the value of the cut-off frequency ω_c and the phase margin. We are looking for a value of K, that yields the prescribed value of the phase margin. There are several ways to determine the value of K. With trial we can use an iterative technique which may converge quite quickly if the steps are chosen appropriately. Another method considers the frequency characteristic of the dynamic components of the regulator $C(z)$. A *CAD* environment may replace these methods, plotting the course of the frequency function of $L(z)$ with unit gain, and then calculating the factor which modifies the gain to reach the required value of the phase margin. This critical step will be demonstrated in Example 13.2.

The last point of the design is checking the behavior of the closed-loop control circuit. The course of the output signal has to be examined also between the sampling points. This can be done by simulation.

Example 13.2 The transfer function of the continuous-time process is

$$P(s) = \frac{e^{-s}}{(1+10s)(1+5s)}. \tag{13.102}$$

Let us design a *PIPD* regulator with sampling time T_s = 1 s. The required type number is $i = 1$ and the phase margin is approximately 60°. First the process is discretized, its pulse transfer function is

$$\begin{aligned} P_{\mathrm{d}}(z) &= z^{-1}\left(1 - z^{-1}\right)\mathcal{Z}\left\{\frac{1}{(1+10s)(1+5s)s}\right\} \\ &= 0.0091\frac{(z+0.9048)}{\left(z - e^{-T_s/10}\right)\left(z - e^{-T_s/5}\right)z} = 0.0091\frac{(z+0.9048)}{(z-0.9048)(z-0.8187)z} \end{aligned} \tag{13.103}$$

Following the design concept given for CT design the breakpoint frequency of the *PI* regulator is chosen to be $\omega_1 = 1/10$, and the breakpoint frequency of the *PD* regulator is chosen to be $\omega_2 = 1/5$. Then

$$C_{\mathrm{PI}}(z) = \frac{z - e^{-1/10}}{z - 1} = \frac{z - 0.9048}{z - 1} \tag{13.104}$$

and

$$C_{\mathrm{PD}}(z) = \frac{z - e^{-1/5}}{z} = \frac{z - 0.8187}{z}. \tag{13.105}$$

The serial connection gives the pulse transfer function of the *PIPD* regulator:

$$C_{\mathrm{PIPD}}(z) = A\frac{z - 0.9048}{z - 1}\frac{z - 0.8187}{z}. \tag{13.106}$$

The loop transfer function is

$$\begin{aligned} L(z) = C_{\mathrm{PIPD}}(z)P_{\mathrm{d}}(z) &= A\frac{z - 0.9048}{z - 1}\frac{z - 0.8187}{z}\frac{0.0091(z+0.9048)}{(z - 0.9048)(z - 0.8187)z} \\ &= A\frac{0.0091(z+0.9048)}{z(z-1)} \end{aligned} \tag{13.107}$$

Analyzing the loop frequency function $L(z)|_{z=e^{j\omega T_s}} = a(\omega)e^{j\varphi(\omega)}$ with $A = 1$, the phase function $\varphi(\omega)$ reaches the value $\varphi = -120°$ when the amplitude is $a_{120} = 0.0661$. As by changing the gain A the phase angle is not modified (only the amplitude function is changed by a factor of A), setting the gain, $A = 1/0.0661 = 15.3$ the phase angle will be $\varphi = -120°$ at $a_{120} = 1$, which means that the phase margin is $\varphi_t = 180° + \varphi(\omega)|_{a=1} = 180° - 120° = 60°$. So the regulator is

$$C_{\mathrm{PIPD}}(z) = 15.13\frac{z - 0.9048}{z - 1}\frac{z - 0.8187}{z}.$$

The unit step responses $u[k]$ and $y(t)$ of the closed-loop are shown in Fig. 13.2. ■

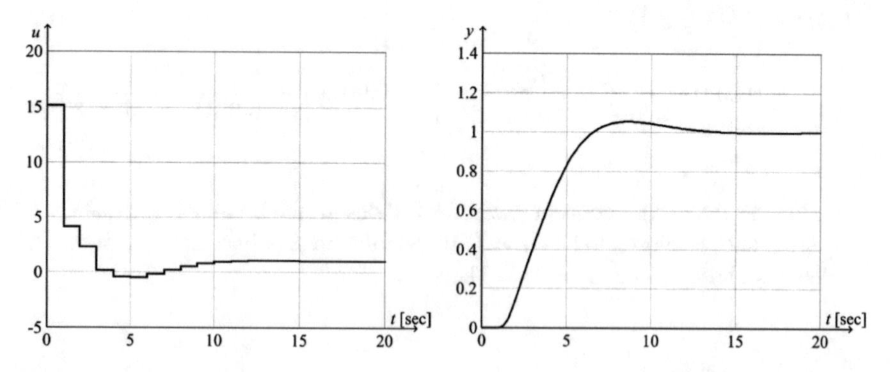

Fig. 13.2 Unit step response and the output signal of a closed-loop *DT* system with *PIPD* regulator

13.3 Design of Discrete-Time Residual Systems

In the case of CT systems, after the regulator design, the loop transfer function often becomes relatively simple. Then the so-called residual systems can be described in analytical form. For sampled data systems—if a DT regulator is designed for the z-transform of the CT process considered together with the zero order holding element—similar simplified examination methods might have been expected. But unfortunately this is not the case, as only the δ-transformation provides the same structure for the DT model as for the CT model. Note that the generally used *SRE* transformation always produces one pole excess. When creating the discrete-time model of a CT process of order n together with a zero order holding element using the z-transformation, the pulse transfer function will contain n poles and generally $(n-1)$ zeros, therefore the analysis of DT residual systems means the consideration of more parameters than in the CT case. The sampling time is a further free parameter. The following example demonstrates these effects.

13.3.1 Continuous-Time Second Order Process with Two Time Lags and Dead-Time

Let us consider the following continuous process containing dead-time given by its transfer function

$$P(s) = \frac{e^{-sT_\mathrm{d}}}{(1+sT_1)(1+sT_2)}, \tag{13.108}$$

It is supposed that the dead-time T_d is an integer multiple of the sampling time T_s. The discrete dead time is expressed as

$$T_\mathrm{d} = dT_\mathrm{s}, \quad (d = 0, 1, 2, \ldots) \tag{13.109}$$

Then the DT model of the continuous process together with the zero order holding element is

$$P_\mathrm{d}(z) = z^{-d}\frac{K_\mathrm{P}(z-z_1)}{(z-p_1)(z-p_2)} = \frac{K_\mathrm{P}(z-z_1)}{z^d(z-p_1)(z-p_2)}. \tag{13.110}$$

Suppose the pulse transfer function of the discrete-time *PID* regulator designed using the pole cancellation technique is

$$C(z) = C_\mathrm{PID}(z) = \frac{K_\mathrm{C}(z-p_1)(z-p_2)}{z(z-1)}. \tag{13.111}$$

Then the loop transfer function is

$$L(z) = C(z)P(z) = \frac{K_\mathrm{C}K_\mathrm{P}(z-z_1)}{z^{d+1}(z-1)} = \frac{K_\mathrm{L}(z-z_1)}{z^{d+1}(z-1)} = \frac{K_\mathrm{I}(z-z_1)}{z^{d+1}(z-1)}. \tag{13.112}$$

and the frequency function of $L(z)$ is

$$L^*(j\omega) = L\left(e^{j\omega T_s}\right) = \frac{K_I\left(e^{j\omega T_s} - z_1\right)}{e^{j(d+1)\omega T_s}\left(e^{j\omega T_s} - 1\right)} = \frac{K_I\left(e^{j\omega T_s} - z_1\right)}{e^{j(d+2)\omega T_s} - e^{j(d+1)\omega T_s}}$$
$$= \frac{K_I[\cos(\omega T_s) + j\sin(\omega T_s) - z_1]}{\cos[(d+2)\omega T_s] + j\sin[(d+2)\omega T_s] - \cos[(d+1)\omega T_s] - j\sin[(d+1)\omega T_s]}. \tag{13.113}$$

The phase angle of $L(j\omega)$ is given by

$$\mathrm{arc}\{L^*(j\omega)\} = \mathrm{arctg}\frac{\sin(\omega T_s)}{\cos(\omega T_s) - z_1} - \mathrm{arctg}\frac{\sin[(d+2)\omega T_s] - \sin[(d+1)\omega T_s]}{\cos[(d+2)\omega T_s] - \cos[(d+1)\omega T_s]}. \tag{13.114}$$

As in the low frequency domain the phase angle of $\frac{1}{z^{(d+1)}(z-1)}$ can be well approximated by

$$\mathrm{arc}\left\{\frac{1}{e^{j(d+1)\omega T_s}\left(e^{j\omega T_s} - 1\right)}\right\} \cong -90^\circ - (d+1)\omega T_s\frac{180^\circ}{\pi}, \tag{13.115}$$

we have that in this range

$$\mathrm{arc}\{L^*(j\omega)\} \cong \mathrm{arctg}\frac{\sin(\omega T_s)}{\cos(\omega T_s) - z_1} - 90^\circ - (d+1)\omega T_s\frac{180^\circ}{\pi}. \tag{13.116}$$

If the control system is designed for the phase margin $\varphi_t = 60^\circ$, then

$$\mathrm{arc}\{L^*(j\omega)\} = -120^\circ \tag{13.117}$$

has to be fulfilled. For this the cut-off frequency ω_c has to be solved from the transcendental equation

$$(d+1)\omega T_s\frac{180^\circ}{\pi} - 30^\circ \cong \mathrm{arctg}\frac{\sin(\omega T_s)}{\cos(\omega T_s) - z_1} \tag{13.118}$$

at $\omega = \omega_c$. Then from

$$|L^*(j\omega)|_{\omega=\omega_c} = \left|\frac{K_I\left(e^{j\omega T_s} - z_1\right)}{e^{jk\omega T_s}\left(e^{j\omega T_s} - 1\right)}\right|_{\omega=\omega_c} = \left|\frac{K_I\left(e^{j\omega T_s} - z_1\right)}{\left(e^{j\omega T_s} - 1\right)}\right|_{\omega=\omega_c} = K_I\left|\frac{\left(e^{j\omega_c T_s} - z_1\right)}{\left(e^{j\omega_c T_s} - 1\right)}\right| = 1 \tag{13.119}$$

$K_{\rm I}$ can be solved for

$$K_{\rm I} = \frac{\left|e^{j\omega_{\rm c}T_{\rm s}} - 1\right|}{\left|e^{j\omega_{\rm c}T_{\rm s}} - z_1\right|}, \tag{13.120}$$

and finally the gain of the regulator is obtained

$$K_{\rm C} = K_{\rm I}/K_{\rm P}. \tag{13.121}$$

It can be seen that especially because of the initial transcendental equation, complicated calculations have to be executed. But with *CAD* facilities, expression

$$L(z) = \frac{K_{\rm I}(z - z_1)}{z^{(d+1)}(z - 1)} = \frac{K_{\rm I}(z - z_1)}{z^k(z - 1)} \tag{13.122}$$

obtained for the loop transfer function can easily be handled. For example, the frequency function belonging to the pulse transfer function $L(z)$ can be determined, and in the required frequency range with a given resolution coherent values of the frequency, the amplitude and the phase can be calculated (just by calling a routine). These values are important for further considerations in the vicinity of the frequency where $\mathrm{arc}\{L(j\omega)\} \cong -120°$. The approximation can be made more accurate with tapering and refining the resolution of the frequency range. The sampling time $T_{\rm s}$ is also a parameter of the calculations (and of the program routine). Its value is also needed in the formal substitution of $z = e^{sT_{\rm s}}|_{s=j\omega}$.

Note that in the case of a system with no dead-time ($d = 0$) and $k = 1$, the initial transcendental equation is

$$\omega T_{\rm s}\frac{180°}{\pi} - 30° \cong \mathrm{arctg}\frac{\sin(\omega T_s)}{\cos(\omega T_s) - z_1}. \tag{13.123}$$

A further remark is that calculating the discrete model of a continuous second order system with two time lags, containing also a zero and dead-time, given by the transfer function

$$P(s) = \frac{(1 + s\tau)e^{-sT_{\rm d}}}{(1 + sT_1)(1 + sT_2)}, \tag{13.124}$$

the pulse transfer function can be obtained also in the form

$$P_{\rm d}(z) = z^{-d}\frac{K_{\rm P}(z - z_1)}{(z - p_1)(z - p_2)} = \frac{K_{\rm P}(z - z_1)}{z^d(z - p_1)(z - p_2)}. \tag{13.125}$$

13.3.2 The Tuschák Method

The discrete regulator design method of Tuschák is based on the fact that also the design methods of DT regulators use essentially pole/zero cancellation methods.

Therefore in most cases good results are obtained if the regulator is designed based on the CT process model, but taking into account the extra phase distortion coming from the sampling. We have already discussed the simplest case of phase distortion, namely the application of the zero order holding element which virtually places an extra dead-time of value $T_s/2$ into the closed-loop, thus the phase characteristic changes unfavorably according to $\varphi_s(j\omega) = \varphi(j\omega) - \omega T_s/2$.

DT models obtained by *SRE* transformation may introduce further undesired distortions. Let us analyze first the *SRE* model of the first order lag element $P(s) = 1/(1+sT)$, where

$$P_d(z) = \frac{b_1 z^{-1}}{1+a_1 z^{-1}} = \frac{b_1}{z+a_1} = \frac{K\left(1-e^{-T_s/T}\right)z^{-1}}{1-e^{-T_s/T}z^{-1}} = \left.\frac{K\left(1-e^{-T_s/T}\right)}{z-e^{-T_s/T}}\right|_{K=1}$$
$$= \frac{1-e^{-T_s/T}}{z-e^{-T_s/T}}. \tag{13.126}$$

Approximating the exponential elements with their TAYLOR expansions,

$$P_d\left(e^{j\omega T_s}\right) \approx \tilde{P}_d(j\omega)$$
$$= \frac{1-\left[1-T_s/T+(T_s/T)^2/2-\cdots\right]}{\left[1+j\omega T_s+(j\omega T_s)^2/2+\cdots\right]-\left[1-T_s/T+(T_s/T)^2/2-\cdots\right]}. \tag{13.127}$$

Neglecting the elements whose degree is higher than two, the following relationship is obtained

$$\tilde{P}_d(j\omega) \approx \frac{1}{1+j\omega T}\frac{1-T_s/2T}{1+(j\omega T-1)T_s/2T} \approx \frac{1}{1+j\omega T}e^{-j\omega T_h^-}, \tag{13.128}$$

where

$$T_h^- = \left.\frac{T_s/2}{1-T_s/2T}\right|_{T_s \ll T} \cong \frac{T_s}{2}. \tag{13.129}$$

Thus in the low frequency domain the additional dead-time is equal to the extra dead-time of $T_s/2$ considered as a consequence of sampling, but for larger sampling times this can be significantly larger, for instance in the case of $T_s = T$, it takes already the value of $T_h^- = T_s$.

In the pulse transfer function, not only do zeros appear correspond to the zeros of the CT process, but also additional zeros appear because of the sampling. Let us consider the DT model of unity gain given by the pulse transfer function $P_d(z) = (z+\gamma)/(1+\gamma)$. The low frequency approximation of its frequency function is

$$P_{\mathrm{d}}\left(e^{j\omega T_{\mathrm{s}}}\right) = \frac{e^{j\omega T_{\mathrm{s}}} + \gamma}{1+\gamma} \approx \frac{1 + j\omega T_{\mathrm{s}} + (j\omega T_{\mathrm{s}})^2/2 + \cdots + \gamma}{1+\gamma}. \tag{13.130}$$

Dividing the numerator and the denominator by $(1+\gamma)$, then neglecting the quadratic and higher degree elements, the approximating frequency function is obtained as

$$\tilde{P}_{\mathrm{d}}(j\omega) \approx 1 + j\omega \frac{T_{\mathrm{s}}}{1+\gamma} \approx e^{j\omega T_{\mathrm{h}}}, \tag{13.131}$$

where the additional negative dead-time (i.e., acceleration) is

$$T_{\mathrm{h}}^{+} = \left. \frac{T_{\mathrm{s}}}{1+\gamma} \right|_{\gamma \approx 0} = T_{\mathrm{s}}. \tag{13.132}$$

If γ is small, then $T_{\mathrm{h}}^{+} \approx T_{\mathrm{s}}$, if it is large, then the value of the additional (positive) dead-time can be smaller than $T_{\mathrm{s}}/2$.

Based on the above considerations, a rough estimate of the entire additional dead-time is given by

$$\tilde{\tilde{T}}_{\mathrm{h}} = \frac{(P-Z)T_{\mathrm{s}}}{2}, \tag{13.133}$$

where P is the number of poles and Z is the number of zeros of the DT model. This does not mean much more than the additional dead-time $T_{\mathrm{s}}/2$ introduced by the zero order holding element (as for *SRE* discretization $P - Z = 1$). Nevertheless it is expedient to calculate all the additional dead-times resulting from the poles and the zeros according to (13.129) and (13.132) and summaring them, to get

$$\tilde{T}_{\mathrm{h}} = \sum_{i=1}^{Z} T_{\mathrm{h},i}^{+} - \sum_{i=1}^{P} T_{\mathrm{h},i}^{-}. \tag{13.134}$$

Example 13.3 Let the transfer function of the CT process be

$$P(s) = \frac{1}{(1+s)(1+5s)(1+10s)}. \tag{13.135}$$

The sampling time is $T_{\mathrm{s}} = 1$. The pulse transfer function is

$$P_{\mathrm{d}}(z) = 0.0024 \frac{(z+0.1903)(z+2.7471)}{(z-0.9048)(z-0.8187)(z-0.3679)}. \tag{13.136}$$

The approximating discrete frequency function in the low frequency domain $(\omega < 1/T_s = 1)$ is calculated as

$$\tilde{P}_d(j\omega) = 0.0024 \frac{(1+0.1903)(1+2.7471)e^{-j\omega\left(\frac{0.5}{1-0.5}+\frac{0.5}{1-0.1}+\frac{0.5}{1-0.05}-1/1.1903-1/3.7471\right)}}{(1-0.9048)(1-0.8187)(1-0.3679)(1+10j\omega)(1+5j\omega)(1+j\omega)} \tag{13.137}$$

or

$$\tilde{P}_d(j\omega) = \frac{e^{-j\omega(2.089-0.84-0.2669)}}{(1+10j\omega)(1+5j\omega)(1+j\omega)} = \frac{e^{-j\omega 0.875}}{(1+10j\omega)(1+5j\omega)(1+j\omega)}. \tag{13.138}$$

As seen here a bit higher value was obtained than 0.5, one-half of the sampling time. ■

13.3.3 Discrete-Time Second Order Process with Time Lag and Dead-Time

The *SRE* discrete-time model of the second order CT time process given by the transfer function (13.108) is of DT form (13.110), which can be written also in the usual form

$$P'_d(z^{-1}) = \frac{b'_1 z^{-1} + b'_2 z^{-2}}{1 + a'_1 z^{-1} + a'_2 z^{-2}} z^{-d} = \frac{b'_o(1+\gamma z^{-1})}{1 + a'_1 z^{-1} + a'_2 z^{-2}} z^{-d'}. \tag{13.139}$$

It is worthwhile to note that using the δ-transformation according to (13.64) one obtains

$$P_d(q^{-1}) = \frac{b''_o q^{-2}}{1 + a''_1 q^{-1} + a''_2 q^{-2}} q^{-d} = \frac{b''_o}{1 + a''_1 q^{-1} + a''_2} q^{-d''}. \tag{13.140}$$

where $d'' = d+2$. So both formulas can be given by the general pulse transfer function $P_d = \mathcal{B}z^{-k}/\mathcal{A}$, only with differing values of $\mathcal{B}$ and k.

The design method using the usual pole cancellation method in the case of the discrete-time *PID* regulator can be described in the simplest way by the following pulse transfer function of the regulator:

$$C(z) = \frac{q_o + q_1 z^{-1} + q_2 z^{-2}}{1 - z^{-1}} G_F(z) = \frac{q_o \mathcal{A}(z)}{1 - z^{-1}} G_F(z), \tag{13.141}$$

where $G_F(z)$ is a serial filter, which forms a part of the regulator. With this compensation the resulting residual loop transfer function is

$$L(z) = \frac{q_o \mathcal{B}}{1 - z^{-1}} G_F(z) z^{-k} = \frac{q_o b_o (1+\gamma z^{-1})}{1 - z^{-1}} = \frac{K_I(1+\gamma z^{-1})}{1 - z^{-1}}. \tag{13.142}$$

For phase margin $\varphi_t = 60°$ the transients of this simple residual closed-loop system are very nice, the overshoot is less than 1–5%. Unfortunately, to calculate the integrating loop gain K_I, the following nonlinear equation has to be solved:

$$x = \frac{1 - 2\mathrm{arctg}\frac{\gamma \sin x}{1+\gamma \cos x}}{2d-1} = g(x). \tag{13.143}$$

Using its solution,

$$K_I = \frac{\sqrt{2(1-\cos x)}}{\sqrt{1+2\gamma\cos x+\gamma^2}}. \tag{13.144}$$

From the last two equations the curves obtained as optimal solutions for setting the parameter K_I as a function of d and γ are shown in Fig. 13.3. For positive values of γ, using the TAYLOR expansion approximations of the functions in the two equations, one obtains

$$K_I = \frac{1}{2k(1+\gamma)-(1-\gamma)}; \quad (\gamma > 0). \tag{13.145}$$

For $\gamma = 0$. which is common when using the δ-transformation (the BÁNYÁSZ-KEVICZKY method),

$$K_I^0 = \frac{1}{2k-1} = \frac{1}{2k-1}\bigg|_{k=d''} = \frac{1}{2(d+2)-1} = \frac{1}{2d+3}; \quad (\gamma \equiv 0). \tag{13.146}$$

It can be simply checked that

$$\frac{1}{T_I^0} = \lim_{T_s \to 0}\frac{K_I^0}{T_s} \approx \lim_{T_s \to 0}\frac{1}{2T_d+3T_s} = \frac{1}{2T_d}, \tag{13.147}$$

which corresponds to the solution (8.23) obtained for the CT residual system.

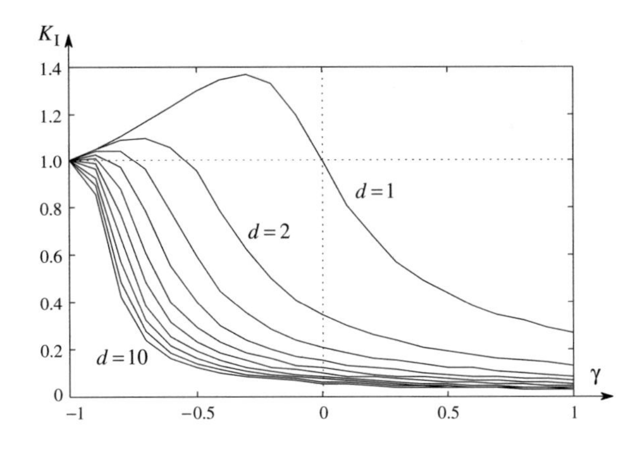

Fig. 13.3 The optimal K_I as a function of d and γ

Let us note that for the domain $\gamma < 0$, the following simple serial filter can be used:

$$G_{\mathrm{F}}(z) = \frac{1}{1 + \gamma z^{-1}}. \tag{13.148}$$

This case corresponds to formula (13.14) of the DT approximate *PID* regulator. This is essentially a zero cancellation technique, which can well be used, if the zero $z_1 = -\gamma$ is stable. If the zero is unstable, then with a pole of value $p_1 = 1/z_1$ the undesired "negative" overshoot in the unit step response of the closed-loop could successfully be decreased. This pole can be placed by the choice of the serial filter

$$G_{\mathrm{F}}(z) = \frac{1+\gamma}{\gamma} \frac{1}{1 + \frac{1}{\gamma} z^{-1} - K_{\mathrm{I}}^{0} \frac{1-\gamma}{\gamma} z^{-d}} \tag{13.149}$$

(the BÁNYÁSZ-KEVICZKY-HETTHÉSSY method).

Chapter 14
State Feedback in Sampled Data Systems

The design methods for controllers based on state feedback in the case of CT processes were discussed in Chap. 9. Next this methodology will be summarized for DT systems. For this purpose consider the state equation of a sampled data linear *LTI* process to be controlled using the results of Sect. 11.4 for the case $d = 0$.

$$\begin{aligned} \boldsymbol{x}[k+1] &= \boldsymbol{F}\boldsymbol{x}[k] + \boldsymbol{g}u[k] \\ y[k] &= \boldsymbol{c}^{\mathrm{T}}\boldsymbol{x}[k] \end{aligned} \tag{14.1}$$

The block scheme represented by the above equations is seen in Fig. 14.1.

Here $u[k]$ and $y[k]$ are the process input and output, respectively, and $\boldsymbol{x}$ denotes the state vector. The equivalent pulse transfer function is now

$$G(z) = \boldsymbol{c}^{\mathrm{T}}(z\boldsymbol{I} - \boldsymbol{F})^{-1}\boldsymbol{g} = \frac{\mathcal{B}(z)}{\det(z\boldsymbol{I} - \boldsymbol{F})} = \frac{\mathcal{B}(z)}{\mathcal{A}(z)} = \frac{b_1 z^{n-1} + \cdots + b_{n-1}z + b_n}{z^n + a_1 z^{n-1} + \cdots + a_{n-1}z + a_n}. \tag{14.2}$$

A classical closed control loop directly applied to the state equation description is shown in Fig. 14.2, where the reference signal is denoted by $r[k]$. The closed-loop is formed by the feedback from the state vector via the linear proportional feedback vector $\boldsymbol{k}^{\mathrm{T}}$ in the form

$$u[k] = k_{\mathrm{r}}r[k] - \boldsymbol{k}^{\mathrm{T}}\boldsymbol{x}[k] \tag{14.3}$$

Based on Fig. 14.2, the state equation of the complete closed-loop system can be written as

$$\begin{aligned} \boldsymbol{x}[k+1] &= \left(\boldsymbol{F} - \boldsymbol{g}\boldsymbol{k}^{\mathrm{T}}\right)\boldsymbol{x}[k] + k_{\mathrm{r}}\boldsymbol{g}r[k] \\ y[k] &= \boldsymbol{c}^{\mathrm{T}}\boldsymbol{x}[k] \end{aligned} \tag{14.4}$$

L. Keviczky et al., *Control Engineering*, Advanced Textbooks in Control and Signal Processing, https://doi.org/10.1007/978-981-10-8297-9_14

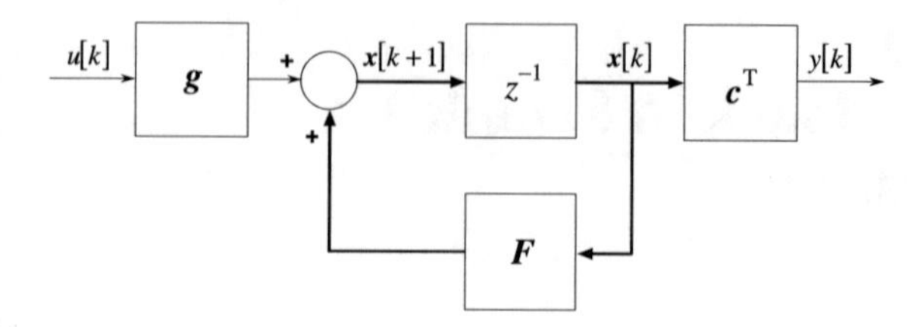

Fig. 14.1 Block scheme of the state equation of the linear time invariant discrete-time system

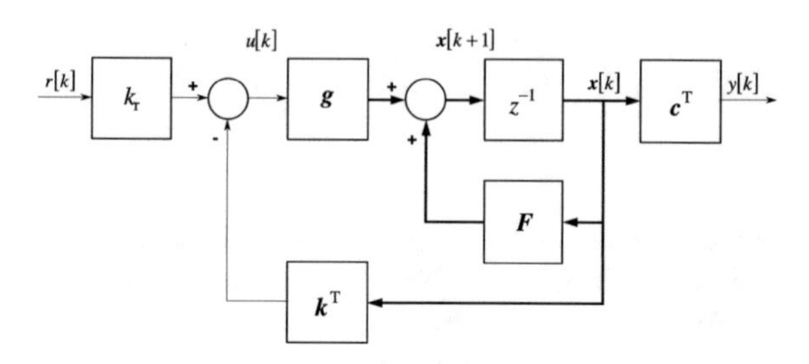

Fig. 14.2 Linear discrete-time control with state feedback

i.e., the dynamics concerning the original system matrix $\boldsymbol{F}$ is modified by the dyadic product $\boldsymbol{g}\boldsymbol{k}^{\mathrm{T}}$ to $\left(\boldsymbol{F}-\boldsymbol{g}\boldsymbol{k}^{\mathrm{T}}\right)$.

The transfer function of the closed control loop is

$$\begin{aligned} T_{\mathrm{ry}}(z) &= \frac{Y(z)}{R(z)} = \boldsymbol{c}^{\mathrm{T}}\left(z\boldsymbol{I}-\boldsymbol{F}+\boldsymbol{g}\boldsymbol{k}^{\mathrm{T}}\right)^{-1}\boldsymbol{g}k_{\mathrm{r}} = \frac{\boldsymbol{c}^{\mathrm{T}}(z\boldsymbol{I}-\boldsymbol{F})^{-1}\boldsymbol{g}k_{\mathrm{r}}}{1+\boldsymbol{k}^{\mathrm{T}}(z\boldsymbol{I}-\boldsymbol{F})^{-1}\boldsymbol{g}} \\ &= \frac{k_{\mathrm{r}}}{1+\boldsymbol{k}^{\mathrm{T}}(z\boldsymbol{I}-\boldsymbol{F})^{-1}\boldsymbol{g}}G(z) = \frac{k_{\mathrm{r}}\mathcal{B}(z)}{\mathcal{A}(z)+\boldsymbol{k}^{\mathrm{T}}\boldsymbol{\Psi}(z)\boldsymbol{g}} \end{aligned} \tag{14.5}$$

which comes from the comparison of the $\mathcal{Z}$-transforms $\boldsymbol{X}(z)=(z\boldsymbol{I}-\boldsymbol{F})^{-1}\boldsymbol{g}U(z)$ [similarly to (3.12)], $U(z)=k_{\mathrm{r}}R(z)-\boldsymbol{k}^{\mathrm{T}}\boldsymbol{X}(z)$ [see (9.3)] and $Y(z)=\boldsymbol{c}^{\mathrm{T}}\boldsymbol{X}(z)$ [see (9.1)], using the matrix inversion lemma (the proof is given in detail in A.9.1 of Appendix A.5 for CT systems). Notice that the state feedback leaves the zeros of the process unchanged, and only the poles of the closed system can be designed by $\boldsymbol{k}^{\mathrm{T}}$.

Introduce the so-called calibration factor k_{r}, by means of which the gain of the T_{ry} can be set to unity, i.e., $T_{\mathrm{ry}}(1)=1$. Obviously the open-loop is not an integrating one, so it can not yield zero error and static gain unity. In order to reach this the process parameters have to be known and the condition

$$k_{\mathrm{r}} = \frac{-1}{\boldsymbol{c}^{\mathrm{T}}\left(\boldsymbol{F} - \boldsymbol{g}\boldsymbol{k}^{\mathrm{T}}\right)^{-1}\boldsymbol{g}} = \frac{\boldsymbol{k}^{\mathrm{T}}\boldsymbol{F}^{-1}\boldsymbol{g} - 1}{\boldsymbol{c}^{\mathrm{T}}\boldsymbol{F}^{-1}\boldsymbol{g}} \tag{14.6}$$

must be fulfilled [see A.9.2 of Appendix A.5]. The above special closed control loop is called state feedback.

14.1 Discrete-Time Pole-Placement State Feedback Regulator

The most natural design method regarding the state feedback is the so-called pole-placement. In this method the feedback vector $\boldsymbol{k}^{\mathrm{T}}$ has to be chosen to provide a prescribed polynomial $\mathcal{R}(z)$ for the characteristic equation of the closed system, e.g. in DT case,

$$\begin{aligned}\mathcal{R}(z) &= z^n + r_1 z^{n-1} + \cdots + r_{n-1}z + r_n = \prod_{i=1}^{n}(z - z_i) = \det\left(z\boldsymbol{I} - \boldsymbol{F} + \boldsymbol{g}\boldsymbol{k}^{\mathrm{T}}\right) \\ &= \mathcal{A}(z) + \boldsymbol{k}^{\mathrm{T}}\Psi(z)\boldsymbol{g}\end{aligned} \tag{14.7}$$

The solution always exists if the process is controllable. If the transfer function of the system to be controlled is known, then it is an exceptional case, because the canonical state equations can be directly written. Based on the controllable controllable canonical forms (3.47) and (11.107) the system matrices can be obtained as

$$\boldsymbol{F}_{\mathrm{c}} = \begin{bmatrix} -a_1 & -a_2 & \ldots & -a_{n-1} & -a_n \\ 1 & 0 & \ldots & 0 & 0 \\ 0 & 1 & \ldots & 0 & 0 \\ \vdots & \vdots & \ddots & \vdots & \vdots \\ 0 & 0 & 0 & 1 & 0 \end{bmatrix}; \ \boldsymbol{c}_{\mathrm{c}}^{\mathrm{T}} = [b_1, b_2, \ldots, b_n]; \tag{14.8}$$

$$\boldsymbol{g}_{\mathrm{c}} = [1, 0, \ldots, 0]^{\mathrm{T}}$$

Taking the special forms of $\boldsymbol{F}_{\mathrm{c}}$ and $\boldsymbol{g}_{\mathrm{c}}$ it can be easily seen that according to the design equation

$$F_c - g_c k_c^T = \begin{bmatrix} -a_1 & -a_2 & \dots & -a_{n-1} & -a_n \\ 1 & 0 & \dots & 0 & 0 \\ 0 & 1 & \dots & 0 & 0 \\ \vdots & \vdots & \ddots & \vdots & \vdots \\ 0 & 0 & 0 & 1 & 0 \end{bmatrix} - \begin{bmatrix} 1 \\ 0 \\ 0 \\ \vdots \\ 0 \end{bmatrix} k_c^T$$
$$= \begin{bmatrix} -r_1 & -r_2 & \dots & -r_{n-1} & -r_n \\ 1 & 0 & \dots & 0 & 0 \\ 0 & 1 & \dots & 0 & 0 \\ \vdots & \vdots & \ddots & \vdots & \vdots \\ 0 & 0 & 0 & 1 & 0 \end{bmatrix} \tag{14.9}$$

the choice

$$k^T = k_c^T = [r_1 - a_1, r_2 - a_2, \dots, r_n - a_n] \tag{14.10}$$

ensures the characteristic equation Eq. (14.7), i.e., the prescribed poles. The value of the calibration factor can be given by a simple computation:

$$k_r = \frac{a_n + (r_n - a_n)}{b_n} = \frac{r_n}{b_n}. \tag{14.11}$$

It can be easily seen from the Eqs. (14.4) and (14.6) that the overall transfer function of the closed-loop system is

$$T_{ry}(z) = \frac{k_r \mathcal{B}(z)}{\mathcal{R}(z)} \tag{14.12}$$

in the case of state feedback pole placement, as was already mentioned in connection with (14.5).

Example 14.1. Consider an unstable process with transfer function

$$G(z) = \frac{-0.2z}{(z-0.8)(z-2)} = \frac{-0.2z^{-1}}{(1-0.8z^{-1})(1-2z^{-1})} = \frac{-0.2z}{z^2 - 2.8z + 1.6} = \frac{-0.2z}{\mathcal{A}(z)},$$

where $\mathcal{A}(z) = (z-0.8)(z-2) = z^2 - 2.8z + 1.6 = z^2 + a_1 z + a_2$. To stabilize the process we should mirror the unstable pole outside the unit circle $p_2^d = 2$ inside the circle, i.e., select $p_2^d = 0.5$. The design polynomial $\mathcal{R}(z) = (z-0.8)(z-0.5) = z^2 - 1.3z + 0.4 = z^2 + r_1 z + r_2$ ensures this goal. So the necessary stabilizing feedback vector is

$$k^T = [\, r_1 - a_1 \quad r_2 - a_2 \,] = [\, -1.3 - (-2.8) \quad 0.4 - (1.6) \,] = [\, 1.5 \quad -1.2 \,].$$

■

The most frequently appearing case of state feedback is when instead of the transfer function, the state space form of the controlled system is given. In connection with (3.67) it has already been discussed that all controllable systems can be written in controllable canonical form by using the transformation matrix $\boldsymbol{T}_{\mathrm{c}} = \boldsymbol{M}_{\mathrm{c}}^{\mathrm{c}}(\boldsymbol{M}_{\mathrm{c}})^{-1}$. This similarity transformation has an effect on the feedback vector, too:

$$\boldsymbol{k}^{\mathrm{T}} = \boldsymbol{k}_{\mathrm{c}}^{\mathrm{T}}\boldsymbol{T}_{\mathrm{c}} = \boldsymbol{k}_{\mathrm{c}}^{\mathrm{T}}\boldsymbol{M}_{\mathrm{c}}^{\mathrm{c}}\boldsymbol{M}_{\mathrm{c}}^{-1} = \boldsymbol{g}_{\mathrm{c}}^{\mathrm{T}}\boldsymbol{M}_{\mathrm{c}}^{-1}\mathcal{R}(\boldsymbol{F}) = [0, 0, \ldots, 1]\boldsymbol{M}_{\mathrm{c}}^{-1}\mathcal{R}(\boldsymbol{F}) \tag{14.13}$$

To compute (14.13), the inverse of the controllability matrix has to be constructed by the system matrices $\boldsymbol{F}$ and $\boldsymbol{g}$, on the one hand. On the other hand, the controllability matrix $\boldsymbol{M}_{\mathrm{c}}^{\mathrm{c}}$ of the controllable canonical form has to be also generated [see (3.61)]. Since this latter depends only on the coefficients a_i in the denominator of the process transfer function, the denominator has to be computed: $\mathcal{A}(z) = \det(z\boldsymbol{I} - \boldsymbol{F})$. The same is true for the computation of $\mathcal{R}(\boldsymbol{F})$ in the second formula. The method of computing the pole placement state feedback vector shown above is named—after its developer—the ACKERMANN method.

Observe that the transformation properties of the CT and DT state equations, their canonical forms and the concepts of controllability and observability are formally completely the same. Deriving from this fact, the state feedback techniques for the control of discrete-time systems also have a great similarity with the CT methods presented above.

14.2 Observer Based Discrete-Time Pole Placement State Feedback Regulator

The method of the state feedback discussed previously requires measuring the state space vector of the state equation describing the process. This is very rarely available, generally only in the case of systems with low order dynamics (for example, mechanical systems described by distance, velocity and acceleration co-ordinates). The usability of the methods depends also on whether measurement or estimation is available on the state vector. For the construction of the state vector, the so-called observer principle has been developed. For this method, the knowledge of the system matrices $\boldsymbol{F}$, $\boldsymbol{g}$ and $\boldsymbol{c}^{\mathrm{T}}$ is necessary, by means of which an exact model of the process is constructed, and applying the same excitation as for the original process, this model (the observer) provides the estimated values $\hat{\boldsymbol{x}}[k]$ and $\hat{y}[k]$ of $\boldsymbol{x}[k]$ and $y[k]$, respectively. The state feedback is performed using $\hat{\boldsymbol{x}}[k]$. The principle is shown in Fig. 14.3.

Strictly speaking, $\hat{\boldsymbol{F}}$, $\hat{\boldsymbol{g}}$ and $\hat{\boldsymbol{c}}^{\mathrm{T}}$ have to be employed in the observer instead of $\boldsymbol{F}$, $\boldsymbol{g}$ and $\boldsymbol{c}^{\mathrm{T}}$. But the particularity of the observer is that besides providing a parallel model, it also constructs an error $\varepsilon[k] = y[k] - \hat{y}[k]$ from the deviation of the original from the estimated output of the process, and feeds it back to the input of

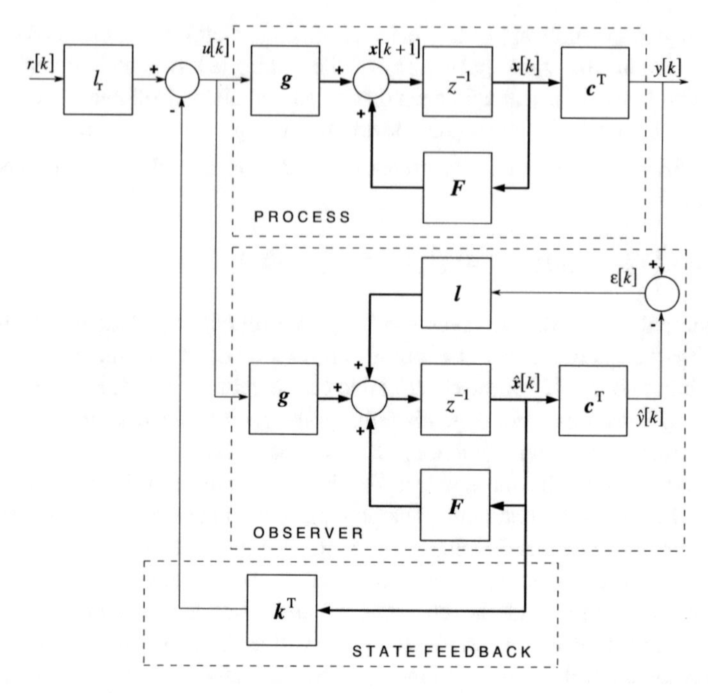

Fig. 14.3 State feedback applying an observer

the observer delay via a proportional feedback vector $\boldsymbol{l}$. This feedback operates until the error exists, i.e., until the outputs of the process and the observer become the same. With the knowledge of the system matrices this operating mode can compensate relatively large errors. It is also seen in the figure that now the state feedback has the form

$$u[k] = k_r r[k] - \boldsymbol{k}^T \hat{\boldsymbol{x}}[k], \tag{14.14}$$

thus $\hat{\boldsymbol{x}}[k]$ appears instead of $\boldsymbol{x}[k]$. After a long and complex derivation, whose details are not discussed here, the transfer function of the complete closed system can be obtained as

$$\begin{aligned} T_{ry}(z) &= \frac{\left[\boldsymbol{c}^T(z\boldsymbol{I}-\boldsymbol{F})^{-1}\boldsymbol{g}\right]\left[1-\boldsymbol{k}^T\left(z\boldsymbol{I}-\boldsymbol{F}+\boldsymbol{g}\boldsymbol{k}^T+\boldsymbol{l}\boldsymbol{c}^T\right)^{-1}\boldsymbol{b}\right]k_r}{1+\left[\boldsymbol{l}^T\left(z\boldsymbol{I}-\boldsymbol{F}+\boldsymbol{g}\boldsymbol{l}^T+\boldsymbol{l}\boldsymbol{c}^T\right)^{-1}\boldsymbol{g}\right]\left[\boldsymbol{c}^T(z\boldsymbol{I}-\boldsymbol{F})^{-1}\boldsymbol{g}\right]} \\ &= \boldsymbol{c}^T\left(z\boldsymbol{I}-\boldsymbol{F}+\boldsymbol{g}\boldsymbol{k}^T\right)^{-1}\boldsymbol{g}k_r = \frac{\boldsymbol{c}^T(z\boldsymbol{I}-\boldsymbol{F})^{-1}\boldsymbol{g}k_r}{1+\boldsymbol{k}^T(z\boldsymbol{I}-\boldsymbol{F})^{-1}\boldsymbol{g}} = \frac{k_r G(z)}{1+\boldsymbol{k}^T(z\boldsymbol{I}-\boldsymbol{F})^{-1}\boldsymbol{g}} = \frac{k_r \mathcal{B}(z)}{\mathcal{R}(z)} \end{aligned} \tag{14.15}$$

which is, perhaps surprisingly, precisely equal to (14.2), i.e., to the case of state feedback without observer. This means that the tracking behavior of the closed system does not depend on the choice of the vector $\boldsymbol{l}$. To examine the operation of the observer, let us construct the vector of the state error

$$\tilde{\boldsymbol{x}}[k] = \boldsymbol{x}[k] - \hat{\boldsymbol{x}}[k] \tag{14.16}$$

and also

$$\tilde{\boldsymbol{x}}[k] = \left(\boldsymbol{F} - \boldsymbol{l}\boldsymbol{c}^{\mathrm{T}}\right)\tilde{\boldsymbol{x}}[k], \tag{14.17}$$

which is very similar to (14.4) without excitation. Very similar methods can be used for the design of observers as were used for state feedback, where the choice of the goal is to ensure the system dynamics (14.17) by the characteristic polynomial

$$\det\left(z\boldsymbol{I} - \boldsymbol{F} + \boldsymbol{l}\boldsymbol{c}^{\mathrm{T}}\right) = \mathcal{F}(z) = z^n + f_1 z^{n-1} + \cdots + f_{n-1}z + f_n \tag{14.18}$$

A solution always exists if the process is observable (This is reasonable if the order of $\mathcal{F}$ is equal to that of $\mathcal{A}$). If the transfer function of the process to be controlled is known, then it is an exceptional case, because then the canonical forms can be directly written. In this case, when the system matrices are based on the observable canonical forms (3.53),

$$\boldsymbol{F}_{\mathrm{o}} = \begin{bmatrix} -a_1 & 1 & 0 & \dots & 0 \\ -a_2 & 0 & 1 & \dots & 0 \\ \vdots & \vdots & \vdots & \ddots & \vdots \\ -a_{n-1} & 0 & 0 & \dots & 1 \\ -a_n & 0 & 0 & \dots & 0 \end{bmatrix}; \; \boldsymbol{c}_{\mathrm{o}}^{\mathrm{T}} = [1, 0, \dots, 0]; \; \boldsymbol{g}_{\mathrm{o}} = [b_1, b_2, \dots, b_n]^{\mathrm{T}} \tag{14.19}$$

Taking the special forms of $\boldsymbol{F}_{\mathrm{o}}$ and $\boldsymbol{c}_{\mathrm{o}}^{\mathrm{T}}$ into account, it is easily seen that according to the design equation

$$\boldsymbol{F}_{\mathrm{o}} - \boldsymbol{l}_{\mathrm{o}}\boldsymbol{c}_{\mathrm{o}}^{\mathrm{T}} = \begin{bmatrix} -a_1 & 1 & 0 & \dots & 0 \\ -a_2 & 0 & 1 & \dots & 0 \\ \vdots & \vdots & \vdots & \ddots & \vdots \\ -a_{n-1} & 0 & 0 & \dots & 1 \\ -a_n & 0 & 0 & \dots & 0 \end{bmatrix} - \boldsymbol{l}_{\mathrm{o}}[1, 0, \dots, 0] = \begin{bmatrix} -f_1 & 1 & 0 & \dots & 0 \\ -f_2 & 0 & 1 & \dots & 0 \\ \vdots & \vdots & \vdots & \ddots & \vdots \\ -f_{n-1} & 0 & 0 & \dots & 1 \\ -f_n & 0 & 0 & \dots & 0 \end{bmatrix} \tag{14.20}$$

the choice

$$\boldsymbol{l} = \boldsymbol{l}_{\rm o} = [f_1 - a_1, f_2 - a_2, \ldots, f_n - a_n]^{\rm T} \tag{14.21}$$

ensures the characteristic equation (14.18), i.e., the prescribed poles.

The general case is now, when the state space equation of the process is given instead of its transfer function. It has already been discussed concerning Eq. (3.79) that all observable systems can be written in observable canonical form by the use of the transformation matrix $\boldsymbol{T}_{\rm o} = \left(\boldsymbol{M}_{\rm o}^{\rm o}\right)^{-1}\boldsymbol{M}_{\rm o}$. This similarity transformation has an effect on the feedback vector, too:

$$\boldsymbol{l} = (\boldsymbol{T}_{\rm o})^{-1}\boldsymbol{l}_{\rm o} = \boldsymbol{M}_{\rm o}^{-1}\boldsymbol{M}_{\rm o}^{\rm o}\boldsymbol{l}_{\rm o}. \tag{14.22}$$

To compute (14.22), the inverse of the observability matrix $\boldsymbol{M}_{\rm o}$ has to be constructed by the general system matrices $\boldsymbol{F}$ and $\boldsymbol{c}^{\rm T}$. On the other hand, the observability matrix $\boldsymbol{M}_{\rm o}^{\rm o}$ of the observable canonical form must be also given (see (3.73). Since this latter one depends only on the coefficients a_i in the denominator of the transfer function of the process, so to its determination the denominator has to be computed: $\mathcal{A}(z) = \det(z\boldsymbol{I} - \boldsymbol{F})$. The computation method of the observer vector shown above is named, after its developer, the ACKERMANN method.

There is an interesting similarity between the design methods of the dynamics of the state feedback and of the observer, a so-called duality, i.e., they correspond to each other under the associations $\boldsymbol{F} \leftrightarrow \boldsymbol{F}^{\rm T}, \boldsymbol{g} \leftrightarrow \boldsymbol{c}^{\rm T}, \boldsymbol{k} \leftrightarrow \boldsymbol{l}^{\rm T}, \boldsymbol{M}_{\rm c}^{\rm c} \leftrightarrow \left(\boldsymbol{M}_{\rm o}^{\rm o}\right)^{\rm T}$.

Based on the state error (14.16) and the equations of the process (14.1), the joint equation of the state feedback and the observer is

$$\begin{bmatrix} \mathbf{x}[k+1] \\ \tilde{\mathbf{x}}[k+1] \end{bmatrix} = \begin{bmatrix} \boldsymbol{F} - \boldsymbol{g}\boldsymbol{k}^{\rm T} & \mathbf{g}\mathbf{k}^{\rm T} \\ \mathbf{0} & \boldsymbol{F} - \boldsymbol{l}\boldsymbol{c}^{\rm T} \end{bmatrix} \begin{bmatrix} \boldsymbol{x}[k] \\ \tilde{\boldsymbol{x}}[k] \end{bmatrix} + \begin{bmatrix} k_{\rm r}\boldsymbol{g} \\ \mathbf{0} \end{bmatrix} r[k]$$
$$e[k] = y[k] - \hat{y}[k] = \mathbf{c}^{\rm T}\tilde{\mathbf{x}}[k] \tag{14.23}$$

Since the right hand side system matrix is upper triangular, the characteristic equation of the closed system is

$$\det\left(z\boldsymbol{I} - \boldsymbol{F} + \boldsymbol{g}\boldsymbol{k}^{\rm T}\right)\det\left(z\boldsymbol{I} - \boldsymbol{F} + \boldsymbol{l}\boldsymbol{c}^{\rm T}\right) = \mathcal{R}(z)\mathcal{F}(z) \tag{14.24}$$

Thus the polynomial is the product of two factors: one is connected to the state feedback, the other is connected to the observer. It is important to remark that in contrast to (14.24), $\mathcal{F}(z)$ does not appear in the transfer function $T_{\rm ry}(z)$ [see (14.12) and (14.15)].

Equation (14.24) representing the observer based state feedback, according to which the characteristic equations of the state feedback and observer are independent, is called the separation principle.

14.3 Two-Step Design Methods Using Discrete-Time State Feedback

It has been shown in the discussion of the state feedback based control, that the most advantageous (favorable) properties of the method are:

- the applicability of the method does not depend on whether the process is stable or unstable
- the tracking behavior does not depend on the applied observer, thus it can be directly designed
- the method is not very sensitive for the exact knowledge of the parameter matrices of the state equation

There are unwanted, unfavorable properties:

- the state feedback is basically a control of 0-type, therefore the remaining error can be eliminated by the calibration factor, which is never very precise using the model of the process
- the state feedback can not change the zeros of the process
- the noise rejection behavior can not be designed directly.

Mainly due to these latter attributes, usually an extra step is included in the design of control systems using state feedback. The necessity of the calibration factor can be easily eliminated by the construction of a cascade integrating controller according to Fig. 14.4.

The joint state equation of the closed system, which now replaces Eq. (14.4), can be written as

$$\begin{aligned}\dot{\boldsymbol{x}}^*[k+1] &= \begin{bmatrix}\dot{\boldsymbol{x}}[k+1]\\ \dot{\delta}[k+1]\end{bmatrix} = \begin{bmatrix}\boldsymbol{F} & \boldsymbol{0}\\ \boldsymbol{c}^{\mathrm{T}} & 0\end{bmatrix}\begin{bmatrix}\boldsymbol{x}[k]\\ \delta[k]\end{bmatrix} + \begin{bmatrix}\boldsymbol{g}\\ 0\end{bmatrix}u[k] + \begin{bmatrix}\boldsymbol{0}\\ -1\end{bmatrix}r[k]\\ &= \left(\boldsymbol{F}^* - \boldsymbol{g}^*\boldsymbol{k}_*^{\mathrm{T}}\right)\boldsymbol{x}^*[k] + \boldsymbol{v}^* r[k]\end{aligned} \tag{14.25}$$

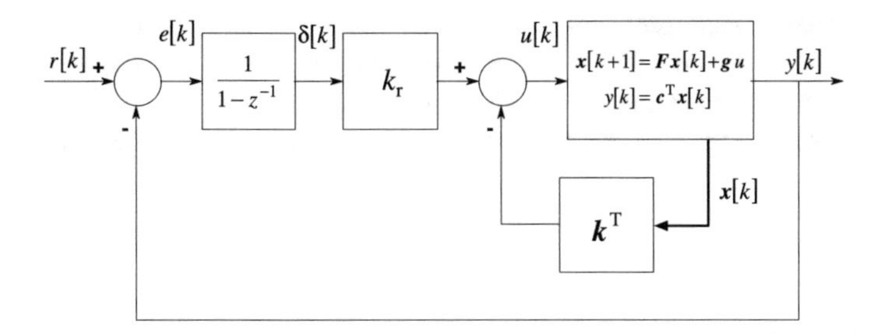

Fig. 14.4 The joint use of the state feedback and the integrating controller

by introducing a new state variable $\delta[k]$, which is the integral of the error $e[k] = r[k] - y[k]$ of the outer loop, where the notations

$$\boldsymbol{F}^* = \begin{bmatrix} \boldsymbol{F} & \boldsymbol{0} \\ c^T & 0 \end{bmatrix}; \quad \boldsymbol{g}^* = \begin{bmatrix} \boldsymbol{g} \\ 0 \end{bmatrix}; \quad \boldsymbol{v}^* = \begin{bmatrix} \boldsymbol{0} \\ -1 \end{bmatrix} \tag{14.26}$$

and the new extended feedback equation

$$u[k] = -\begin{bmatrix} \boldsymbol{k}^{\mathrm{T}} & k_{\mathrm{r}} \end{bmatrix} \begin{bmatrix} \boldsymbol{x}[k] \\ \delta[k] \end{bmatrix} = -\boldsymbol{k}_*^{\mathrm{T}} \boldsymbol{x}^*[k] = \frac{k_{\mathrm{r}}}{1 - z^{-1}} e[k] - \boldsymbol{k}^{\mathrm{T}} \boldsymbol{x}[k] \tag{14.27}$$

are taken into account.

Equation (14.27) clearly shows the integrating effect. The item $\boldsymbol{k}^{\mathrm{T}}\boldsymbol{x}[k]$, however, can be considered as a generalization of the derivative effect.

Thus the closed-loop control having also an integrator can be described by a state equation which has its dimension higher by one than the earlier one, where now k_{r} has also to be determined besides $\boldsymbol{k}^{\mathrm{T}}$. For the design of the extended system the characteristic polynomial $\mathcal{R}^*(z)$ having order greater by one has to be prescribed, then the design Eq. (14.13) of the ACKERMANN method can be directly applied here too. If the process is not given in the transfer function form, then the general state equation has to be rewritten first into a controllable canonical form, as was shown in (10.13).

Notice that the extended task can not be solved sequentially, i.e., by determining first the $\boldsymbol{k}^{\mathrm{T}}$ belonging to $\mathcal{R}(z)$, and then k_{r} based on $\mathcal{R}^*(z) = \mathcal{R}(z)(z - z_{n+1})$. The task has to be solved in one step for $\boldsymbol{k}_*^{\mathrm{T}}$ on the basis of $\mathcal{R}^*(z)$.

An integrating effect can also be included by designing the state feedback for a modified process $G^*(z) = zG(z)/(z-1)$ instead of the transfer function $G(z)$. Note that the feedback vectors obtained for the earlier case and for this latter approach are not the same!

Obviously, besides the *I*-controller, a higher order regulator can also be applied. The solution of the pole placement, however, can not be obtained automatically by the ACKERMANN method, and may lead to a complicated system of non-linear equations.

In the case of state feedback applying observer an *I* or higher order regulator, instead of the regulator of 0-type, can also be applied in the error feedback of the observer using the methods shown above.

The unchanged zeros of the process can be compensated by a serial compensator

$$K_{\mathrm{s}}(z) = G_{\mathrm{s}}(z) \frac{\mathcal{N}(z)}{\mathcal{B}_+(z)} \tag{14.28}$$

where it is assumed, according to the method applied in Chap. 7, that the numerator of the process is $\mathcal{B}(z) = \mathcal{B}_+(z)\mathcal{B}_-(z)$. Here $\mathcal{B}_+$ contains the stable zeros and $\mathcal{B}_-$ the unstable zeros. For realizability, $\mathcal{N}(z)/\mathcal{B}_+(z)$ has to be proper, thus only as many zeros can be placed in the transfer function of the closed system as there are stable zeros in the process.

Finally the loop transfer function has the form

$$T_{\mathrm{ry}}(z) = \frac{\mathcal{N}(z)}{\mathcal{R}(z)} k_{\mathrm{r}} G_{\mathrm{s}}(z) \mathcal{B}_{-}(z) \tag{14.29}$$

where the effect of the invariant $\mathcal{B}_{-}(z)$ can be attenuated optimally by the filter $G_{\mathrm{s}}(z)$. In many cases a simple, but not optimal $G_{\mathrm{s}}(z) = 1$ is chosen.

A favorable design of the disturbance rejection feature can be reached by applying a *YP* controller in the outer cascade loop. This can be done since the state feedback is capable of stabilizing any process, even an unstable one. In general, the control of an unstable process has two steps. In the first step the process is stabilized, then in the second step, via a second outer loop, the required quality goals can be ensured even by a *TDOF* structure.

A stabilizing controller using state feedback can be applied only to delay-free processes. If the process has a significant delay, then the only possibility is to switch to a sampled-data control using the general polynomial method [see Chap. 15].

14.4 Discrete-Time *LQ* State Feedback Regulator

With the method presented in the previous section, arbitrary (stabilizing) pole-placement can be performed via the so-called state feedback from the state vector of the process. A further optimality task can also be solved by the technique of state feedback. The goal of this task to control optimally the DT *LTI* process (11.33–11.34) by the minimization of a complicated optimality criterion

$$I = \frac{1}{2} \sum_{k=0}^{\infty} \left\{ \boldsymbol{x}^{\mathrm{T}}[k] \boldsymbol{W}_{\mathrm{x}} \boldsymbol{x}[k] + W_{\mathrm{u}} u^2[k] \right\} \tag{14.30}$$

Here $\boldsymbol{W}_{\mathrm{x}}$ is a real symmetric positive semi-definite matrix, weighting the DT state vector and W_{u} is a positive scalar, weighting the DT actuator signal. The solution optimizing the criterion is a state feedback in the form

$$u[k] = -\boldsymbol{k}_{\mathrm{LQ}}^{\mathrm{T}} \boldsymbol{x}[k] \tag{14.31}$$

[see (9.3)], where $\boldsymbol{k}_{\mathrm{LQ}}^{\mathrm{T}}$ is the feedback vector

$$\boldsymbol{k}_{\mathrm{LQ}}^{\mathrm{T}} = \frac{1}{W_{\mathrm{u}}} \boldsymbol{g}^{\mathrm{T}} \boldsymbol{P} \tag{14.32}$$

Here the symmetric positive semi-definite matrix $\boldsymbol{P}$ is the solution of the algebraic Riccati equation

$$\boldsymbol{PF} + \boldsymbol{F}^{\mathrm{T}}\boldsymbol{P} - \frac{1}{W_{\mathrm{u}}}\boldsymbol{Pgg}^{\mathrm{T}}\boldsymbol{P} = -\boldsymbol{W}_{\mathrm{x}} \tag{14.33}$$

The (algebraic) Riccati equation is nonlinear in $\boldsymbol{P}$, therefore it does not have an explicit algebraic solution. The CAD systems used in control engineering, however, have several numerical algorithms for the solution of the above equation. This controller is called an *LQ* (*Linear Quadratic*: *Linear regulator—Quadratic criterion*) regulator.

The state equation of the closed system provided by the *LQ* regulator has the form

$$\boldsymbol{x}[k+1] = \left(\boldsymbol{F} - \boldsymbol{g}\boldsymbol{k}_{\mathrm{LQ}}^{\mathrm{T}}\right)\boldsymbol{x}[k]; \quad \bar{\boldsymbol{F}} = \boldsymbol{F} - \boldsymbol{g}\boldsymbol{k}_{\mathrm{LQ}}^{\mathrm{T}} \tag{14.34}$$

(The derivation of the *LQ* regulator for CT systems can be found in A.9.6 of Appendix A.5, the derivation of the DT controller can be done with a very similar analogy).

If the transfer function of the process is known, then the controllable canonical form can be easily written in analogy with the CT Eq. (9.10) for the special $\boldsymbol{F}_{\mathrm{c}}$ and $\boldsymbol{g}_{\mathrm{c}}$ formed by (14.8), according to the design algorithm (14.10) of the classical DT state feedback. The feedback vector $\boldsymbol{k}_{\mathrm{LQ}}^{\mathrm{T}}$ comes from the *LQ* control design (from the solution of the Riccati equation). So by turning back the derivation of (14.10), the coefficients of the characteristic polynomial $\mathcal{R}(s)$ of the closed-loop system are given by

$$[r_1, r_2, \ldots, r_n]^{\mathrm{T}} = \mathbf{k}_{\mathrm{LQ}}^{\mathrm{T}} + [a_1, a_2, \ldots, a_n]^{\mathrm{T}} \tag{14.35}$$

In the case of *LQ* control it is also possible to apply an observer for the determination of the state vector.

Notice that the state feedback vector $\boldsymbol{k}_{\mathrm{LQ}}^{\mathrm{T}}$ also leaves the zeros of the process unchanged.

Chapter 15
General Polynomial Method for the Design of Discrete-Time Controllers

Unfortunately the application of the *DE* to CT processes cannot handle a time-delay, since the method can be used only for polynomials. Time-delay systems can be stabilized only in the discrete-time case. Assume that the pulse transfer function of the process is

$$G\left(z^{-1}\right) = G_+\left(z^{-1}\right)\bar{G}_-\left(z^{-1}\right) = G_+\left(z^{-1}\right)G_-\left(z^{-1}\right)z^{-d}, \text{ or } G = G_+\bar{G}_- = \\ = G_+G_-z^{-d} \tag{15.1}$$

where G_+ is stable and its inverse is also stable (*SIS*: *Stable Inverse Stable*). $\overline{G}_-$ is unstable and its inverse is also unstable (*UIU*: *Unstable Inverse Unstable*). G_- is also *UIU*. Here, in general, the inverse of the time-delay part cannot be realized, because it would be an ideal predictor. Thus a reasonable factorization of the process is

$$G = \frac{\mathcal{B}}{\mathcal{A}}z^{-d} = \frac{\mathcal{B}_+\mathcal{B}_-}{\mathcal{A}_+\mathcal{A}_-}z^{-d} = \left(\frac{\mathcal{B}_+}{\mathcal{A}_+}\right)\left(\frac{\mathcal{B}_-}{\mathcal{A}_-}\right)z^{-d} = G_+G_-z^{-d} \tag{15.2}$$

Here $\mathcal{A}_+$ contains the stable poles, $\mathcal{A}_-$ the unstable ones. Similarly, $\mathcal{B}_+$ includes the stable zeros, $\mathcal{B}_-$ the unstable ones. The general design *DE* for discrete systems is simply obtained from (10.14) by formally changing $\mathcal{B}_-$ to $\mathcal{B}_-z^{-d}$. The new form of (10.14) becomes

$$\begin{array}{llcll} (\mathcal{A}_+\mathcal{A}_-) & (\mathcal{B}_+\mathcal{X}_\mathrm{d}\mathcal{X}') & + & \left(\mathcal{B}_+\mathcal{B}_-z^{-d}\right) & (\mathcal{A}_+\mathcal{Y}_\mathrm{d}\mathcal{Y}') & = \mathcal{R}' = \mathcal{A}_+\mathcal{B}_+\mathcal{R} \\ \mathcal{A} & \mathcal{X} & + & \mathcal{B} & \mathcal{Y} & = \mathcal{R}' \end{array} \tag{15.3}$$

L. Keviczky et al., *Control Engineering*, Advanced Textbooks in Control and Signal Processing, https://doi.org/10.1007/978-981-10-8297-9_15

The modified *DE* is

$$\begin{array}{llllll} (\mathcal{A}_{-}\mathcal{X}_{\mathrm{d}}) & \mathcal{X}' & + & (\mathcal{B}_{-}z^{-d}\mathcal{Y}_{\mathrm{d}}) & \mathcal{Y}' & = \mathcal{R}' \\ \mathcal{A}' & \mathcal{X}' & + & \mathcal{B}' & \mathcal{Y}' & = \mathcal{R}' \end{array} \tag{15.4}$$

where $\mathcal{A}' = \mathcal{A}_{-}\mathcal{X}_{\mathrm{d}}$ and $\mathcal{B}' = \mathcal{B}_{-}z^{-d}\mathcal{Y}_{\mathrm{d}}$ are known and the controller is obtained again as

$$C = \frac{\mathcal{Y}}{\mathcal{X}} = \frac{\mathcal{A}_{+}\mathcal{Y}_{\mathrm{d}}\mathcal{Y}'}{\mathcal{B}_{+}\mathcal{X}_{\mathrm{d}}\mathcal{X}'} = \frac{\left(\frac{\mathcal{Y}_{\mathrm{d}}}{\mathcal{R}}\right)\mathcal{Y}'\mathcal{A}}{\mathcal{B}_{+}\left(1 - \frac{\mathcal{Y}_{\mathrm{d}}}{\mathcal{R}}\mathcal{Y}'\mathcal{B}_{-}z^{-d}\right)} = \frac{P'_{\mathrm{w}}\mathcal{Y}'}{1 - P'_{\mathrm{w}}\mathcal{Y}'\mathcal{B}_{-}z^{-d}}\frac{\mathcal{A}}{\mathcal{B}_{+}} \tag{15.5}$$

The Youla-regulator is integrating if a unit gain is ensured for the reference model: $R_{\mathrm{n}}(\omega = 0) = R_{\mathrm{n}}(z = 1) = 1$. This cannot be automatically guaranteed for the stabilizing controller coming from the *DE*. This solution is guaranteed if $\mathcal{X}_{\mathrm{d}}$ brings the pole $z = 1$ into the denominator. To solve the *DE* Eq. (15.4), the equation has to be formed in powers of z.

What was discussed in detail in the Chap. 10 relating to the *DE* will not be repeated here. Note that the transfer characteristic of the whole control loop is

$$y = T_{\mathrm{r}}y_{\mathrm{r}} + Sy_{\mathrm{n}} = R_{\mathrm{r}}G_{\mathrm{r}}\mathcal{B}_{-}z^{-d}y_{\mathrm{r}} + \left(1 - R'_{\mathrm{n}}\mathcal{Y}'\mathcal{B}_{-}z^{-d}\right)y_{\mathrm{n}} \tag{15.6}$$

It can be clearly seen that the filter G_{r} can be chosen arbitrarily and can be optimized to attenuate the effect of $\mathcal{B}_{-}$. Unfortunately the same statement cannot be made about the optimization of the disturbance rejection. Here $\mathcal{Y}'$ comes from the modified *DE* (15.4), so it cannot be chosen arbitrarily, therefore the attenuation of the effect of $\mathcal{B}_{-}$ cannot be solved as easily, as was seen with the Youla-parameterization and tracking properties (15.6).

Example 15.1 Let the controlled system be a first order $(n = 1)$, unstable DT process

$$G(z^{-1}) = \frac{\mathcal{B}(z^{-1})}{\mathcal{A}(z^{-1})} = \frac{-0.2z^{-1}}{1 - 1.2z^{-1}} = \frac{-0.2}{z - 1.2} \tag{15.7}$$

whose pole $p = 1.2$ is outside the unit circle. Determine the controller $C = \mathcal{Y}/\mathcal{X}$ which stabilizes the process by prescribing the characteristic polynomial $\mathcal{R}(z) = z - 0.2 = 0$. The controller is sought in the form of $n - 1 = 0$ order, which can be reached by the structure

$$C = \frac{\mathcal{Y}}{\mathcal{X}} = \frac{K}{1} = K \tag{15.8}$$

i.e. by a proportional controller. Based on (15.4), we have

$$\begin{gathered}\mathcal{A}\mathcal{X}+\mathcal{B}\mathcal{Y}=\mathcal{R}\\(z-1.2)-0.2K=z-0.2\end{gathered}\tag{15.9}$$

from which $C=K=-5$ is obtained for the controller. It can be checked by simple computation that the pulse transfer function of the closed system is

$$T=\frac{1}{z-0.2}=\frac{z^{-1}}{1-0.2z^{-1}},\tag{15.10}$$

thus the unstable pole has been successfully allocated to the prescribed place inside the unit circle, by means of which the system is stabilized. The static transfer of the closed-loop is not unity, because the controller is proportional and not integrating. To get better control, it is reasonable to apply a further outer cascade loop, as was seen with state feedback control. ■

Example 15.2 Let the controlled system be a first order ($n=1$), stable DT process

$$G\left(z^{-1}\right)=\frac{\mathcal{B}(z^{-1})}{\mathcal{A}(z^{-1})}=\frac{0.2z^{-1}}{1-0.8z^{-1}}=\frac{0.2}{z-0.8}\tag{15.11}$$

and the goal is to make it faster. Assuming an *ODOF* system, our design goal is expressed by the reference model

$$R_{\rm r}=R_{\rm n}=\frac{0.8z^{-1}}{1-0.2z^{-1}}=\frac{0.8}{z-0.2}\tag{15.12}$$

Now the Youla-regulator is of integrating type, i.e.,

$$C_{\rm opt}=C_{\rm id}=\frac{R_{\rm n}G_{+}^{-1}}{1-R_{\rm n}}=\frac{1}{1-\frac{0.8z^{-1}}{1-0.2z^{-1}}}\frac{0.8z^{-1}}{1-0.2z^{-1}}\frac{1-0.8z^{-1}}{0.2z^{-1}}=4\frac{1-0.8z^{-1}}{1-z^{-1}}\tag{15.13}$$

(because the zero of the denominator is $z=1$), and the transfer function of the closed system is

$$T=\frac{0.8z^{-1}}{1-0.2z^{-1}}=\frac{0.8}{z-0.2}\tag{15.14}$$

whose static transfer is unity corresponding to a control of 1-type.

Based on (15.12), the characteristic equation for the design by *DE* is $\mathcal{R}(z)=z-0.2=0$. Now the controller is also sought in the form of order $n-1=0$, thus according to (15.8), proportional controller is applied. Equation (15.9) becomes

$$\begin{aligned} \mathcal{AX} + \mathcal{BY} &= \mathcal{R} \\ (z - 0.8) + 0.2K &= z - 0.2 \end{aligned} \tag{15.15}$$

from which the controller is $C = K = 3$. It can be checked by simple computation that the overall transfer function of the closed-loop system is now

$$T = \frac{0.6}{z - 0.2} = \frac{0.6z^{-1}}{1 - 0.2z^{-1}} \tag{15.16}$$

The prescribed pole 0.2 is successfully allocated, but the loop is of 0-type, therefore the gain of T is 0.75. The above two examples represent well the practice, i.e., for stable systems the YOULA-parameterization has to be applied, while for stabilizing unstable systems, the application of *DE*, or the state feedback discussed in this chapter can provide the solution. ■

Example 15.3 Let the controlled system be a first order ($n = 1$), unstable, time-delay DT process

$$P(z^{-1}) = \frac{\mathcal{B}(z^{-1})}{\mathcal{A}(z^{-1})} = \frac{-0.2z^{-1}}{1 - 1.2z^{-1}} z^{-1} = \frac{-0.2z^{-2}}{1 - 1.2z^{-1}} = \frac{-0.2}{z(z - 1.2)} \tag{15.17}$$

whose pole $p = 1.2$ is outside the unit circle. Observe that this formally corresponds to a second order process because of the time-delay. Therefore the stabilizing controller $C = \mathcal{Y}/\mathcal{X}$ is sought in a first order form with three parameters

$$C = \frac{\mathcal{Y}}{\mathcal{X}} = \frac{y_o z + y_1}{z + x_1} = \frac{y_o + y_1 z^{-1}}{1 + x_1 z^{-1}} \tag{15.18}$$

Because of realizability conditions it is reasonable to select a stable third degree characteristic polynomial $\mathcal{R}(z) = z(z - 0.2)^2 = z(z^2 - 0.4z + 0.04)$ for the controller design. The number of unknown parameters is three and the relevant *DE* is

$$\begin{aligned} \mathcal{AX} + \mathcal{BY} &= \mathcal{R} \\ (z^2 - 1.2z)(z + x_1) - 0.2(y_o z + y_1) &= z(z^2 - 0.4z + 0.04) \end{aligned} \tag{15.19}$$

and solving the equation for y_o, y_1 and x_1 we get

$$C = \frac{-5z}{z + 0.8} = \frac{-5}{1 + 0.8z^{-1}} \tag{15.20}$$

It can be checked easily that the overall transfer function of the closed-loop system is

$$T = \frac{z}{z(z-0.2)^2} = \frac{z^{-1}}{(1-0.2z^{-1})^2}z^{-1} = \frac{z^{-1}}{1-0.4z^{-1}+0.04^{-2}}z^{-1} \tag{15.21}$$

The prescribed double poles at 0.2 have been successfully allocated, but the control loop is of 0-type, thus the gain of T is 1.5625. Thus a controller having a relatively simple structure could solve a difficult problem, i.e., it can stabilize an unstable time-delay process. ■

Chapter 16
Outlook

The goal of this chapter is to illustrate some further subjects in control engineering. In the previous sections single variable (*SISO*), linear systems with constant parameters (*LTI*) were considered. The systems in practice, however, are usually nonlinear, multivariable and have varying parameters. It is not surprising, that the solution of these kinds of problems needs higher level control engineering theory. Neither does this chapter deal with all these subjects, instead it gives a short summary of four areas, which belong to the modern theory of *SISO* systems. These are:

- Norms of control engineering signals and systems
- Methods of numerical optimization
- Introduction to system identification
- Iterative and adaptive control schemes.

16.1 Norms of Control Engineering Signals and Operators

A norm in a complex linear space is interpreted as a real number, called the norm of $\boldsymbol{x}$ and denoted by $\|\boldsymbol{x}\|$, which can be applied to any vector $\boldsymbol{x}$ of the space, and which satisfies the relationships below

$\|\boldsymbol{x}\| > 0$ if $\boldsymbol{x} \neq 0$, and $\|0\| = 0$.
$\|a\boldsymbol{x}\| = |a|\|\boldsymbol{x}\|$ for an arbitrary complex number a
$\|\boldsymbol{x}+\boldsymbol{y}\| \leq \|\boldsymbol{x}\| + \|\boldsymbol{y}\|$, which is the so-called triangle inequality.

The same concept exists regarding the linear vector-spaces of dimension n, and formally the same is valid for functions, too.

The quality of the control—as was seen in the previous sections—is connected with the error signal, or to the sensitivity function. The error signal is a function of time, but the sensitivity function is a complex frequency function, thus they are all functions. Their magnitude somehow has to be defined, because their value at a given frequency does not characterize the whole function, not speaking about their

L. Keviczky et al., *Control Engineering*, Advanced Textbooks in Control and Signal Processing, https://doi.org/10.1007/978-981-10-8297-9_16

magnitude. A mathematical notion, the above norm, is used for characterizing the magnitude of a function. Next, some basic norms will be presented, whose definitions will explain their meaning.

16.1.1 Norms of Signals

$$\mathcal{L}_1 \text{ norm:} \quad \|u(t)\|_1 = \int_{-\infty}^{\infty} |u(t)| \mathrm{d}t \tag{16.1}$$

$$\mathcal{L}_2 \text{ norm:} \quad \|u(t)\|_2 = \sqrt{\int_{-\infty}^{\infty} |u(t)|^2 \mathrm{d}t} \tag{16.2}$$

$$\mathcal{L}_\infty \text{ norm:} \quad \|u(t)\|_\infty = \max_t |u(t)| \tag{16.3}$$

In the practice usually input functions ($u(t) \equiv 0$, if $t < 0$) are investigated, where the lower limit of their integral is zero.

From the integrals of errors (integral criteria) discussed in Chap. 4, $I_3 = IAE = \|e(t)\|_1$ is the $\mathcal{L}_1$ norm of $e(t)$, $I_2 = \|e(t)\|_2^2$ is the square of the $\mathcal{L}_2$ norm. The relationships are quite obvious, nevertheless the integral criteria are considered rather engineering quality measures, but the norms are strict mathematical definitions.

For non-final-time signals there is the definition of power as

$$\text{pow}[u(t)] = \sqrt{\lim_{T\to\infty} \frac{1}{2T} \int_{-T}^{T} |u(t)|^2 \mathrm{d}t} \tag{16.4}$$

Note that final-time, constrained signals have only energy, their power is zero. Thus if $\|u(t)\|_2 < \infty$ then $\text{pow}[u(t)] = 0$.

The simplest inequalities regarding these norms are

$$\text{pow}[u(t)] \leq \|u(t)\|_\infty; \quad \text{if} \quad \|u(t)\|_\infty < \infty \tag{16.5}$$

$$\|u(t)\|_2 \leq \sqrt{\|u(t)\|_\infty \|u(t)\|_1}; \quad \text{if} \quad \|u(t)\|_\infty < \infty \text{ and } \|u(t)\|_1 < \infty \tag{16.6}$$

16.1.2 Operator Norms

Using the frequency function $H(j\omega)$ of an *LTI* system having a stable transfer function $H(s)$ the following norms can be defined.

$$\mathcal{H}_2 \text{ norm:} \quad \|H(j\omega)\|_2 = \sqrt{\frac{1}{2\pi}\int_{-\infty}^{\infty} |H(j\omega)|^2 \mathrm{d}\omega} \tag{16.7}$$

$$\mathcal{H}_\infty \text{ norm:} \quad \|H(j\omega)\|_\infty = \max_{\omega}|H(j\omega)| \tag{16.8}$$

These operator norms are usually called system norms.

The computation of the $\mathcal{H}_2$ norm can be performed on the basis of the PARSEVAL-theorem.

$$\|H(j\omega)\|_2^2 = \frac{1}{2\pi}\int_{-\infty}^{\infty} |H(j\omega)|^2 \mathrm{d}\omega = \frac{1}{2\pi j}\oint H(-s)H(s)\mathrm{d}s = \sum \mathrm{Res}[H(-s)H(s)], \tag{16.9}$$

where the residues of $[H(-s)H(s)]$ have to be taken into consideration on the left half plane. The expression (16.9) can be used only (i.e., $\mathcal{H}_2$ is finite), if $H(s)$ is strictly proper and has no pole on the imaginary axis. It is worth noting that

$$\|H(j\omega)\|_2 = \sqrt{\int_{-\infty}^{\infty} |w(t)|^2 \mathrm{d}t} = \sqrt{\int_{0}^{\infty} |w(t)|^2 \mathrm{d}t}, \tag{16.10}$$

where $w(t)$ is the weighting function of the system having transfer function $H(s)$.

If the system $H(s)$ is given in state-space form $(\boldsymbol{A}, \boldsymbol{b}, \boldsymbol{c}^{\mathrm{T}})$, then the $\mathcal{H}_2$ norm can be computed by the following expression

$$\|H(j\omega)\|_2 = \sqrt{\boldsymbol{c}^{\mathrm{T}}\boldsymbol{L}\boldsymbol{c}}, \tag{16.11}$$

where

$$\boldsymbol{L} = \int_{0}^{\infty} e^{\boldsymbol{A}t}\boldsymbol{b}\boldsymbol{b}^{\mathrm{T}}e^{\boldsymbol{A}^{\mathrm{T}}t}\mathrm{d}t. \tag{16.12}$$

Instead of the computation of the integral (16.12), $\boldsymbol{L}$ can be simply determined by solving the system of linear equations for $\boldsymbol{L}$:

$$\boldsymbol{A}\boldsymbol{L} + \boldsymbol{L}\boldsymbol{A}^{\mathrm{T}} = -\boldsymbol{b}\boldsymbol{b}^{\mathrm{T}} \tag{16.13}$$

(see A.16.1 in Appendix A.5). Equation (16.13) can also be solved by the conventional solution technique for systems of linear equations if the unknown column vectors of $\boldsymbol{L}$ are collected into one column vector.

The computation of the $\mathcal{H}_\infty$ norm is not easy, though its geometrical interpretation is very simple: it is the farthest distance of the NYQUIST diagram of $H(j\omega)$ from the origin. Since $H(s)$ and $H(j\omega)$ are usually rational functions, the possible places of the extrema of the absolute value (the necessary condition) are derived from the zeros of the first order derivative. This equation, however, yields a high order system of polynomial equations even for a low order process, whose solution requires numerical techniques. That is why, instead of an analytical solution, numerical methods are used directly to determine the maximum of $|H(j\omega)|$. The $\mathcal{H}_\infty$ norm is finite if $H(s)$ is proper and has no pole on the imaginary axis or in the right half plane.

(The computation of the $\mathcal{H}_\infty$ norm for error-function operators can be performed by the NEVANLINNA-PICK approximation procedure, but its discussion goes beyond the content of this textbook.)

The most important inequality regarding the $\mathcal{H}_\infty$ norm is

$$\|H_1(j\omega)H_2(j\omega)\|_\infty \leq \|H_1(j\omega)\|_\infty \|H_2(j\omega)\|_\infty. \tag{16.14}$$

Keeping the former notations, let $u(t)$ be the input and $y(t)$ the output of the system with transfer function $H(s)$. The most important relationships of the signals and norms of the system are for stable processes:

$$\|y(t)\|_2 \leq \|H(j\omega)\|_\infty \|u(t)\|_2, \tag{16.15}$$

therefore it can be stated that the $\mathcal{H}_\infty$ norm is the upper limit of the gain of the $\mathcal{L}_2$ norm. Based on the inequality

$$\|y(t)\|_\infty \leq \|w(t)\|_1 \|u(t)\|_\infty \tag{16.16}$$

it can be simply seen that the $\mathcal{L}_1$ norm of the weighting function is the upper limit of the gain of the $\mathcal{L}_\infty$ norm. Thus the upper limit of the maximum of the unit step response $y(t) = v(t)$ (if $u(t) = 1(t)$) is equal to the integral of the absolute value of the weighting function. Similar relations are valid for the following inequality

$$\|y(t)\|_\infty \leq \|H(j\omega)\|_2 \|u(t)\|_2. \tag{16.17}$$

It comes from the comparison of (16.16) and (16.17) that

$$\|y(t)\|_\infty \leq \min\{\|w(t)\|_1 \|u(t)\|_\infty; \|H(j\omega)\|_2 \|u(t)\|_2\}, \tag{16.18}$$

where a more strict condition is applied. Thus the $\mathcal{H}_\infty$, $\mathcal{H}_2$ and $\mathcal{L}_1$ norms, for certain signals, can correspond to the upper limit of the gain.

Similar relationships can be formulated for the power of the input and output signals:

$$\mathrm{pow}[y(t)] \leq \|H(j\omega)\|_\infty \mathrm{pow}[u(t)], \tag{16.19}$$

by means of which we obtain

$$\mathrm{pow}[y(t)] \leq \|H(j\omega)\|_\infty \|u(t)\|_\infty. \tag{16.20}$$

From the comparison of these two latter inequalities, it follows that

$$\begin{aligned} \mathrm{pow}[y(t)] &\leq \min\{\|H(j\omega)\|_\infty \mathrm{pow}[u(t)]; \|H(j\omega)\|_\infty \|u(t)\|_\infty\} \\ &= \|H(j\omega)\|_\infty \min\{\mathrm{pow}[u(t)]; \|u(t)\|_\infty\} \end{aligned} \tag{16.21}$$

where the more strict condition is applied.

It was shown in Chap. 7 that the optimality of *YP* controllers applied for stable processes can be reached via the optimal choice of the embedded filters $G_{\mathbf{x}}|_{\mathbf{x}=\mathrm{r,n}}$ (transfer functions). Their optimality for the error transfer functions $R_{\mathbf{x}}(1 - G_{\mathbf{x}}P_{-}e^{-sT_d})|_{\mathbf{x}=\mathrm{r,n}}$ can be ensured by the minimization of the operator norms $\mathcal{H}_2$ and $\mathcal{H}_\infty$.

16.2 Basic Methods of the Numerical Optimization

The optimization problems of control engineering can usually be formulated by seeking the minimum of a scalar-vector function $f(\boldsymbol{x})$. The function to be minimized may be e.g., an integral criterion, a signal or operator norm, the vector components of the vector space of the searching are the parameters of the controller. Basically two main groups of extremum seeking methods can be distinguished depending on whether only the value of the function can be computed, or also its first and second order derivatives can be determined at a point $\boldsymbol{x}$.

16.2.1 Direct Seeking Methods

In the case of direct seeking (*DS*) methods only the value of the function $f(\boldsymbol{x})$ can be computed at a given point of seeking the minimum. The most effective *DS* method is the so-called adaptive simplex method of Nelder and Mead. In an n-dimensional space, a simplex is a shape given by $(n+1)$ points. Thus in two-dimensional space it is a triangle, in the three-dimensional case it is a tetrahedron. Find the minimum of $f(\boldsymbol{x})$ in a two dimensional space. First consider the simplex *ABC* shown in Fig. 16.1. Compute the values of $f(\boldsymbol{x})$ at the three points of the simplex. Based on the values $f(\boldsymbol{x}_A)$, $f(\boldsymbol{x}_B)$ and $f(\boldsymbol{x}_C)$, let us arrange in order of their magnitudes the corresponding coordinate vectors of the three points. Assume that the biggest value is obtained at the point $f(\boldsymbol{x}_A = \boldsymbol{x}_o)$. Mirror this point $\boldsymbol{x}_o$ to the center point of the opposite—less by one order—shape, i.e., now via the middle point $\boldsymbol{x}_1$ of the line *BC* to the point $\boldsymbol{x}_2$. Then continue this procedure (stepping) in the obtained direction until the values of $f(\boldsymbol{x}_i)$ increase. Assume that at point $\boldsymbol{x}_4$ the value is $f(\boldsymbol{x}_4) > f(\boldsymbol{x}_3)$, i.e., the minimum seeking algorithm does not give a better

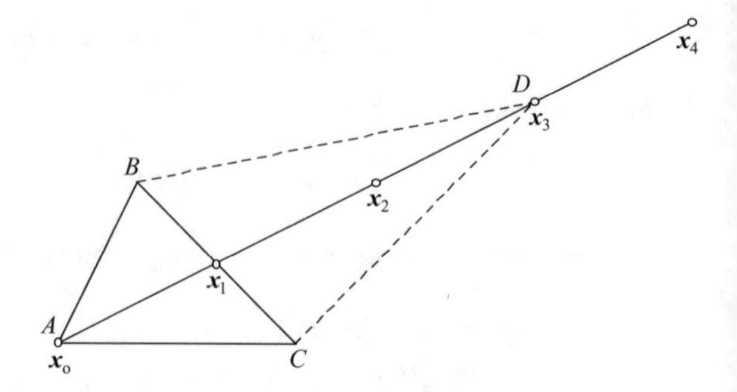

Fig. 16.1 Scheme of the adaptive simplex method

point. This means that the first point of the new simplex will be $\boldsymbol{x}_3$ and the simplex will be given by the triangle *BCD*. Then the point $\boldsymbol{x}_B$ belonging to the second biggest function value $f(\boldsymbol{x}_B)$ has to be mirrored on the middle point of line *CD*, then the seeking steps have to be continued in this direction. If all the points of the original simplex have already been mirrored, then we get into a completely new simplex whose form follows the form of the function $f(\boldsymbol{x})$ to a certain extent. The procedure is continued until mirroring all points of the simplex only a worse point is found, i.e., a bigger value of $f(\boldsymbol{x})$ is obtained. This case is called the limiting (or boundary) simplex. Then the sought minimum is inside this simplex.

The method is continued by formulating a new simplex with half size edges based on the worst point, i.e., by shrinking the simplex. The algorithm is started again from this shrunk simplex. The search method is stopped when the size of the limit simplex in each coordinate direction is within a certain accuracy threshold (the convergence limit).

The advantage of the simplex method is that it can easily handle both explicit constraints

$$\boldsymbol{x}_{\min} \leq \boldsymbol{x} \leq \boldsymbol{x}_{\max} \tag{16.22}$$

and so-called k implicit constraints, such as

$$g_j(\boldsymbol{x}) \leq 0 \quad j = 1, \ldots, k. \tag{16.23}$$

To achieve this, the starting point $\boldsymbol{x}_o$ has to fulfill the above conditions, then during the stepping the above restrictions are handled as if a bigger $f(\boldsymbol{x})$ had been obtained, thus the seeking in that direction has to be stopped.

The adaptive simplex method is able to find the minimum of a function of even a very special form with acceptable efficiency. Of course, it can determine only the minimum of a unimodal function, i.e., when $f(\boldsymbol{x})$ has only one extremum, or it can seek for a local minimum in a given region.

If the task is such that several extrema can be expected, i.e., $f(\boldsymbol{x})$ is multimodal, then the adaptive complex method can be applied. The "complex" is defined by a shape (set) given by $N > n+1$ points in an n-dimensional space. Usually N is much bigger than n, and the algorithm has to be started with an equally distributed point set in the search space. The algorithm operates similarly to the adaptive simplex method, but now the given point has to be mirrored via the geometrical center of all the other $N-1$ points, then the stepping has to be continued in this direction.

16.2.2 Gradient Based Methods

In the cases when the first and second derivates of the function $f(\boldsymbol{x})$ can be computed, then algorithms faster than the *DS* methods can also be constructed. The general canonical form of the methods using the gradient is the following iterative algorithm:

$$\boldsymbol{x}_{i+1} = \boldsymbol{x}_i - \boldsymbol{G}(\boldsymbol{x}_i)\frac{\mathrm{d}f(\boldsymbol{x}_i)}{\mathrm{d}\boldsymbol{x}}. \tag{16.24}$$

The gradient methods can be basically distinguished by how to choose the weighting matrix $\boldsymbol{G}(\boldsymbol{x}_i)$ (Note that each version of the algorithms approximates the gradient $\mathrm{d}f(\boldsymbol{x}_i)/\mathrm{d}\boldsymbol{x}$ in a different way).

From the different ways of choosing the weighting matrix $\boldsymbol{G}(\boldsymbol{x}_i)$, the first is the so-called gradient method:

$$\boldsymbol{G}(\boldsymbol{x}_i) = G(\boldsymbol{x}_i) = \left[\frac{\mathrm{d}f^{\mathrm{T}}(\boldsymbol{x}_i)}{\mathrm{d}\boldsymbol{x}}\boldsymbol{H}(\boldsymbol{x}_i)\frac{\mathrm{d}f(\boldsymbol{x}_i)}{\mathrm{d}\boldsymbol{x}}\right]^{-1}, \tag{16.25}$$

where $\boldsymbol{H}(\boldsymbol{x}_i)$ is the HESSIAN matrix (see (A.1.31)) of the function $f(\boldsymbol{x})$ at the point $\boldsymbol{x}_i$. It is interesting that now $G(\boldsymbol{x}_i)$ is a scalar. This method uses a second order approximation in the direction opposite to the gradient, and puts the next iteration point at the minimum of the parabola taken in this direction. The significant disadvantage of this method is that in the case of "curving" valleys, it slows down because it cannot follow precisely the deepness shape of the valley.

The next method is the NEWTON-RAPHSON method (sometimes it is also called the GAUSS-NEWTON method), where

$$\boldsymbol{G}(\boldsymbol{x}_i) = [\boldsymbol{H}(\boldsymbol{x}_i)]^{-1}. \tag{16.26}$$

This method fits a general quadratic surface (multidimensional ellipsoid) at an iteration point and puts the next iteration point at the calculable extremum of this shape.

The above two methods using gradients have also very clear geometrical interpretations, the other methods can be considered as different combinations of these.

The gradient methods are much more effective than the *DS* ones for so-called "well behaved" functions, but for exceptional functions, e.g., having the form of a banana, they slow down. Their further disadvantage is that they are not very effective in the case of constraints, because they usually shrink to the trajectory point crossing the constraining surface. In this case, certain techniques use the solution to push the iteration off this surface and the search starts again.

There are several procedures and software programs available for all the above methods in the different program packages and in an object oriented *CAD* environment.

It has been noted in Sect. 4.8 in connection with the square error area that its minimization generally provides an optimal step response function having a relatively high overshoot. Therefore it seems reasonable to construct the optimization task which performs this minimization of the integral criterion $I_2 = f(\boldsymbol{x})$ under the restriction for the overshoot $\sigma = g(\boldsymbol{x}) \leq 1.05$. This task guarantees a "nice" step response function with a small overshoot.

Example 16.1 The expression for the so-called "function of banana" frequently used in optimization tasks is

$$f(\boldsymbol{x}) = 100\left(x_2 - x_1^2\right)^2 + (1 - x_1)^2. \tag{16.27}$$

The function in 3D is shown in Fig. 16.2, whose minimum is at the point $\boldsymbol{x} = [1, 1]$.

The operation of the adaptive simplex method is illustrated in Fig. 16.3. The procedure starts from the point $\boldsymbol{x} = [-1.9, 2]$ and after 210 iterations it finds the minimum (i.e., it computes the function's value at 210 points).

Figure 16.4 shows the operation of the gradient method, more exactly its inability to find the minimum after computing the function's value at 210 points, and the gradients at 200 points, but it stopped at the beginning of the valley.

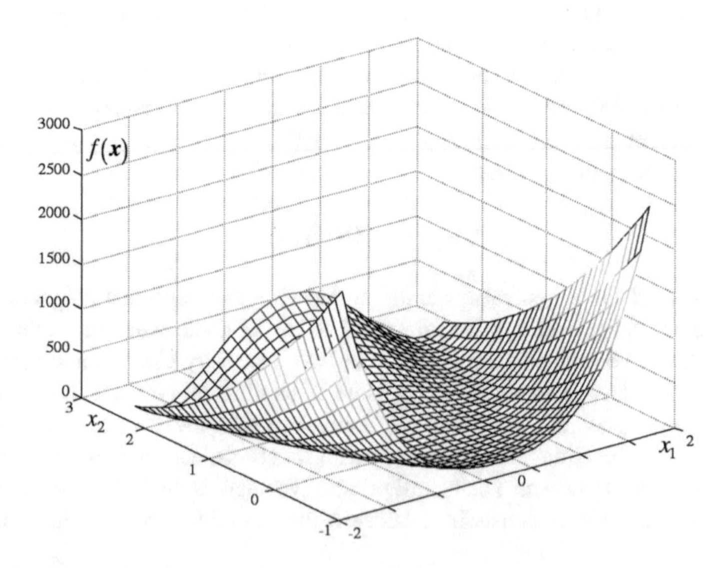

Fig. 16.2 The so-called "function of banana" applied quite often in optimization tasks

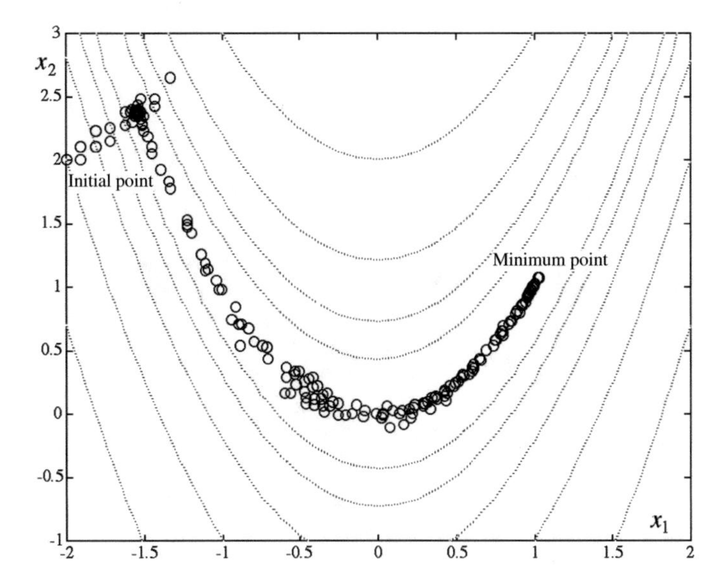

Fig. 16.3 Optimum seeking by simplex method

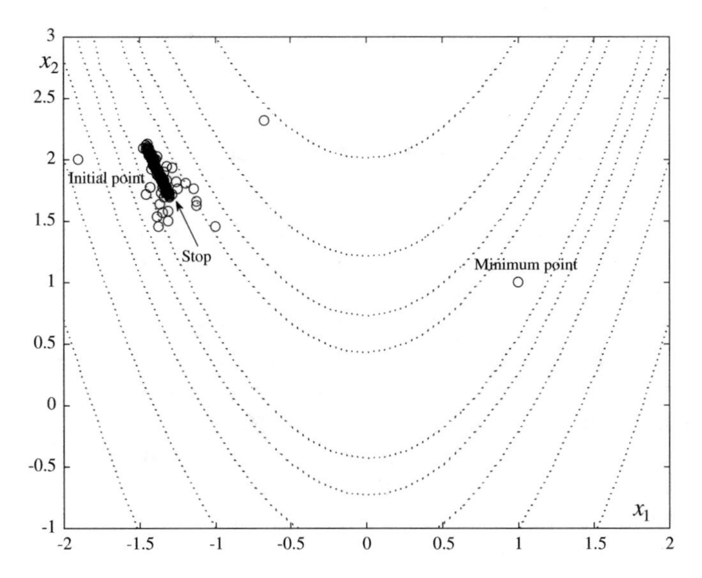

Fig. 16.4 The inability of the gradient method to find the minimum of (16.27)

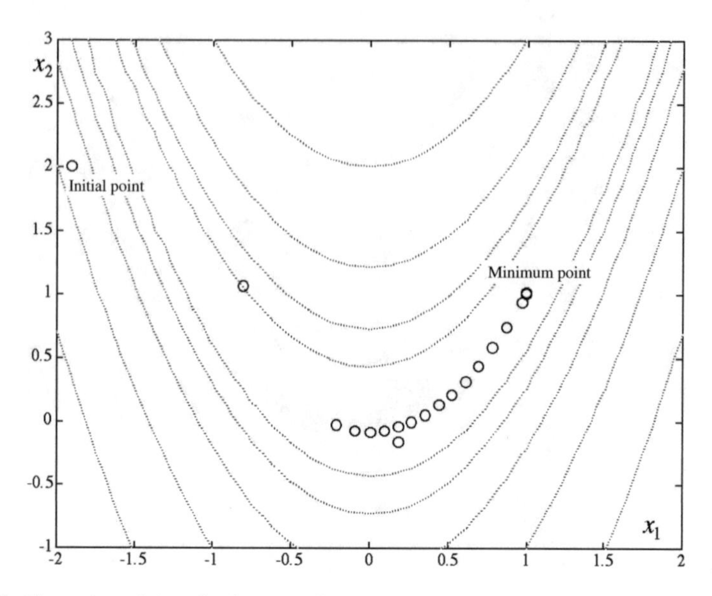

Fig. 16.5 Illustration of the effectiveness of the NEWTON-RAPHSON method

The effectiveness of the NEWTON-RAPHSON method is demonstrated in Fig. 16.5: the method found the minimum after 21 iterations. ■

16.3 Introduction to Process Identification

It has been seen in Sect. 2.4 that one of the basic tasks in control engineering is process identification, when the model $\hat{P}$ of the process P to be controlled is determined from the measurements of the input and output signals. Process identification, starting with simple grapho-analytical methods, has today become an independent (autonomous) discipline; its methods and results can be found in several books. With the spread of modern computational techniques, almost standard tools are available to solve the most important tasks. Here only some of the topmost methods are discussed, just to illustrate the applied algorithms and techniques.

Process identification methods substantially differ from each other, depending on the task to be solved, i.e., whether the static characteristics or the dynamic model of the process has to be determined.

16.3.1 Identification of Static Processes

Assume that the static characteristic of the process is a line $p_o + p_1 u$, which can be measured with measurement error e

$$y = p_o + p_1 u + e. \tag{16.28}$$

The input signal u is measured without error, or it is a known signal put into the system (active experiment). The input and output signals are measured jointly at N points. These values are approximated by the linear model

$$\hat{y} = \hat{p}_o + \hat{p}_1 u = \boldsymbol{f}^{\mathrm{T}}(u)\hat{\boldsymbol{p}}; \quad \boldsymbol{f}^{\mathrm{T}}(u) = [1 \quad u]; \quad \hat{\boldsymbol{p}} = [\hat{p}_o \quad \hat{p}_1]^{\mathrm{T}} \tag{16.29}$$

as seen in Fig. 16.6. Here $\boldsymbol{f}(u)$ is called the vector of function components.

If the additive measurement error has zero average, then the so-called *Least Squares (LS)* method provides the unbiased estimation of the process parameters. The *LS* method takes the sum of the squares of the differences between the measured value and the model output at each point and optimizes it according to the criterion

$$V(\hat{\boldsymbol{p}}, N) = \frac{1}{2}\sum_{j=1}^{N}\left[y_j - \boldsymbol{f}^{\mathrm{T}}(u_j)\hat{\boldsymbol{p}}\right]^2 = \frac{1}{2}[\boldsymbol{y} - \boldsymbol{F}_{\mathrm{u}}\hat{\boldsymbol{p}}]^{\mathrm{T}}[\boldsymbol{y} - \boldsymbol{F}_{\mathrm{u}}\hat{\boldsymbol{p}}], \tag{16.30}$$

where

$$\boldsymbol{F}_{\mathrm{u}} = \begin{bmatrix} 1 & u_1 \\ 1 & u_2 \\ \vdots & \vdots \\ 1 & u_N \end{bmatrix} = \begin{bmatrix} \boldsymbol{f}^{\mathrm{T}}(u_1) \\ \boldsymbol{f}^{\mathrm{T}}(u_2) \\ \vdots \\ \boldsymbol{f}^{\mathrm{T}}(u_N) \end{bmatrix} \quad \text{and} \quad \boldsymbol{y} = \begin{bmatrix} y_1 \\ y_2 \\ \vdots \\ y_N \end{bmatrix}. \tag{16.31}$$

The system of vector equations for the N samples is

$$\boldsymbol{y} = \boldsymbol{F}_{\mathrm{u}}\boldsymbol{p} + \boldsymbol{e}, \tag{16.32}$$

where $\boldsymbol{e} = [e_1 e_1 \ldots e_N]^{\mathrm{T}}$ is the vector of the measurement errors. The parameter estimation minimizing the sum of squares according to A.16.2 in Appendix A.5 is

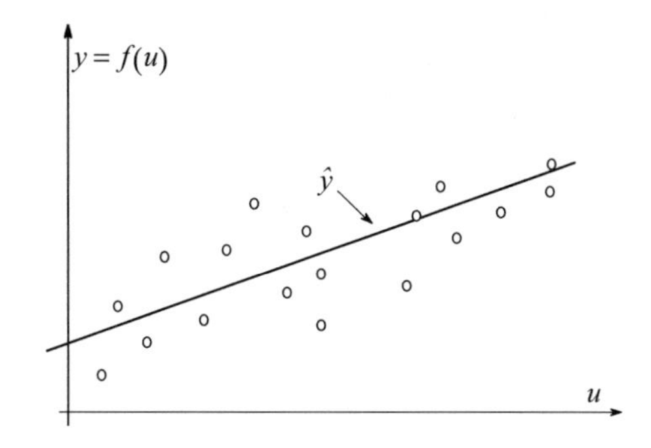

Fig. 16.6 Identification of linear static model

$$\hat{\boldsymbol{p}} = \left[\boldsymbol{F}_{\mathrm{u}}^{\mathrm{T}}\boldsymbol{F}_{\mathrm{u}}\right]^{-1}\boldsymbol{F}_{\mathrm{u}}^{\mathrm{T}}\boldsymbol{y}. \tag{16.33}$$

This estimation is unbiased, i.e., $E\{\hat{\boldsymbol{p}}\} = \boldsymbol{p}$, thus the expected value of $\hat{\boldsymbol{p}}$ is the unknown original parameter vector $\boldsymbol{p}$. If e has a normal distribution, then the $\hat{\boldsymbol{p}}$ obtained by the *LS* estimation has minimum variance and is the *best* estimator of $\boldsymbol{p}$.

In many cases the input signal u is not known in advance, just measured (passive experiment). If u is a random signal, then in order to get an unbiased estimation by the *LS* method, the independence of the signals e and u has to be assumed.

The computation of the solution (16.33) can be made easier by taking the following relationships into account

$$\boldsymbol{F}_{\mathrm{u}}^{\mathrm{T}}\boldsymbol{F}_{\mathrm{u}} = \sum_{j=1}^{N}\boldsymbol{f}(u_j)\boldsymbol{f}^{\mathrm{T}}(u_j) \quad \text{and} \quad \boldsymbol{F}_{\mathrm{u}}^{\mathrm{T}}\boldsymbol{y} = \sum_{j=1}^{N}\boldsymbol{f}(u_j)y_j. \tag{16.34}$$

Assume that the static characteristic of the process is a parabola $p_{\mathrm{o}} + p_1 u + p_2 u^2$, and the additive measurement error is e.

$$y = p_{\mathrm{o}} + p_1 u + p_2 u^2 + e. \tag{16.35}$$

The input signal u is assumed to be measured without error. The input and output signals are measured jointly at N points, and the following nonlinear (quadratic) model is fitted to the measured values, as seen in Fig. 16.7. Now introduce

$$\hat{y} = \hat{p}_{\mathrm{o}} + \hat{p}_1 u + \hat{p}_2 u^2 = \boldsymbol{f}^{\mathrm{T}}(u)\hat{\boldsymbol{p}}; \quad \boldsymbol{f}^{\mathrm{T}}(u) = \begin{bmatrix}1 & u & u^2\end{bmatrix}; \quad \hat{\boldsymbol{p}} = \begin{bmatrix}\hat{p}_{\mathrm{o}} & \hat{p}_1 & \hat{p}_2\end{bmatrix}^{\mathrm{T}}. \tag{16.36}$$

Observe that the model $\hat{y} = \boldsymbol{f}^{\mathrm{T}}(u)\hat{\boldsymbol{p}}$ is still linear in the parameters. Thus the *LS* method can be applied unchanged if the matrix $\boldsymbol{F}_{\mathrm{u}}$ is formulated from the function component vector $\boldsymbol{f}(u)$ according to the quadratic model (16.36):

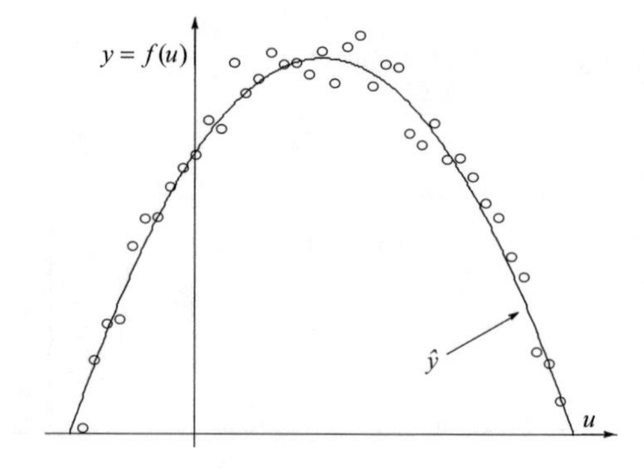

Fig. 16.7 Identification of a nonlinear (quadratic) static model

$$\boldsymbol{F}_{\mathrm{u}} = \begin{bmatrix} 1 & u_1 & u_1^2 \\ 1 & u_2 & u_2^2 \\ \vdots & \vdots & \vdots \\ 1 & u_N & u_N^2 \end{bmatrix} = \begin{bmatrix} \boldsymbol{f}^{\mathrm{T}}(u_1) \\ \boldsymbol{f}^{\mathrm{T}}(u_2) \\ \vdots \\ \boldsymbol{f}^{\mathrm{T}}(u_N) \end{bmatrix} \tag{16.37}$$

and $\hat{\boldsymbol{p}}$ is computed again by Eq. (16.33).

Observe that a relatively wide class of functions can be written in a form that is linear in its parameters.

If the static characteristic is nonlinear, then an extremum seeking method is used to minimize $V(\hat{\boldsymbol{p}}, N)$, which, e.g., can be chosen from those discussed in Sect. 16.2.

16.3.2 Identification of Dynamic Processes

Nowadays the identification of dynamic processes exclusively means the determination of a discrete time (*DT*) model. It has been shown in Sect. 11.4 that a *DT* system given by the so-called filter form

$$y[k] = G(z^{-1})u[k] = \frac{\mathcal{B}(z^{-1})z^{-d}}{\mathcal{A}(z^{-1})}u[k] = \frac{\mathcal{B}(z^{-1})z^{-d}}{1+\tilde{\mathcal{A}}(z^{-1})}u[k] \tag{16.38}$$

can be written in a form linear in parameters as

$$\begin{aligned} y[k] &= \mathcal{B}(z^{-1})z^{-d}u[k] - \tilde{\mathcal{A}}(z^{-1})y[k] = \boldsymbol{f}^{\mathrm{T}}(u, y, k)\boldsymbol{p}_{\mathrm{ba}} \\ &= b_1 u[k-d-1] + b_1 u[k-d-2] + \cdots + b_n u[k-d-n] - a_1 y[k-1] - \cdots - a_n y[k-n] \end{aligned} \tag{16.39}$$

where

$$\begin{aligned} \boldsymbol{f}^{T}(u, y, k) &= [u[k-d-1]\; u[k-d-2]\; \ldots\; u[k-d-n]\; -y[k-1]\; \ldots\; -y[k-n]] \\ \boldsymbol{p}_{\mathrm{ba}} &= [b_1\; b_1\; \ldots\; b_n\; a_1\; \ldots\; a_n] \end{aligned} \tag{16.40}$$

This technique, by means of which the difference equation of the dynamic *DT* systems is made "quasi-linear", opens the possibility of formulating further process identification algorithms similar to the *LS* method. The measurement noise problems of *DT* systems, however, should be discussed in a basically different way than for static characteristics. The measuring situation is illustrated in Fig. 16.8.

Here the measurement of $u[k]$ is assumed to be without error, but the noiseless output signal $v[k]$ of the system is assumed to be measured with an additive measurement error $y_{\mathrm{n}}[k]$. This output noise $y_{\mathrm{n}}[k]$ derives from an independent, zero mean, so-called white noise via the noise model $\mathcal{C}(z^{-1})/\mathcal{D}(z^{-1})$.

This task essentially requires the identification of two models: the process model and the noise model. The task can be drastically simplified by certain assumptions made regarding the noise model. If the noise model has the form $1/\mathcal{A}(z^{-1})$, i.e.,

Fig. 16.8 Measured signals of a linear dynamic discrete-time system

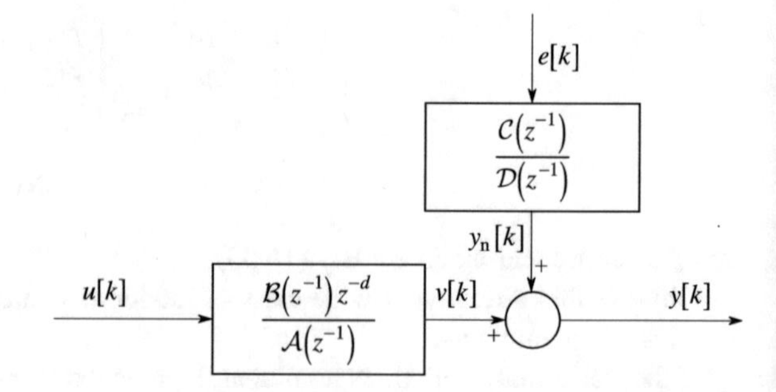

$$y[k] = \frac{\mathcal{B}(z^{-1})z^{-d}}{\mathcal{A}(z^{-1})}u[k] + \frac{1}{\mathcal{A}(z^{-1})}e[k] \tag{16.41}$$

then the original process can be rewritten as

$$y[k] = \boldsymbol{f}^{\mathrm{T}}(u, y, k)\boldsymbol{p}_{\mathrm{ba}} + e[k]. \tag{16.42}$$

Observe that this form essentially corresponds to Eqs. (16.28) and (16.35) seen in the identification of the static characteristics, thus the *LS* method can be directly applied if the matrix $\boldsymbol{F}(u)$ is constructed from $\boldsymbol{f}^{\mathrm{T}}(u, y, k)$ instead of $\boldsymbol{f}(u)$, and $\hat{\boldsymbol{p}}_{\mathrm{ba}}$ is the parameter vector. Let us create first the vector

$$\boldsymbol{y} = [y[1] \quad y[2] \quad \ldots \quad y[N]]^{\mathrm{T}} \tag{16.43}$$

and the matrix

$$\boldsymbol{F}_{\mathrm{uy}} = \begin{bmatrix} \boldsymbol{f}^{\mathrm{T}}(u, y, 1) \\ \boldsymbol{f}^{\mathrm{T}}(u, y, 2) \\ \vdots \\ \boldsymbol{f}^{\mathrm{T}}(u, y, N) \end{bmatrix} \tag{16.44}$$

The parameter estimation by the *LS* method has also the form of (16.33)

$$\hat{\boldsymbol{p}}_{\mathrm{ba}} = \left[\boldsymbol{F}_{\mathrm{uy}}^{\mathrm{T}}\boldsymbol{F}_{\mathrm{uy}}\right]^{-1}\boldsymbol{F}_{\mathrm{uy}}^{\mathrm{T}}\boldsymbol{y}. \tag{16.45}$$

This estimation is asymptotically unbiased, i.e., $\mathrm{plim}_{N\to\infty}\{\hat{\boldsymbol{p}}\} = \boldsymbol{p}$, thus the probabilistic limit value of $\hat{\boldsymbol{p}}$ is the unknown original parameter vector $\boldsymbol{p}$. The independence of $e[k]$ has to be assumed, because $\boldsymbol{f}^{\mathrm{T}}(u, y, k)$ has measured values depending on $e[k]$. If $e[k]$ has a normal distribution, then the $\hat{\boldsymbol{p}}$ resulting from the *LS* estimation is the minimum variance (best) estimator of $\boldsymbol{p}$.

Unfortunately the form of the noise model $1/\mathcal{A}(z^{-1})$ is very special, so it cannot be used generally. The noise model $\mathcal{C}(z^{-1})/\mathcal{A}(z^{-1})$ can be considered as a more general form, when

$$y[k] = \frac{\mathcal{B}(z^{-1})z^{-d}}{\mathcal{A}(z^{-1})}u[k] + \frac{\mathcal{C}(z^{-1})}{\mathcal{A}(z^{-1})}e[k] = \frac{\mathcal{B}(z^{-1})z^{-d}}{1+\tilde{\mathcal{A}}(z^{-1})}u[k] + \frac{1+\tilde{\mathcal{C}}(z^{-1})}{\mathcal{A}(z^{-1})}e[k]. \tag{16.46}$$

This form preserves the generality of the noise model $\mathcal{C}(z^{-1})/\mathcal{D}(z^{-1})$, but it contains a great number of redundant parameters because of bringing the fractions to a common denominator. The quasi-linearization by (16.36) can be easily performed here, too.

$$\begin{aligned} y[k] &= \mathcal{B}(z^{-1})z^{-d}u[k] - \tilde{\mathcal{A}}(z^{-1})y[k] + \tilde{\mathcal{C}}(z^{-1})e[k] + e[k] \\ &= b_1u[k-d-1] + b_2u[k-d-2] + \cdots + b_nu[k-d-n] - a_1y[k-1] - \cdots - a_ny[k-n] \\ &\quad + c_1e[k-1] + \cdots + c_ne[k-n] + e[k] = \boldsymbol{f}^{\mathrm{T}}(u,y,e,k)\boldsymbol{p}_{\mathrm{bac}} + e[k] \end{aligned} \tag{16.47}$$

The most important disadvantage of this form of the model is that the past values of $e[k]$ in the vector $\boldsymbol{f}(u,y,e,k)$ are not known. But if the estimation $\hat{\boldsymbol{p}}_{\mathrm{bac}}$ of the $\boldsymbol{p}_{\mathrm{bac}}$ is known then an estimation $\hat{e}[k]$ of the source noise can always be computed in the form

$$\hat{e}[k] = y[k] - \boldsymbol{f}^{\mathrm{T}}(u,y,\hat{e},k)\hat{\boldsymbol{p}}_{\mathrm{bac}}, \tag{16.48}$$

where now the computed (estimated) value $\hat{e}[k]$ is in $\boldsymbol{f}(u,y,\hat{e},k)$. Creating the matrix

$$\boldsymbol{F}_{\mathrm{uy\hat{e}}} = \begin{bmatrix} \boldsymbol{f}^{\mathrm{T}}(u,y,\hat{e},1) \\ \boldsymbol{f}^{\mathrm{T}}(u,y,\hat{e},2) \\ \vdots \\ \boldsymbol{f}^{\mathrm{T}}(u,y,\hat{e},N) \end{bmatrix} \tag{16.49}$$

formally again an *LS* estimation is obtained based on (16.35) and (16.43) in the form

$$\hat{\boldsymbol{p}}_{\mathrm{bac}} = \left[\boldsymbol{F}^{\mathrm{T}}_{\mathrm{uy\hat{e}}}\boldsymbol{F}_{\mathrm{uy\hat{e}}}\right]^{-1}\boldsymbol{F}^{\mathrm{T}}_{\mathrm{uy\hat{e}}}\boldsymbol{y}. \tag{16.50}$$

Since the series $\hat{e}[k], (k = 1,\ldots,N)$, is always computed for a given $\hat{\boldsymbol{p}}_{\mathrm{bac}}$, here only an iteration method can be realized, i.e., the series $\hat{e}[k]$ has to be computed after each estimation step. The iteration is continued until the difference between the consecutively estimated parameter vectors becomes less than a given error. (This iteration is called a relaxation-type one.) The solution (16.50) belonging to Eq. (16.47) is called the *Extended Least Squares* (*ELS*) method. Several other versions of this method are known, using different noise models, which has resulted in a huge number of methods in the literature.

Theoretically the most accurate method can be obtained by minimizing the loss function

$$V(\hat{\boldsymbol{p}}_{\text{bac}},N)=\frac{1}{2}\sum_{j=1}^{N}\left\{y[j]-\boldsymbol{f}^{\text{T}}(u,y,\hat{e},j)\hat{\boldsymbol{p}}_{\text{bac}}\right\}^{2}=\frac{1}{2}\left[\boldsymbol{y}-\boldsymbol{F}_{\text{uy}\hat{e}}\hat{\boldsymbol{p}}_{\text{bac}}\right]^{\text{T}}\left[\boldsymbol{y}-\boldsymbol{F}_{\text{uy}\hat{e}}\hat{\boldsymbol{p}}_{\text{bac}}\right] \tag{16.51}$$

in the space of the parameter vector $\hat{\boldsymbol{p}}_{\text{bac}}$, which means a general minimum seeking problem. Even for different noise models, the first and second order derivates of $V(\hat{\boldsymbol{p}}_{\text{bac}},N)$ with respect to the parameters can be relatively easily computed, so effective minimum seeking algorithms can be constructed this way. The methods directly minimizing (16.50) are called *Maximum Likelihood* (*ML*) methods. This method requires zero average, normal, white noise for $e[k]$.

Those methods which use simultaneously available N data-pairs of the input and output signals are called "off-line" or "batch" methods. All the above methods belong to this category.

There are measurement situations when the model obtained formerly by an estimation method is modified (renewed) after getting a new measured data-pair. These methods are called "on-line" or "recursive" identification methods.

16.3.3 Discrete-Time to Continuous-Time Transformation

It has been seen during the discussion of the basic discrete-time process identification methods that these methods—deriving from their character—provide the operators of models $\hat{G}(z^{-1})$ or $\hat{G}(q^{-1})$ constructed by the estimated parameters $\hat{\boldsymbol{p}}_{\text{ba}}$ of the pulse transfer function $G(z^{-1})$ or pulse transfer operator $G(q^{-1})$ of the process. (From the process identification point of view there is no importance attached to these notations and meanings.) Here $\hat{G}(z^{-1})$ is used. In many cases, however, the model $\hat{P}(s)$ of the original CT system is required as a result of the identification. This conversion, i.e., the equivalence at the sampling points, can be solved only by assuming a holding term of a given type. In connection with Eqs. (11.30) and (11.31) it has been already shown that in the case of a zero order hold, thus applying an *SRE* transformation, the parameter matrices of the DT state equations are

$$\boldsymbol{F}=e^{\boldsymbol{A}T_{\text{s}}}\quad\text{and}\quad\boldsymbol{g}=\boldsymbol{A}^{-1}\left(e^{\boldsymbol{A}T_{\text{s}}}-\boldsymbol{I}\right)\boldsymbol{b}. \tag{16.52}$$

Formally, the parameter matrices of the *SRE* equivalent CT systems can be obtained by the reverse of the equations

$$\boldsymbol{A}=\frac{1}{T_{\text{s}}}\ln(\boldsymbol{F})\quad\text{and}\quad\boldsymbol{b}=\frac{1}{T_{\text{s}}}\ln(\boldsymbol{F})(\boldsymbol{F}-\boldsymbol{I})^{-1}\boldsymbol{g}. \tag{16.53}$$

Here $\ln(\boldsymbol{F})$ is the logarithm of the matrix $\boldsymbol{F}$, which is defined and computed by the definitions valid for matrix functions (see $e^{\boldsymbol{A}}$ in Chap. 3). Based on the above the algorithm of the discrete-continuous transformation is:

1. Based on the estimated $\hat{\boldsymbol{p}}_{\mathrm{ba}}$, construct the state-space description of a DT model by the controllable canonical form $\hat{\boldsymbol{F}}_{\mathrm{c}};\hat{\boldsymbol{g}}_{\mathrm{c}};\hat{\boldsymbol{c}}_{\mathrm{c}}$ or an observable $\hat{\boldsymbol{F}}_{\mathrm{o}};\hat{\boldsymbol{g}}_{\mathrm{o}};\hat{\boldsymbol{c}}_{\mathrm{o}}$ canonical form.
2. Using the transformation Eq. (16.53), compute the state-space form $\hat{\boldsymbol{A}};\hat{\boldsymbol{b}};\hat{\boldsymbol{c}}$ of the CT model. This step results in parameter matrices of general form having n^2+2n parameters.
3. Transform the CT state-space model $\hat{\boldsymbol{A}};\hat{\boldsymbol{b}};\hat{\boldsymbol{c}}$ to a controllable or observable canonical form by either the transformation matrix $\boldsymbol{T}_{\mathrm{c}}=\boldsymbol{M}_{\mathrm{c}}^{\mathrm{c}}(\boldsymbol{M}_{\mathrm{c}})^{-1}$ or $\boldsymbol{T}_{\mathrm{o}}=(\boldsymbol{M}_{\mathrm{o}}^{\mathrm{o}})^{-1}\boldsymbol{M}_{\mathrm{o}}$ from which the parameters of the transfer function $\hat{P}(s)$ of the CT model sought can be easily determined from the non-redundant structure corresponding to the canonical form. Note that in the case of the canonical form, it is not necessary to compute the whole matrix $\hat{\boldsymbol{A}}_{\mathrm{c}}$ or $\hat{\boldsymbol{A}}_{\mathrm{o}}$, it is sufficient to compute the first row or column.

The above transformation techniques are the most compact ones, but of course, there are different ways to solve the problem. The same accurate result can be obtained by decomposing $\hat{G}(z^{-1})$ into partial fractions and then the discrete-continuous transformation can be made term-by-term.

16.3.4 Recursive Parameter Estimation

First consider the recursive version of the *LS* method. Assume that *N* data-pairs are processed and the *LS* estimate is available in the form

$$\hat{\boldsymbol{p}}[N]=\left\{\boldsymbol{F}^{\mathrm{T}}[N]\boldsymbol{F}[N]\right\}^{-1}\boldsymbol{F}^{\mathrm{T}}[N]\boldsymbol{y}_N. \tag{16.54}$$

If we want to modify our estimate obtained by (16.54) using the new data-pairs $u[N+1]$ and $y[N+1]$ measured in the $[N+1]$-th time instant, then it can be computed by the following recursive relationships

$$\hat{\boldsymbol{p}}[N+1]=\hat{\boldsymbol{p}}[N]+\boldsymbol{R}[N+1]\boldsymbol{f}(N+1)\left\{y[N+1]-\boldsymbol{f}^{\mathrm{T}}(N+1)\hat{\boldsymbol{p}}[N]\right\} \tag{16.55}$$

and

$$\boldsymbol{R}[N+1]=\boldsymbol{R}[N]-\frac{\boldsymbol{R}[N]\boldsymbol{f}(N+1)\boldsymbol{f}^{\mathrm{T}}(N+1)\boldsymbol{R}[N]}{1+\boldsymbol{f}^{\mathrm{T}}(N+1)\boldsymbol{R}[N]\boldsymbol{f}(N+1)} \tag{16.56}$$

(see A.16.3 in Appendix A.5). Here $\boldsymbol{f}(N+1)$ means a general function independently of whether the method is applied to a static or dynamic process model. The so-called convergence matrix $\boldsymbol{R}[N]$ is

$$\boldsymbol{R}[N] = \left\{ \sum_{j=1}^{N} \boldsymbol{f}(j)\boldsymbol{f}^{\mathrm{T}}(j) \right\}^{-1} = \left\{ \boldsymbol{F}^{\mathrm{T}}[N]\boldsymbol{F}[N] \right\}^{-1}. \tag{16.57}$$

The equation-pair (16.55) and (16.56) belong to the family of the so-called learning, adaptive estimation algorithms, which are included in the canonical equation of the general *Stochastic Approximation* (*SA*):

$$\hat{\boldsymbol{p}}[k+1] = \hat{\boldsymbol{p}}[k] + \boldsymbol{R}[k+1]\frac{\mathrm{d}V(\hat{\boldsymbol{p}}, k)}{\mathrm{d}\hat{\boldsymbol{p}}}. \tag{16.58}$$

These *SA* algorithms differ from each other in the choice of the convergence matrix $\boldsymbol{R}[k]$ and the way to compute the gradient. Here only the best-known method has been discussed.

If the parameters of the process are varying, then it might be necessary to forget in a certain sense the validity of the former model and take into account the new measurements with higher importance. To solve this so-called "forgetting" problem, assume that the past is forgotten by using the following matrix

$$\boldsymbol{F}[N+1] = \begin{bmatrix} \lambda \boldsymbol{F}[N] \\ \boldsymbol{f}^{\mathrm{T}}(N+1) \end{bmatrix} \text{instead of } \boldsymbol{F}[N+1] = \begin{bmatrix} \boldsymbol{F}[N] \\ \boldsymbol{f}^{\mathrm{T}}(N+1) \end{bmatrix}$$

where the forgetting factor is $0 \leq \lambda \leq 1$. If $\lambda =$ constant, then it is enough to use the following convergence matrix

$$\boldsymbol{R}[N+1] = \frac{1}{\lambda^2}\left\{ \boldsymbol{R}[N] - \frac{\boldsymbol{R}[N]\boldsymbol{f}(N+1)\boldsymbol{f}^{\mathrm{T}}(N+1)\boldsymbol{R}[N]}{\lambda^2 + \boldsymbol{f}^{\mathrm{T}}(N+1)\boldsymbol{R}[N]\boldsymbol{f}(N+1)} \right\} \tag{16.59}$$

instead of (16.56).

A constant forgetting factor, however, may cause problems, if the new measurements do not have significantly new information, since this algorithm forgets exponentially the old information, and so $\boldsymbol{R}[N]$ may become singular. Therefore the choice of the corresponding forgetting strategy is the most critical part of the adaptive estimation method.

Note that Eqs. (16.55), (16.56) and (16.60) are usually called naive programming formulas. By means of them the method can be simply presented but numerically they behave badly. They are mostly used for purposes of demonstration or, simulation. In practice the canonical, diagonal form of $\boldsymbol{R}[N]$ and its recursive forms are used: this solution works best from the numerical point of view. This method uses the so-called GIVENS transformation.

16.3.5 Model Validation

During process identification the determination of a model of acceptable correctness (accuracy) can be made only by an iterative process. Its main steps are presented in Fig. 16.9.

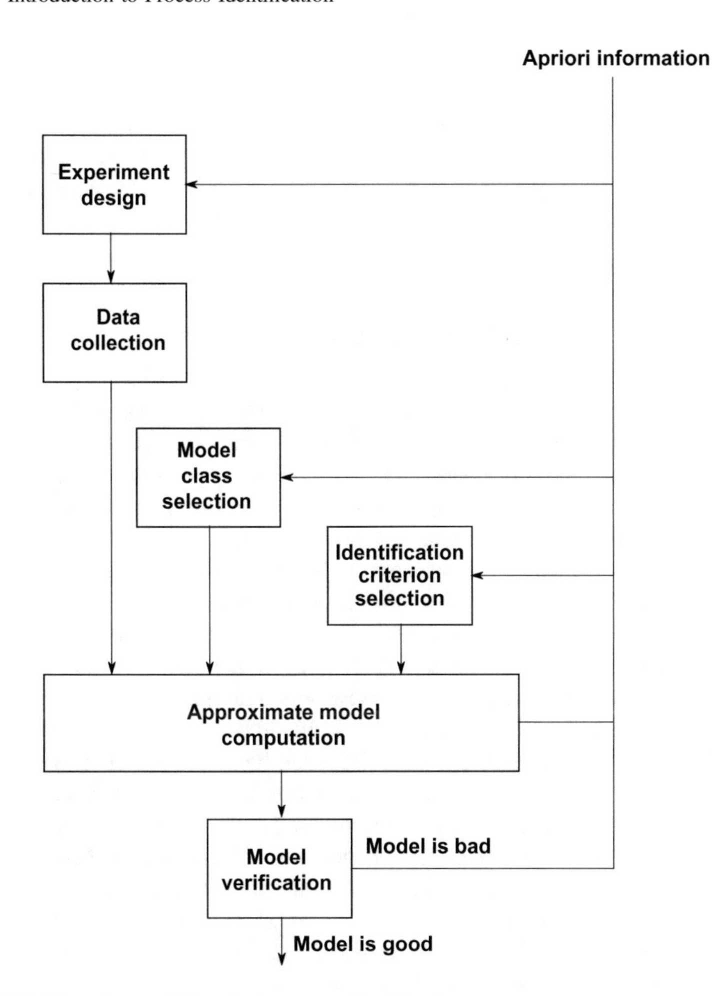

Fig. 16.9 The scheme of the whole process identification

The identification starts using certain preliminary information. First the so-called "design of experiments" is performed. In this step the optimal allocation of the measurement points is determined for static modeling. For dynamic modeling, however, one assumes that optimal input signals representing the significant frequency region are generated. This latter is called input design. The accuracy of the final model depends significantly on this step, therefore several theories deal with the optimal solution of this task [see, for example, Sect. 7.5].

The effect of the optimal measurement points or input signal for the process is realized by active experiment designs, and the data are collected during the experiments.

Based on the preliminary information, the class of the model and the identification criterion are chosen, then the determination of an approximate model is performed (parameter estimation).

Then the model output and the measured output of the process obtained for the same input are compared. From the deviations, qualitative, goodness of fit measures can be constructed for checking the acceptability of the model (model validation).

If the accuracy of the model is not satisfactory then the iteration is continued with a newer experiment design. The procedure is stopped when the accuracy of the model is acceptable.

16.4 Iterative and Adaptive Control Schemes

In the previous sections some off-line and on-line methods of process identification have been discussed. Not only can the process model be improved by repeated experiments, but also the controller, if a controller is designed and realized based on the model, applied in a closed-loop with the computed optimal parameters. This joint task is required, on the one hand, at the initial tuning of the regulator, and on the other hand, in the case of a process with slowly varying parameters, in the continuous adaptation of the regulator (adaptive control).

In the case of modern, microprocessor based compact controllers, nowadays there is embedded possibility for a certain kind of automatic tuning. The commercial controllers usually apply the ÅSTRÖM-type relay-tuning (see Sect. 8.3).

The more demanding optimal controller is based on the iterative strategy of a certain learning-adaptive version of joint identification-control (*simultaneous identification and control).* This strategy assumes that the identification is performed without opening the closed control loop, i.e., under normal operation conditions. The identification is usually off-line, i.e., it is based on the simultaneous processing of N data pairs. Based on the obtained model an optimal controller is determined and this controller is used in the next off-line experiment. By this technique the optimality of the controller can be gradually improved as the model becomes more and more accurate, while the normal operation of the process is hardly disturbed.

Certainly it is also possible to improve the parameter estimation of the process applying a recursive parameter estimating technique in every sampling instant, and the optimal controller output is applied to the process only delayed by the computation time of the optimal controller. In the case of today's fast operating computing equipment, this solution, to a very good approximation, can be considered as simultaneous processing in the case of significantly slower processes. This strategy is called adaptive control. The determination of a reliable controller is not an easy task. A recursive parameter estimation algorithm is required which does not forget the learned model if the new measurements do not have significant new information. If the quantity of the new information is considerable, then it is able to follow the slowly varying parameters by due forgetting strategies.

In the case of certain, so-called predictive controllers, the process model is not identified directly, but finally in the algorithms, the determination of the process model is always present.

Appendix

A.1 Mathematical Summary

A.1.1 Some Basic Theorems of Matrix Algebra

The following scheme is called a matrix

$$\mathbf{A} = \begin{bmatrix} a_{11} & a_{12} & \dots & a_{1n} \\ a_{21} & a_{22} & \dots & a_{2n} \\ \vdots & \vdots & \ddots & \vdots \\ a_{m1} & a_{m2} & \dots & a_{mn} \end{bmatrix}, \tag{A.1.1}$$

where the values a_{ij} are the elements of the matrix. If its elements are real, then the matrix is called a real matrix, if they are complex, then the matrix is called a complex matrix. In general, a matrix has m rows and n columns. The dimension (the size) of the matrix is $m \times n$. A matrix of type $m \times n$ is rectangular; an $n \times n$ matrix is called square (quadratic) matrix, an $m \times 1$ matrix is a column matrix (column vector), a $1 \times n$ matrix is a row matrix (row vector), a 1×1 matrix is called a scalar.

Matrices are usually denoted by *bold* (*fat*) capital letters, the column and row vectors are denoted by *bold* lower case letters. The determinant of the square matrix $\mathbf{A}$ is denoted by $|\mathbf{A}|$ (or written as $\det(\mathbf{A})$).

The transpose of the matrix $\mathbf{A}$ is denoted by $\mathbf{A}^{\mathrm{T}}$, and it means the result of the mirroring of its elements for the main diagonal.

$$\mathbf{A}^{\mathrm{T}} = \begin{bmatrix} a_{11} & a_{21} & \dots & a_{m1} \\ a_{12} & a_{22} & \dots & a_{m1} \\ \vdots & \vdots & \ddots & \vdots \\ a_{1n} & a_{2n} & \dots & a_{mn} \end{bmatrix}. \tag{A.1.2}$$

L. Keviczky et al., *Control Engineering*, Advanced Textbooks in Control and Signal Processing, https://doi.org/10.1007/978-981-10-8297-9

If $\boldsymbol{A}$ is an $m \times n$ matrix, then its transpose is an $n \times m$ matrix, and it is trivial that $(\boldsymbol{A}^{\mathrm{T}})^{\mathrm{T}} = \boldsymbol{A}$. If $\boldsymbol{A}^{\mathrm{T}} = \boldsymbol{A}$ then it is called mirror matrix.

A vector is usually considered a column matrix, and a row matrix is denoted as the transpose of a column matrix, e.g.,

$$\boldsymbol{x} = \begin{bmatrix} x_1 \\ \vdots \\ x_n \end{bmatrix} = [x_1, \ldots, x_n]^{\mathrm{T}} = [\boldsymbol{x}^{\mathrm{T}}]^{\mathrm{T}}. \tag{A.1.3}$$

The elements of the zero matrix, or zero vector, are all zeros. The diagonal matrix has elements different from zero only along the main diagonal, i.e.,

$$\boldsymbol{D} = \mathbf{diag}[a_{11}, a_{22}, \ldots, a_{nn}]. \tag{A.1.4}$$

If in a diagonal matrix all the diagonal elements are unity, then the matrix is called the unit matrix: $\boldsymbol{I} = \mathbf{diag}[1, 1, \ldots, 1]$.

Two matrices are equal if all the corresponding elements are equal. The sum of two or more matrices of the same type is obtained by summing the corresponding elements. The multiplication of a matrix by a scalar is obtained by multiplying each element of the matrix by the scalar. The most characteristic case is the multiplication of two matrices, e.g., when a matrix $\boldsymbol{A}$ of type $m \times l$ is multiplied by a matrix $\boldsymbol{B}$ of type $l \times n$,

$$\boldsymbol{C} = \boldsymbol{AB}, \tag{A.1.5}$$

where

$$c_{ij} = \sum_{k=1}^{l} a_{ik} b_{kj}; \quad \begin{cases} i = 1, 2, \ldots, m \\ j = 1, 2, \ldots, n \end{cases} \tag{A.1.6}$$

i.e., the element in the i-th row and j-th column of the matrix $\boldsymbol{C}$ of type $m \times n$ is obtained by multiplying the i-th row of $\boldsymbol{A}$ by the j-th column of $\boldsymbol{B}$. (The number l of columns of $\boldsymbol{A}$ must be equal to the number l of the rows of $\boldsymbol{B}$.) Matrix multiplication is associative and distributive, but, in general, is not commutative: $\boldsymbol{AB} \neq \boldsymbol{BA}$. If $\boldsymbol{AB} = \boldsymbol{BA}$, then in this case the matrices are interchangeable (commutative). Note that the determinant of the square product matrix $|\boldsymbol{C}| = \det(\boldsymbol{C})$ is obtained by multiplying the determinants $|\boldsymbol{A}|$ and $|\boldsymbol{B}|$ of the factor matrices, i.e., $|\boldsymbol{C}| = |\boldsymbol{A}||\boldsymbol{B}|$.

The scalar product of two vectors having the same dimension can be expressed as the product of matrices, by

$$\boldsymbol{a} \cdot \boldsymbol{b} = \boldsymbol{a}^{\mathrm{T}} \boldsymbol{b} = \boldsymbol{b}^{\mathrm{T}} \boldsymbol{a} = \boldsymbol{b} \cdot \boldsymbol{a}. \tag{A.1.7}$$

If the scalar product of two different, non-zero vectors is zero, then the two vectors are called orthogonal.

The following expression represents a very important rule

$$[\boldsymbol{AB}]^{\mathrm{T}} = \boldsymbol{B}^{\mathrm{T}}\boldsymbol{A}^{\mathrm{T}}. \tag{A.1.8}$$

The inverse of a square, regular (nonsingular, i.e., its determinant is non-zero) matrix is a matrix, for which the following expression is valid.

$$\boldsymbol{A}^{-1}\boldsymbol{A} = \boldsymbol{A}\boldsymbol{A}^{-1} = \boldsymbol{I}. \tag{A.1.9}$$

The inverse of $\boldsymbol{A}$ is given by the rule

$$\boldsymbol{A}^{-1} = \frac{\mathbf{adj}(\boldsymbol{A})}{|\boldsymbol{A}|}. \tag{A.1.10}$$

Here $|\boldsymbol{A}|$ is the (non-zero) determinant of $\boldsymbol{A}$, and the adjunct matrix $\mathbf{adj}(\boldsymbol{A})$ of $\boldsymbol{A}$ is obtained by mirroring the matrix whose elements are sub-determinants of appropriate sign belonging to each element of $\boldsymbol{A}$. Since the rule $|\boldsymbol{AB}| = |\boldsymbol{A}||\boldsymbol{B}|$ is valid, therefore, according to $1 = |\boldsymbol{I}| = |\boldsymbol{A}^{-1}\boldsymbol{A}| = |\boldsymbol{A}^{-1}||\boldsymbol{A}|$, $\boldsymbol{A}$ has an unambiguous inverse only if $|\boldsymbol{A}| \neq 0$, i.e., the matrix $\boldsymbol{A}$ is non-singular. It is obvious that

$$[\boldsymbol{A}^{-1}]^{-1} = \boldsymbol{A} \quad \text{and} \quad [\boldsymbol{A}^{-1}]^{\mathrm{T}} = [\boldsymbol{A}^{\mathrm{T}}]^{-1}. \tag{A.1.11}$$

Furthermore if $\boldsymbol{A}$ and $\boldsymbol{B}$ are regular square matrices, then

$$[\boldsymbol{AB}]^{-1} = \boldsymbol{B}^{-1}\boldsymbol{A}^{-1}. \tag{A.1.12}$$

The matrices $s\boldsymbol{I} - \boldsymbol{A}$, or $\boldsymbol{A} - s\boldsymbol{I}$, are called the characteristic matrices of the square matrix $\boldsymbol{A}$, and the equation $\mathcal{A}(s) = |s\boldsymbol{I} - \boldsymbol{A}| = 0$ is called the characteristic equation. The roots $\lambda_i (i = 1, 2, \ldots, n)$ of the characteristic equation are the eigenvalues of the matrix $\boldsymbol{A}$. Due to the main pivot theorem the eigenvectors $\boldsymbol{v}_i (i = 1, 2, \ldots n)$ of $\boldsymbol{A}$ fulfill the following vector equations:

$$\boldsymbol{A}\boldsymbol{v}_i = \lambda_i \boldsymbol{v}_i \quad (i = 1, 2, \ldots, n). \tag{A.1.13}$$

This is the definition of the eigenvectors. If the vectors $\boldsymbol{v}_j$ are linearly independent, then the matrix $\boldsymbol{A}$ has a simple structure, if the vectors are not independent, then the matrix is called deteriorated.

The CAYLEY-HAMILTON theorem has significant importance in the matrix theory: any matrix $\boldsymbol{A}$ satisfies its own characteristic equation, i.e., $\mathcal{A}(\boldsymbol{A}) = 0$. (Here in the scalar polynomial equation $\mathcal{A}(s) = 0$, s^i is replaced by $\boldsymbol{A}^i$ $(i = 1, 2, \ldots, n)$, while s^0 is by $\boldsymbol{A}^0 = \boldsymbol{I}$, so finally a matrix polynomial equation is obtained.)

In many cases it is necessary to express the inner structure of a matrix, therefore so-called block-matrices are applied, e.g.,

$$M = \begin{bmatrix} A & B \\ C & D \end{bmatrix}. \tag{A.1.14}$$

According to the matrix multiplication rules

$$\begin{bmatrix} A & B \\ C & D \end{bmatrix}\begin{bmatrix} E & F \\ G & H \end{bmatrix} = \begin{bmatrix} AE+BG & AF+BH \\ CE+DG & CF+DH \end{bmatrix}. \tag{A.1.15}$$

The determinant of a quasi-diagonal matrix is

$$\begin{vmatrix} A & B \\ O & D \end{vmatrix} = \det\begin{bmatrix} A & B \\ O & D \end{bmatrix} = \det(A)\det(D) = |A||D|. \tag{A.1.16}$$

The product ab^{T} is called the dyadic product. The inverse of the matrix A extended by the addition of a dyadic product can be given very simply, if the inverse of A is known:

$$\left(A + ab^{\mathrm{T}}\right)^{-1} = A^{-1} - \frac{\left(A^{-1}a\right)\left(b^{\mathrm{T}}A^{-1}\right)}{1 + b^{\mathrm{T}}A^{-1}a} \tag{A.1.17}$$

A.1.2 Some Basic Formulas of Vector Analysis

In vector analysis for Euclidean space there are scalar-scalar functions

$$f = f(x), \tag{A.1.18}$$

so-called scalar-vector functions

$$f = f(\boldsymbol{x}), \tag{A.1.19}$$

and vector-vector functions

$$\boldsymbol{f} = \boldsymbol{f}(\boldsymbol{x}). \tag{A.1.20}$$

(All these are the special cases of the most general but very rare matrix-matrix functions $F = F(X)$.) In many cases multivariable scalar-scalar, scalar-vector or vector-vector functions occur, e.g.,

$$f = f(x, u); \quad f = f(\boldsymbol{x}, \boldsymbol{u}); \quad \boldsymbol{f} = \boldsymbol{f}(\boldsymbol{x}, \boldsymbol{u}) \tag{A.1.21}$$

or functions containing independent variables (time or a parameter) also appear

$$f = f(x, u, t); \quad f = f(\boldsymbol{x}, \boldsymbol{u}, t); \quad \boldsymbol{f} = \boldsymbol{f}(\boldsymbol{x}, \boldsymbol{u}, t). \tag{A.1.22}$$

Certain rules for differentiation are very important. The derivative with respect to a scalar is very simple, e.g.,

$$\frac{\mathrm{d}\boldsymbol{x}(t)}{\mathrm{d}t} = \left[\frac{\mathrm{d}x_1}{\mathrm{d}t}, \frac{\mathrm{d}x_2}{\mathrm{d}t}, \ldots, \frac{\mathrm{d}x_n}{\mathrm{d}t}\right]^{\mathrm{T}} = [\dot{x}_1, \dot{x}_2, \ldots, \dot{x}_n] = \dot{\boldsymbol{x}} \tag{A.1.23}$$

$$\frac{\mathrm{d}\boldsymbol{A}(t)}{\mathrm{d}t} = \begin{bmatrix} \dot{a}_{11} & \dot{a}_{12} & \ldots & \dot{a}_{1n} \\ \dot{a}_{21} & \dot{a}_{22} & \ldots & \dot{a}_{2n} \\ \vdots & \vdots & \ddots & \vdots \\ \dot{a}_{m1} & \dot{a}_{m2} & \ldots & \dot{a}_{mn} \end{bmatrix} = \dot{\boldsymbol{A}} \tag{A.1.24}$$

The gradient of a scalar-vector function is a column vector

$$\mathbf{grad}[f(\boldsymbol{x})] = \frac{\mathrm{d}f(\boldsymbol{x})}{\mathrm{d}\boldsymbol{x}} = \left[\frac{\mathrm{d}f(\boldsymbol{x})}{\mathrm{d}x_1} \quad \frac{\mathrm{d}f(\boldsymbol{x})}{\mathrm{d}x_2} \quad \ldots \quad \frac{\mathrm{d}f(\boldsymbol{x})}{\mathrm{d}x_n}\right]^{\mathrm{T}}, \tag{A.1.25}$$

which means the application of a multivariable differential-operator

$$\frac{\mathrm{d}}{\mathrm{d}\boldsymbol{x}} = \left[\frac{\mathrm{d}}{\mathrm{d}x_1} \quad \frac{\mathrm{d}}{\mathrm{d}x_2} \quad \ldots \quad \frac{\mathrm{d}}{\mathrm{d}x_n}\right]^{\mathrm{T}} \tag{A.1.26}$$

thus

$$\mathbf{grad}[f(\boldsymbol{x})] = \frac{\mathrm{d}}{\mathrm{d}\boldsymbol{x}} f(\boldsymbol{x}) = \frac{\mathrm{d}f(\boldsymbol{x})}{\mathrm{d}\boldsymbol{x}}. \tag{A.1.27}$$

The JACOBIAN matrix is

$$\boldsymbol{J} = \boldsymbol{J}(\boldsymbol{f}, \boldsymbol{x}) = \begin{bmatrix} \frac{\mathrm{d}f_1}{\mathrm{d}x_1} & \frac{\mathrm{d}f_1}{\mathrm{d}x_2} & \ldots & \frac{\mathrm{d}f_1}{\mathrm{d}x_n} \\ \frac{\mathrm{d}f_2}{\mathrm{d}x_1} & \frac{\mathrm{d}f_2}{\mathrm{d}x_2} & \ldots & \frac{\mathrm{d}f_2}{\mathrm{d}x_n} \\ \vdots & \vdots & \ddots & \vdots \\ \frac{\mathrm{d}f_m}{\mathrm{d}x_1} & \frac{\mathrm{d}f_m}{\mathrm{d}x_2} & \ldots & \frac{\mathrm{d}f_m}{\mathrm{d}x_n} \end{bmatrix}. \tag{A.1.28}$$

Avoiding complicated notations, the JACOBIAN matrix is symbolically denoted by

$$J(f,x) = \frac{\mathrm{d}f(x)}{\mathrm{d}x^{\mathrm{T}}} \tag{A.1.29}$$

and its transpose is

$$J^{\mathrm{T}}(f,x) = \frac{\mathrm{d}f^{\mathrm{T}}(x)}{\mathrm{d}x}. \tag{A.1.30}$$

Thus the transpose of the gradient vector is

$$\mathbf{grad}^T[f(x)] = \left[\frac{\mathrm{d}f(x)}{\mathrm{d}x}\right]^{\mathrm{T}} = \frac{\mathrm{d}f(x)}{\mathrm{d}x^{\mathrm{T}}} = J(f,x). \tag{A.1.31}$$

The second order derivatives of a scalar-vector function can be arranged into the HESSIAN matrix

$$H = H(f,x) = \begin{bmatrix} \frac{\mathrm{d}^2 f}{\mathrm{d}x_1^2} & \frac{\mathrm{d}^2 f}{\mathrm{d}x_1 \mathrm{d}x_2} & \cdots & \frac{\mathrm{d}^2 f}{\mathrm{d}x_1 \mathrm{d}x_n} \\ \frac{\mathrm{d}^2 f}{\mathrm{d}x_2} & \frac{\mathrm{d}^2 f}{\mathrm{d}x_2^2} & \cdots & \frac{\mathrm{d}^2 f}{\mathrm{d}x_2 \mathrm{d}x_n} \\ \vdots & \vdots & \ddots & \vdots \\ \frac{\mathrm{d}^2 f}{\mathrm{d}x_n} & \frac{\mathrm{d}^2 f}{\mathrm{d}x_n \mathrm{d}x_2} & \cdots & \frac{\mathrm{d}^2 f}{\mathrm{d}x_n^2} \end{bmatrix}. \tag{A.1.32}$$

A.2 Signals and Systems

The general topics of signals and systems directly connected to control engineering have been discussed in the main sections of this textbook. For completeness there are, however, some special fields whose effect and availability has to be known, but they cannot be connected directly to control engineering. From the subject of an excitation with special periodic signals, only the standard sine excitation was discussed for the better understanding of the frequency functions.

Dynamics of linear processes with periodic excitation

Let $u(t)$ be a function of time with a period T_{p}, i.e., $u(t+T_{\mathrm{p}}) = u(t)$. Introduce the notation $u_{\mathrm{A}}(t)$ for denoting the basic function (or truncated function) determining the periodic signal, which in the time domain $0<t<T_{\mathrm{p}}$ is equal to $u(t)$, but otherwise is zero.

$$u_{\mathrm{A}}(t) = [1(t) - 1(t - T_{\mathrm{p}})]u(t) = \begin{cases} u(t); & 0<t<T_{\mathrm{p}} \\ 0; & t \leq 0;\ t > T_{\mathrm{p}} \end{cases}. \tag{A.2.1}$$

The periodic function $u(t)$ can be obviously constructed by repeated shifts and sums of the basic function $u_{\mathrm{A}}(t)$, according to the definition

$$u(t) = \sum_{j=0}^{\infty} u_{\mathrm{A}}\left(t - jT_{\mathrm{p}}\right) = 1_{\mathrm{A}}(t)u(t), \tag{A.2.2}$$

where $1_{\mathrm{A}}(t)$ is called the repetitive operator. Determine the LAPLACE transform of the basic function $u_{\mathrm{A}}(t)$, i.e., the function $U_{\mathrm{A}}(s)$. Due to the shift theorem

$$\mathcal{L}\left\{u_{\mathrm{A}}\left(t - jT_{\mathrm{p}}\right)\right\} = e^{-jsT_{\mathrm{p}}} U_{\mathrm{A}}(s) \tag{A.2.3}$$

and applying it to (A.2.2), the LAPLACE transform of the periodic signal $u(t)$ is

$$U(s) = \mathcal{L}\{u(t)\} = \mathcal{L}\{1_{\mathrm{A}}(t)u(t)\} = \sum_{j=0}^{\infty} e^{-jsT_{\mathrm{p}}} U_{\mathrm{A}}(s) = U_{\mathrm{A}}(s) \sum_{j=0}^{\infty} e^{-jsT_{\mathrm{p}}}. \tag{A.2.4}$$

Notice that here the summing equation for the geometric series can be applied

$$U(s) = \mathcal{L}\{1_{\mathrm{A}}(t)u(t)\} = \frac{U_{\mathrm{A}}(s)}{1 - e^{-sT_{\mathrm{p}}}}. \tag{A.2.5}$$

If the LAPLACE transform $U(s)$ of a signal can be written in the form of (A.2.5), then using the basic function $u_{\mathrm{A}}(t) = \mathcal{L}^{-1}\{U_{\mathrm{A}}(s)\}$, the time function of the periodic signal can be easily determined. The condition $u_{\mathrm{A}}(t) = 0$ for $t > T_{\mathrm{p}}$ must be fulfilled.

Next the system dynamics, i.e., the process response is investigated when a periodic signal is put to as input of an *LTI* system. The response can be gotten by the LAPLACE transform of the process output if the system is originally free of energy. The LAPLACE transform of the output by using the conventional transfer function notation $H(s) = \mathcal{B}(s)/\mathcal{A}(s)$ is

$$Y(s) = U(s)H(s) = \frac{U_{\mathrm{A}}(s)}{1 - e^{-sT_{\mathrm{p}}}} \frac{\mathcal{B}(s)}{\mathcal{A}(s)}. \tag{A.2.6}$$

In general $Y(s)$ is not the transform of a periodic signal, since the condition $\mathcal{L}^{-1}\{Y(s)\} = \mathcal{L}^{-1}\{U_{\mathrm{A}}(s)H(s)\} = 0$ is not fulfilled for $t > T_{\mathrm{p}}$. $H(s)$ is always (except for the case of dead-time) a rational function, but this cannot be said about $U_{\mathrm{A}}(s)$. Decompose the function $Y(s)$ into the sum of a periodic and a non-periodic function, i.e.,

$$Y(s) = \frac{U_{\mathrm{A}}(s)}{1 - e^{-sT_{\mathrm{p}}}} \frac{\mathcal{B}(s)}{\mathcal{A}(s)} = \frac{Y_{\mathrm{A}}(s)}{1 - e^{-sT_{\mathrm{p}}}} + \frac{\mathcal{C}(s)}{\mathcal{A}(s)}, \tag{A.2.7}$$

where $y_A(t) = \mathcal{L}^{-1}\{Y_A(s)\}$, $y_A(t > T_p) = 0$, and $\mathcal{C}(s)$ are unknown polynomials. From this equation the basic function of the periodic output component can be expressed as

$$Y_A(s) = \frac{\mathcal{B}(s)U_A(s) - (1 - e^{-sT_p})\mathcal{C}(s)}{\mathcal{A}(s)} = H(s)U_A(s) - \left(1 - e^{-sT_p}\right)\frac{\mathcal{C}(s)}{\mathcal{A}(s)}. \tag{A.2.8}$$

By transforming back $Y_A(s)$, zero has to be obtained for the time $t > T_p$. Using these conditions $\mathcal{C}(s)$ can be determined. Apply the expansion theorem and assuming single poles we get

$$y_A(t) = \sum_{i=1}^{n} \frac{\mathcal{B}(p_i)U_A(p_i) - (1 - e^{-p_iT_p})\mathcal{C}(p_i)}{\mathcal{A}'(p_i)} e^{p_it} = 0; \quad t > T_p. \tag{A.2.9}$$

Since the factors e^{p_it} cannot be zero, therefore the function $y_A(t)$ can be zero for all time points $t > T_p$ only if the coefficients of all n factors are zero, i.e.,

$$\mathcal{C}(p_i) = \frac{\mathcal{B}(p_i)U_A(p_i)}{1 - e^{-p_iT_p}} = \alpha_i; \quad i = 1, \ldots, n. \tag{A.2.10}$$

This condition, at the same time, gives the solution for the coefficients of the unknown $\mathcal{C}(s)$, since n independent linear equations can be formulated.

$$1 + c_1p_i + c_2p_i^2 + \cdots + c_np_i^n = \alpha_i; \quad i = 1, \ldots, n. \tag{A.2.11}$$

The coefficients come from the solution of these equations whose compact form is

$$\begin{bmatrix} c_1 \\ c_2 \\ \vdots \\ c_n \end{bmatrix} = \begin{bmatrix} p_1 & p_1^2 & \cdots & p_1^n \\ p_2 & p_2^2 & \cdots & p_2^n \\ \vdots & \vdots & \ddots & \vdots \\ p_n & p_n^2 & \cdots & p_n^n \end{bmatrix}^{-1} \begin{bmatrix} \alpha_1 - 1 \\ \alpha_2 - 1 \\ \vdots \\ \alpha_n - 1 \end{bmatrix}. \tag{A.2.12}$$

Based on (A.2.7) the complete time function of the process output is

$$y(t) = 1_A(t)y_A(t) + y_{tr}(t), \tag{A.2.13}$$

where $y_{tr}(t)$ is the so-called non-periodic transient factor

$$y_{tr}(t) = \mathcal{L}^{-1}\left\{\frac{\mathcal{C}(s)}{\mathcal{A}(s)}\right\} = \sum_{i=1}^{n} \frac{\mathcal{C}(p_i)}{\mathcal{A}'(p_i)} e^{p_it} = \sum_{i=1}^{n} \frac{\mathcal{B}(p_i)}{\mathcal{A}'(p_i)} \frac{U_A(p_i)}{1 - e^{-p_iT_p}} e^{p_it}. \tag{A.2.14}$$

Since the expansion theorem requires only the substitution values of $\mathcal{C}(p_i)$, it is not necessary to solve the system of equations (A.2.12).

Based on (A.2.8) the basic function $y_A(t)$ of the output signal is

$$y_A(t) = \mathcal{L}^{-1}\{H(s)U_A(s)\} - \mathcal{L}^{-1}\left\{\frac{\mathcal{C}(s)}{\mathcal{A}(s)}\right\}; \quad 0<t<T_p, \tag{A.2.15}$$

which is obtained by the inverse LAPLACE transform. (Here the effect of e^{-sT_p} in (A.2.8) does not have to be taken into account, because the response is out of the basic period.) Applying the expansion theorem yields

$$\begin{aligned} y_A(t) &= \mathcal{L}^{-1}\left\{\frac{\mathcal{B}(s)}{\mathcal{A}(s)}U_A(s)\right\} - \sum_{i=1}^{n}\frac{\mathcal{C}(p_i)}{\mathcal{A}'(p_i)}e^{p_i t} \\ &= \mathcal{L}^{-1}\left\{\frac{\mathcal{B}(s)}{\mathcal{A}(s)}U_A(s)\right\} - \sum_{i=1}^{n}\frac{\mathcal{B}(p_i)}{\mathcal{A}'(p_i)}\frac{U_A(p_i)}{1-e^{-p_i T_p}}e^{p_i t}; \quad 0<t<T_p. \end{aligned} \tag{A.2.16}$$

Note that $y_A(t) \neq \mathcal{L}^{-1}\{H(s)U_A(s)\}$, thus $Y_A(s) \neq H(s)U_A(s)$.

The process output of an *LTI* process excited by a periodic signal has two factors: a periodic signal and a transient signal. After the transient is died out only the periodic component remains. These two components appear even if the initial energy content of the process is zero (the initial state vector in the state equation is zero), i.e., the above two components must not be mistaken for the factors obtained from the solution of the homogeneous (un-excited) and inhomogeneous (excited) state equations. The above components of the response obtained for a periodic excitation appear even if the initial condition is not a zero vector. Thus, in the general case, the process response has three components.

A.3 Standard Control Engineering Signals and Notations

A.3.1 Standard Notations in Control Engineering

The design, installation, operation and maintenance of process control systems require the cooperation of the participants who are working on the solution of the task. In order to achieve this, it is required to use common notation in the documentation of each piece of equipment of the different process control functions. In the documentation, the notation of the process control equipment refers to its technical character and how it is connected to the process. Standard graphical and alphanumeric notation helps the engineers and, technicians to interpret the design documentation.

The notation systems and standard protocols may differ in different branches of the industry (chemical, energy, agriculture, etc.).

The standard DIN (Deutsches Institut für Normung) 19227 contains several graphical symbols for sensors, controllers, actuators, and control equipment. Further recommendations can be found in standards DIN 1946, 2429, 2481, 19239 and 30600.

The instrumentation and control functions are usually represented by a circle or oval curve containing letters and numbers. The letters refer to the character of the physical quantity and the control function, the numbers give the place of the equipment in the process (e.g., serial number of the valve, motor or sensor).

In the instrumentation designs [see Fig. A.3.1] the first letter of the text in the circle refers to the character of the measured or controlled quantity, e.g., the meaning of some of the first letters are: E—electrical signal, F—flowing quantity, G—movement or position, L- level, P- pressure, Q- composition or other material character (frequently it is denoted by A, too), S- speed, T- temperature, V- viscosity. The second letter means the control function, e.g., T- sensing, C- control. For example the text LC in the circle means level controller. Further letters can refer to further functions, e.g., to alarm, security operation, computer connection, transducers, etc. Figure A.3.1 illustrates the composition control of the liquid in the mixer tank and the standard notations of the valve, composition sensor and controller.

There is an other standard, KKS (Kraftwerk Kennzeichen System), which has been developed in the German electrical industry and primarily used by European firms. This notation fits the functional structure of the technology. The process control functions and notations fit with the mechanical and electrical power transmission functions and notations. The unified notations of the equipments make it possible to identify the technological units in a decomposed part of a complex technology. For example, the notation 03GCR31AA101 for a valve means that it is

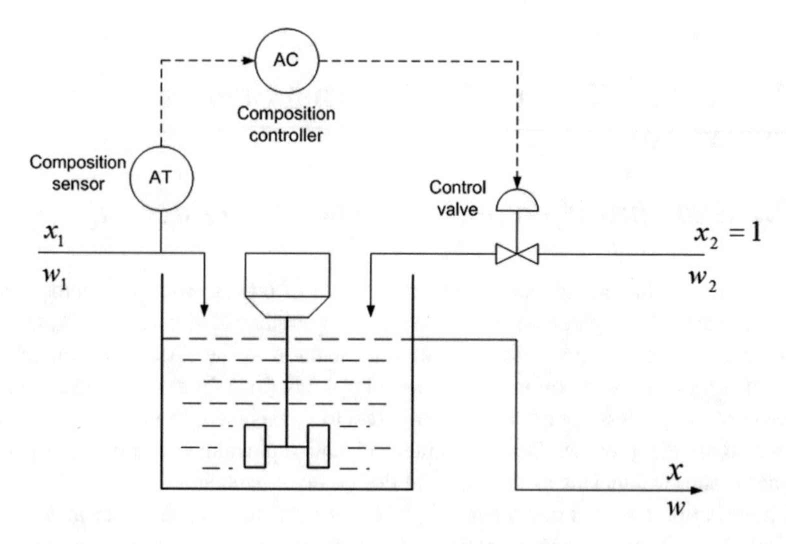

Fig. A.3.1 Typical notations applied in the instrumentation designs

Table A.3.1 Most generally used names in control engineering

Control	Disturbance, noise
Open-loop control	Manipulated variable
Closed-loop control, feedback control	Output signal, controlled variable
Process, plant	Reference signal, set-point
Sensor	Error signal
Actuator	Control signal
Controller, regulator, control algorithm	Measured output, sensor output

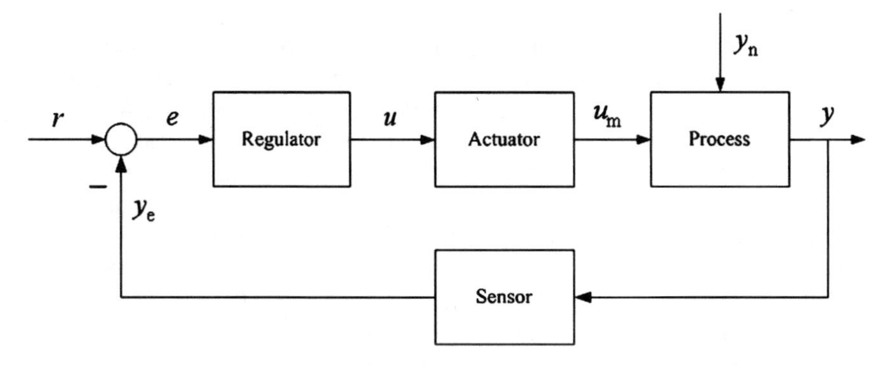

Fig. A.3.2 Operational scheme of the control loop

in system 03, GCR means the subtechnology, 31 is the serial number of the pipeline. AA means to which equipment this valve is connected, 101 is the serial number of the valve. This detailed notation makes it possible to identify unambiguously the equipments. Each technology has its own system identification notations.

A.3.2 The Names of the Most Important Signals in Control Systems

Table A.3.1 contains certain names most generally used in control engineering.

The operation scheme of a control loop is shown in Fig. A.3.2. The dynamics of the actuator and sensor are usually included in the dynamics of the controlled system. The joint scheme is shown by Fig. A.3.3.

A.4 Computer-Aided Design (*CAD*) Systems

Nowadays the design of complex systems is inconceivable without computers. The fast computers, the sophisticated developing environments and the well elaborated design algorithms make it possible to design and simulate simple, precise and

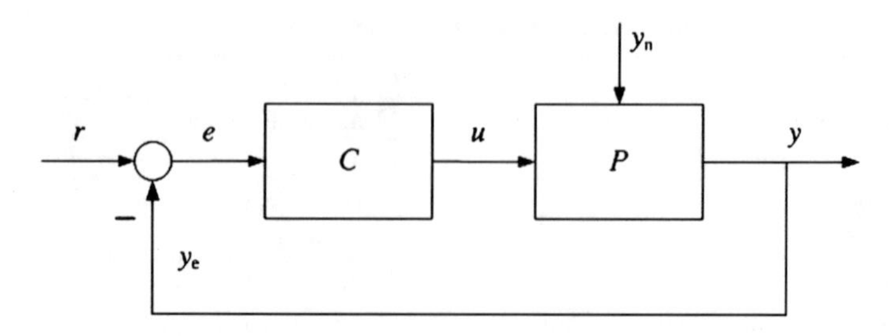

Fig. A.3.3 The joint scheme of the control loop

flexible control systems. The design consists of two phases: the design of the controllers and the simulation of the system. The control parameters are determined on the basis of the quality requirements. During the simulation the operation of the system is investigated for given parameters on the basis of a criterion or visual performance. A graphical presentation is more and more in the front, because this technique makes possible a fast, precise and information rich presentation.

There are several program packages available for the design of control systems. These can be classified in two groups. The first group contains the packages for general mathematical computations which might be extended for the design of controllers. The other group contains the industrial control systems whose main goal is to perform the control or to solve special control tasks.

A.4.1 Mathematical Program Packages

The most well-known control design packages were primarily developed for general mathematical computations, but later they were extended by special tools for helping the design procedures. There are several program packages, however, which originally were not designated for the design of controllers, but later, due to their mathematical and graphical capabilities, were applied for design, too.

MATLAB®

The program package **MATLAB®** has been elaborated for scientific and engineering computations, simulation and graphical presentation. It provides a strong background to the solution of differential equations, handling matrix algebra and the solution of other mathematical problems and to the presentation of the results in good quality and also graphically. The extended application of **MATLAB®** derives from the fact that its command set can be extended by toolboxes. A toolbox is actually a function library developed for supporting different subject areas. **MATLAB®** has very good graphical capabilities, and relatively complex design tasks can be performed within an acceptable running time range. The programming of **MATLAB®** is interactive, which means that it performs the commands row by row without translation. Its speed is based on coding the critical program parts in a lower level language, generally in C or C++, and on direct access to the system matrix structure.

The essence of the matrix programming is that the matrix operations are performed automatically by the triggered functions for all elements of the matrix instead of special embedded cycles. **MATLAB®** supports several mathematical operations, procedures (e.g., handling of complex numbers, computation of inverse and eigenvalues of a matrix, Fourier transformation, convolution computation and determination of the roots of equations). **MATLAB®** does not support directly symbolic computations but makes that possible by the *Symbolic Math Toolbox*. The *Symbolic Toolbox* is based on **MAPLE®** but it has an interface to **MATLAB®**.

MATLAB® is primarily used in the engineering environment. If a new algorithm or theory appears then they are immediately developed in the form of toolboxes or function libraries in order to investigate and compare them with other methods.

The *Control System Toolbox* contains functions for the design and simulation of control systems. The controller can be given in transfer function or state space form. It is able to investigate continuous and discrete systems in the time and frequency domains. It can handle single and multi input-multi output, linear and nonlinear systems. The toolboxes are open, they can be easily extended with other functions and algorithms. The *Control System Toolbox* can be well used with other toolboxes, e.g., with the *Fuzzy Logic Toolbox*, *Model Predictive Control Toolbox*, *Nonlinear Control Design Blockset, System Identification Toolbox* and *Robust Control Toolbox*.

SIMULINK® is a graphical program package for modeling and simulation of dynamic systems. The simulation is interactive, therefore the effect of changing the parameters can be well presented. In **SIMULINK®** the dynamic system is given by a block-diagram, the different blocks can be copied from a library. **SIMULINK®** is able to simulate linear and nonlinear systems in the continuous, discrete and hybrid domains. **SIMULINK®** simulates the models by integrating ordinary differential equations. It can use several integrating methods. The result of the simulation can be further used by **MATLAB®** for data processing or graphical presentation. The graphical abilities of **SIMULINK®** facilitate significantly the design and simulation of the controllers.

MATHEMATICA®

MATHEMATICA® is an interactive system for mathematical computations. It supports numerical and symbolic computations and also includes a high level programming language which makes it possible for the user to develop new procedures. **MATHEMATICA®** is one of the most effective systems for general mathematical computations, which has roughly two million users all over the world.

Starting from the 60s there have been programs for special computations, but **MATHEMATICA®** with its completely new approach made it possible to handle uniformly the different fields of technical computation. Appearing in 1988 it brought significant change in the usage of the computers in several fields. The program was developed by the research group of *Wolfram Research* led by Stephen

WOLFRAM. The key development was to develop a new symbolic computer language which made first possible to handle a wide range of objects necessary for technical computations by a few basic categories (primitives). Among the developers and users a high number of mathematicians and research engineers can be found. It is very popular in education, nowadays several hundreds of textbooks are based on it and it is a very important tool among students worldwide. It is very useful in writing complex studies, reports, because it provides a uniform environment for computation, modeling, text editing and graphical presentation. One of its disadvantages is that its learning curve is quite steep, the acquirement of its basic operation is not easy. Its most important advantage is its openness, it can be easily extended to new subject areas, as, e.g., to applied mathematics, informatics, control engineering, economics, sociology, etc.

In **MATHEMATICA®** the basic arithmetic operations can be performed. It can also handle complex numbers. Its most important data structure is the list, which practically corresponds to a set. The lists can be defined as embedded, and different operations can be accomplished on them, e.g., unification, cut, adding a term and deleting a term, etc. The matrices are the special forms of the lists. The typical matrix operations can also be performed, like inversion and eigenvalue computations.

Due to its symbolic capabilities it can be well used for algebraic transformations. Several such transformations can be made very easily which are difficult to compute by hand, e.g., simplification of fractions, series expansions, decomposition into partial fractions, solving equations, minimum seeking, differentiation, and integration.

In **MATHEMATICA®** the functions are formal transformation rules. Any kind of object can appear as the input or output of a function. The function may consist of mathematical commands, program control commands (e.g., if, then, for) or it can be written even in another programming language (e.g., FORTRAN, C).

Due to its graphical capabilities the data can be presented in one, two or three dimensions.

MAPLE®

MAPLE® is a general computer algebraic system for solving mathematical problems and presenting technical figures with excellent quality. It is easy to learn and anybody can perform complex mathematical computations after a very short time. **MAPLE®** contains also high level programming languages by means of which the users can define their own procedures. Its main feature is providing symbolic computations, algebraic transformations, series expansions, integration and differential computations. It can be used in several areas of mathematics, e.g., for solving linear algebraic, statistical and group theoretical tasks. The commands can be performed interactively or in a group (*batch mode*). It can be well used in education and for development. Its capabilities can be extended by adding outer functions. It contains more than 2500 functions for different subject areas. Several of them were developed by external, independent companies, firms and research institutes. The most frequently used function libraries, toolboxes are:

- *Global Optimization Toolbox*
- *Database Integration Toolbox*
- *Fuzzy Sets*
- **MAPLE**®*Professional Math Toolbox* for **LabVIEW**®
- *Analog Filter Design Toolbox*
- *ICP* for **MAPLE**® (*Intelligent Control and Parameterization*: it makes possible the design of automatic, intelligent and robust controllers)

Its mathematical capabilities and the *ICP toolbox* provides the opportunity to solve control engineering tasks but in spite of this it is mainly used by statisticians and mathematicians and less by control engineers.

SysQuake®

SysQuake® is a very similar system to **MATLAB**® concerning its commands. It has been developed for solving design tasks interactively directly on the screen. By its help, e.g., by directly changing the place of the poles and zeros, the breakpoint frequencies, the controller or process parameters, several system attributes (Bode diagram, Nyquist diagram, root-locus, transfer functions of the closed-loop signals) can be followed simultaneously in the design procedure. The software tools for man-machine interaction can be easily realized in object-oriented structures.

A.4.2 Industrial Control Systems

Nowadays, industrial control systems have special *CAD* tools. Sometimes these do not provide a wide range of design possibilities: they are usually restricted only to those algorithms ensuring the operation of a given system. In many cases this means only a simple *PID* controller whose parameters can be set in a simulation environment. The industrial control systems are usually able to perform certain kind of automatic design, e.g., in the case adaptive systems where the parameters of the controller are automatically set based on the system's behavior. Several significant industrial companies have serious system and control design background. They can be sorted according to their functions:

- Firms producing integrated control systems, *Rockwell*, *Honeywell*. They perform the control of the whole factories, like *Rockwell Automation Ltd.*
  Rockwell Software: their program package enables the integrated control of the whole factories including automation tasks.
- Robot manufacturing firms: *Fanuk, Panasonic, ABB*. Nowadays ready made robots perform a certain part of the automated manufacturing.
- PLC producing firms: *Siemens*, *Allen Bradly* (*Rockwell*), *Toshiba*. The PLC (Programmable Logic Controller) is one of the main elements of the industrial process control systems.
- Firms producing data collecting and measurement systems, like: *National Instruments, Siemens*, etc.

Among the above firms several have also some additional activities. They generally develop program systems which can be used only for their machines and equipments. From the great number of industrial systems perhaps only the **LabVIEW®** program package developed by National Instruments is widely used and has become an accepted developing environment by other firms as well.

LabVIEW®

LabVIEW® provides a graphical developing environment for data collection, signal processing, and data presentation. It makes possible flexible, high level programming without the complexity of programming languages. It has all the programming tools (e.g., handling of data structures, cycles and events) which are given in classical programming languages, but in a simpler environment. **LabVIEW®** has also an embedded translator whose efficiency is comparable to a C translator concerning the speed and memory requirements.

The effectiveness and popularity of **LabVIEW®** is due to the fact that it has several (presently about 50) program libraries, toolkits available for developers. These include different virtual tools, sample programs and documentation fitting well with the developing environments and applications. These functions are designed and optimized for such special demands which comprise a wide range of fields, from signal processing, communication to the data structure. The main toolkits are the following:

- *Application Deployment & Targeting Modules*
- *Software Engineering & Optimization Tools*
- *Data Management and Visualization*
- *Real-Time and FPGA Deployment*
- *Embedded System Deployment*
- *Signal Processing and Analysis*
- *Automated Testing*
- *Image Acquisition and Machine Vision*
- *Control Design & Simulation*
- *Industrial Control*

The *Control Design Toolkit* is able to design and analyze controllers in the **LabVIEW®** environment. The main features of the *Control Design Toolkit are:*

- The **LabVIEW®** *Control Design Toolkit* can design and analyze the controllers in the **LabVIEW®** environment. It provides interactive graphical design, e.g. by the help of root-locus.
- The process and the controller can be given in transfer function and state-space forms.
- These modules are integrated with the **LabVIEW®** *Simulation Module*.
- The behavior of the system can be investigated by several tools, e.g. step response function, Bode diagram, allocation of zeros and poles, etc.

LabVIEW® ensures an integrated environment for data collection, identification, controller design and simulation. The system's behavior can be graphically investigated, while its parameters can be adjusted.

A.5 Proofs and Derivations (By Chapters)

A.2.1

It is very simple to determine the Bode diagram of

$$H(s) = 1 + sT; \quad H(j\omega) = 1 + j\omega T = |H(j\omega)|e^{j\varphi(\omega)}. \tag{A.2.1}$$

The dependence of its absolute value and phase angle on the frequency is

$$|H(j\omega)| = \sqrt{1+\omega^2T^2} = \left[10\lg\left(1+\omega^2T^2\right)\right]\mathrm{dB}; \quad \varphi(\omega) = \mathrm{arctg}\,\omega T. \tag{A.2.2}$$

Investigating the asymptotic behavior of the functions we get

$$H(j\omega) \approx 1; \quad |H(j\omega)| \approx 0\,\mathrm{dB}; \quad \varphi(\omega) \approx 0, \quad \text{if} \quad \omega \ll \omega_1 = 1/T \tag{A.2.3}$$

and

$$\begin{aligned} &H(j\omega) \approx j\omega \mathrm{T}, \;\; |\mathrm{H}(\mathrm{j}\omega)| \approx (20\lg\omega + 20\lg\mathrm{T})\mathrm{dB}; \\ &\varphi(\omega) \approx 90^\circ, \quad \text{if} \quad \omega \gg \omega_1 = 1/\mathrm{T} \end{aligned} \tag{A.2.4}$$

If logarithmic scaling is applied for the frequency axis then both asymptotes of the amplitude are straight lines. On the frequency axis there are two points at a distance of a decade, for which $\omega_2 = 10\omega_1$, i.e., $\lg\omega_2 = 1 + \lg\omega_1$. Thus in logarithmic scale the decade means constant distance. So in the region $\omega \gg \omega_1$ the asymptote of the curve is a line having slope of 20 dB/decade, which cuts the 0 dB axis at ω_1 (the brake frequency). Here the actual value is

$$|H(j\omega_1)| = (20\lg 2)\,\mathrm{dB} = 3\,\mathrm{dB} \quad \text{and} \quad \varphi(\omega_1) = \mathrm{arctg}1 = 45^\circ \tag{A.2.5}$$

The tangents of the functions are

$$\frac{\mathrm{d}|H(j\omega)|}{\mathrm{dlg}\omega} = 10\frac{\mathrm{dlg}(1+\omega^2T^2)}{\mathrm{d}\omega}\frac{\mathrm{d}\omega}{\mathrm{dlg}\omega} = 10\frac{2\omega T^2}{1+\omega^2T^2}\omega\,\mathrm{dB/decade} \tag{A.2.6}$$

$$\frac{\mathrm{d}\varphi(\omega)}{\mathrm{dlg}\omega} = \frac{\mathrm{darctg}\omega T}{\mathrm{d}\omega}\frac{\mathrm{d}\omega}{\mathrm{dlg}\omega} = \frac{T}{1+\omega^2T^2}\frac{\omega}{\lg e}\frac{180^\circ}{\pi}\mathrm{degree/decade} \tag{A.2.7}$$

and their slopes at the break frequency

$$\left.\frac{\mathrm{d}|H(j\omega)|}{\mathrm{dlg}\omega}\right|_{\omega_1} = 10\,\mathrm{dB/decade} \tag{A.2.8}$$

$$\left.\frac{\mathrm{d}\varphi(\omega)}{\mathrm{dlg}\omega}\right|_{\omega_1} = 66\,\mathrm{degree/decade} \tag{A.2.9}$$

A.3.1

The solution of the state equation can be given by (3.18). To prove it let us differentiate the equation

$$\frac{\mathrm{d}\boldsymbol{x}(t)}{\mathrm{d}t} = \frac{\mathrm{d}}{\mathrm{d}t}\left[e^{\boldsymbol{A}t}\boldsymbol{x}(0)\right] + \frac{\mathrm{d}}{\mathrm{d}t}\left[\int_0^t e^{\boldsymbol{A}(t-\tau)}\boldsymbol{bu}(\tau)\mathrm{d}\tau\right], \tag{A.3.1}$$

where

$$\frac{\mathrm{d}}{\mathrm{d}t}\left[e^{\boldsymbol{A}t}\boldsymbol{x}(0)\right] = \boldsymbol{A}e^{\boldsymbol{A}t}\boldsymbol{x}(0) \tag{A.3.2}$$

and

$$\begin{aligned}\frac{\mathrm{d}}{\mathrm{d}t}\left[\int_0^t e^{\boldsymbol{A}(t-\tau)}\boldsymbol{bu}(\tau)\mathrm{d}\tau\right] &= \int_0^t \frac{\mathrm{d}}{\mathrm{d}t}\left[e^{\boldsymbol{A}(t-\tau)}\boldsymbol{bu}(\tau)\right]\mathrm{d}\tau + \frac{\mathrm{d}t}{\mathrm{d}t}\left[e^{\boldsymbol{A}(t-\tau)}\boldsymbol{bu}(\tau)\right]_{\tau=t} \\ &\quad - \frac{\mathrm{d}0}{\mathrm{d}t}\left[e^{\boldsymbol{A}(t-\tau)}\boldsymbol{bu}(\tau)\right]_{\tau=0} = \int_0^t \boldsymbol{A}e^{\boldsymbol{A}(t-\tau)}\boldsymbol{bu}(\tau)\mathrm{d}\tau + \boldsymbol{bu}(\tau)\end{aligned} \tag{A.3.3}$$

where the expressions $\mathrm{d}t/\mathrm{d}t = 1$, $\mathrm{d}0/\mathrm{d}t = 0$ and $\left.e^{\boldsymbol{A}(t-\tau)}\right|_{\tau=t} = 1$ are taken into consideration. Thus the derivative of (3.18) is

$$\frac{\mathrm{d}\boldsymbol{x}(t)}{\mathrm{d}t} = \boldsymbol{A}e^{\boldsymbol{A}t}\boldsymbol{x}(0) + \int_0^t \boldsymbol{A}e^{\boldsymbol{A}(t-\tau)}\boldsymbol{bu}(\tau)\mathrm{d}\tau + \boldsymbol{bu}(t) = \boldsymbol{Ax}(t) + \boldsymbol{bu}(t). \tag{A.3.4}$$

A.3.2

In the case of zero initial conditions (i.e. $\boldsymbol{x}(0) = \boldsymbol{0}$) and $d = 0$, the impulse response of a system to the excitation $u(t) = \delta(t)$ can be computed from (3.18)

$$\begin{aligned} \boldsymbol{x}(t) &= \int_0^t e^{\boldsymbol{A}(t-\tau)} \boldsymbol{b} \delta(\tau) \mathrm{d}\tau = e^{\boldsymbol{A}t} \left[\int_0^t e^{-\boldsymbol{A}\tau} \delta(\tau) \mathrm{d}\tau \right] \boldsymbol{b} = e^{\boldsymbol{A}t} \left[e^{-\boldsymbol{A}\tau} \delta(\tau) \right]_0^t \boldsymbol{b} \\ &= e^{\boldsymbol{A}t} \left[-e^{-\boldsymbol{A}t} \delta(t) + e^{-\boldsymbol{A}0} \delta(0) \right] \boldsymbol{b} = e^{\boldsymbol{A}t} \boldsymbol{b} \\ w(t) &= y(t) = \boldsymbol{c}^{\mathrm{T}} \boldsymbol{x}(t) = \boldsymbol{c}^{\mathrm{T}} e^{\boldsymbol{A}t} \boldsymbol{b} \end{aligned} \tag{A.3.5}$$

which is equal to (3.25) which was obtained in the operator domain.

A.3.3

One of the most important theorems in matrix theory is the CAYLEY-HAMILTON Theorem. A matrix fulfills its own characteristic equation, i.e., the equation $\mathcal{A}(\boldsymbol{A}) = 0 = \det(s\boldsymbol{I} - \boldsymbol{A}) = 0$ which is formally the same as

$$\mathcal{A}(\boldsymbol{A}) = \boldsymbol{0} \tag{A.3.6}$$

[see Appendix A.1]. Equation (A.3.7) is satisfied also by the matrix polynomial $\mathcal{P}(\boldsymbol{A})$ of matrix $\boldsymbol{A}$, but also by any such matrix function $\boldsymbol{F}(\boldsymbol{A})$ whose associated function $f(s)$ is analytical (regular) in a certain region around the origin of the s-plane. Let the basic matrix be $\boldsymbol{F}(\boldsymbol{A}) = e^{\boldsymbol{A}\tau}$, then based on the above expressions we get

$$e^{\boldsymbol{A}\tau} = \alpha_{\mathrm{o}}(\tau)\boldsymbol{I} + \alpha_1(\tau)\boldsymbol{A} + \cdots + \alpha_{\mathrm{n}-1}(\tau)\boldsymbol{A}^{n-1}. \tag{A.3.7}$$

A.5.1

The NYQUIST stability criterion can be derived from the CAUCHY argument principle of the theory of complex functions.

The argument principle

Let Γ be a closed curve, not cutting itself, in the complex plane, which surrounds the region D. Consider the function $f(z)$ of the complex variable z. Suppose the function $f(z)$ has P poles and Z zeros in the domain D. All poles and zeros are taken into account with their multiplicity. In all the other points of the domain the function is analytic (thus at these points it is differentiable).

Due to the argument principle, going round the curve anti-clockwise, the angle change $\Delta_\Gamma \arg f(z)$ of the function $f(z)$ is $2\pi(Z - P)$,

$$\frac{1}{2\pi}\Delta_\Gamma \arg f(z) = \frac{1}{2\pi j}\int_\Gamma \frac{f'(z)}{f(z)}\mathrm{d}z = Z - P. \tag{A.5.1}$$

Proof Assume that $f(z)$ has a zero of multiplicity m at the point $z = \alpha$. In the vicinity of the zero the function $f(z)$ can be written as: $f(z) = (z-\alpha)^m g(z)$, where $g(z)$ is an analytic function. Constitute the expression $f'(z)/f(z)$:

$$\frac{f'(z)}{f(z)} = \frac{m}{z-\alpha} + \frac{g'(z)}{g(z)}. \tag{A.5.2}$$

The second term on the right hand side of (A.5.2) is analytic at $z = \alpha$. The numerator of the first term gives the residue.

In (A.5.1) the integral around the closed curve is the sum of the residues, considering the zeros and poles it is $Z - P$. Otherwise, taking into account that

$$\frac{f'(z)}{f(z)} = \frac{\mathrm{d}}{\mathrm{d}z}\ln f(z) \tag{A.5.3}$$

the following relationship can be derived:

$$\begin{aligned}\int_\Gamma \frac{f'(z)}{f(z)}\mathrm{d}z &= \int_\Gamma \mathrm{d}(\ln f(z)) = \int_\Gamma \mathrm{d}(\ln\{|f(z)|\exp(j\arg f(z))\}) \\ &= \int_\Gamma \mathrm{d}\ln|f(z)| + j\int_\Gamma \mathrm{d}(\arg f(z)) = j\Delta_\Gamma \arg f(z) = 2\pi j(Z-P)\end{aligned} \tag{A.5.4}$$

This proves the argument principle given by (A.5.1), thus

$$\frac{1}{2\pi}\Delta_\Gamma \arg f(z) = Z - P. \tag{A.5.5}$$

The Nyquist stability criterion

Investigate the stability of a closed control loop havingnegative feedback. The characteristic equation is

$$1 + L(s) = 0 \tag{A.5.6}$$

where $L(s)$ is the transfer function of the open loop.

Consider the closed curve on the complex plane shown in Fig. 5.17. If $L(s)$ has poles also on the imaginary axis, then pass around them at a small radius according to Fig. 5.18. The characteristic polynomial can also be written in the form of (5.31) as

$$1+L(s) = 1+\frac{\mathcal{N}(s)}{\mathcal{D}(s)} = \frac{\mathcal{D}(s)+\mathcal{N}(s)}{\mathcal{D}(s)} = k\frac{(s-z_1)(s-z_2)\ldots(s-z_n)}{(s-p_1)(s-p_2)\ldots(s-p_n)}. \quad \text{(A.5.7)}$$

Let the characteristic polynomial be the function $f(z)$ to be used for a mapping. Mapping the curve of Fig. 5.17 by the characteristic polynomial, the argument principle can be applied. Since the curve of Fig. 5.17 is passed around clockwise, the number of times R the mapped curve encircles the origin is

$$\frac{1}{2\pi}\Delta_\Gamma \arg f(z) = R = P - Z. \quad \text{(A.5.8)}$$

To ensure stability, the characteristic equation must not have roots in the right half-plane, thus the condition for stability is

$$Z = 0 \quad \text{(A.5.9)}$$

and from this,

$$R = P. \quad \text{(A.5.10)}$$

This means that the control system is stable if the curve mapping the curve of Fig. 5.17 by the characteristic polynomial encircles the origin anti-clockwise as many times as there are the unstable, right half-plane poles of the open-loop.

Mapping the curve $L(s)$ instead of the characteristic polynomial we get the so-called complete Nyquist curve. Investigating its windings around the point $-1+0j$, the system is stable if the condition (A.5.10) is fulfilled.

A.9.1

Use the notation introduced in (3.13)

$$\boldsymbol{\Phi}(s) = (s\boldsymbol{I}-\boldsymbol{A})^{-1} = \frac{\mathbf{adj}(s\boldsymbol{I}-\boldsymbol{A})}{\det(s\boldsymbol{I}-\boldsymbol{A})} = \frac{\mathbf{adj}(s\boldsymbol{I}-\boldsymbol{A})}{\mathcal{A}(s)} = \frac{\boldsymbol{\Psi}(s)}{\mathcal{A}(s)} \quad \text{(A.9.1)}$$

to simplify the complex form $\boldsymbol{c}^{\mathrm{T}}\left(s\boldsymbol{I}-\boldsymbol{A}+\boldsymbol{b}\boldsymbol{k}^{\mathrm{T}}\right)^{-1}\boldsymbol{b}$ and use the matrix inversion lemma

$$\left(s\boldsymbol{I}-\boldsymbol{A}+\boldsymbol{b}\boldsymbol{k}^{\mathrm{T}}\right)^{-1} = \left[\boldsymbol{\Phi}^{-1}(s)+\boldsymbol{b}\boldsymbol{k}^{\mathrm{T}}\right]^{-1} = \boldsymbol{\Phi}(s) - \boldsymbol{\Phi}(s)\boldsymbol{b}\left[1+\boldsymbol{k}^{\mathrm{T}}\boldsymbol{\Phi}(s)\boldsymbol{b}\right]^{-1}\boldsymbol{k}^{\mathrm{T}}\boldsymbol{\Phi}(s) \quad \text{(A.9.2)}$$

by means of which

$$\boldsymbol{c}^{\mathrm{T}}\left(s\boldsymbol{I}-\boldsymbol{A}+\boldsymbol{b}\boldsymbol{k}^{\mathrm{T}}\right)^{-1}\boldsymbol{b} = \boldsymbol{c}^{\mathrm{T}}\boldsymbol{\Phi}(s)\boldsymbol{b} - \frac{\boldsymbol{c}^{\mathrm{T}}\boldsymbol{\Phi}(s)\boldsymbol{b}\boldsymbol{k}^{\mathrm{T}}\boldsymbol{\Phi}(s)\boldsymbol{b}}{1+\boldsymbol{k}^{\mathrm{T}}\boldsymbol{\Phi}(s)\boldsymbol{b}} = \frac{\boldsymbol{c}^{\mathrm{T}}\boldsymbol{\Phi}(s)\boldsymbol{b}}{1+\boldsymbol{k}^{\mathrm{T}}\boldsymbol{\Phi}(s)\boldsymbol{b}}. \quad \text{(A.9.3)}$$

So (9.5) can be further modified

$$T_{\mathrm{ry}}(s)=\frac{\boldsymbol{c}^{\mathrm{T}}\boldsymbol{\Phi}(s)\boldsymbol{b}k_{\mathrm{r}}}{1+\boldsymbol{k}^{\mathrm{T}}\boldsymbol{\Phi}(s)\boldsymbol{b}}. \tag{A.9.4}$$

Note that here

$$\boldsymbol{c}^{\mathrm{T}}\boldsymbol{\Phi}(s)\boldsymbol{b}=\boldsymbol{c}^{\mathrm{T}}(s\boldsymbol{I}-\boldsymbol{A})^{-1}\boldsymbol{b}=P(s)=\frac{\mathcal{B}(s)}{\mathcal{A}(s)} \tag{A.9.5}$$

by means of which

$$\begin{aligned}T_{\mathrm{ry}}(s)&=\frac{\boldsymbol{c}^{\mathrm{T}}\boldsymbol{\Phi}(s)\boldsymbol{b}k_{\mathrm{r}}}{1+\boldsymbol{k}^{\mathrm{T}}\boldsymbol{\Phi}(s)\boldsymbol{b}}=\frac{k_{\mathrm{r}}}{1+\boldsymbol{k}^{\mathrm{T}}\frac{\boldsymbol{\Psi}(s)}{\mathcal{A}(s)}\boldsymbol{b}}P(s)=\frac{k_{\mathrm{r}}}{1+\boldsymbol{k}^{\mathrm{T}}\frac{\boldsymbol{\Psi}(s)}{\mathcal{A}(s)}\boldsymbol{b}}\frac{\mathcal{B}(s)}{\mathcal{A}(s)}\\&=\frac{k_{\mathrm{r}}\mathcal{B}(s)}{\mathcal{A}(s)+\boldsymbol{k}^{\mathrm{T}}\boldsymbol{\Psi}(s)\boldsymbol{b}}\end{aligned} \tag{A.9.6}$$

A.9.2

The static unit gain of the transfer function$T_{\mathrm{ry}}(s)$ of the closed system can be ensured by the scaling factor k_{r}. From the condition

$$T_{\mathrm{ry}}(s)\big|_{s=0}=\boldsymbol{c}^{\mathrm{T}}\left(-\boldsymbol{A}+\boldsymbol{b}\boldsymbol{k}^{\mathrm{T}}\right)^{-1}\boldsymbol{b}k_{\mathrm{r}}=1 \tag{A.9.7}$$

it is obtained that

$$k_{\mathrm{r}}=-1/\boldsymbol{c}^{\mathrm{T}}\left(\boldsymbol{A}-\boldsymbol{b}\boldsymbol{k}^{\mathrm{T}}\right)^{-1}\boldsymbol{b}. \tag{A.9.8}$$

Applying the matrix inversion lemma in the denominator,

$$\left(\boldsymbol{A}-\boldsymbol{b}\boldsymbol{k}^{\mathrm{T}}\right)^{-1}=\boldsymbol{A}^{-1}+\boldsymbol{A}^{-1}\boldsymbol{b}\left[1-\boldsymbol{k}^{\mathrm{T}}\boldsymbol{A}^{-1}\boldsymbol{b}\right]^{-1}\boldsymbol{k}^{\mathrm{T}}\boldsymbol{A}^{-1}, \tag{A.9.9}$$

we get that

$$\begin{aligned}\boldsymbol{c}^{\mathrm{T}}\left(\boldsymbol{A}-\boldsymbol{b}\boldsymbol{k}^{\mathrm{T}}\right)^{-1}\boldsymbol{b}&=\boldsymbol{c}^{\mathrm{T}}\boldsymbol{A}^{-1}\boldsymbol{b}+\frac{\boldsymbol{c}^{\mathrm{T}}\boldsymbol{A}^{-1}\boldsymbol{b}\boldsymbol{k}^{\mathrm{T}}\boldsymbol{A}^{-1}\boldsymbol{b}}{1-\boldsymbol{k}^{\mathrm{T}}\boldsymbol{A}^{-1}\boldsymbol{b}}=\frac{\boldsymbol{c}^{\mathrm{T}}\boldsymbol{A}^{-1}\boldsymbol{b}\left(1+\boldsymbol{k}^{\mathrm{T}}\boldsymbol{A}^{-1}\boldsymbol{b}-\boldsymbol{k}^{\mathrm{T}}\boldsymbol{A}^{-1}\boldsymbol{b}\right)}{1-\boldsymbol{k}^{\mathrm{T}}\boldsymbol{A}^{-1}\boldsymbol{b}}\\&=\frac{\boldsymbol{c}^{\mathrm{T}}\boldsymbol{A}^{-1}\boldsymbol{b}}{1-\boldsymbol{k}^{\mathrm{T}}\boldsymbol{A}^{-1}\boldsymbol{b}}\end{aligned} \tag{A.9.10}$$

So the other form of (A.9.8) is

$$k_r = \frac{-1}{c^T\left(A - bk^T\right)^{-1}b} = \frac{k^T A^{-1} b - 1}{c^T A^{-1} b}. \tag{A.9.11}$$

A.9.3

As was seen in the derivation of (9.10), the pole allocating state feedback vector $k^T = k_c^T$ can easily be computed from the controllable canonical form. It was discussed in connection with Eq. (3.67) that all controllable systems can be rewritten into controllable canonical form by the transformation matrix $T_c = M_c^c (M_c)^{-1}$. From this we can get the similarity transformation (9.13) of the feedback vector

$$k^T = k_c^T T_c = k_c^T M_c^c M_c^{-1}. \tag{A.9.12}$$

Instead of the relatively complicated transformation matrix T_c, another simpler method is also available. Find the matrix T of the similarity transformation by the following expressions

$$A_c = TAT^{-1} \quad \text{and} \quad b_c = Tb \tag{A.9.13}$$

The similarity transformation of the matrix A can also be expressed in the form

$$A_c T = TA. \tag{A.9.14}$$

Introducing the notation t_i^T for the rows of the matrix T we can write that

$$A_c T = \begin{bmatrix} -a_1 & -a_2 & \dots & -a_{n-1} & -a_n \\ 1 & 0 & \dots & 0 & 0 \\ 0 & 1 & \dots & 0 & 0 \\ \vdots & \vdots & \ddots & \vdots & \vdots \\ 0 & 0 & 0 & 1 & 0 \end{bmatrix} \begin{bmatrix} t_1^T \\ t_2^T \\ t_3^T \\ \vdots \\ t_n^T \end{bmatrix} = TA = \begin{bmatrix} t_1^T \\ t_2^T \\ t_3^T \\ \vdots \\ t_n^T \end{bmatrix} A = \begin{bmatrix} t_1^T A \\ t_2^T A \\ t_3^T A \\ \vdots \\ t_n^T A \end{bmatrix}. \tag{A.9.15}$$

Executing the operations we get that

$$A_c T = \begin{bmatrix} -a_1 t_1^T - a_2 t_2^T \cdots - a_{n-1} t_{n-1}^T - a_n t_n^T \\ t_1^T \\ t_2^T \\ \vdots \\ t_{n-1}^T \end{bmatrix} = \begin{bmatrix} t_1^T A \\ t_2^T A \\ t_3^T A \\ \vdots \\ t_n^T A \end{bmatrix} = \begin{bmatrix} t_n^T A^{n-1} \\ t_n^T A^{n-2} \\ t_n^T A^{n-3} \\ \vdots \\ t_n^T \end{bmatrix} A = TA. \tag{A.9.16}$$

As a consequence of the equality of the two sides the following recursive relationship holds between the row vectors $\boldsymbol{t}_i^{\mathrm{T}}$, if $\boldsymbol{t}_n^{\mathrm{T}}$ is known

$$\boldsymbol{t}_{i-1}^{\mathrm{T}} = \boldsymbol{t}_i^{\mathrm{T}}\boldsymbol{A}; \quad i = n, n-1, \ldots, 2 \tag{A.9.17}$$

or in another form,

$$\boldsymbol{t}_{i-1}^{\mathrm{T}} = \boldsymbol{t}_n^{\mathrm{T}}\boldsymbol{A}^{n-i+1}; \quad i = n, n-1, \ldots, 2. \tag{A.9.18}$$

Thus the transformation matrix is

$$\boldsymbol{T}_{\mathrm{c}} = \boldsymbol{T} = \begin{bmatrix} \boldsymbol{t}_n^{\mathrm{T}}\boldsymbol{A}^{n-1} \\ \boldsymbol{t}_n^{\mathrm{T}}\boldsymbol{A}^{n-2} \\ \boldsymbol{t}_n^{\mathrm{T}}\boldsymbol{A}^{n-3} \\ \vdots \\ \boldsymbol{t}_n^{\mathrm{T}} \end{bmatrix}. \tag{A.9.19}$$

Similarly, based on (A.9.13) and (A.9.16) we get that

$$\boldsymbol{b}_{\mathrm{c}} = \boldsymbol{T}\boldsymbol{b} = \begin{bmatrix} \boldsymbol{t}_1^{\mathrm{T}} \\ \boldsymbol{t}_2^{\mathrm{T}} \\ \boldsymbol{t}_3^{\mathrm{T}} \\ \vdots \\ \boldsymbol{t}_n^{\mathrm{T}} \end{bmatrix} \boldsymbol{b} = \begin{bmatrix} \boldsymbol{t}_n^{\mathrm{T}}\boldsymbol{A}^{n-1} \\ \boldsymbol{t}_n^{\mathrm{T}}\boldsymbol{A}^{n-2} \\ \boldsymbol{t}_n^{\mathrm{T}}\boldsymbol{A}^{n-3} \\ \vdots \\ \boldsymbol{t}_n^{\mathrm{T}} \end{bmatrix} \boldsymbol{b} = \begin{bmatrix} \boldsymbol{t}_n^{\mathrm{T}}\boldsymbol{A}^{n-1}\boldsymbol{b} \\ \boldsymbol{t}_n^{\mathrm{T}}\boldsymbol{A}^{n-2}\boldsymbol{b} \\ \boldsymbol{t}_n^{\mathrm{T}}\boldsymbol{A}^{n-3}\boldsymbol{b} \\ \vdots \\ \boldsymbol{t}_n^{\mathrm{T}}\boldsymbol{b} \end{bmatrix}, \tag{A.9.20}$$

whose transposed form is

$$\boldsymbol{b}_{\mathrm{c}}^{\mathrm{T}} = \boldsymbol{t}_n^{\mathrm{T}}\left[\boldsymbol{b} \quad \boldsymbol{A}\boldsymbol{b} \quad \ldots \quad \boldsymbol{A}^{n-2}\boldsymbol{b} \quad \boldsymbol{A}^{n-1}\boldsymbol{b}\right] = \boldsymbol{t}_n^{\mathrm{T}}\boldsymbol{M}_{\mathrm{c}}, \tag{A.9.21}$$

where $\boldsymbol{M}_{\mathrm{c}}$ is the controllability matrix. From this,

$$\boldsymbol{t}_n^{\mathrm{T}} = \boldsymbol{b}_{\mathrm{c}}^{\mathrm{T}}(\boldsymbol{M}_{\mathrm{c}})^{-1}. \tag{A.9.22}$$

Thus $\boldsymbol{t}_n^{\mathrm{T}}$ is the first row of the inverse of the controllability matrix, since

$$\boldsymbol{b}_{\mathrm{c}}^{\mathrm{T}} = [1, 0, \ldots, 0]. \tag{A.9.23}$$

Consider the transpose of the feedback vector (A.9.12)

$$\boldsymbol{k}^{\mathrm{T}}=\boldsymbol{k}_{\mathrm{c}}^{\mathrm{T}}\boldsymbol{T}_{\mathrm{c}}=[r_1-a_1,r_2-a_2,\ldots,r_n-a_n]\begin{bmatrix}\boldsymbol{t}_n^{\mathrm{T}}\boldsymbol{A}^{n-1}\\ \boldsymbol{t}_n^{\mathrm{T}}\boldsymbol{A}^{n-2}\\ \boldsymbol{t}_n^{\mathrm{T}}\boldsymbol{A}^{n-3}\\ \vdots\\ \boldsymbol{t}_n^{\mathrm{T}}\end{bmatrix}. \tag{A.9.24}$$

Executing the operation we get the equation

$$\boldsymbol{k}^{\mathrm{T}}=\boldsymbol{t}_n^{\mathrm{T}}\sum_{i=1}^{n}r_i\boldsymbol{A}^{n-i}-\boldsymbol{t}_n^{\mathrm{T}}\sum_{i=1}^{n}a_i\boldsymbol{A}^{n-i} \tag{A.9.25}$$

then adding $\boldsymbol{A}^n$ to both sums we get a very interesting form,

$$\boldsymbol{k}^{\mathrm{T}}=\boldsymbol{t}_n^{\mathrm{T}}\mathcal{R}(\boldsymbol{A})-\boldsymbol{t}_n^{\mathrm{T}}\mathcal{A}(\boldsymbol{A}). \tag{A.9.26}$$

Due to the CAYLEY-HAMILTON theorem all square matrices satisfy their characteristic polynomial, therefore $\mathcal{A}(\boldsymbol{A})=\mathcal{R}(\boldsymbol{A})=0$. The final form of (A.9.26) is

$$\boldsymbol{k}^{\mathrm{T}}=\boldsymbol{t}_n^{\mathrm{T}}\mathcal{R}(\boldsymbol{A}). \tag{A.9.27}$$

This latter equation is called the ACKERMANN formula. The expression (A.9.12) can be evaluated much easier by computational methods than by (A.9.27).

A.9.4

Based on the diagonal canonical form, from the basic relationship (9.7) of the pole allocation we can get by equivalent rewriting that

$$\mathcal{R}(s)-\mathcal{A}(s)=\frac{\mathcal{B}(s)}{\boldsymbol{c}_{\mathrm{d}}^{\mathrm{T}}(s\boldsymbol{I}-\boldsymbol{A}_{\mathrm{d}})^{-1}\boldsymbol{b}^{\mathrm{d}}}\boldsymbol{k}_{\mathrm{d}}^{\mathrm{T}}(s\boldsymbol{I}-\boldsymbol{A}_{\mathrm{d}})^{-1}\boldsymbol{b}^{\mathrm{d}}=\mathcal{A}(s)\boldsymbol{k}_{\mathrm{d}}^{\mathrm{T}}(s\boldsymbol{I}-\boldsymbol{A}_{\mathrm{d}})^{-1}\boldsymbol{b}^{\mathrm{d}}, \tag{A.9.28}$$

which yields

$$\frac{\mathcal{R}(s)}{\mathcal{A}(s)}=1+\boldsymbol{k}_{\mathrm{d}}^{\mathrm{T}}(s\boldsymbol{I}-\boldsymbol{A}_{\mathrm{d}})^{-1}\boldsymbol{b}^{\mathrm{d}}. \tag{A.9.29}$$

Decomposing the left side into partial fractions, and taking the diagonal character of the system into account, it can be seen that

$$\frac{\mathcal{R}(s)}{\mathcal{A}(s)}=1+\sum_{i=1}^{n}\frac{k_i^{\mathrm{d}}b_i^{\mathrm{d}}}{s-\lambda_i}=1+\sum_{i=1}^{n}\frac{k_i^{\mathrm{d}}\beta_i}{s-\lambda_i}. \tag{A.9.30}$$

Applying the expansion theory valid for the simple poles of the partial fractions, thus multiplying both sides with $(s-\lambda_i)$ and substituting $s=\lambda_i$, we get the expression

$$k_i^d b_i^d = \prod_{j=1}^{n}(\lambda_i-\mu_j)\Big/\prod_{\substack{j=1\\ i\neq j}}^{n}(\lambda_i-\lambda_j) \tag{A.9.31}$$

This procedure has to be performed for all the poles.

A.9.5

Taking the matrix inversion identity (A.1.17) in Appendix A.1 into account the following steps of the rewriting can be easily followed:

$$\begin{aligned} T_{\mathrm{ry}}(s) &= \frac{\left[\boldsymbol{c}^{\mathrm{T}}(s\boldsymbol{I}-\boldsymbol{A})^{-1}\boldsymbol{b}\right]\left[1-\boldsymbol{k}^{\mathrm{T}}\left(s\boldsymbol{I}-\boldsymbol{A}+\boldsymbol{b}\boldsymbol{k}^{\mathrm{T}}+\boldsymbol{l}\boldsymbol{c}^{\mathrm{T}}\right)^{-1}\boldsymbol{b}\right]k_{\mathrm{r}}}{1+\left[\boldsymbol{k}^{\mathrm{T}}\left(s\boldsymbol{I}-\boldsymbol{A}+\boldsymbol{b}\boldsymbol{k}^{\mathrm{T}}+\boldsymbol{l}\boldsymbol{c}^{\mathrm{T}}\right)^{-1}\boldsymbol{b}\right]\left[\boldsymbol{c}^{\mathrm{T}}(s\boldsymbol{I}-\boldsymbol{A})^{-1}\boldsymbol{b}\right]} \\ &= \boldsymbol{c}^{\mathrm{T}}\left(s\boldsymbol{I}-\boldsymbol{A}+\boldsymbol{b}\boldsymbol{k}^{\mathrm{T}}\right)^{-1}\boldsymbol{b}k_{\mathrm{r}} = \frac{\boldsymbol{c}^{\mathrm{T}}(s\boldsymbol{I}-\boldsymbol{A})^{-1}\boldsymbol{b}k_{\mathrm{r}}}{1+\boldsymbol{k}^{\mathrm{T}}(s\boldsymbol{I}-\boldsymbol{A})^{-1}\boldsymbol{b}} \\ &= \frac{k_{\mathrm{r}}P(s)}{1+\boldsymbol{k}^{\mathrm{T}}(s\boldsymbol{I}-\boldsymbol{A})^{-1}\boldsymbol{b}} = \frac{k_{\mathrm{r}}\boldsymbol{B}(s)}{\boldsymbol{R}(s)} \end{aligned} \tag{A.9.32}$$

A.9.6

The so-called *LQ* controller, discussed in 9.5, is a special case of a generally formulated optimization problem. In the general case the task is to determine the control signal $u(t)$ of the system given by the state equation

$$\dot{\boldsymbol{x}}(t) = \frac{\mathrm{d}\boldsymbol{x}(t)}{\mathrm{d}t} = \boldsymbol{f}[\boldsymbol{x}(t), u(t)], \tag{A.9.33}$$

which minimizes the general integral criterion

$$I = \frac{1}{2}\int_0^{T_{\mathrm{f}}} F[\boldsymbol{x}(t), u(t)]\mathrm{d}t = I[u(t)]. \tag{A.9.34}$$

The solution is provided by the so-called minimum principle, by means of which the so-called HAMILTON function

$$H(t) = F[\boldsymbol{x}(t), u(t)] + \boldsymbol{\lambda}(t)^{\mathrm{T}} \boldsymbol{f}[\boldsymbol{x}(t), u(t)] \tag{A.9.35}$$

has to be constructed, for which the following necessary conditions of the extremum values

$$\frac{\mathrm{d}H(t)}{\mathrm{d}u(t)} = 0; \quad \frac{\mathrm{d}H(t)}{\mathrm{d}\boldsymbol{x}(t)} = -\frac{\mathrm{d}\boldsymbol{\lambda}(t)}{\mathrm{d}t} = -\dot{\boldsymbol{\lambda}}(t) \tag{A.9.36}$$

must be fulfilled. (The sufficient condition of the minimum is that $\partial^2 H/\partial u^2 > 0$.) The HAMILTON function and the necessary condition for the minimum (A.9.36) corresponds formally to the LAGRANGE method of the conditional optimum (thus $\boldsymbol{\lambda}$ is the co-vector of the method), since the minimum of $I[u(t)]$ has to be reached under the condition (A.9.33). (Note that in the state space arbitrary motion is not allowed, only those corresponding to (A.9.33).) For the solution it is usually assumed that $\boldsymbol{\lambda}(t) = \boldsymbol{P}(t)\boldsymbol{x}(t)$, i.e., it can be derived from the state vector by a linear transformation, so

$$\dot{\boldsymbol{\lambda}}(t) = \dot{\boldsymbol{P}}(t)\boldsymbol{x}(t) + \boldsymbol{P}\dot{\boldsymbol{x}}(t). \tag{A.9.37}$$

If the upper limit of the integral is infinity ($T_{\mathrm{f}} = \infty$) then $\boldsymbol{P}(t) = \boldsymbol{P}$ is constant, so $\dot{\boldsymbol{P}} = \boldsymbol{0}$, thus

$$\boldsymbol{\lambda}(t) = \boldsymbol{P}\boldsymbol{x}(t) \quad \text{and} \quad \dot{\boldsymbol{\lambda}}(t) = \boldsymbol{P}\dot{\boldsymbol{x}}(t). \tag{A.9.38}$$

The *LQ* regulator of the *LTI* process has to solve the task

$$I = \frac{1}{2}\int_0^{\infty} \left[\boldsymbol{x}^{\mathrm{T}}(t)\boldsymbol{W}_{\mathrm{x}}\boldsymbol{x}(t) + W_{\mathrm{u}}u^2(t)\right]\mathrm{d}t = \min_{u(t)} \tag{A.9.39}$$

under the condition of linear system dynamics

$$\dot{\boldsymbol{x}}(t) = \boldsymbol{A}\boldsymbol{x}(t) + \boldsymbol{b}u(t). \tag{A.9.40}$$

The HAMILTON function now is

$$H(t) = \frac{1}{2}\left[\boldsymbol{x}^{\mathrm{T}}(t)\boldsymbol{W}_{\mathrm{x}}\boldsymbol{x}(t) + W_{\mathrm{u}}u^2(t)\right] + \boldsymbol{\lambda}^{\mathrm{T}}[\boldsymbol{A}\boldsymbol{x}(t) + \boldsymbol{b}u(t)], \tag{A.9.41}$$

whose second order derivate is $\partial^2 H/\partial u^2 = W_{\mathrm{u}} > 0$, so the necessary condition is, at the same time sufficient, too. The necessary condition, on the one hand, is

$$\frac{\mathrm{d}H(t)}{\mathrm{d}u(t)} = W_\mathrm{u}u(t) + \boldsymbol{\lambda}^\mathrm{T}\boldsymbol{b} = W_\mathrm{u}u(t) + \boldsymbol{b}^\mathrm{T}\boldsymbol{\lambda} = 0 \tag{A.9.42}$$

from which the optimal control is

$$u(t) = -\frac{1}{W_\mathrm{u}}\boldsymbol{b}^\mathrm{T}\boldsymbol{\lambda}(t) = -\frac{1}{W_\mathrm{u}}\boldsymbol{b}^\mathrm{T}\boldsymbol{P}\boldsymbol{x}(t) = -\boldsymbol{k}_\mathrm{LQ}^\mathrm{T}\boldsymbol{x}(t) \tag{A.9.43}$$

On the other, the matrix $\boldsymbol{P}$ in Eq. (A.9.43) has to be determined. For this, consider the complete state equation of the closed system

$$\dot{\boldsymbol{x}} = \boldsymbol{A}\boldsymbol{x} - \boldsymbol{b}\boldsymbol{k}_\mathrm{LQ}^\mathrm{T}\boldsymbol{x} = \left(\boldsymbol{A} - \boldsymbol{b}\boldsymbol{k}_\mathrm{LQ}^\mathrm{T}\right)\boldsymbol{x} = \left(\boldsymbol{A} - \frac{1}{W_\mathrm{u}}\boldsymbol{b}\boldsymbol{b}^\mathrm{T}\boldsymbol{P}\right)\boldsymbol{x} = \bar{\boldsymbol{A}}\boldsymbol{x}, \tag{A.9.44}$$

which has the same form as for the state feedback. Thus the *LQ* regulator is a state feedback controller. Based on Eqs. (A.9.36) and (A.9.44) the co-vector is

$$\dot{\boldsymbol{\lambda}} = \boldsymbol{P}\dot{\boldsymbol{x}} = \left(\boldsymbol{P}\boldsymbol{A} - \frac{1}{W_\mathrm{u}}\boldsymbol{P}\boldsymbol{b}\boldsymbol{b}^\mathrm{T}\boldsymbol{P}\right)\boldsymbol{x} = \boldsymbol{P}\bar{\boldsymbol{A}}x, \tag{A.9.45}$$

which has to satisfy the equation

$$\begin{aligned}\dot{\boldsymbol{\lambda}}(t) &= -\frac{\mathrm{d}H(t)}{\mathrm{d}\boldsymbol{x}(t)} = -\boldsymbol{W}_\mathrm{x}\boldsymbol{x}(t) - \boldsymbol{A}^\mathrm{T}\boldsymbol{\lambda}(t) = -\boldsymbol{W}_\mathrm{x}\boldsymbol{x}(t) - \boldsymbol{A}^\mathrm{T}\boldsymbol{P}\boldsymbol{x}(t)\\ &= -\left(\boldsymbol{W}_\mathrm{x} + \boldsymbol{A}^\mathrm{T}\boldsymbol{P}\right)\boldsymbol{x}(t)\end{aligned} \tag{A.9.46}$$

coming from the necessary condition (A.9.36). Comparing the last two equations, the following equality

$$\boldsymbol{P}\boldsymbol{A} - \frac{1}{W_\mathrm{u}}\boldsymbol{P}\boldsymbol{b}\boldsymbol{b}^\mathrm{T}\boldsymbol{P} = -\left(\boldsymbol{W}_\mathrm{x} + \boldsymbol{A}^\mathrm{T}\boldsymbol{P}\right) \tag{A.9.47}$$

is obtained for symmetric $\boldsymbol{P}$. By rewriting we get the so-called nonlinear algebraic Riccati matrix equation

$$\boldsymbol{P}\boldsymbol{A} + \boldsymbol{A}^\mathrm{T}\boldsymbol{P} - \frac{1}{W_\mathrm{u}}\boldsymbol{P}\boldsymbol{b}\boldsymbol{b}^\mathrm{T}\boldsymbol{P} = \boldsymbol{W}_\mathrm{x}, \tag{A.9.48}$$

which has no explicit algebraic solution, but there are several fast numerical methods available for its computation.

The joint state equation of the state vector and co-vector of the closed system can be easily written using Eqs. (A.9.44) and (A.9.46)

$$\begin{bmatrix} \dot{\boldsymbol{x}} \\ \dot{\boldsymbol{\lambda}} \end{bmatrix} = \begin{bmatrix} \boldsymbol{A} & \frac{1}{W_u}\boldsymbol{b}\boldsymbol{b}^{\mathrm{T}}\boldsymbol{P} \\ \boldsymbol{W}_x & -\boldsymbol{A}^{\mathrm{T}} \end{bmatrix} \begin{bmatrix} \boldsymbol{x} \\ \boldsymbol{\lambda} \end{bmatrix}. \tag{A.9.49}$$

Note that if the upper limit of the integral is finite ($T_f < \infty$) then $\boldsymbol{P} = \boldsymbol{P}(t)$ depends on the time and the RICCATI matrix equation has to be solved in advance for the domain $0 \le t \le T_f$.

Next it will be shown that the solution $\boldsymbol{P}$ of the RICCATI matrix equation has exceptional meaning. Substitute Eq. (A.9.41) of the optimal control into the criterion (A.9.37)

$$I = \frac{1}{2}\int_0^\infty \left[\boldsymbol{x}^{\mathrm{T}}(t)\boldsymbol{W}_x\boldsymbol{x}(t) + W_u u(t)\right]\mathrm{d}t = \frac{1}{2}\int_0^\infty \left\{\boldsymbol{x}^{\mathrm{T}}(t)\boldsymbol{W}_x\boldsymbol{x}(t) + W_u\left[-\boldsymbol{k}_{\mathrm{LQ}}^{\mathrm{T}}\boldsymbol{x}(t)\right]^2\right\}\mathrm{d}t$$
$$= \frac{1}{2}\int_0^\infty \left[\boldsymbol{x}^{\mathrm{T}}(t)\boldsymbol{W}_x\boldsymbol{x}(t) + \frac{1}{W_u}\boldsymbol{x}^{\mathrm{T}}(t)\boldsymbol{P}^{\mathrm{T}}\boldsymbol{b}\boldsymbol{b}^{\mathrm{T}}\boldsymbol{P}\boldsymbol{x}(t)\right]\mathrm{d}t = \frac{1}{2}\int_0^\infty \left[\boldsymbol{x}^{\mathrm{T}}(t)\bar{\boldsymbol{W}}_x\boldsymbol{x}(t)\right]\mathrm{d}t \tag{A.9.50}$$

where

$$\bar{\boldsymbol{W}}_x = \boldsymbol{W}_x + \frac{1}{W_u}\boldsymbol{P}^{\mathrm{T}}\boldsymbol{b}\boldsymbol{b}^{\mathrm{T}}\boldsymbol{P}. \tag{A.9.51}$$

The solution for the closed system of (A.9.44) without excitation is

$$\boldsymbol{x}(t) = e^{\bar{\boldsymbol{A}}t}\boldsymbol{x}(0) \tag{A.9.52}$$

so the criterion (A.9.50) for the case without excitation is

$$I = \frac{1}{2}\int_0^\infty \left[\boldsymbol{x}^{\mathrm{T}}(0)e^{\bar{\boldsymbol{A}}^{\mathrm{T}}t}\bar{\boldsymbol{W}}_x e^{\bar{\boldsymbol{A}}t}\boldsymbol{x}(0)\right]\mathrm{d}t = \frac{1}{2}\boldsymbol{x}^{\mathrm{T}}(0)\boldsymbol{P}\boldsymbol{x}(0), \tag{A.9.53}$$

where it is assumed that

$$\boldsymbol{P} = \int_0^\infty e^{\bar{\boldsymbol{A}}^{\mathrm{T}}t}\bar{\boldsymbol{W}}_x e^{\bar{\boldsymbol{A}}t}\mathrm{d}t. \tag{A.9.54}$$

To prove this, carry out the integration

$$\boldsymbol{P} = \int_0^\infty e^{\bar{\boldsymbol{A}}^{\mathrm{T}}t}\bar{\boldsymbol{W}}_{\mathrm{x}}e^{\bar{\boldsymbol{A}}t}\mathrm{d}t = \left[e^{\bar{\boldsymbol{A}}^{\mathrm{T}}t}\bar{\boldsymbol{W}}_{\mathrm{x}}\bar{\boldsymbol{A}}^{-1}e^{\bar{\boldsymbol{A}}t}\right]_0^\infty - \int_0^\infty \bar{\boldsymbol{A}}^{\mathrm{T}}e^{\bar{\boldsymbol{A}}^{\mathrm{T}}t}\bar{\boldsymbol{W}}_{\mathrm{x}}\bar{\boldsymbol{A}}^{-1}e^{\bar{\boldsymbol{A}}t}\mathrm{d}t. \tag{A.9.55}$$

Furthermore, if $\bar{\boldsymbol{A}}$ is stable, then

$$\boldsymbol{P} = -\bar{\boldsymbol{W}}_{\mathrm{x}}\bar{\boldsymbol{A}}^{-1} - \bar{\boldsymbol{A}}^{\mathrm{T}}\left(\int_0^\infty e^{\bar{\boldsymbol{A}}^{\mathrm{T}}t}\bar{\boldsymbol{W}}_{\mathrm{x}}e^{\bar{\boldsymbol{A}}t}\mathrm{d}t\right)\bar{\boldsymbol{A}}^{-1} = -\bar{\boldsymbol{W}}_{\mathrm{x}}\bar{\boldsymbol{A}}^{-1} - \bar{\boldsymbol{A}}^{\mathrm{T}}\boldsymbol{P}\bar{\boldsymbol{A}}^{-1}. \tag{A.9.56}$$

Here, it has been used that $\bar{\boldsymbol{A}}^{-1}e^{\bar{\boldsymbol{A}}t} = e^{\bar{\boldsymbol{A}}t}\bar{\boldsymbol{A}}^{-1}$. Finally the equation

$$\boldsymbol{P}\bar{\boldsymbol{A}} + \bar{\boldsymbol{A}}^{\mathrm{T}}\boldsymbol{P} = -\bar{\boldsymbol{W}}_{\mathrm{x}} \tag{A.9.57}$$

is obtained, which is called the Lyapunov equation. The equation is only virtually linear in $\boldsymbol{P}$, since $\bar{\boldsymbol{W}}_{\mathrm{x}}$ and $\bar{\boldsymbol{A}}$ also depend on $\boldsymbol{P}$. Rewriting the equation we get again the algebraic Riccati matrix equation (A.9.48). By this, on the one hand, the relationship (A.9.54) is proved for $\boldsymbol{P}$, on the other hand the meaning of $\boldsymbol{P}$ is also shown: namely that it is the quadratic cost function matrix associated with the control ensuring the minimum of the criterion (A.9.37) for the case without excitation.

It is also interesting to investigate how the Hamilton function itself changes in time. Determine the time derivates

$$\frac{\mathrm{d}H}{\mathrm{d}t} = \left[\frac{\mathrm{d}H}{\mathrm{d}\boldsymbol{x}}\right]^{\mathrm{T}}\frac{\mathrm{d}\boldsymbol{x}}{\mathrm{d}t} + \left[\frac{\mathrm{d}H}{\mathrm{d}u}\right]^{\mathrm{T}}\frac{\mathrm{d}u}{\mathrm{d}t} + \left[\frac{\mathrm{d}H}{\mathrm{d}\boldsymbol{\lambda}}\right]^{\mathrm{T}}\frac{\mathrm{d}\boldsymbol{\lambda}}{\mathrm{d}t} \tag{A.9.58}$$

Since based on (A.9.33) and (A.9.35) it can be stated that

$$\frac{\mathrm{d}H}{\mathrm{d}\boldsymbol{\lambda}} = \frac{\mathrm{d}H}{\mathrm{d}\boldsymbol{x}} = \boldsymbol{f} \tag{A.9.59}$$

and taking Eq. (A.9.36) into account, we get

$$\left[\frac{\mathrm{d}H}{\mathrm{d}\boldsymbol{x}}\right]^{\mathrm{T}}\frac{\mathrm{d}\boldsymbol{x}}{\mathrm{d}t} + \left[\frac{\mathrm{d}H}{\mathrm{d}\boldsymbol{\lambda}}\right]^{\mathrm{T}}\frac{\mathrm{d}\boldsymbol{\lambda}}{\mathrm{d}t} = 0. \tag{A.9.60}$$

Thus finally

$$\frac{\mathrm{d}H}{\mathrm{d}t} = 0, \tag{A.9.61}$$

i.e., the Hamilton function is constant (assuming that neither the control, nor the state vector have restrictions). Thus in the case without limitation, the Hamilton function is time invariant and it is invariant also for the input (see the necessary condition (A.9.36) for the extremum).

A.11.1

Based on the transfer function of the zero order hold

$$W_{\mathrm{ZOH}}(s) = \frac{1 - e^{-sT_s}}{s} \tag{A.11.1}$$

the frequency function of the holding element is

$$\begin{aligned} W_{\mathrm{ZOH}}(s)\big|_{s=j\omega} &= \frac{1 - e^{-j\omega \mathrm{T_s}}}{j\omega} = \frac{2e^{-j\omega \mathrm{T_s}/2}\left(e^{j\omega \mathrm{T_s}/2} - e^{-j\omega \mathrm{T_s}/2}\right)}{2j\omega} \\ &= T_s \frac{\sin(\omega \mathrm{T_s}/2)}{\omega \mathrm{T_s}/2} e^{-j\omega \mathrm{T_s}/2} \end{aligned} \tag{A.11.2}$$

Based on the above the absolute value function of the zero order hold can be written as

$$|W_{\mathrm{ZOH}}(j\omega)| = T_s \left| \frac{\sin(\omega T_s/2)}{\omega T_s/2} \right| \tag{A.11.3}$$

and its phase function is

$$\angle\{W_{\mathrm{ZOH}}(j\omega)\} = \angle\{\sin(\omega T_s/2)\} - \omega T_s/2 \tag{A.11.4}$$

Both components of the frequency function are drawn for the choice $T_s = 1$ s in Figs. A.11.1 and A.11.2. It can be seen that the absolute value function becomes

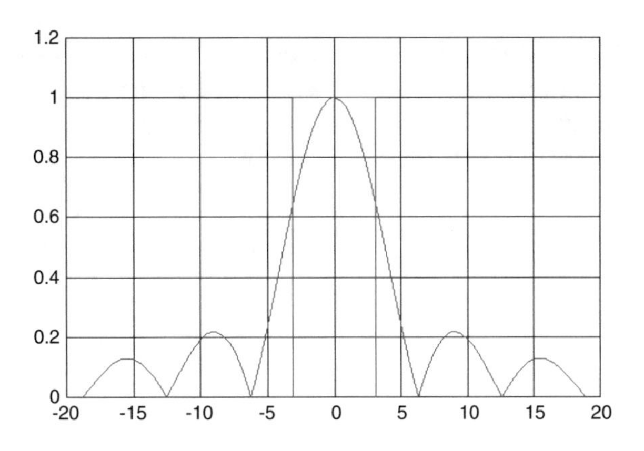

Fig. A.11.1 The absolute value of the frequency function of the zero order hold

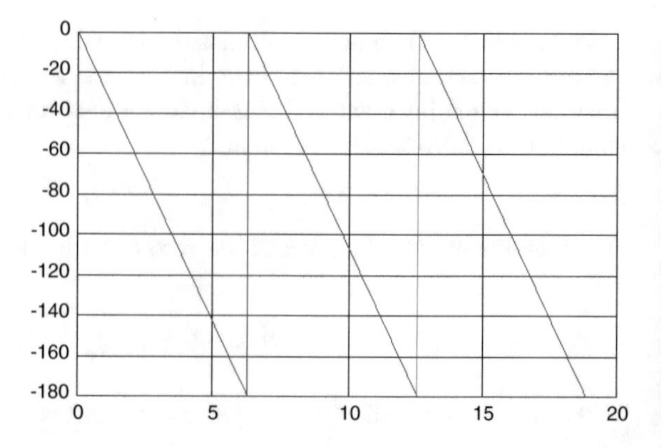

Fig. A.11.2 The phase function of the frequency function of the zero order hold

zero at the sampling frequency $\omega_s = 2\pi/T_s = 6.28\,\text{rad/s}$ and at its integer multiples, and the phase function has a linear character, its value corresponds to a delay term of $T_h = T_s/2$, at the singular points the phase changes by $\pm 180°$.

In Fig. A.11.1 the characteristic of the active linear filter in the region $\omega \leq \omega_s/2 = \omega_{max}$ is also shown. It can be seen, that the amplitude distortion can be neglected only in the lowest frequency region.

A.11.2

Let us start from the expression

$$\mathbf{Z}\{f[k]\} = F(z) = \sum_{k=0}^{\infty} f[k]z^{-k} = f[0] + f[1]z^{-1} + f[2]z^{-2} + \cdots + f[k]z^{-k} + \cdots \tag{A.11.1}$$

defining the z-transform of a discrete signal $f[k] (k = 0, 1, 2, \ldots)$ as an infinite geometric progression. Multiplying both sides by the factor z^{k-1}

$$F(z)z^{k-1} = f[0]z^{k-1} + f[1]z^{k-2} + f[2]z^{k-3} + \cdots + f[k]z^{-1} + \cdots \tag{A.11.2}$$

is obtained, which is actually the Laurent series of the expression $z^{k-1}F(z)$ at $z = 0$. Consider now a circle C around the origin of the complex plane, which includes all the poles of $z^{k-1}F(z)$. Since in the above expression the coefficient of z^{-1} is $f[k]$, and at the same time, this coefficient is the residue of $z^{k-1}F(z)$, we obtain

$$f[k] = \mathcal{Z}^{-1}\{F(z)\} = \frac{1}{2\pi j} \oint_C F(z)z^{k-1} \mathrm{d}z. \tag{A.11.3}$$

A.16.1

Due to the definition of $\mathcal{H}_2$ and the PARSEVAL theorem

$$\|H(j\omega)\|_2 = \sqrt{\int_0^\infty \|w(t)\|^2 \mathrm{d}t} = \sqrt{\int_0^\infty \boldsymbol{c}^\mathrm{T} e^{\boldsymbol{A}t} \boldsymbol{b}\boldsymbol{b}^\mathrm{T} e^{\boldsymbol{A}^\mathrm{T} t} \boldsymbol{c} \mathrm{d}t}. \tag{A.16.1}$$

Thus

$$\|H(j\omega)\|_2^2 = \boldsymbol{c}^\mathrm{T} \left[\int_0^\infty e^{\boldsymbol{A}t} \boldsymbol{b}\boldsymbol{b}^\mathrm{T} e^{\boldsymbol{A}^\mathrm{T} t} \mathrm{d}t \right] \boldsymbol{c}. \tag{A.16.2}$$

Introduce the notation

$$\boldsymbol{L} = \int_0^\infty e^{\boldsymbol{A}t} \boldsymbol{b}\boldsymbol{b}^\mathrm{T} e^{\boldsymbol{A}^\mathrm{T} t} \mathrm{d}t \tag{A.16.3}$$

so finally the $\mathcal{H}_2$ norm can be computed as

$$\|H(j\omega)\|_2 = \sqrt{\boldsymbol{c}^\mathrm{T} \boldsymbol{L} \boldsymbol{c}}. \tag{A.16.4}$$

Differentiate the integral in the A.16.3

$$\frac{\mathrm{d}}{\mathrm{d}t} e^{\boldsymbol{A}t} \boldsymbol{b}\boldsymbol{b}^\mathrm{T} e^{\boldsymbol{A}^\mathrm{T} t} = \boldsymbol{A} e^{\boldsymbol{A}t} \boldsymbol{b}\boldsymbol{b}^\mathrm{T} e^{\boldsymbol{A}^\mathrm{T} t} + e^{\boldsymbol{A}t} \boldsymbol{b}\boldsymbol{b}^\mathrm{T} e^{\boldsymbol{A}^\mathrm{T} t} \boldsymbol{A}^\mathrm{T} \tag{A.16.5}$$

then integrate both sides of the equation over the domain $[0, \infty]$:

$$\left[e^{\boldsymbol{A}t} \boldsymbol{b}\boldsymbol{b}^\mathrm{T} e^{\boldsymbol{A}^\mathrm{T} t} \right]_0^\infty = \boldsymbol{A} \left[\int_0^\infty e^{\boldsymbol{A}t} \boldsymbol{b}\boldsymbol{b}^\mathrm{T} e^{\boldsymbol{A}^\mathrm{T} t} \mathrm{d}t \right] + \left[\int_0^\infty e^{\boldsymbol{A}t} \boldsymbol{b}\boldsymbol{b}^\mathrm{T} e^{\boldsymbol{A}^\mathrm{T} t} \mathrm{d}t \right] \boldsymbol{A}^\mathrm{T}, \tag{A.16.6}$$

from which by simple computation and considering (A.16.3) we get the system of linear equations

$$-\boldsymbol{b}\boldsymbol{b}^\mathrm{T} = \boldsymbol{A}\boldsymbol{L} + \boldsymbol{L}\boldsymbol{A}^\mathrm{T} \tag{A.16.7}$$

for $\boldsymbol{L}$.

A.16.2

Rewriting the criterion (16.27) in detail the form

$$V(\hat{\boldsymbol{p}}) = \frac{1}{2}\left[\boldsymbol{y}^{\mathrm{T}}\boldsymbol{y} - 2\boldsymbol{y}^{\mathrm{T}}\boldsymbol{F}(\boldsymbol{u})\hat{\boldsymbol{p}} + \hat{\boldsymbol{p}}^{\mathrm{T}}\boldsymbol{F}^{\mathrm{T}}(\boldsymbol{u})\boldsymbol{F}(\boldsymbol{u})\hat{\boldsymbol{p}}\right] \tag{A.16.8}$$

is obtained, and making its gradient equal to zero yields

$$\frac{\mathrm{d}V(\hat{\boldsymbol{p}})}{\mathrm{d}\hat{\boldsymbol{p}}} = -\boldsymbol{F}^{\mathrm{T}}(\boldsymbol{u})\boldsymbol{y} + \boldsymbol{F}^{\mathrm{T}}(\boldsymbol{u})\boldsymbol{F}(\boldsymbol{u})\hat{\boldsymbol{p}} = \boldsymbol{0}. \tag{A.16.9}$$

Solving the equation for $\hat{\boldsymbol{p}}$, the best parameter estimator is obtained in the form

$$\hat{\boldsymbol{p}} = \left[\boldsymbol{F}^{\mathrm{T}}(\boldsymbol{u})\boldsymbol{F}(\boldsymbol{u})\right]^{-1}\boldsymbol{F}^{\mathrm{T}}(\boldsymbol{u})\boldsymbol{y}. \tag{A.16.10}$$

A.16.3

Assume that processing N data pairs the off-line LS estimation of the parameters is available as

$$\hat{\boldsymbol{p}}[N] = \left[\boldsymbol{F}^{\mathrm{T}}[N]\boldsymbol{F}[N]\right]^{-1}\boldsymbol{F}^{\mathrm{T}}[N]\boldsymbol{y}_N, \tag{A.16.11}$$

then compute the LS estimation for the $(N+1)$-th point

$$\begin{aligned}\hat{\boldsymbol{p}}[N+1] &= \left[\boldsymbol{F}^{\mathrm{T}}[N+1]\boldsymbol{F}[N+1]\right]^{-1}\boldsymbol{F}^{\mathrm{T}}[N+1]\boldsymbol{y}_{N+1} \\ &= \left\{\begin{bmatrix}\boldsymbol{F}[N] \\ \boldsymbol{f}^{\mathrm{T}}[N+1]\end{bmatrix}^{\mathrm{T}}\begin{bmatrix}\boldsymbol{F}[N] \\ \boldsymbol{f}^{\mathrm{T}}[N+1]\end{bmatrix}\right\}^{-1}\begin{bmatrix}\boldsymbol{F}[N] \\ \boldsymbol{f}^{\mathrm{T}}[N+1]\end{bmatrix}^{\mathrm{T}}\begin{bmatrix}\boldsymbol{y}_N \\ y[N+1]\end{bmatrix} \\ &= \left\{\boldsymbol{F}^{\mathrm{T}}[N]\boldsymbol{F}^{\mathrm{T}}[N] + \boldsymbol{f}[N+1]\boldsymbol{f}^{\mathrm{T}}[N+1]\right\}^{-1}\left\{\boldsymbol{F}^{\mathrm{T}}[N]\boldsymbol{y}_N + \boldsymbol{f}[N+1]y[N+1]\right\}\end{aligned} \tag{A.16.12}$$

For the solution it is required to compute the inverse of the matrix extended by the dyadic product

$$\boldsymbol{R}[N+1] = \left\{\boldsymbol{F}^{\mathrm{T}}[N+1]\boldsymbol{F}^{\mathrm{T}}[N+1]\right\}^{-1} = \left\{\boldsymbol{F}^{\mathrm{T}}[N]\boldsymbol{F}^{\mathrm{T}}[N] + \boldsymbol{f}[N+1]\boldsymbol{f}^{\mathrm{T}}[N+1]\right\}^{-1}. \tag{A.16.13}$$

According to A.1.17 in Appendix A.1,

$$\left\{\boldsymbol{F}^{\mathrm{T}}[N+1]\boldsymbol{F}^{\mathrm{T}}[N+1]\right\}^{-1} = \left\{\boldsymbol{F}^{\mathrm{T}}[N]\boldsymbol{F}^{\mathrm{T}}[N]\right\}^{-1} - \frac{\left\{\boldsymbol{F}^{\mathrm{T}}[N]\boldsymbol{F}^{\mathrm{T}}[N]\right\}^{-1}\boldsymbol{f}[N+1]\boldsymbol{f}^{\mathrm{T}}[N+1]\left\{\boldsymbol{F}^{\mathrm{T}}[N]\boldsymbol{F}^{\mathrm{T}}[N]\right\}^{-1}}{1+\boldsymbol{f}^{\mathrm{T}}[N+1]\left\{\boldsymbol{F}^{\mathrm{T}}[N]\boldsymbol{F}^{\mathrm{T}}[N]\right\}^{-1}\boldsymbol{f}[N+1]}. \tag{A.16.14}$$

Using the definition of $\boldsymbol{R}[N]$ by (16.54) we get

$$\boldsymbol{R}[N+1] = \boldsymbol{R}[N] - \frac{\boldsymbol{R}[N]\boldsymbol{f}(N+1)\boldsymbol{f}^{\mathrm{T}}(N+1)\boldsymbol{R}[N]}{1+\boldsymbol{f}^{\mathrm{T}}(N+1)\boldsymbol{R}[N]\boldsymbol{f}(N+1)} \tag{A.16.13}$$

Substituting the above recursive equation of the convergence matrix into (A.16.12), the recursive equation of the parameter estimation is obtained as

$$\hat{\boldsymbol{p}}[N+1] = \hat{\boldsymbol{p}}[N] + \boldsymbol{R}[N+1]\boldsymbol{f}(N+1)\left\{y[N+1] - \boldsymbol{f}^{\mathrm{T}}(N+1)\hat{\boldsymbol{p}}[N]\right\}. \tag{A.16.14}$$

The term "recursive" comes from the fact that the renewal equations of both $\boldsymbol{R}[N]$ and $\hat{\boldsymbol{p}}[N]$ can be computed from the previous values by adding a new term.

Authors

László Keviczky

Ruth Bars

Jenő Hetthéssy

Csilla Bányász

L. Keviczky et al., *Control Engineering*, Advanced Textbooks in Control and Signal Processing, https://doi.org/10.1007/978-981-10-8297-9

Pictures of Some of the Scientists Cited in This Book

L. Keviczky et al., *Control Engineering*, Advanced Textbooks in Control and Signal Processing, https://doi.org/10.1007/978-981-10-8297-9

References

1. Åström KJ (2002) Control system design. Lecture notes. University of California, Berkeley
2. Åström KJ, Wittenmark B (1984) Computer controlled systems. Theory and design. Prentice Hall, Englewood Cliffs
3. Åström KJ, Hägglund T (2006) Advanced PID control. ISA—Instrumentation, Systems and Automation Society, USA
4. Åström KJ, Murray RM (2008) Feedback systems: an introduction for scientists and engineers. Princeton University Press, Princeton. (http://www.cds.caltech.edu/~murray/amwiki/index.php/)
5. Bosgra OH, Kwakernaak H, Meinsma G (2004) Design methods for control systems. Winter Course, Dutch Institute of Systems and Control, The Netherlands
6. Csáki F (1966) Szabályozások dinamikája. Akadémiai Kiadó, Budapest
7. Csáki F (1973) State-space methods for control systems. Akadémiai Kiadó, Budapest
8. Fodor Gy (1967) Lineáris rendszerek analízise. Műszaki Könyvkiadó, Budapest
9. Horowitz IM (1963) Synthesis of feedback systems. Academic Press, New York
10. Kailath T (1980) Linear systems. Prentice-Hall, Englewood Cliffs
11. Keviczky L, Bányász Cs (2015) Two-degree of freedom control systems: Youla parameterization approach. Elsevier, New York
12. Lantos B (2001) Irányítási rendszerek elmélete és tervezése I (Egyváltozós szabályozások). Akadémiai Kiadó, Budapest
13. Levine WS (ed) (1996) The control handbook. CRC, Boca Raton
14. Mikles J, Fikar M (2004) Process modeling, identification and control 2 (identification and optimal control). STU Press, Bratislava
15. Nof YS (ed) (2009) Springer handbook of automation. Springer, Berlin
16. Ogata K (1987) Discrete-time control systems. Prentice-Hall, Englewood Cliffs
17. Szilágyi B (1981) Folyamatszabályozás, *A szabályozás egyensúlyi helyzete* (4. füzet). Tankönyvkiadó, Budapest
18. Tuschák R (1994) Szabályozástechnika. Műegyetemi Kiadó, Budapest

L. Keviczky et al., *Control Engineering*, Advanced Textbooks in Control and Signal Processing, https://doi.org/10.1007/978-981-10-8297-9

Index

L. Keviczky et al., *Control Engineering*, Advanced Textbooks in Control and Signal Processing, https://doi.org/10.1007/978-981-10-8297-9

Printed in the United States
By Bookmasters